K3: A New Problem List in Low-Dimensional Topology

Mathematical Surveys and Monographs
Volume 295

K3: A New Problem List in Low-Dimensional Topology

R. İnanç Baykur
Robion C. Kirby
Daniel Ruberman
Editors

Carolyn Abbott, John A. Baldwin, Thomas Koberda,
Marc Lackenby, Tye Lidman, Robert Lipshitz,
Ciprian Manolescu, Jeffrey Meier, Maggie Miller,
Priyam Patel, Mark Powell, Nick Salter,
and Radmila Sazdanović
Associate Editors

AMERICAN MATHEMATICAL SOCIETY
Providence, Rhode Island

2020 *Mathematics Subject Classification.* Primary 57Kxx, 57Mxx, 57Nxx, 57Rxx, 57Sxx, 53Cxx, 14Dxx, 32Sxx, 20Fxx.

For additional information and updates on this book, visit
www.ams.org/bookpages/surv-295

Library of Congress Cataloging-in-Publication Data

Cataloging-in-Publication Data has been applied for by the AMS.
See http://www.loc.gov/publish/cip/.

∞ The paper used in this book is acid-free and falls within the guidelines
established to ensure permanence and durability.
Visit the AMS home page at `https://www.ams.org/`

10 9 8 7 6 5 4 3 2 1   31 30 29 28 27 26

Contents

Introduction

Low-dimensional topology has been a highly active field over the last half-century, with remarkable achievements and breakthroughs coming at a staggering pace. The subject, comprising—at least—knot theory and the study of manifolds of dimensions up to four, has been repeatedly transformed by new ideas from hyperbolic geometry, differential geometry, mathematical physics, contact and symplectic geometry, geometric analysis, and other areas of mathematics, as well as advances in traditional topological methods. This book is a compendium of problems in low-dimensional topology, each presented with some background and references. It is aimed at graduate students and more experienced researchers alike, highlighting that low-dimensional topology is a problem-driven subject.

In the rest of this introduction, we give a little background on and history of the subject and of the role that earlier problem lists have played in its development.

Because of the difficulty of visualizing higher-dimensional spaces, the basic constructions arising from the works of Riemann and Poincaré were from the earliest days of topology much easier to picture in lower dimensions. So, it is perhaps surprising that a systematic theory of manifold topology was developed first in dimensions five and higher. This was largely due to the rapid development of homotopy theory, handlebody theory, smoothing theory, and surgery theory in the 1950s and 1960s. Spectacular advances coming from (and inspiring the development of) these tools include the discovery and classification of exotic spheres, the solution of the generalized Poincaré conjecture, and comprehensive classification theories for many classes of manifolds. The classification of piecewise linear (PL) structures on homotopy tori, in particular, was crucial in showing that the tools that had been so effective in the study of PL and smooth manifolds of dimension at least five could finally be applied to topological manifolds. See [**CRR00**, **CRR01**] for a collection of survey articles on this era in high-dimensional topology.

In the 1970s, it became clear that these methods broke down in lower dimensions. Knot theory, in conjunction with 2- and 3-dimensional topology, received increasing attention, with some interactions amongst these subjects starting to be understood. Dimension four was a complete mystery, with only a few isolated results. Many topologists were attracted to the wide variety of open problems in the area, and by the end of the decade there was a sense that a new field was opening up.

A fundamental contribution to the progress and unity of the field was played by two problem lists [**Kir78**, **Kir97**] compiled by Rob Kirby, the second representing a large expansion and update of the first[1]. It is hard to overstate the influence of these documents: especially for a young researcher, the lists of problems could be taken

[1]We refer to [**Kir78**] as **K1** and [**Kir97**] as **K2** throughout; the present volume is known to its contributors as **K3**.

as definitions of the field. To this day, it is considered a signature achievement to solve a *Kirby problem* and those who do so will usually point it out in their papers. Those lists were, as is the current effort, a product of the community; many mathematicians contributed problems and others detailed write-ups that provided background and inspiration for early-career mathematicians. The nature of the subject was already changing rapidly around the time of **K1** in the late 1970s, with Thurston's revolutionary ideas about hyperbolic 3-manifolds and mapping class groups and the development of new tools to visualize smooth 4-manifolds. Just a few years later the breakthroughs of Freedman on topological 4-manifolds and Donaldson on smooth 4-manifolds showed how dramatically different dimension four was from any other dimension. This period also saw Jones's surprising discovery of his polynomial invariant for links and the multitude of quantum invariants that followed it.

Since the publication of **K2**, low-dimensional topology has continued to expand rapidly. That list was compiled just as Seiberg-Witten gauge theory was introduced in the mid-1990s, an event that had a huge effect on 4-manifold theory, especially in conjunction with rapid developments in symplectic and contact topology. The subsequent explosion of invariants of 3-manifolds and links from Floer theory and categorification, and Perelman's work on Ricci flow, resulted in the solution of many important problems and the expansion of the field in unexpected directions. It has been clear for at least a decade that a new problem list was needed to guide the field and to give a fresh source of research ideas for a broad range of mathematicians. This book represents our effort to create such a list.

The **K1** and **K2** lists were compiled initially from problem sessions held at conferences, coupled with direct solicitation of problems over the succeeding months and years. The great expansion of the field meant that this model would be difficult to reproduce, and we decided to proceed in a more focused manner. We organized a one-week workshop at the American Institute of Mathematics (AIM) in the fall of 2023, hosting about 35 senior, mid- and early-career mathematicians, for extensive discussions about directions in various subfields and to write problems. Nearly half of the problems in the book were first drafted during this intense yet amazingly productive week. Many others were solicited by participants from researchers not at the workshop; the diversity of participants at the workshop was helpful both for the workshop's productivity and for gathering a wide range of problems at this second stage. We followed the basic structure of **K1** and **K2**, with five broad subject areas: (I) *Knot theory*; (II) *Surfaces*; (III) *3-manifolds*; (IV) *4-manifolds*; and (V) *Miscellaneous* for problems that are related to low-dimensional topology but did not readily fit into one of the other four chapters. Of course, many of the problems, such as those concerning knot concordance, cross these artificial boundaries. We tried to keep problems on a given topic together, rather than spreading them across chapters.

Following the workshop, we recruited an editorial board from workshop attendees, with some additional members. The board was divided into five groups corresponding to the areas described above. Each group reviewed the problems in its area and solicited new problems from a wide variety of experts in the field. The response from the wider community was generous and brought many interesting new problems for us to consider. Each problem, once agreed upon as suitable, was sent to at least one referee and revised (or rejected as unsuitable) accordingly. Although most problems bear the names of the people who proposed them for **K3** and of those who wrote them up, the referees received the problems without indication of authorship. Because problems were short (mostly less than a page) each one received careful attention, and we received many valuable comments and corrections from referees and the broader community. The final editing was done by a smaller subset of the editorial board, listed below.

The structure of the book. Like its predecessors, *K3: A new problem list in low-dimensional topology* consists of a sequence of problems, each with a bit of background material and discussion to help make the problem accessible to our readers. The list contains nearly 400 problems, many of which include multiple subproblems, questions, and conjectures—altogether presenting close to a thousand questions interspersed throughout the text. The problems are placed within five separate chapters, while keeping in mind that the nature of low-dimensional topology means that many topics could have been placed in more than one chapter. Each chapter is sorted into thematically related topics and has an overall introduction laying out the main themes. It should be understood that the bibliography represents a compromise between completeness (which would drastically increase the length of the book) and being useful to a wide variety of researchers. The citations, especially those of preprints, are not meant as endorsements but as pointers to work related to each topic. Conversely, our description of unpublished results, including arxiv preprints, as "announced" or "claimed" is not intended to cast doubt on them, but merely to acknowledge the important role of peer review.

The **K2** iteration of the problem list contained the first list, **K1**, together with extensive updates on progress over the intervening 20 years. Repeating this structure seemed impractical; it would have essentially tripled the size of this book. Hence, we chose to have mostly newly written problems, with a few major, unsolved problems from the earlier lists included for inspiration and completeness. It would have been unconscionable, for instance, to have a problem list without the smooth 4-dimensional Poincaré Conjecture, so it is here. Some unsolved problems from **K2** appear in an updated form to take account of progress over the intervening years; this is always noted in the current write-up. The problems from **K2** that are updated in this volume are listed in a table at the end of the book, with the problem number in this book. A separate effort is underway to update **K2** with brief progress reports on the problems therein.

A note on attributions. Many of problems presented here had a somewhat complex route to their final form, having been suggested by workshop participants, solicited by the editors, or suggested by a referee as an elaboration of a submitted problem. Some of our problems are conjectures made in print, but the actual origin of many problems is imprecise. Where possible, at the end of each problem we indicate the **Scribe**, i.e., the person or people who wrote at least the initial draft of what now appears in the list. Scribes had some leeway in the depth of discussion of a problem and on the choice of references, although both of these could be amended following suggestions of the referees and editors. Most problems also have a **Proposer**, whose contribution ranged from suggesting that **K3** should have a problem on a certain topic to providing a full draft of the problem. When no Proposer or Scribe is listed, either the problem was proposed/written communally or the information of who proposed/wrote it was not recorded. Final responsibility for the problems rests with the editors.

For named conjectures, such as the 4D Poincaré Conjecture, the problem statement is preceded by the name.

Editors and contributors. The editorial board evolved over the year and a half of editing and included thirteen associate editors, who were primarily responsible for specific chapters but also made various contributions across chapters. Chapter editors are listed at the beginning of each chapter. The final editing team was responsible for arranging the problems within each chapter, formatting them in a standard style, and carrying out related editorial tasks.

Editorial Board: Carolyn Abbott, John Baldwin, İnanç Baykur, Robion Kirby, Thomas Koberda, Marc Lackenby, Tye Lidman, Robert Lipshitz, Ciprian Manolescu, Jeffrey Meier, Maggie Miller, Priyam Patel, Mark Powell, Nick Salter, Daniel Ruberman, and Radmila Sazdanović.

Final Editing Team: John Baldwin, İnanç Baykur, Robert Lipshitz, Mark Powell, Daniel Ruberman, and Nick Salter. Arunima Ray generously volunteered to create the index.

Proposers: Here we list all the names of Proposers for **K3** problems, according to the definition above, extracted from the individual problems. This list, and the list of Scribes below, was not easy to track, and we apologize for any inadvertent omissions or errors.

Carolyn Abbott, Ian Agol, Jørgen Andersen, Tarik Aougab, Dave Auckly, John Baldwin, Richard H. Bamler, İnanç Baykur, Simon Blatt, Hans Boden, Joe Boninger, Danny Calegari, Jason Cantarella, Jae Choon Cha, Julian Chaidez, Jon Chaika, Abhijit Champanerkar, John Crisp, Daniel Cristofaro-Gardiner, Irving Dai, Robert Edwards, Yasha Eliashberg, John Etnyre, Benson Farb, Ronald Fintushel, Michael Freedman, David Gabai, David Gay, Marco Golla, Cameron Gordon, Joshua Greene, Joel Hass, Matthew Hedden, Jennifer Hom, Jim Hoste, Daniel Kasprowski, Mikhail Khovanov, Sanghyun Kim, Seungwon Kim, Thomas Koberda, Ilya Kofman, Hokuto Konno, Robin Koytcheff, Slava Krushkal, Marc Lackenby, Tao Li, Anatoly Libgober, Tye Lidman, Francesco Lin, Jianfeng Lin, Robert Lipshitz, Charles Livingston, Kathryn Mann, Ciprian Manolescu, Bruno Martelli, Jeffrey Meier, Maggie Miller, Allison Moore, Anubhav Mukherjee, András Némethi, Yi Ni, Luis Paris, Priyam Patel, Lisa Piccirillo, Olga Plamenevskaya, Mark Powell, Jessica Purcell, Arunima Ray, Alan Reid, Daniel Ruberman, Dan Rutherford, Osamu Saeki, Nick Salter, Martin Scharlemann, Henrik Schumacher, Laura Starkston, Ronald Stern, Matthew Stoffregen, Alexander Stoimenow, Clifford Taubes, Abigail Thompson, Bülent Tosun, Jeremy Van Horn-Morris, Mehdi Yazdi, and Alex Zupan.

Scribes: Here we list those who had a hand in writing the problems, either an initial or subsequent draft. We do not list the editors (who rewrote many of the problems.)

Ian Agol, Tarik Aougab, Michel Boileau, Francis Bonahon, Joe Boninger, Danny

Calegari, Jason Cantarella, Julian Chaidez, Daren Chen, Anthony Conway, Daniel Cristofaro-Gardiner, Irving Dai, Renaud Detcherry, Robert Edwards, John Etnyre, Charles Frohman, David Futer, David Gabai, David Gay, Cameron Gordon, Eugene Gorsky, Josh Greene, Matthew Hedden, Joel Hass, Kristen Hendricks, Jennifer Hom, Jim Hoste, Stanislav Jabuka, Sanghyun Kim, Seungwon Kim, Alexandra Kjuchukova, Hokuto Konno, Robin Koytcheff, Robert Kusner, Tao Li, Tian-Jun Li, Francesco Lin, Jianfeng Lin, Charles Livingston, Sara Maloni, Bruno Martelli, András Némethi, Lenny Ng, Yi Ni, Tim Perutz, Lisa Piccirillo, Olga Plamenevskaya, Jessica Purcell, Arunima Ray, Alan Reid, Joshua Sabloff, Dan Rutherford, Osamu Saeki, Ramanujan Santharoubane, Martin Scharlemann, Steven Sivek, Laura Starkston, Abigail Thompson, Bülent Tosun, Giancarlo Urzúa, Jeremy Van Horn-Morris, Helen Wong, Mehdi Yazdi, Ian Zemke, Jonathan Zung, and Alex Zupan.

Acknowledgments. We thank the American Institute of Mathematics for hosting the workshop *K3: A new problem list in low-dimensional topology* in the fall of 2023, where this project kicked off, and for their flexibility in adapting to the specific needs of this endeavor. We thank the participants of the AIM workshop for their active engagement and contributions, as well as the many other colleagues listed above who submitted problems to our list, helped with scribing, or refereed several hundred submitted problems.

We are especially grateful to the **K3** chapter editors, who played leading roles in soliciting and handling problems and who have continued their invaluable service as the editorial board of **K3**. This book would not have been possible without their thoughtful input and diligent work.

We are also deeply thankful to the broader geometry and topology community for fostering the rich and collaborative mathematical culture we have tried to reflect in this problem list. Numerous colleagues—too many to name—have motivated the selection of problems curated here through their work and conversations with each of us. We are indebted to all those who showed enthusiasm and excitement for a new problem list in low-dimensional topology.

R. İnanç Baykur was partially supported by NSF grants DMS-2005327 and DMS-2506431, and Simons Foundation Travel Grant. Daniel Ruberman was partially supported by NSF grant DMS-1952790.

CHAPTER 1

Knot theory

Chapter 1 Editors: Tye Lidman, Jeffrey Meier, Daniel Ruberman, and Radmila Sazdanović

Knots and links, as mathematical objects, were initially studied by a variety of combinatorial and topological methods. While more modern tools such as hyperbolic geometry, gauge theory, and quantum topology have resulted in a great deal of progress, many older problems remain unsolved and many new ones have arisen.

This chapter presents a selection of problems that touch on classical topics in knot theory, on algorithmic and combinatorial questions, and on the mathematics of knots as physical objects. A major focus is on new invariants such as Floer and Khovanov homology theories, and on interactions between knot theory and 3- and 4-dimensional topology. The chapter has several sections on concordance, and on Legendrian knot theory, both of which exemplify such interactions.

Most terminology is taken from standard sources such as the books of Burde-Zieschang-Heusener [**BZH14**], Lickorish [**Lic97**], and Rolfsen [**Rol90**]. Unless otherwise specified (as in Problems 1.61 and 1.81 where wild knots are discussed) our convention is that knots are smooth embeddings. On the other hand, surfaces bounded by knots or links in 4-manifolds may be smooth or topological (and then locally flat or not), so we will specify the category in each case. Some further definitions are given in the introduction to each section.

1.1. Classical knot theory

The problems in this section are concerned with classical invariants, dating to the earliest days of knot theory. We start with some definitions.

(1) The *crossing number* $c(K)$ of a knot K in S^3 is the minimal number of crossings in a diagram for K, the minimum being taken over all possible diagrams of K.
(2) The *unknotting number* $u(K)$ of a knot K in S^3 is the minimum, taken over all diagrams of K, of the number of crossings that must be changed to obtain a diagram of the unknot.
(3) A knot $P \subset S^1 \times D^2$ is called a *satellite operator*. Its winding number $w(P)$ is the homology class (considered up to sign) $[P] \in H_1(S^1 \times D^2) \cong \mathbb{Z}$. Given a knot K, the satellite knot $P(K)$ (or K_P) is the image of P under an identification of $S^1 \times D^2$ with a neighborhood of K that gives the 0-framing of K.
(4) Conway mutation [**Con70**] is the operation of removing a portion of a link that hits a 2-sphere in 4 points, and replacing it after a 180° rotation around an axis (interchanging the points in pairs).

(5) A link L is *fibered* if its exterior $S^3\backslash\nu(L)$ has the structure of a bundle over S^1 with fiber a compact surface with boundary.

PROBLEM 1.1. *Is the crossing number additive under connected sum:*

$$c(K_1\#K_2) = c(K_1) + c(K_2)?$$

REMARKS. This is Problem 1.65 in [**Kir97**]. Many special cases are known. The most famous case is due to Kauffman [**Kau87**], Murasugi [**Mur87**], and Thistlewaite [**Thi87**], who showed the conjecture for alternating knots (and more generally adequate knots, defined in [**LT88**]). Lackenby [**Lac09**] has shown in the general case that if $K_1, \ldots, K_n$ are knots, then

$$\frac{c(K_1) + \cdots + c(K_n)}{152} \leqslant c(K_1\# \cdots \#K_n).$$

Scribed by: I. Zemke

PROBLEM 1.2.

(a) *Show that if P is a nontrivial satellite operator and K_P is a nontrivial satellite of a knot K, then*

$$c(K_P) \geqslant c(K),$$

where c denotes the crossing number.

(b) *More generally, show that if P is a satellite operator then*

$$c(K_P) \geqslant c(K) \cdot w(P)^2,$$

where $w(P)$ denotes the geometric winding number.

REMARKS.

(1) Part (a) is Problem 1.67 in [**Kir97**]. Part (b) appears as the Satellite Crossing Number Conjecture in [**BM24a**].

(2) Note that a special case of this problem is whether

$$c(K_1\#K_2) \geqslant c(K_1),$$

which is a weaker form of Problem 1.1.

It is reasonable to expect equality in part (a) only when P is the core of the solid torus.

Lackenby [**Lac14**] has shown the inequality

$$c(K_P) \geqslant c(K)/10^{13}.$$

Kalfagianni-Lee [**KL23a**] have shown that if K is a nontrivial adequate knot that has an adequate diagram with writhe 0, then

$$c(Wh_{\pm}(K)) = 4c(K) + 2,$$

where $Wh_{\pm}(K)$ denote the Whitehead doubles of K. Kalfagianni-McConkey [**KM24**] have shown that if K is adequate, then the cable knot $K_{p,q}$ satisfies

$$c(K_{p,q}) \geqslant q^2 \cdot c(K) + 1$$

for any coprime p, q. Here q, denotes the longitudinal winding of the cabling pattern. A recent preprint of Baker–Motegi claims that the more general conjecture is true if one puts sufficiently many full twists in a pattern with geometric winding number at least 2 [**BM24a**].

Scribed by: J. Hom, I. Zemke

PROBLEM 1.3. *How does unknotting number behave under connected sum and mutation?*

(a) *Does the connected sum of n nontrivial knots have unknotting number at least n?*
(b) *Is unknotting number invariant under mutation?*

REMARKS.

(1) Problem 1.69 (B) from [**Kir97**] asks if unknotting number is additive; a recent preprint [**BH25**] claims an example where it fails to be additive, where the knots in question are the $(2, 7)$ torus knot and its mirror.
(2) Question (a) is [**Kir97**, Problem 1.69 (A)], and is still open. It has a positive answer for $n = 2$ by Scharlemann [**Sch85b**].
(3) With regard to Question (b), Gordon–Luecke [**GL06**] show that having unknotting number one is invariant under mutation.

Scribed by: C. Gordon, J. Meier

PROBLEM 1.4. *Let P be a nontrivial satellite pattern with winding number $w(P) \neq 0$. Then for any nontrivial knot K and its satellite K_P, one has*

$$u(K_P) \geqslant w(P) + 1.$$

REMARKS.

(1) This problem appears as [**HLP22**, Conjecture 1.10].
(2) One might guess that $w(P)^2$ is a lower bound. However, it can be shown that
$$u((T_{2,3})_{2,1}) \leqslant 3,$$
where $K_{p,q}$ denotes the (p, q) cable of the knot K.
(3) It follows from [**ST88**] that if $w(P) \neq 0$, then
$$u(K_P) \geqslant 2.$$
Problem 1.4 would be a generalization of this result. More recently, Hom, Lidman and Park [**HLP22**] have claimed that if K is a nontrivial knot, then
$$u(K_{p,q}) \geqslant p.$$
In the above, p indicates the winding number of the satellite operation.

Scribed by: I. Zemke

PROBLEM 1.5. *Is there a relationship between genus and unknotting number for specific classes of knots? Here are two instances of classes of knots for which there might be an interesting answer.*

(a) *Conjecture (Murasugi-Przytycki). For every positive fibered knot K, the genus of K equals its unknotting number.*
(b) *Question (Stoimenow). For every alternating fibered knot, is the genus greater than or equal to the unknotting number?*

REMARKS.

(1) There are examples of knots where the genus of a knot is greater than the unknotting number, and vice versa. However, there could be an inequality in some particular classes of knots. In part (a) there is evidence for an inequality one direction; in a different setting, part (b), the evidence points to an inequality in the other direction.
(2) The conjecture in (a) appears at the end of [**MP89**]. By work of Rudolph [**Rud93, Rud99**] it is now known that genus $\leqslant$ unknotting number for a positive knot. For a positive braid closure, manipulations with braid group relations [**BW84**] allow one to construct an unknotting sequence of length = genus, so that genus = unknotting number for a positive braid knot. However, such arguments are not available for the slightly more general case of a positive fibered knot.
(3) A Knotinfo [**LM23b**] search shows that this holds for alternating fibered knots of at most 13 crossings, with strict inequality holding in all but 13 of the 2106 such knots.

Proposed for K3 by: A. Stoimenow
Scribed by: D. Ruberman

PROBLEM 1.6. *Suppose that V_1 and V_2 are S–equivalent Seifert forms. Does there exist a* **fixed** *knot K bounding Seifert surfaces F_1 and F_2 for which the associated Seifert forms are V_1 and V_2?*

REMARKS.

(1) A Seifert form of a knot is a bilinear integral form V for which $V - V^*$ is unimodular (with $V^*(\alpha, \beta) = V(\beta, \alpha)$). Every Seifert form V arises as the Seifert form of a Seifert surface F for some knot K. If K bounds Seifert surfaces F_1 and F_2, then the associated Seifert forms are S–equivalent. See [**Tro73**] or [**Kaw96**, Chapter 5] for more details.
(2) One way to think about this problem is as a comparison of two flavors of S–equivalence: that of S–equivalence of Seifert forms, versus that of S–equivalence of knots, with two knots being S–equivalent if they possess Seifert forms that are S–equivalent [**Kea04, NS03**]. It is shown in [**NS03**] that two knots are S–equivalent if and only if they are related by a finite number of *doubled-delta moves*. These moves do not preserve the knot type in general: for example, it is shown in [**NS03**] that a knot is S–equivalent to the unknot if and only if it has Alexander polynomial is equal to 1.

(3) Consider the notation, defined for a Seifert form V and for a knot K:

$$[V]_S = \{\tilde{V} \mid \tilde{V} \text{ is a Seifert form } S\text{–equivalent to } V\},$$
$$[K]_S = \{\tilde{K} \mid \tilde{K} \text{ is a knot } S\text{–equivalent to } K\}.$$

Since for a given knot K, any two of its Seifert forms V_1 and V_2 are S–equivalent, it follows that $V_1, V_2 \in [V]_S$, where V is any Seifert form of K. Thus, the set of all Seifert forms of K maps to a subset of the S–equivalence class of any one of its Seifert forms. Given this, the question asked in this problem, can be rephrased as follows.

QUESTION. For a given S–equivalence class $[V]_S$ of a Seifert form V, does there exist a ***fixed*** knot K such that the obvious map

$$\{\tilde{V} \mid \tilde{V} \text{ is a Seifert form of } K\} \mapsto [V]_S$$

is surjective?

The key here being the existence of a ***single*** knot K for which the indicated map is surjective, as every element of $[V]_S$ arises as the Seifert form of ***some*** knot; cf. [**AFMW23**, page 20].

(4) In [**AFMW23**], the authors pose the genus-one version of this problem as Problem 7.7, and claim a complete answer under the additional hypothesis that Seifert surfaces be disjoint. They find a necessary and sufficient condition for a pair of 2×2 matrices to be Seifert forms for a pair of disjoint Seifert surfaces of the same knot, and they give an explicit construction realizing infinitely many pairs of disjoint Seifert surfaces with the same boundary for any pair satisfying the condition.

Proposed for K3 by: C. Livingston
Scribed by: S. Jabuka, J. Meier

PROBLEM 1.7. *Show that the sequence of absolute values of the coefficients of the Alexander polynomial of a link are:*

(a) *concave (Fox's trapezoidal conjecture), or*
(b) *for an alternating link, log concave (Stoimenow's strong Fox conjecture).*
(c) *For a positive link, is the sequence of the coefficients of the Conway polynomial strongly log concave?*

REMARKS.

(1) Let K be a link with Alexander polynomial (up to a factor of t^r, $r \in \frac{1}{2}\mathbb{Z}$)

$$\Delta_K(t) = \sum_{i=0}^{n} a_i t^i,$$

with $a_0, a_n \neq 0$, and Conway polynomial

$$\nabla_K(z) = z^{c(K)-1} \sum_{i=0}^{m} c_i z^{2i}$$

defined by $\nabla_K(t - t^{-1}) = t^{-n}\Delta_K(t^2)$, where $c(K)$ denotes the number of components of K. Recall that $a_i = (-1)^n a_{n-i}$; that for a non-split alternating link, $a_i a_{i+1} < 0$, $0 \leqslant i < n$; and that for a positive link, $c_i > 0$, $0 \leqslant i \leqslant m$.

For the Strong Fox Conjecture (3) below, recall that a sequence of positive numbers $(a_i)_{0 \leqslant i \leqslant n}$, is called *log concave* if $a_{i-1}a_{i+1} \leqslant a_i^2$ for all i with $0 < i < n$. That is, the sequence $(\log(a_i))$ is concave. For the final conjecture (4), recall that a sequence of positive numbers $(c_i)_{0 \leqslant i \leqslant m}$, is called *strongly log concave* if $c_{i-1}c_{i+1} < c_i^2$ for all i with $0 < i < m$.

(2) (Fox's trapezoidal conjecture [**Fox62**, Problem 12]) Let K be an alternating knot with Alexander polynomial as above. Then

$$|a_0| \leqslant |a_1| \leqslant \cdots \leqslant |a_{\frac{n}{2}}| \geqslant |a_{\frac{n}{2}+1}| \geqslant \cdots \geqslant |a_n|.$$

Moreover, if $|a_i| = |a_{i+1}|$ for some $i < \frac{n}{2}$, then

$$|a_i| = |a_{i+1}| = \cdots = |a_{n-i}|. \tag{1}$$

See for example [**Mur85, AJK24**] for some progress on Fox's conjecture.

(3) (Strong Fox Conjecture) Let K be an alternating link with Alexander polynomial as above. Then the sequence of absolute values of the coefficients $(|a_i|)$ is log concave. The equality $a_{i-1}a_{i+1} = a_i^2$ for some $i < n/2$ holds only if $a_{i-1} = -a_i = a_{i+1}$. Moreover, the length of this 'trapezoid top', i.e. the length of the longest string of equalities as in Equation (1) is bounded above by the absolute value of the signature. Conjecture (3) first appeared in [**Sto05**]. It is called the Strong Fox Conjecture in [**Ban22**]. Log-concavity is a ubiquitous phenomenon appearing in many fields of mathematics, for example, [**AHK17**] and [**Sta89**].

The Strong Fox Conjecture was proved for 2-bridge knots by Banfield [**Ban22**] and for special alternating links by Hafner–Meszaros–Vidinas [**HMV24**]. The conjecture, including the statement regarding equality, was verified by Stoimenow [**Sto16**, Section 9.2] for alternating knots of genus at most 4 and for certain plumbed links [**Sto21**, Theorem 5.1]

(4) Conjecture (Stoimenow): Let K be a positive link with Conway polynomial as above. Then the sequence of coefficients (c_i) is strongly log concave.

Conjecture (4) was proved for special alternating links by Stoimenow [**Sto05**, Theorem 1].

Proposed for K3 by: A. Stoimenow
Scribed by: D. Chen, D. Ruberman

PROBLEM 1.8. *Which multi-variable Laurent polynomials arise as the multi-variable Alexander polynomial of a link in the 3-sphere or, more generally, a homology 3-sphere?*

REMARKS.

(1) A link in a homology sphere Y has a covering corresponding to the homomorphism $\pi_1(Y \backslash L) \twoheadrightarrow \mathbb{Z}^n$ arising from the isomorphism $H^1(Y \backslash L) \cong \mathbb{Z}^n$, where n is the number of components of the link L. The homology of this covering gives rise to a multi-variable Alexander polynomial

$\Delta_L(t_1, .., t_n) \in \mathbb{Z}[t_1^{\pm 1}, ..t_n^{\pm 1}]$, well defined up to multiplication by units, which are all of the form $\pm t_1^{\nu_1} \cdots t_n^{\nu_n}$

(2) The problem has been solved for the single-variable Alexander polynomial of knots. If K is a knot in a homology sphere, then:
- $\Delta_K(t) = \Delta_K(t^{-1})$
- $\Delta_K(1) = 1$.

Moreover, any polynomial satisfying these two properties is realized by a knot in the 3-sphere (or any homology sphere) [**Sei35, Lev65, Rol90**]. Torres [**Tor53**] found analogous (necessary but not sufficient–see [**Hil81, Pla86**]) properties of the multi-variable Alexander polynomial of a link. It is perhaps surprising that while there is no known characterization of the Alexander polynomial of two-component links, Bailey found a characterization [**Bai77**] (discussed in [**Pla86**]) of their Alexander modules. A general reference for the ideas here is Hillman's book [**Hil12**].

(3) The problem can be refined by focusing on specific families of links.

QUESTION (i). Which polynomials arise for a given family of knots or links?

Some partial results are known.
- *Torus links.* If $T_{p,q}$ is the $(p,q)-$torus link, let $d = \gcd(p,q)$. Then
$$\Delta_{T_{p,q}}(t_1, t_2, \ldots, t_d) = \frac{[(t_1 t_2 \ldots t_d)^{pq/d} - 1]^d [(t_1 t_2 \ldots t_d) - 1]}{[(t_1 t_2 \ldots t_d)^{p/d} - 1][(t_1 t_2 \ldots t_d)^{q/d} - 1]}.$$
See [**Rol90**] for knots and [**EN85**, Theorem 12.1]; compare [**Ore21**, Corollary 8.2] for links with $d > 1$.
- *Two-bridge links.* Many authors have found necessary (but still insufficient) conditions on the Alexander polynomial of a 2-bridge knot or link. Results of Koseleff and Pecker [**KP15**] subsume most earlier results. Given a 2-variable polynomial, Hoste [**Hos20**], gives an algorithm to decide if the polynomial is the Alexander polynomial of a 2-component, 2-bridge link.
- *Alternating links.* The Alexander polynomials of alternating knots have special properties [**Cro59, Mur58**], and their coefficients are conjectured to have various concavity properties. See Problem 1.7 for an extensive discussion of these. It would be of interest to extend such properties to the case of alternating links.

(4) Similarly, any 3-manifold with $H^1(X) \cong \mathbb{Z}^n$ has an associated multi-variable Alexander polynomial $\Delta_X \in \mathbb{Z}[H^1(X)] \cong \mathbb{Z}[t_1^{\pm 1}, ..t_n^{\pm 1}]$.

QUESTION (ii). Which Laurent polynomials arise as the multi-variable Alexander polynomial of a 3-manifold?

Scribed by: M. Hedden, J. Hoste

1.2. Knots and 3-manifolds

Knots and links give rise to 3-manifolds in several ways. The simplest is that their complements are open 3-manifolds and their exteriors are 3-manifolds with boundary a union of tori. Hence the fundamental group of the complement, known as the *knot or link group*, is a knot invariant.

One can fill in the boundaries of a link exterior with solid tori. This process, which depends on a choice of slope (the simple closed curve on the torus that bounds a disk in the solid torus) for each solid torus, is known as *Dehn surgery*. We follow the convention that the closed curve representing p times the meridian plus q times the longitude corresponds to slope p/q. Every closed, connected, orientable 3-manifold can be obtained in this way from a link exterior in S^3, and the study of 3-manifolds that are obtained by Dehn surgery on a *knot* has a rich history; see [**Gor09**] for an introduction and survey.

One can also fill in the boundary of an arbitrary 3-manifold M with $\partial M = T^2$, where there may not be a standard parameterization of the boundary. In this setting, the slope r is just a homology class in ∂M. We denote by $M(r)$ the result of gluing the solid torus to M with meridional disk glued to the slope r, and call it the *Dehn filling*.

Additionally, one can take branched covers of S^3 branched along a link L. As with Dehn surgery, every 3-manifold can be obtained in this way. The simplest version of this construction is the n-fold cyclic branched cover of L, denoted $\Sigma_n(L)$.

PROBLEM 1.9. *If Dehn surgery on a knot K gives a lens space, then K is a Berge knot.*

REMARKS.

(1) This appears as [**Kir97**, Problem 1.78]. Let $(\Sigma; H_1, H_2)$ denote the genus-two Heegaard splitting of S^3 (cf. Section 3.3). A knot $K \subset S^3$ is a *Berge knot* if it lies on Σ and there are properly embedded disks $D_i \subset H_i$ such that $|K \cap D_i| = 1$. These knots are also called *doubly primitive*; see [**Gor09**, Section 3.2].

(2) The conjecture, often referred to as the *Berge Conjecture*, is known for small order lens spaces [**Gor09**]. For example, if surgery on K gives $L(p, q)$ with $|p| \leqslant 5$, then K is the unknot or the trefoil [**KMOS07**]. In general, if a lens space were to arise as surgery on a non-Berge knot, that lens space must also arise as surgery on a Berge knot [**Gre13b**]. The putative knot necessarily has the same knot Floer homology as that corresponding Berge knot.

(3) From its description, any Berge knot is necessarily a tunnel number one knot. The Berge conjecture has been verified for all tunnel number one knots in the preprint [**LMP25**] and all non-hyperbolic knots [**Mos71**, **BL89**].

(4) If surgery on a non-torus knot produces a lens space, then it must be an integral surgery by the Cyclic Surgery Theorem [**CGLS87**]. The size of the surgery coefficient is also heavily constrained. If $S^3_p(K) = L(p, q)$ for a nontrivial, non-trefoil knot K, $2g(K) - 1 \leqslant |p| - 2\sqrt{4(|p| + 1)/5}$ [**Gre13b**]. If K is a counterexample to the Berge conjecture, then $|p| < 4g(K) - 1$ [**Bak06**].

(5) Using Heegaard Floer homology, it is also known that a knot with a lens space surgery must be fibered and, after possibly mirroring, strongly quasi-positive [**Ni07, Hed10**]. The knot Floer homology of the dual knot $\widehat{K}$ in $L(p,q)$ has smallest possible knot Floer homology, $\dim \widehat{HFK}(\widehat{K}) = \dim \widehat{HF}(L(p,q))$ [**Ras07, Hed11, Gre13b**]. Such a knot is called *Floer simple*, and one proposed strategy is to prove that Floer simple knots are one-bridge with respect to the standard genus one Heegaard splitting of $L(p,q)$ [**BGH08**], as Berge proved that such knots are necessarily doubly primitive [**Ber18**].

Scribed by: T. Lidman

PROBLEM 1.10 (Generalized Property R Conjecture). *Let $L \subset S^3$ be an n-component link such that 0-framed Dehn surgery on L results in $\#^n(S^1 \times S^2)$. Is there a sequence of handleslides converting L into an n-component unlink?*

REMARKS.

(1) An n-component link $L \subset S^3$ whose 0-surgery is $\#^n(S^1 \times S^2)$ is called an *R-link*. An R-link is said to have *generalized property R* if it can be converted by handleslides to an unlink. (Note that all framings and linking numbers are zero.)

Gabai proved [**Gab87a**] that the only 1-component R-link is the unknot, solving the Property R Conjecture. The Generalized Property R Conjecture (GPRC) posits that all R-links have generalized property R [**Kir97**, Problem 1.82] and was studied in [**GST10**].

(2) There are two ways to weaken the GPRC that are of interest. Let $L \sqcup L'$ denote the split union of links L and L'.

CONJECTURE (Stable GPRC). *If L is an R-link, then there is an unlink U such that $L \sqcup U$ can be converted by handleslides to an unlink.*

Many potential counterexamples to the Stable GPRC are given in [**GST10, MZ22**].

(3) A *Hopf pair* is a canceling 1-handle/2-handle pair, which can be represented in the Kirby calculus by a Hopf link with a dotted-circle component and a 0-framed component [**GS99**, Section 5.4].

CONJECTURE (Weak GPRC). *If L is an R-link, then there is an unlink U and a split union H of Hopf pairs such that $L \sqcup U \sqcup H$ can be converted by handleslides to the split union $U' \sqcup H'$ of an unlink and a split union of Hopf pairs.*

A sufficient condition for a 2-component link to satisfy the Weak GPRC is given in [**MZ22**, Theorem 1.1].

(4) There are connections between these conjectures and other important problems. Given an R-link L, one can construct a homotopy 4-ball B_L by attaching 0-framed 2-handles to $S^3 \times I$ along L and capping off with a 4-dimensional 1-handlebody. Turning this handle-decomposition upside down gives a balanced presentation P_L of the trivial group.

Any homotopy 4-sphere that can be built without 1-handles is $B_L \cup B^4$ for some R-link L. If an R-link L satisfies the Weak GPRC, then B_L is diffeomorphic to B^4. So, the Weak GPRC implies the Smooth Poincaré Conjecture in dimension 4 for homotopy 4-spheres built without 1-handles.

The Andrews–Curtis Conjecture (ACC–see Problem 5.10) posits that any balanced presentation of the trivial group can be converted to the trivial presentation through balanced presentations. If the GPRC is true, then so is the ACC. There is also a stable version of the ACC that is implied by the Stable GPRC [**GST10, MZ22**].

If L is an R-link, then L bounds a collection of homotopy-ribbon disks in the homotopy 4-ball B_L. Thus, such links provide potential counterexamples to the Slice-Ribbon Conjecture (Problem 1.50) but are ribbon if they satisfy the Stable GPRC [**AT16b**]; see [**MZ25**, Proposition 5.2]. If K is a fibered, homotopy-ribbon knot, then the closed monodromy of K admits an extension across a handlebody [**CG83a**]. Any such extension gives rise to a link L lying on the fiber of K such that L and $K \cup L$ are both R-links, and $K \cup L$ is ribbon if L (equivalently, $K \cup L$) satisfies the Stable GPRC.

Scribed by: J. Meier

PROBLEM 1.11 (Cabling conjecture). *Let $K \subset S^3$ be a knot and $r \in \mathbb{Q}$. If r-framed Dehn surgery on K is not prime, then K is a nontrivial cable of a knot $J \subset S^3$ and r is the slope of the cabling annulus.*

REMARKS.

(1) The conjecture is due to González-Acuña and Short [**GAnS86**]. If $K = J_{p,q}$ is a cable knot, then there is an annulus in $S^3 \backslash \nu(K)$ that has boundary equal to two parallel curves of slope pq. This annulus is called the *cabling annulus* and pq is referred to as the slope of the cabling annulus. Using a cut and paste argument, one can show that $S^3_{pq}(K) = S^3_{q/p}(J) \# S^3_{p/q}(U)$, where U is the unknot. Therefore, if $p \neq 1$, then $S^3_{pq}(K)$ is not prime.

(2) Many special cases of the conjecture are known. Gordon and Luecke [**GL87**] prove that if r is a reducible slope, then r is an integer and the surgery contains a lens space summand. The conjecture is known for satellite knots [**Sch90**], strongly invertible knots [**EMn92**], alternating knots [**MT92**] and symmetric knots [**LZ94, HS98, HM97b**].

(3) Greene [**Gre15**] has shown that if r-framed surgery on K gives a connected sum of nontrivial lens spaces, then K is a (p, q)-cable of either the unknot or a torus knot and that $r = pq$.

(4) Dunfield [**Dun20**] observed that a similar conjecture may hold for knots in $S^1 \times S^2$. He conjectures that there are no hyperbolic knots in $S^1 \times S^2$ that have non-prime Dehn surgery.

(5) A weaker version of the cabling conjecture is the *two summands conjecture*, which says that if $S^3_r(K)$ is non-prime, then it cannot have more than two prime summands. Note that it follows from work of Howie [**How02**],

Sayari [**Say09**] and Valdez-Sánchez [**VS99**] that $S^3_r(K)$ can never contain more than three irreducible summands.

Scribed by: I. Zemke

PROBLEM 1.12. *This problem presents several variations on the Cosmetic Surgery Conjecture, discussed in turn below.*

(a) (*Cosmetic Surgery Conjecture*) *Two surgeries along inequivalent slopes are never purely cosmetic.*
(b) (*Oriented knot complement conjecture*) *If K_1 and K_2 are knots in a closed, oriented 3-manifold N whose complements are homeomorphic via an orientation-preserving homeomorphism, then there exists a homeomorphism of N taking K_1 to K_2.*
(c) *Let $K \subset S^3$ be a nontrivial knot, then K admits no purely cosmetic surgeries.*
(d) *If $K \subset S^3$ is not the torus knot $T_{2g+1,2}$, then K admits no chirally cosmetic surgeries with inequivalent slopes.*

REMARKS.

(1) This is an update to [**Kir97**, Problem 1.81], contributed by S. Bleiler. Parts (a) and (b) appear as Conjectures 6.1 and 6.2 of [**Gor91**] respectively.
(2) For the rest of this problem, M denotes an irreducible oriented 3-manifold with torus boundary that is not the solid torus. Two fillings along distinct slopes r_1 and r_2 are called *purely cosmetic* if there is an orientation-preserving homeomorphism between $M(r_1)$ and $M(r_2)$, and *chirally cosmetic* if there is an orientation-reversing homeomorphism between these fillings. (The terminology would be purely or chirally cosmetic *surgery* if M is a knot complement.) Two slopes are said to be *equivalent* if there exists a homeomorphism (either preserving or reversing the orientation) of M taking one slope to the other.
(3) When [**Kir97**, Problem 1.81] was written, several examples of chirally cosmetic surgeries on inequivalent slopes were known. More recently, Ichihara and Jong [**IJ18**], with help from Masai, found a hyperbolic one-cusped manifold admitting a pair of chirally cosmetic surgeries with inequivalent slopes, such that the resulting manifold is also hyperbolic. This gives a counterexample to [**Kir97**, Problem 1.81B].

    Dunfield found that the manifold $o9_{39009}$ from Burton's census [**Bur14**] had Dehn fillings $(-1, 3)$ and $(-3, 2)$, which are isometric by an orientation reversing isometry, but where the core curves of the Dehn fillings are geodesics of different lengths. This provides a counterexample to [**Kir97**, Problem 1.81C].
(4) There are different versions of the Cosmetic Surgery Conjecture (a) in the literature. A stronger version of the conjecture, which is still open, asserts that if there exist purely cosmetic surgeries on M, then there is an orientation-preserving homeomorphism of M sending one slope to the other. This version would imply that M does not admit purely cosmetic surgeries if M is hyperbolic. If there exist two slopes that are related

by an orientation-reversing homeomorphism, and the surgeries on these two slopes are amphichiral, then this would be a counterexample to the stronger version of the conjecture, but not to the version of the conjecture stated in this problem.

(5) Lackenby [**Lac97**] proved that, for "most" surgeries on null-homotopic knots in manifolds with $b_1 > 0$, the original manifold, the knot, and the slope are determined by the resulting manifold from the surgery.

When M is hyperbolic, there is an explicit constant $C = C(M)$, satisfying that if the normalized lengths of two inequivalent slopes r_1, r_2 are both greater than C, then this pair is not (purely or chirally) cosmetic [**FPS22, FPS25**].

Using code that is publicly available on GitHub [**FPS24**], Conjecture (a) has been verified for all the 59,107 one-cusped manifolds in the SnapPy census. This code can also be used to investigate Conjecture (c) above for knots in S^3.

(6) If M is the irreducible complement of a null-homologous knot in a 3-manifold of the form $Z\#S^1 \times S^2$, then the slopes of any pair of cosmetic surgeries must be $\{\pm r\}$ [**Ni13**]. A similar result holds when M is the complement of a knot in a closed 3-manifold whose Thurston norm is strictly smaller than that of M [**Ni11**].

(7) The Cosmetic Surgery Conjecture (a) is equivalent to the Oriented Knot Complement Conjecture (b), which is an analogue of the Knot Complement Theorem [**GL89**].

In [**Kir97**, Problem 1.81D], the homeomorphism of N is required to be orientation-preserving. Since our definition of equivalent slopes does not require the homeomorphism to preserve orientation, we do not require the homeomorphism of N to be orientation-preserving in the problem statement here.

Gordon and Luecke [**GL89**] proved Conjecture (b) for $N = S^3$ and $S^1 \times S^2$. It also holds for certain L-spaces [**Gai18, Rav22**]. Cremaschi–Yarmola proved it for knots in an orientable circle bundle over a genus $g \geqslant 2$ surface [**CY24**].

(8) When M is the complement of a knot $K \subset S^3$, the Cosmetic Surgery Conjecture is equivalent to saying that if K admits purely cosmetic surgeries with slopes r_1, r_2 then K is amphichiral and $r_1 = -r_2$. The conclusion that $r_1 = -r_2$ was proved in [**NW15**]. Conjecture (c) is stronger than this special case of Conjecture (a), because there may exist an amphichiral knot in S^3 such that a nonzero surgery on it is also amphichiral. This would be a counterexample to Conjecture (c), but not to Conjecture (a).

Futer–Purcell–Schleimer [**FPS25**] verified Conjecture (c) for knots with up to 19 crossings.

Hanselman [**Han23**] proved that if K admits a pair of purely cosmetic surgeries, then the pair of slopes is either $\{\pm 2\}$ or $\{\pm \frac{1}{q}\}$ for an integer q that is explicitly determined from the knot Floer homology of K, and the former case happens only when $g(K) = 2$. Daemi–Lidman–Miller Eisemeier [**DLME24**] subsequently announced that they have eliminated the case of $\{\pm \frac{1}{q}\}$. As a result, any pair of purely cosmetic slopes must be $\{\pm 2\}$, and the knot has genus 2 and trivial Alexander polynomial.

(9) Although [**Kir97**, Problem 1.81 (B)] has been disproved, no such examples have been found for knots in S^3. This is the motivation for Conjecture (d).
(10) There are many known constraints on chirally cosmetic surgeries; see [**IIS22**] for an overview. Ozsváth and Szabó [**OS11**] proved that if K admits a pair of (purely or chirally) cosmetic surgeries with slopes r_1, r_2, then either r_1, r_2 have opposite signs or the surgery is an L-space. Futer–Purcell–Schleimer [**FPS25**] have shown that Conjecture (d) holds for all hyperbolic knots with up to 15 crossings.
(11) Somewhat weaker versions of Conjecture (b), in which the knots are required to be null-homotopic, may be found in [**Kir97**, Problem 1.80 (B) and (C)]. Using standard arguments in Heegaard Floer homology, one can verify the conjecture in [**Kir97**, Problem 1.80(C)] when Y is an L-space. See, for example, [**Gai18, Rav22**].
(12) A list of problems about (mostly chirally) cosmetic surgeries on knots in S^3 can be found in [**Ito22**].

Proposed for K3 by: C. Gordon
Scribed by: Y. Ni

PROBLEM 1.13. *Let K be a null-homotopic knot in a 3-manifold Y, and let $Y_0(K)$ be the manifold obtained by 0-surgery on K.*

(a) *Conjecture: Let F be a Seifert surface for K, and let $\widehat{F} \subset Y_0(K)$ be the closed surface obtained by capping off ∂F with a disk. If $Y_0(K)$ is a surface bundle over S^1 with fiber in the homology class $[\widehat{F}]$, then K is a fibered knot with fiber in the homology class $[F]$.*
(b) *Conjecture: If F is taut in the exterior of K in Y, then $\widehat{F}$ is also taut in $Y_0(K)$.*

REMARKS.

(1) Conjecture (a) is an updating of [**Kir97**, Problem 1.80C], and is due to Boileau. Conjecture (b) is due to Gabai.
(2) Let M be a compact, oriented, connected 3-manifold, let $F \subset M$ be a properly embedded connected surface. We say that F is *taut* if it realizes the Thurston norm in the homology class $[F] \neq 0 \in H_2(M, \partial M)$, or if F is a sphere.
(3) In the original version of Conjecture (a), the surgery slope can be any rational number, and there is no restriction on the homology class of the fiber. However, there are simple counterexamples to the original question [**Ni09**]. More precisely, there exist nonfibered null-homotopic knots such that every integral surgery on the knot yields a manifold that fibers over S^1.
(4) It is known that $Y_0(K)$ does not contain a homologically essential 2-sphere if $Y \backslash K$ does not contain such 2-spheres and K is nontrivial [**Gab87a, Lac97, HL22, Ni23b**].

Conjecture (a) and Conjecture (b) are known for knots in S^3 [**Gab87b**], L-spaces [**Ni07**, **AN09**], and certain reducible manifolds [**Gab87b**, **Ni13**]. They are also known for knots obtained from the unknot by crossing changes of the same sign [**Ni24**].

Scribed by: Y. Ni, M. Boileau

PROBLEM 1.14. *For which nonzero $r \in \mathbb{Q}$ is it true that for every nontrivial knot $K \subset S^3$ there is a homomorphism*

$$\pi_1(S^3_r(K)) \to \mathrm{SU}(2)$$

with nonabelian image?

REMARKS.

(1) If this statement is true for some $r \in \mathbb{Q}$, then it's true for $-r$; hence, we assume $r > 0$. Kronheimer–Mrowka proved [**KM04**] that there is a homomorphism with nonabelian image for all $r \in (0, 2]$. Work of Baldwin–Sivek [**BS23a**] together with announced results of Baldwin–Li–Sivek–Ye [**BLSY24**], and Farber–Reinoso–Wang [**FRW24**] proves the same for all $r \in (2, 5)$ with prime power numerators.

(2) Let $r = p/q > 0$ with p and q relatively prime, and assume that $\Delta_K(\zeta^2) \neq 0$ for any pth root of unity ζ. Then it is further known [**BS23a**] that if $\pi_1(S^3_r(K))$ only admits abelian SU(2)-representations, then K is fibered, strongly quasipositive, and $r \geqslant 2g(K) - 1$.

(3) Note that for most $r \in \mathbb{Z}$, there is a torus knot whose r–surgery is SU(2)-abelian. Problem 3.51 deals with a related problem about the characterization of rational homology spheres whose fundamental groups have only Abelian SU(2) representations.

Scribed by: J. Baldwin

PROBLEM 1.15.

(a) *Are there integral homology spheres with arbitrarily large (integral) Dehn surgery number? Are there irreducible examples? Does the connected sum of n nontrivial homology spheres have Dehn surgery number at least n?*

(b) *Given a 3-manifold Y with a smooth orientation-preserving $\mathbb{Z}/n\mathbb{Z}$ symmetry ϕ, Sakuma* [**Sak01**] *(generalizing* [**PS01**]*) proved that Y admits a surgery description compatible with the symmetry. Define the equivariant (integral) Dehn surgery number of (Y, ϕ) as the minimal number of connected components of any such link. Do there exist (Y, ϕ) whose equivariant (integral) Dehn surgery number is arbitrarily larger than the (integral) Dehn surgery number?*

(c) *Following the remarks below, we can consider the following specific case: Does there exist a manifold with a 2-fold symmetry whose equivariant (integral) Dehn surgery number is larger than the (integral) Dehn surgery number?*

REMARKS.

(1) Any oriented connected closed 3-manifold Y is obtained by (integral) Dehn surgery on a link in the 3-sphere, and the minimal number of connected components of any such link is called the (integral) Dehn surgery number of Y.
(2) Part (a) is a version of Problem 3.102 from [**Kir97**], asked by Auckly. The integral Dehn surgery case is interesting to consider from the perspective of 4-manifolds bounded by Y, as it is related to the minimum number of handles needed to describe the 4-manifold.

    There are many examples of integral homology spheres with Dehn surgery number at least 2, the earliest due to [**GL87**], hyperbolic examples due to [**Auc97**], and Seifert fibered examples due to [**HKL16b**].
(3) Regarding part (b), it should be possible to show that the equivariant surgery number does not agree with the surgery number. Consider $Y = S^3_2(4_1)$, which is a Seifert fibered space (see, for example, [**BW01**, Theorem 1.1(4))]). Since Y is Seifert fibered, it has a $\mathbb{Z}/n\mathbb{Z}$ action, for all n, while 4_1 does not. It remains to be shown that there is no other knot K for which some $2/q$-surgery is Y; a plausible approach is to use the mapping cone formula for Heegaard Floer homology.

    Finding the examples as in part (b) for which the equivariant surgery number is arbitrarily larger than the Dehn surgery number is, as in part (a), more challenging.
(4) Lastly, we point out that this problem is closely related to Problem 3.39.

Scribed by: J. Hom

PROBLEM 1.16.

(a) *Given a knot $K \subset S^3$ determine all knots $K' \subset S^3$ for which the branched double covers of S^3 along K and K' are homeomorphic.*
(b) *Find a set of moves on links in S^3 so that two links are related by these moves if and only if they have the same branched double cover.*

REMARKS.

(1) These are Problems 3.25 and 1.22 in [**Kir97**], respectively. See Problem 1.17 for a related problem about alternating links. Let $\Sigma(K)$ denote the branched double cover of S^3 along K.
(2) It follows from the Orbifold Geometrization Theorem that there are finitely many K' with $\Sigma(K') \cong \Sigma(K)$. See [**Pao05**] for a discussion.
(3) Mecchia and Zimmerman showed that if $\Sigma(K)$ is hyperbolic, then there are at most eight other K' with $\Sigma(K') \cong \Sigma(K)$ [**MZ04**]. Kawauchi showed that this bound is sharp, exhibiting infinitely many 9-tuples of distinct hyperbolic knots in S^3 with homeomorphic branched double covers [**Kaw06**]. However, his construction is very non-explicit, and it would be interesting to have some explicit constructions. Mecchia further described how hyperbolic knots with the same double branched cover can be understood in terms of isometries of the cover [**Mec01**]
(4) If K is a 2-bridge knot and $\Sigma(K') \cong \Sigma(K)$ then $K' = K$ [**HR85**].

(5) If K and K' are related by Conway mutation then $\Sigma(K') \cong \Sigma(K)$. There are also non-mutant knots with the same branched double covers, including $T(3,7)$ and $P(-2,3,7)$. In this example, the Khovanov homologies have the same total rank, but Watson [**Wat10**] gives non-mutant knots with homeomorphic branched double covers the total rank of whose Khovanov homologies differ, showing that the rank of Khovanov homology is not an invariant of branched double covers. See also [**Pao01**].

(6) Higher order cyclic branched covers are generally much better knot invariants. For example, work of Zimmerman [**Zim98**] shows that if $n \geqslant 3$ and $K \subset S^3$ is hyperbolic, then the n-fold cyclic branched cover of K distinguishes K from other knots so long as K does not admit periodic symmetries of order n, and the latter condition is satisfied whenever $n > 2g(K) + 1$. See [**Pao05**] for further discussion.

(7) Extending work of Montesinos [**Mon85**], Piergallini constructed a complete set of moves relating links with the same simple, irregular 3-fold branched covers [**Pie91**].

Scribed by: J. Baldwin, T. Lidman

PROBLEM 1.17. *Can an alternating link and a non-alternating link have homeomorphic branched double covers?*

REMARKS.

(1) The answer is conjectured to be no in [**Gre13a**, Conjecture 1.4] since Greene shows that two alternating links have homeomorphic branched double covers if and only if they are Conway mutants. Note that Viro proved that mutant links have homeomorphic branched double-covers [**Vir76**, Theorem 1], and Menasco proved that mutation preserves the property of being alternating [**Men84**, Proof of Theorem 3(b)].

(2) The problem is related to Problems 1.22 and 3.25 of [**Kir97**] which ask for a kind of classification of pairs of knots with homeomorphic branched double covers.

(3) If a Seifert-fibered space is the branched double-cover of a pair of distinct links, then the pair consists of either a non-alternating torus link and a non-alternating Montesinos link, or else two mutant Montesinos links [**MR02**, Introduction].

(4) Dunfield investigated knots with at most 16 crossings whose branched double-covers are hyperbolic and of small enough volume to appear in the Hodgson-Weeks census of closed hyperbolic 3-manifolds. He reports 3765 non-alternating knots with such branched covers but only 178 alternating knots, and no manifold occurs as the branched double-cover of both kinds of knots (private correspondence from 2013).

(5) While there exist many constructions of pairs of non-mutant links with homeomorphic branched double-covers, these constructions typically appear to produce non-alternating examples (cf. [**MW86**, **MZ04**]).

Scribed by: J. Greene

PROBLEM 1.18.

(a) *(Meridional Rank Conjecture) Is the meridional rank* $\mu(L)$ *of every link* L *equal to its bridge number* $b(L)$*?*
(b) *Given two knots* $K_1, K_2 \subset S^3$ *such that* $\pi_1(S^3\backslash K_1)$ *and* $\pi_1(S^3\backslash K_2)$ *have the same finite quotients, does* $\mu(K_1) = \mu(K_2)$*? In particular, is it possible to explicitly detect the meridional rank* $\mu(K)$ *of a knot* $K \subset S^3$ *from the finite quotients of* $\pi_1(S^3\backslash K)$*?*
(c) *Given two links* $L_1, L_2 \subset S^3$*, if there exists a surjective homomorphism* $\pi_1(S^3\backslash L_1) \twoheadrightarrow \pi_1(S^3\backslash L_2)$ *that sends the meridians of* L_1 *to the meridians of* L_2*, do the bridge numbers satisfy* $b(L_1) \geqslant b(L_2)$*?*
(d) *Given two knots* $K_1, K_2 \subset S^3$*, if there exists a surjective homomorphism* $\pi_1(S^3\backslash K_1) \twoheadrightarrow \pi_1(S^3\backslash K_2)$ *(which may not send meridians to meridians):*
    (i) *Do the meridional ranks satisfy* $\mu(K_1) \geqslant \mu(K_2)$*?*
    (ii) *Do the bridge numbers satisfy* $b(K_1) \geqslant b(K_2)$*?*
(e) *Let* $K \subset S^3$ *be a hyperbolic knot. Is every subgroup of* $\pi_1(S^3\backslash K)$ *generated by at most* $b(K) - 1$ *meridians free? See* [**BJW18**].

REMARKS.

(1) Problem 1.11 of [**Kir97**] poses Question (a) for knots, and was proposed by Cappell and Shaneson.
(2) The *meridional rank* $\mu(L)$ of a link L is the minimal number of meridians needed to generate the fundamental group of its complement. It is bounded above by the bridge number $b(L)$.
(3) Question (b) has a negative answer for links: let W_n denote the result of applying $|n| \geqslant 2$ twists to the Whitehead link W_0. Then W_n and W_0 have homeomorphic exteriors, but $\mu(W_0) = b(W_0) = 2$, while $\mu(W_n) = b(W_n) = 3$.

    Note that the homeomorphism between the exterior of W_n and W_0 discussed above does not send meridians to meridians. Under the meridian-preserving hypothesis, the epimorphism of groups in Question (c) implies $\mu(L_1) \geqslant \mu(L_2)$, so a negative answer to Question (c) implies a negative answer to the Meridional Rank Conjecture.
(4) In the case of knots, Question (d) represents a strengthening of Question (c).
(5) Different answers to the two items under Question (d) would imply a negative answer to the Meridional Rank Conjecture. Note that there are many epimorphisms between knot groups that are not meridian-preserving [**GAn75**]. For closed orientable manifolds, the answer to the analogous question with respect to the Heegaard genus is negative if one does not assume the epimorphism to be induced by a non-zero degree map (see [**BF18**]). The non-zero degree condition, in the knot complement case, implies that the epimorphism is meridian preserving.
(6) In Question 4 the bridge number is defined using bridge trisections (see [**MZ17a, JMMZ22**] and [**AAD**$^+$**23**, Question 6.1]; cf. Problem 4.111) or, equivalently, using Morse position.

    By [**JP25**], the meridional rank of 2-spheres in S^4 can collapse under connected sum; therefore, either additivity of bridge number or the equality of bridge number and meridional rank (or both) fails for 2-knots.

(7) Problem 4.117 describes a four-dimensional analog of the Meridional Rank Conjecture.

Scribed by: A. Kjuchukova, D. Ruberman

PROBLEM 1.19. *Let* $Y = Y_1 \# Y_2$ *be a connected sum of 3-manifolds with* $Y_i \neq S^3$*, for* $i = 1, 2$*. Let* $\Phi \colon Y \to Y$ *be a Dehn twist around the connected sum separating 2-sphere, and assume* Y_1 *and* Y_2 *are such that* Φ *is not isotopic to the identity. Does there exist a knot* $K \subseteq Y$ *such that* K *and* $\Phi(K)$ *are not ambiently isotopic?*

REMARKS. Aceto-Bregman-Davis-Park-Ray [**ABD⁺20**] studied the analogous question for knots in prime 3-manifolds. They showed that the Gluck twist on $S^1 \times S^2$ (see Section 4.1 for the definition) can be detected by a knot that is equivalent but not isotopic to its image under the Gluck twist. They also showed that for irreducible 3-manifolds, diffeomorphisms that preserve free homotopy classes of loops are isotopic to the identity. Hence the key remaining question is whether Dehn twists around connected sum spheres can be detected by knots that are equivalent but not isotopic.

Scribed by: M. Powell

1.3. Heegaard Floer and instanton knot homology

Floer homologies have become pervasive in low-dimensional topology over the past few decades. Floer theory, in different manifestations, produces invariants of many objects: knots, mapping classes of surfaces, 3-manifolds, 4-manifolds, etc. While there have been numerous applications of these theories, understanding structural properties of these invariants has proved fruitful as well. This section focuses on a few problems on Floer homology theories for knots.

For concreteness, we discuss basic principles for the 'hat' Heegaard knot Floer homology for knots in the 3-sphere, denoted $\widehat{HFK}$, which produces an abelian group out of a bridge decomposition of a knot, through an analysis of certain Lagrangian tori in a symmetric product of a surface [**OS04c**, **Ras03**]. This invariant admits a bigrading, $\widehat{HFK}(S^3, K) = \oplus_{m,a} \widehat{HFK}_m(S^3, K, a)$, and categorifies the symmetrized Alexander polynomial, e.g.

$$\sum_{m,a} (-1)^m \dim \widehat{HFK}_m(S^3, K, a) = \Delta_K(t).$$

This invariant strengthens what the Alexander polynomial can see: knot Floer homology detects the Seifert genus [**OS04b**], i.e. $\widehat{HFK}(S^3, K, n) = 0$ for $|n| > g(K)$ and $\widehat{HFK}(S^3, K, g(K)) \neq 0$, and fiberedness since $\widehat{HFK}(S^3, K, g(K)) = \mathbb{Z}$ if and only if K is a fibered knot [**Ghi08**, **Ni07**]. A further benefit is that knot Floer homology (with $\mathbb{F}_2$-coefficients) is functorial under surface cobordisms between knots [**Juh16**, **Zem19b**], and hence has applications to problems in knot concordance. Knot Floer homology extends to knots in other 3-manifolds and is in fact a refinement of Heegaard Floer homology for 3-manifolds, the simplest version is denoted $\dim \widehat{HF}$. For more background on the invariants of 3-manifolds, see

Section 3.6. There is also a version with richer structure, HFK^-. For surveys on knot Floer homology, see for example [**Man16a, Hom21**]. There are many other incarnations of Floer homology theories for knots, including ones coming from Yang-Mills gauge theory (instanton knot Floer homology), Seiberg-Witten gauge theory (monopole knot Floer homology), and symplectic field theory (embedded contact knot homology), which share many structural properties (or are isomorphic) with $\widehat{HFK}$.

PROBLEM 1.20.

(a) *Are there any null-homologous Floer minimal knots with irreducible complements other than the Borromean knots B_g, $g \geqslant 0$, in any 3-manifolds?*

It is reasonable to make the following more concrete conjectures.

(b) *Conjecture (Ni): If $b_1(Y) < 2$, then the only null-homologous Floer minimal knot in Y is the unknot.*

(c) *Conjecture (Ni): If the Thurston norm of Y is not identically zero, then Y does not contain any null-homologous Floer minimal knot with irreducible complement.*

(d) *Conjecture (Hedden, Rasmussen): The only Floer minimal knots in lens spaces are simple knots.*

REMARKS.

(1) Let Y be a closed, oriented, connected 3-manifold, and let $K \subset Y$ be a rationally null-homologous knot. It is known that

$$\mathrm{rank}\widehat{HFK}(Y, K) \geqslant \mathrm{rank}\widehat{HF}(Y).$$

We say K is **Floer minimal** if equality in the above inequality holds. Clearly, unknots are Floer minimal. Moreover, the Borromean knot [**OS04c**, Figure 16] $B_g \subset \#^{2g}(S^1 \times S^2)$ is Floer minimal. If the p-surgery on a knot $L \subset S^3$ is an L-space and $p \geqslant 2g(L)$, then the dual knot in the L-space is Floer minimal [**Hed11, Ras07**].

It is known that the only null-homologous Floer minimal knot in Y is the unknot, when Y is an L-space [**NW14**], and when Y fibers over S^1 with fiber of genus > 1 [**Ni14**].

(2) When Y is the lens space $L(p, q)$, there is a standard genus-1 Heegaard diagram with p intersection points. Any two base points z, w will specify a Floer minimal knot. Such knots are called *simple*. Hedden [**Hed11**] and Rasmussen [**Ras07**] proved that Conjecture (d) would imply the Berge Conjecture (Problem 1.9 and [**Kir97**, Problem 1.78]).

Proposed for K3 and scribed by: Y. Ni

PROBLEM 1.21.

(a) *For a given positive integer g, are there only finitely many L-space knots of genus g?*

A related but more general question is:

(b) *Question (Hedden): For which knots $L \subset S^3$ are there only finitely many knots $K \subset S^3$ with $\widehat{HFK}(S^3, K) \cong \widehat{HFK}(S^3, L)$?*

(c) *In some special cases, one expects stronger results saying that a class of knots is characterized by its knot Floer homology. We state 3 conjectures, ordered by their strength.*

(i) *For a knot* $K \subset S^3$, *if* $\widehat{HFK}(S^3, K) \cong \widehat{HFK}(S^3, T_{2g+1,2})$, *then* $K = T_{2g+1,2}$.

(ii) *For a knot* $K \subset S^3$, *if* $\widehat{HFK}(S^3, K) \cong \widehat{HFK}(S^3, T_{p,q})$, *then* K *is an iterated torus knot.*

(iii) *Conjecture (Ni): For an L-space knot* $K \subset S^3$, *if all roots of* $\Delta_K(t)$ *lie on the unit circle, then* K *is an iterated torus knot.*

REMARKS.

(1) A knot $K \subset S^3$ is called an *L-space knot* if there exists a positive integer p such that $S^3_p(K)$, the p-surgery on K is an L-space (see Problem 3.48).

(2) The answer to Question (a) is "yes" when $g \leqslant 2$ [**Ghi08, FRW24**].

If K is an L-space knot, then K is fibered and strongly quasipositive [**Ni07, Hed10**],

$$\widehat{HFK}(S^3, K, g(K) - 1) \cong \mathbb{Z}_{(-1)}, \tag{2}$$

where $\mathbb{Z}_{(-1)}$ means a copy of $\mathbb{Z}$ supported in the Maslov grading -1 [**OS05b**]. (The condition (2) implies that the monodromy of the fibration is freely isotopic to a diffeomorphism without fixed point [**BHS25, Ni23a, GS22a**].) However, these properties are not enough to guarantee the finiteness. For each $g \geqslant 2$, Misev [**Mis21**] constructed infinitely many strongly quasipositive fibered knots with the same Seifert form as the torus knot $T_{2g+1,2}$. By taking cables of these knots, the condition (2) can also be satisfied.

(3) The finiteness in Question (b) does not hold for all knots L: see [**HW18**, Acknowledgements], [**MS15**], and [**Wan22**, Theorem 1.3]. In particular, Hedden and Watson showed that, given any non-trivial band sum of two unknots, altering the number of twists in the band gives rise to an infinite family of distinct knots with the same knot Floer homology [**HW18**, Theorem 1].

(4) Conjecture (i) in part (c) is known for $g = 1, 2$ [**Ghi08, FRW24**]. Conjecture (iii) would yield a positive answer to Problem 1.22; see [**BBG19a**].

Scribed by: Y. Ni

PROBLEM 1.22. *If* K *is a hyperbolic L-space knot, show that its branched cover* $\Sigma_2(K)$ *is not an L-space.*

REMARKS. This is a conjecture of A. Moore. The corresponding statement for higher order covers $\Sigma_n(K)$ was shown by Boileau-Boyer-Gordon [**BBG19a**] for $n \geqslant 4$ and by [**FRW24**] for $n = 3$. Boyer-Gordon-Hu [**BGH25**, Corollary 7.3] have announced that $\Sigma_n(K)$ has a left-orderable group for each $n \geqslant 2$, so in particular, a positive solution would be in accordance with the L-space conjecture, Problem 3.48.

Proposed for K3 by: A. Moore

PROBLEM 1.23. *Let K be a cubic graph embedded in the plane, and let* Tait(K) *be the number of Tait colorings of K.*

(a) *Is* $\dim J^\sharp(K) = \text{Tait}(K)$*?*
(b) *Is* $J^\flat(K)$ *nonzero when K is bridgeless?*

REMARKS.

(1) Part (a) is Conjecture 1.2 in [**KM19a**].
(2) A Tait coloring of a cubic (also called trivalent) graph is a labeling of the edges of the graph by three colors such that edges of all of the colors are incident at each vertex. In [**KM19a**], Kronheimer–Mrowka used gauge theory to assign to a web $K \subset \mathbb{R}^3$ a finite-dimensional vector space $J^\sharp(K)$ over $\mathbb{Z}/2\mathbb{Z}$, and proved that $J^\sharp(K)$ is zero if and only if K has an embedded bridge. They also proposed a combinatorial counterpart $J^\flat(K)$, which Khovanov–Robert [**KR21**] showed is well-defined for planar webs $K \subset \mathbb{R}^2$. Work of Kronheimer–Mrowka [**KM19b**] and Khovanov–Robert [**KR21**] implies that

$$\dim J^\flat(K) \leqslant \text{Tait}(K) \leqslant \dim J^\sharp(K)$$

for any planar web K. The Four Color Theorem is equivalent to the statement that every bridgeless cubic planar graph has a Tait coloring, and a new proof would therefore follow from an affirmative answer to either of the questions in this problem.
(3) Boozer [**Boo23**] gave numerical evidence suggesting that the inequality $\dim J^\flat(K) \leqslant \text{Tait}(K)$ may sometimes be strict. In [**Boo25**], he also announced that $J^\flat$ and $J^\sharp$ do not coincide as functors.

Scribed by: J. Baldwin

1.4. Jones polynomial and related invariants

The Jones polynomial was discovered in 1984 [**Jon85**] and has been intensely studied since then with exciting results, including proofs of the classical Tait Conjectures [**Kau87, Mur87, Thi87, Thi88, MT93**]. Despite heavy development and investigation, many aspects of this invariant remain mysterious and connections to other, more geometric, aspects of low-dimensional topology are generally unknown. Much of the interest in the Jones polynomial is motivated by its algebraic structure and connections to theoretical physics and hyperbolic geometry. See [**Lê16**] for a thorough introduction to the Jones polynomial and its quantum topological generalizations, especially the colored Jones polynomial and skein modules, and connections to other aspects of knot theory, such as alternating knots and the A-polynomial (via the AJ-conjecture).

PROBLEM 1.24 (Jones Unknot Detection).

(a) *Is there a nontrivial knot with the same Jones polynomial as the unknot?*
(b) *Does there exist a nontrivial knot whose colored Jones polynomials are all trivial?*
(c) *Is there a nontrivial knot with trivial HOMFLYPT polynomial? (A knot with trivial HOMFLYPT polynomial would have trivial Jones polynomial.)*

REMARKS.

(1) Question (a) was first proposed by V. Jones; see [**Kir97**, Problems 1.88(a) and (c)] and [**AALM00**].
(2) Computer calculations by Sikora and Tuzun [**TS21**] have shown there is no nontrivial knot with up to 24 crossings with trivial Jones polynomial.
(3) As observed in [**MM01**], a positive answer to Question (b) would follow from the Volume Conjecture stated in Problem 1.27.
(4) By tabulating links, Thistlethwaite found links with two or more components with trivial Jones polynomials [**Thi01**]. This was extended in [**EKT03**] to infinite families of links with trivial Jones polynomials. But these examples do not have trivial HOMFLYPT polynomials.
(5) Kronheimer and Mrowka have proven that a knot has trivial Khovanov homology if and only if it is trivial [**KM11**].

Scribed by: J. Hoste, H. Wong

PROBLEM 1.25.

(a) *Conjecture: The noncommutative A-ideal of a knot K is exactly the annihilator of the colored Jones polynomial (the infinite dimensional vector whose entries are the colored Jones polynomials of K).*
(b) *(AJ Conjecture) Let $A(K; L, M)$ be the A-polynomial of the knot K, and $\alpha(K, q; M, L)$ a polynomial of minimal degree in L and co-prime coefficients that annihilates the colored Jones polynomial $J_n(K; q)$. Then*

$$\alpha(K, -1; M, L) = A(K; L, M)$$

up to a factor depending on M.

REMARKS.

(1) The version of Conjecture (a) for $\mathfrak{sl}_2$ appears in [**FGL02**]; the full version is stated in [**Sik08**]. The AJ conjecture (b) is due to Garoufalidis [**Gar04**].
(2) The noncommutative A-ideal of $S^3 - K$ is the left ideal in the Kauffman bracket skein algebra of the torus that annihilates the empty skein in the knot complement. [**FGL02**] shows that a series built from the colored Jones polynomial $J_n(K; q)$ annihilates the noncommutative A-ideal, which leads to recursive relationships satisfied by $J_n(K; q)$.

    The noncommutative A-ideal is computed for 3_1 [**Gel02**], $(2, 2p + 1)$ torus knots [**GS03**], and 4_1 [**GS04**]. Demonstrating the nontriviality of the noncommutative A-ideal remains a significant unresolved problem.
(3) The A-polynomial of a knot K [**CL98**] is the defining polynomial for the character variety formed by $\mathrm{SL}_2(\mathbb{C})$-representations of $\pi_1(\partial(S^3 - K))$ that extend to $\pi_1(S^3 - K)$. See [**GL05**] for details on the definition

of $\alpha(K, q; M, L)$, which comes from linear recursion relations satisfied by $J_n(K;q)$. The AJ Conjecture is proved for certain 2-bridge knots [**Lô6, LZ17**] and pretzel knots [**LT15**], and a diagrammatic approach is suggested in [**DG20**]. See also [**Guk05**] for the physics context of the AJ conjecture.

(4) The two Conjectures are closely intertwined, and the AJ Conjecture is an important step towards understanding the relationship between classical and quantum invariants.

Scribed by: C. Frohman, H. Wong

PROBLEM 1.26 (Jones Slope Conjecture). *For a knot K, the Jones slopes $js(K)$ are the set of cluster points in*

$$\left\{\frac{4}{n^2}\deg_+(J_n(K;q)\right\}_{n\in\mathbb{N}} \bigcup \left\{\frac{4}{n^2}\deg_-(J_n(K;q)\right\}_{n\in\mathbb{N}},$$

where $\deg_+$ *denotes the highest degree and* $\deg_-$ *the lowest degree.*

REMARKS.

(1) The Jones slopes of a knot K are a subset of the boundary slopes of incompressible surfaces, boundary-incompressible orientable surfaces in $S^3 - K$.

(2) This conjecture was originally stated by Garoufalidis [**Gar11**]. There is a stronger version of the conjecture due to Kalfagianni and Tran [**KT15**]. Garoufalidis proved his conjecture for up to 10 crossings [**Gar11**], and the Slope Conjecture or its stronger form has been proved for various families of knots [**FKP11, GvdV16, LvdV16, Lee22, BMT21, BMT20, MT17, LYL19, GLVDV20**].

Scribed by: C. Frohman, H. Wong

PROBLEM 1.27 (Kashaev–Murakami–Murakami Volume Conjecture). *For a link $L \subset S^3$,*

$$\frac{1}{2\pi}\mathrm{Vol}_{\mathrm{hyp}}(S^3 - L) = \lim_{n\to\infty}\frac{1}{n}\log\left|J_n(L, e^{\frac{2\pi i}{n}})\right|.$$

REMARKS. Here, $J_n(L,q)$ is the nth colored Jones polynomial of L in the variable q, and $\mathrm{Vol}_{\mathrm{hyp}}(S^3 - L)$ is the hyperbolic volume of the link complement. This problem was originally proposed for a complex-valued invariant for triangulated links in S^3 [**Kas97**], which was later shown to be determined by the colored Jones polynomial of the link complement [**MM01**]. Note that this conjecture would imply that the collection of colored Jones polynomials would characterize the unknot; see Problem 1.24(b). The conjecture has been verified for some knots, including knots with up to seven crossings [**Oht16, OY18, Oht17**], torus knots [**KT00**], Whitehead doubles of $(2, b)$ torus knots or links [**Zhe07**], Whitehead chains, [**vdV08**], and certain cables of 4_1 [**LT10, MT25**].

Note that there is also a stronger complexified version involving the Chern–Simons invariant. See [**Mur07, MMO$^+$02**].

Scribed by: C. Frohman, H. Wong

PROBLEM 1.28. *Let L be a link in the thickened annulus $S^1 \times I \times I$.*

(a) Wrapping conjecture*: $w(L)$ is equal to the maximal nonzero annular degree of the annular Kauffman bracket of L.*
(b) Categorifed wrapping conjecture*: $w(L)$ is equal to the maximal nonzero annular grading of the annular Khovanov homology of L.*

REMARKS.

(1) The *wrapping number* $w(L)$ of L is the minimum number of times L intersects a meridional disk.
(2) The wrapping number $w(L)$ is less than or equal to the maximal nonzero annular degree of the annular Kauffman bracket of L, as well as the maximal nonzero annular grading of the annular Khovanov homology of L. The problem is asking whether these upper bounds are necessarily achieved; that is, whether either of these invariants detects the wrapping number.
(3) The wrapping conjecture is due to Hoste–Przytycki [**HP95a**], who proved it for any annular link with a $\pm$-adequately wrapped diagram.
(4) The categorified wrapping conjecture is due to Grigsby, who proved with Ni that it holds for string links [**GN14**]. This conjecture would follow from the wrapping conjecture since the graded Euler characteristic of annular Khovanov homology is equal to the annular Kauffman bracket. It is known via spectral sequences relating annular Khovanov homology with Floer-theoretic invariants [**GW10, Xie21, XZ25**] that annular Khovanov homology is nontrivial in the annular grading given by the *generalized Thurston norm* of the meridional disk. The difficulty is that the generalized meridional Thurston norm can be arbitrarily smaller than the wrapping number; see Martin's work for this and related progress [**Mar23**].

Scribed by: J. Baldwin

1.5. Khovanov and related homologies

In his seminal paper [**Kho00**] Khovanov categorified the Jones polynomial in the form of a bigraded homology theory, now known as Khovanov link homology, a breakthrough that led to a number of categorifications in knot theory. It was noted early on that Heegaard Floer knot homology (which categorifies the Alexander polynomial, and is discussed in Section 1.3) has many parallels with Khovanov homology. Applications of Khovanov homology include Rasmussen's integer-valued s-invariant which was used to give a combinatorial proof of Milnor's conjecture on the unknotting number of torus knots [**Ras10**] and is part of a strategy to disprove the smooth Poincaré Conjecture [**FGMW10**]. Khovanov homology detects the unknot [**KM11**], a question that is still open for the Jones polynomial, and provides a lower bound on the unknotting number [**AD19**].

Khovanov link homology, also known as the $\mathfrak{sl}_2$-link homology, has inspired categorifications associated with the representation theory of higher $\mathfrak{sl}_n$s, as well as other quantum invariants such as the triply graded HOMFLYPT (Khovanov-Rozansky) homology [**KR08b**] (which categorifies the HOMFLYPT polynomial), categorifications associated to other representations of Lie algebras, and skein algebras. These invariants have rich algebraic structure and exciting connections to other fields, such as representation theory, graph theory, and symplectic geometry, and hence have been a major focus of the field. For a brief overview on these invariants, see [**Ras21, KL23b**]. The problems in this section focus on these aspects, while many others in the list also explore connections with other topics in low-dimensional topology.

PROBLEM 1.29. *Is the first inequality below true? If so, is the second?*

(a) (*Vol-Det Conjecture*) *For any alternating hyperbolic knot* $K \subset S^3$,
$$\mathrm{vol}(K) < 2\pi \log \det(K),$$
where vol is the hyperbolic volume of K *and det is the determinant.*

(b) *For any hyperbolic knot* $K \subset S^3$,
$$\mathrm{vol}(K) < 2\pi \log \operatorname{rank} \overline{Kh}(K),$$
where $\overline{Kh}$ *is the reduced Khovanov homology of* K*?*

REMARKS.

(1) The proposed inequalities relate quantum topology and hyperbolic geometry in the style of the Volume Conjecture [**Kas97, MM01**] (see Problem 1.27), but may be more approachable.
(2) A correlation between determinant and volume for alternating hyperbolic knots was first noticed by Dunfield, and Champanerkar, Kofman and Purcell conjectured in [**CKP16**] that (1) holds. They moreover showed that the bound is sharp, and conjectured the second more general inequality (b).
(3) Burton verified (a) for several families of knots including 2-bridge knots, closed alternating 3-braids, and an infinite family of alternating 4-braids [**Bur18**].
(4) Champanerkar, Kofman and Lalin proved (a) for other infinite families of alternating knots using number theoretic methods in [**CKL19**], and gave related conjectures involving other volume-type quantities and the Mahler measures of certain graph-theoretic polynomials. See also [**CK25**].

Proposed for K3 by: A. Champanerkar, I. Kofman, J. Purcell
Scribed by: J. Baldwin, J. Boninger

PROBLEM 1.30. *Does the Khovanov homology of every nontrivial knot contain 2-torsion?*

REMARKS. This is a conjecture due to Shumakovitch [**Shu14**] originally posted in 2004. It has been verified for some infinite families of knots [**Shu14, AP04, PS14, DM25b, DM25a**]. Moreover, Gujral–Wang have shown [**GW25**] that if K is a knot with no 2-torsion in its Khovanov homology, and J is obtained from K by a proper rational tangle replacement, then

$$\mathrm{rank}\overline{Kh}(K) \leqslant \mathrm{rank}\overline{Kh}(J),$$

where $\overline{Kh}$ refers to reduced Khovanov homology. It follows in particular that Shumakovitch's conjecture holds for knots with unknotting number 1.

If proven, the conjecture would offer an alternative proof that Khovanov homology detects the unknot.

Proposed for K3 by: A. Shumakovitch

PROBLEM 1.31.

(a) *Compute the Khovanov homology for all torus knots $T(m,n)$.*
(b) *Compute the Khovanov–Rozansky $\mathfrak{gl}(N)$ homology for all torus knots.*

REMARKS.

(1) The data for the Khovanov homology of $T(m,n)$ is available at [**LM23b**] for small m,n. Khovanov–Rozansky homology data is available at [**Lew23a, Lew23b**].
(2) The homology is known to have a lot of torsion [**MPS$^+$18**]. Computing Khovanov homology with $\mathbb{Z}_2$ coefficients and $\mathfrak{gl}(N)$ homology with $\mathbb{Z}_N$ coefficients might be easier than rational coefficients [**GOR13**]. On the other hand, an open conjecture [**PS14, MPS$^+$18**] states that the Khovanov homology of $T(m,n)$ for m prime and $n > m$ has m-torsion.
(3) The triply graded Khovanov–Rozansky homology of torus knots (with integer coefficients) is computed in [**Mel22**]. By [**Ras15**] there is a spectral sequence from triply graded homology to Khovanov (resp. Khovanov–Rozansky) homology, but the differentials are not known explicitly.
(4) By [**Sto09**], there is a well-defined limit of the Khovanov homology of $T(m,n)$ as $m \to \infty$, also known as $T(n,\infty)$. Rozansky [**Roz14**] shows that the Khovanov homology of $T(n,\infty)$ provides a categorified equivalent of the Jones–Wenzl projector [**Wen87**]. The triply graded homology of $T(n,\infty)$ is computed in [**Hog18**], and precise conjectures for the Khovanov homology of $T(n,\infty)$ are discussed in [**GOR13, GL15**].
(5) In contrast, the Heegaard Floer homology of $T(m,n)$ is well known [**OS05b**].

Scribed by: E. Gorsky

PROBLEM 1.32.

(a) *Recover the Jones polynomial of links $L \subset \mathbb{R}^3$ by counting solutions to the Kapustin–Witten equations on $\mathbb{R}^3 \times \mathbb{R}_+$ with Nahm pole boundary conditions;*
(b) *Recover the Khovanov homology of links $L \subset \mathbb{R}^3$ by counting solutions to the Haydys–Witten equations on $\mathbb{R} \times \mathbb{R}^3 \times \mathbb{R}_+$ with Nahm pole boundary conditions;*

(c) *Use the Kapustin–Witten or Haydys–Witten equations to construct invariants of links in other 3-manifolds.*

REMARKS.

(1) Khovanov homology [**Kho00**] is a link invariant defined combinatorially, whose Euler characteristic is the Jones polynomial. It has many properties similar to Floer homologies (e.g., functoriality under cobordisms) and, as such, it is natural to ask if it has an interpretation in terms of gauge theory. Such an interpretation was proposed by Witten in [**Wit12**], and is summarized in points (a) and (b) of the problem.

(2) The Kapustin-Witten (KW) equations are a set of gauge invariant PDE's in four dimensions introduced in [**KW07**]. The Nahm pole boundary conditions for these equations are described in [**Wit12**]; they have a more special form along the link $L \times \{0\} \subset \mathbb{R}^3 \times \{0\}$. Gaiotto and Witten [**GW12**] sketched an approach to showing that the count of solutions to the KW equations with Nahm pole boundary conditions gives the coefficients of the Jones polynomial. However, there is still work to be done to make this mathematically rigorous.

(3) The Haydys-Witten (HW) equations were introduced in [**Wit12**] and [**Hay15**]. Witten conjectured that, if one forms a complex whose generators are the KW solutions and whose differentials count HW solutions (with appropriate limits and boundary conditions), then its homology is Khovanov homology.

(4) Part of the appeal of these gauge-theoretic ideas is that they may provide a definition of the Jones polynomial/Khovanov homology manifestly independent of the link diagram. Moreover, we can write the same equations with $\mathbb{R}^3$ replaced by another 3-manifold, leading to part (c) of the problem. Optimistically, the counts of KW solutions could give the class of the link L in the Kauffman bracket skein module of the 3-manifold [**Prz91, Tur88**], or be related to the $\widehat{Z}$ invariants predicted by physicists [**GPV17, GPPV20**]; see Problem 3.68. More ambitiously, the count of HW solutions should give analogues of Khovanov homology for links in 3-manifolds. At the moment, Khovanov homology is only defined in a combinatorial, concrete way for links in S^3, interval bundles over surfaces [**APS04**], $\mathbb{RP}^3$ [**Man07, Gab13**], and connected sums of $S^1 \times S^2$ [**Roz10, Wil21**]. If indeed the Haydys-Kapustin-Witten program can be carried through, it would also be important to compare the result with the Khovanov lasagna skein modules of Morrison-Walker-Wedrich [**MWW22**].

(5) Mathematicians have made progress on understanding the analytic properties of the KW and HW equations; see [**MW14**], [**MW20**], [**Tau13**], [**LT20**], [**He19**], [**HM19a**], [**Tau18**], [**Tau19**]. Some of the compactness results (e.g. [**Tau18**, Theorem B]) only hold when the 3-manifold has positive Ricci curvature, which suggests that part (c) should first be attempted under this assumption.

Proposed for K3 and scribed by: C. Manolescu

PROBLEM 1.33. *Describe topological necessary or sufficient conditions for a link to have KR-parity. For example:*

(a) *Are all links with KR-parity positive? Quasipositive?*
(b) *Do all algebraic links have KR-parity?*

REMARKS.

(1) We say that a link has *KR-parity* if its triply graded Khovanov–Rozansky homology is concentrated in homological degrees of the same parity. The motivation for this property comes from the work of Hogancamp and Elias [**EH19**], who developed a recursive procedure for computing the Khovanov–Rozansky homology of some links; their procedure relies on the link having KR-parity.
(2) By [**Mel22, HM19b**] all torus links have KR-parity.
(3) If a link has KR-parity, then all coefficients in the HOMFLY polynomial $P(a, q)$ should have the same sign (depending on conventions, one might need to change the variable a to $-a$).
(4) There are examples of positive links that do not have KR-parity, for example the knot 10_{139} which is the closure of the braid $\sigma_1\sigma_2(\sigma_1\sigma_2\sigma_2\sigma_1)^2$.
(5) Let FT_i denote the full twist on the first i strands. It is conjectured [**GHSR20, OR23, Tur24**] that the closure of the braid $\mathrm{FT}_2^{d_2}\mathrm{FT}_3^{d_3}\cdots\mathrm{FT}_n^{d_n}$ has KR-parity as long as all $d_i \geqslant 0$.

Scribed by: E. Gorsky

PROBLEM 1.34.

(a) *Khovanov and Rozansky* [**KR08b**] *used braid presentations to define a triply graded link homology theory whose Euler characteristic is the HOMFLYPT polynomial. Find a description of this homology in terms of arbitrary planar diagrams of the link, and determine whether some flavor of this theory is functorial with respect to link cobordisms.*
(b) *Do the same for Cautis' link homologies from* [**Cau17**], *which are also defined in terms of braid presentations.*

REMARKS.

(1) There are several versions of the HOMFLYPT link homology in [**KR08b**]: unreduced, middle, reduced, and totally reduced; see [**Ras15**]. Reduction refers to choosing a basepoint and taking a mapping cone in the chain complex, which results in a smaller homology.

Only the totally reduced theory (where a basepoint is chosen on each link component) has a chance at functoriality. For the original unreduced theory, the homology of the unknot is the product of the exterior and polynomial algebras on one generator each, and this is infinite-dimensional over the ground field (i.e., over the homology of the empty link); thus, it cannot admit a functorial extension unless suitably modified. The totally reduced theory assigns finite-dimensional homology groups to all links and has a chance at being functorial for link cobordisms (decorated by paths). A similar kind of functoriality appears in knot Floer homology [**Zem19b**].

(2) The definition of HOMFLYPT homology in [**KR08b**] starts with a complex associated to a braid diagram. If one constructs the same complex from an arbitrary planar diagram, the resulting homology does not behave well with respect to the Reidemeister II(b) move; see [**Web07**, Section 3.1] and [**Abe17**].
(3) HOMFLYPT homology can also be interpreted as the Hochschild homology of the Rouquier complexes formed from Soergel bimodules [**Kho07**]. Functoriality of Rouquier complexes under braid cobordisms was proved in [**EK10**].
(4) In [**Cau17**], Cautis used categorical $\mathfrak{sl}_n$ actions to define triply-graded link invariants that categorify the HOMFLYPT polynomial of links colored by arbitrary partitions. He also defined a finite dimensional categorification of the $\mathfrak{sl}(n)$ link polynomial colored by symmetric powers of the standard representation. See also [**QRS18**] for a combinatorial definition of Cautis' homology, and [**RW20**] for the equivariant version.

Proposed for K3 by: M. Khovanov
Scribed by: C. Manolescu

PROBLEM 1.35.

(a) *Is symplectic Khovanov homology isomorphic to Khovanov homology, over* $\mathbb{Z}$*?*
(b) *Give a construction of odd symplectic Khovanov homology* $\mathrm{Kh}^{\mathrm{odd}}_{symp}(K)$.

REMARKS. Symplectic Khovanov homology, introduced in [**SS06**], is a candidate Floer-theoretic construction of Khovanov homology. Abouzaid and Smith [**AS19**] showed that it is isomorphic to ordinary Khovanov homology over $\mathbb{Q}$ (or more generally over fields of characteristic 0), but their proof does not work with integer coefficients.

It is not obvious from the construction how to produce a version similarly related to odd Khovanov homology.

Scribed by: K. Hendricks, C. Manolescu

PROBLEM 1.36. *Categorify the* ($\mathfrak{sl}(2)$, $\mathfrak{sl}(N)$, *HOMFLYPT) skein algebras for surfaces.*

REMARKS.

(1) The skein algebra of a surface, introduced by Przytycki-Sikora [**PS00b**] and Turaev [**Tur91**], encodes curves on the surface up to the Kauffman bracket skein relations.
(2) The skein algebra of the annulus is categorified by the annular Khovanov (or Khovanov–Rozansky) homology [**GW10, GLW18, QR18**].
(3) There are several constructions of Khovanov-type invariants of links in thickened surfaces [**APS04, QW21**]. However, it is not known if any of these can support a categorical analogue of the skein product.

(4) The HOMFLYPT skein algebra of the torus is closely related to the elliptic Hall algebra [**MS17b**]. Categorification of the elliptic Hall algebra is an important problem in geometric representation theory [**Neg22**].

Scribed by: E. Gorsky, C. Manolescu

PROBLEM 1.37.

(a) *For every link $L \subset \mathbb{R}^3$, every simple Lie algebra $\mathfrak{g}$, and every coloring of the components of L with irreducible representations of $\mathfrak{g}$, construct a bigraded link homology theory that is functorial under link cobordisms, and whose Euler characteristic is the Reshetikhin–Turaev link invariant associated to $\mathfrak{g}$ and L.*

(b) *For each of the theories above, define an odd version (with the same mod 2 reduction) or explain the obstruction to its existence.*

REMARKS.

(1) Link homologies corresponding to $\mathfrak{sl}(n)$ with its standard vector representation were constructed by Khovanov [**Kho00**] for $n = 2$ and Khovanov–Rozansky [**KR08a**] for all n. Generalizations to exterior products of the standard representations of $\mathfrak{sl}(n)$ were constructed in [**Wu14**], [**Yon11**], [**QR16**], and were proved to be functorial under link cobordisms in [**ETW18**].

(2) Webster [**Web17**] has a categorification of the Reshetikhin–Turaev link invariants for any simple Lie algebra $\mathfrak{g}$ and any labeling of the link components by irreducible representations of $\mathfrak{g}$. His homology is bigraded. For most $\mathfrak{g}$ and labelings by irreducible representations, his homology of the unknot is infinite-dimensional over the ground field and thus cannot extend to a functorial link homology theory. Such a theory should at least assign a commutative Frobenius algebra to the unknot, necessarily finite-dimensional over the ground field. (A naive guess is that Webster homology extends to a functorial theory, possibly up to overall scaling, if and only if all representations are minuscule.) It is also possible that a good functorial theory would be more complicated algebraically, e.g. be defined over a dg ring, or have more interesting gradings.

(3) The odd version of Khovanov homology was defined by Ozsváth, Rasmussen, and Szabó in [**ORS13**]; it is isomorphic to Khovanov homology over $\mathbb{Z}/2$ but different over $\mathbb{Z}$. The odd version appears naturally in relation to Floer homologies of the double branched cover of the link; see [**OS05c**], [**Sca15**]. Its functoriality (up to sign) was studied by Migdail and Wehrli [**MW24**].

(4) Mikhaylov and Witten [**MW15**] studied odd Khovanov homology from the perspective of physics. They argued that there should be odd versions for the link homologies associated to $\mathfrak{so}(2n+1)$, but not for $\mathfrak{su}(n)$; except for $n = 2$, when $\mathfrak{su}(2) = \mathfrak{so}(3)$. The odd link homologies for $\mathfrak{so}(2n+1)$ for $n \geqslant 2$ have not yet been constructed mathematically.

(5) Ellis and Lauda [**EL16**] gave an odd categorification of quantum $\mathfrak{sl}(2)$, using a covering Kac-Moody algebra. They pointed out that the theory of covering Kac-Moody algebras only exists in finite type for $\mathfrak{so}(2n+1)$.

(6) Khovanov, Putyra and Vaz [**KPV24**] constructed odd analogues of Soergel bimodules and Rouquier complexes, and pointed out an obstruction to defining an odd HOMFLYPT homology.

Proposed for K3 by: M. Khovanov
Scribed by: C. Manolescu

1.6. Knot concordance

The study of knot concordance dates from the investigations of Fox and Milnor [**FM57, FM66**] into singularities of surfaces embedded in 4-space. Thus, although the subject matter came from knot theory, concordance was intimately tied up from the outset with 4-dimensional topology. We have divided the problems in this area, into three sections; by necessity the boundaries between these are a bit blurred. The first is concerned with knot concordance groups and variations (doubly slice knots, knots with symmetries). The second, Section 1.7, contains problems about ribbon knots and the relation of ribbon concordance, and the third, Section 1.8, about related 4-dimensional questions.

Some definitions that are used in the three sections follow.

(1) Two knots K_0 and K_1 are *smoothly* (resp., *topologically*) *concordant* if there is a smoothly (resp., topologically locally flatly) embedded cylinder $C \subset S^3 \times [0,1]$ such that $C \cap (S^3 \times \{i\}) = K_i$ for $i = 0, 1$.
(2) A knot is *smoothly* (resp. *topologically*) *slice* if it is smoothly (resp., topologically) concordant to the unknot. Equivalently, a slice knot is one that bounds a disk in B^4 (in the appropriate category).
(3) More generally, the *smooth* (resp., *topological*) *slice genus* of a knot K is the minimum genus of a smoothly (resp., topologically locally flatly) embedded surface in B^4 bounded by K and is denoted $g_4(K)$ (resp., $g_4^{top}(K)$).
(4) Let $\mathcal{C}$ (resp. $\mathcal{C}^{\mathrm{TOP}}$) denote the set of equivalence classes of knots modulo smooth (resp., topological) concordance. This is an abelian group under the binary operation of connected summation and is referred to as the *smooth* (resp., *topological*) *concordance group.*

See the introduction to Section 1.7 for definitions relating to ribbon knots and ribbon concordance.

PROBLEM 1.38. *What is the structure of the smooth knot concordance group?*

(a) *Is there a torsion element of the smooth concordance group $\mathcal{C}$ having order other than two? Is there a torsion element in $\mathcal{C}_{TS}$, the smooth concordance group of topologically slice knots, having order other than two?*
(b) *Is there a knot K having concordance order two in $\mathcal{C}$ which is not concordant to a strongly negatively amphichiral knot?*
(c) *Do there exist infinitely divisible elements in $\mathcal{C}$, that is, does there exist K such that for infinitely many $n \in \mathbb{N}$, there is a J such that $[K] = n[J]$?*

REMARKS.

(1) This is [**Kir97**, Problem 1.32]. Fox and Milnor's original work on the concordance group shows there exist elements of order two [**FM66**]; Hedden, Kim, and Livingston [**HKL16a**] show the existence of an infinite subgroup of topologically slice knots of order two. Many elements of infinite order are known; for example, $\mathcal{C}_{TS}$ contains subgroup isomorphic to $\mathbb{Z}^\infty$ [**End95**], and in fact a direct summand isomorphic to $\mathbb{Z}^\infty$ [**DHST21, OSS17b**].
(2) Any infinitely divisible element of $\mathcal{C}$ must be sent to zero under any integer-valued homomorphism from the concordance group, such as the Oszváth-Szabó τ-invariant or the Rasmussen s-invariant.

Proposed for K3 by: C. Gordon
Scribed by: K. Hendricks

PROBLEM 1.39.

(a) *Do the algebraic knots freely generate a subgroup of the smooth concordance group $\mathcal{C}$?*
(b) *Do the algebraic knots freely generate a subgroup of the topological concordance group $\mathcal{C}^{\mathrm{TOP}}$?*

REMARKS.

(1) This question is originally due to Rudolph [**Rud76**], although it is not specified whether $\mathcal{C}$ or $\mathcal{C}^{\mathrm{TOP}}$ is intended. Litherland showed that torus knots are independent in the smooth concordance group [**Lit79**]; his proof yields the topological version as well. Litherland's theorem uses Levine-Tristram signatures, but various authors have shown that there exist connected sums of algebraic knots and their mirrors that are algebraically slice but not smoothly slice [**LM83, HKL12**] implying that Litherland's strategy is not sufficient to answer the question above.
(2) This problem is especially interesting in light of its connection to the slice-ribbon conjecture (Problem 1.50): a result of Miyazaki shows that non-trivial linear combinations of iterated torus knots are not ribbon [**Miy94**, Corollary 8.4], so the slice-ribbon conjecture implies Rudolph's conjecture. Indeed, Baker [**Bak16**] and Abe-Tagami [**AT16a**] observed that the slice-ribbon conjecture implies a stronger form of Rudolph's conjecture, to wit, that the set of prime fibered strongly quasi-positive knots is linearly independent in the smooth concordance group. Large families of knots satisfying both Rudolph's conjecture and this strengthening were exhibited by Conway, Kim, and Politarczyk [**CKP23**].
(3) In a related direction, Baker showed that the slice-ribbon conjecture implies that if two fibered knots supporting the tight contact structure on S^3 are concordant, then they are in fact isotopic [**Bak16**, Corollary 4]. Abe and Tagami observed that it would then follow that the set of fibered

knots supporting the tight contact structure on S^3 is linearly independent in the smooth concordance group and contains the algebraic knots [**AT16a**, Observation 1.3]. This can be thought of as a generalization of Rudolph's question.

Scribed by: K. Hendricks

PROBLEM 1.40. *A satellite operator $P \subset S^1 \times D^2$ induces an operation P on the concordance group $\mathcal{C}$* [**Gor75**].

(a) *Let P be a winding number one satellite operator with $P(U) \simeq U$. If $P(K) \simeq U$, does this mean that $K \simeq U$?*
(b) *Is the Whitehead doubling operator injective on $\mathcal{C}$? Can we exhibit any (winding-number-zero) operator that is injective on $\mathcal{C}$?*
(c) *Conjecture (Hedden; see* [**HPC21**]*). The only homomorphisms on $\mathcal{C}$ induced by satellite operators are the zero map, the identity, and the involution induced by orientation reversal.*
(d) *Conjecture (Hedden, Pinzón Caicedo) Let P be a non-constant satellite operator with winding number zero. Does $\langle \mathrm{im} P \rangle$ necessarily have infinite rank? Some special cases are:*
    (i) *Does there always exist a fixed knot K such that $\langle \{P(nK)\}_{n\in\mathbb{Z}} \rangle$ has infinite rank?*
    (ii) *If $\{nK\}_{n\in\mathbb{Z}}$ is any rank-one subgroup of $\mathcal{C}$, does $\langle \{P(nK)\}_{n\in\mathbb{Z}} \rangle$ have infinite rank?*

REMARKS.

(1) Satellite operators are defined in the Introduction to this section.
(2) The overall conjecture (d) is easily verified in the case of non-zero winding number. It was shown in [**HK12b**] that Whitehead doubling has infinite rank, with many other patterns treated in [**HPC21**]. The authors subsequently posed the general conjecture, as well as subconjecture (i).
(3) In a positive answer to [**DHMS24**], subconjecture (i) was announced for the Whitehead doubling operation. Indeed, K was taken to be the (right-handed) trefoil; this was previously unknown. The authors gave a general class of patterns (as well as a flexible Floer-theoretic condition on K) for subconjecture (i) to hold; they subsequently made the stronger subconjecture (ii). Roughly speaking, the latter indicates that any (non-constant, winding number zero) pattern *expands rank* in the knot concordance group.

Scribed by: I. Dai

PROBLEM 1.41. *This problem is concerned with the stable 4-genus $g_s(K)$ of a knot K, defined below.*

(a) *Is there a knot K such that $g_s(K) \in \mathbb{Q}\backslash\mathbb{Z}$?*
(b) *Is there a knot K such that $g_s(K) \notin \mathbb{Q}$?*

(c) *Is there a knot K such that $0 < g_s(K) < 1/2$?*

(d) [**Liv10**, Question 1] *Does K represent a torsion element in the concordance group if and only if $g_s(K) = 0$?*

REMARKS.

(1) The stable 4-genus of a knot K is defined by

$$g_s(K) = \lim_{n\to\infty} \frac{g_4(nK)}{n}.$$

(2) One instance under which the stable 4-genus $g_s(K)$ of a knot K may become strictly less than $g_4(K)$, is when for some $m \geqslant 2$, the knot mK has $g_4(mK) < mg_4(K)$, for in this case

$$g_s(K) = \lim_{l\to\infty} \frac{g_4(lmK)}{lm} \leqslant \lim_{l\to\infty} \frac{lg_4(mK)}{lm} = \frac{g_4(mK)}{m} < g_4(K).$$

Examples of such knots are exhibited in [**Liv10**]. For instance, for $m = 2$, an infinite family of knots K is presented with $g_4(K) = 1$ and $g_4(2K) \leqslant 1$, leading to $g_s(K) \leqslant \frac{1}{2}$. Thus, the strict inequality $g_s(K) < g_4(K)$ may point to interesting phenomena with regards to oriented surfaces in the 4-ball bounded by connected sums of K.

(3) Positive answers to the first two parts of this problem automatically imply the inequality $g_s(K) < g_4(K)$, but the problem goes beyond by asking how "exotic" $g_s(K)$ can become.

In [**Liv10**], it is shown that for any $\epsilon > 0$, there exists a knot K_ϵ with stable 4-genus near $\frac{1}{2}$, in the sense of this double inequality:

$$\frac{1-\epsilon}{2} \leqslant g_s(K_\epsilon) \leqslant \frac{1}{2}.$$

Similarly, in [**Ilt22**] it is shown that if K_n denotes the twist knot with n full twists, where n is such that the Pell equation $x^2 - (4n+1)y^2 = -1$ has integral solutions in x and y, then

$$\frac{1}{2} - \frac{6}{2n+7} \leqslant g_s(K_n) \leqslant \frac{1}{2}.$$

The above bounds can be interpreted as suggesting that $\frac{1}{2}$ may have a special significance with regards to the stable 4-genus, motivating part (c).

(4) The lower bound of $0 < g_s(K)$ is required in part (c) of the question to exclude knots K that represent torsion elements in the concordance group; any such knot will have stable 4-genus equal to 0. This leads to part (d), which suggests a connection with torsion in the concordance group (See Problem 1.38).

Proposed for K3 by: C. Livingston
Scribed by: S. Jabuka, J. Meier

PROBLEM 1.42. *Do there exist algebraically concordant Seifert forms V_1 and V_2 for which there do not exist concordant knots K_1 and K_2 with Seifert forms V_1 and V_2, respectively?*

REMARKS.

(1) Seifert forms V_1 and V_2 are algebraically concordant if $V_1 \oplus -V_2$ is metabolic. This problem is open regardless of whether one works with topological or smooth concordance of knots.

(2) As an example, the Seifert forms
$$V_1 = \begin{pmatrix} 3 & 2 \\ 1 & 3 \end{pmatrix} \quad \text{and} \quad V_2 = \begin{pmatrix} 1 & 2 \\ 1 & 9 \end{pmatrix}$$
are algebraically concordant: the set of vectors $\{(0,-3,6,-1),(2,-1,0,1)\}$ is a basis for a metabolizer of $V_1 \oplus -V_2$. Does there exist a pair of concordant knots K_1 and K_2 having these Seifert forms? Conjecture 1.11 of [**Liv01**] posits that any knots K_1 and K_2 having as Seifert forms these particular V_1 and V_2, respectively, cannot be concordant.

Proposed for K3 by: C. Livingston
Scribed by: J. Meier

PROBLEM 1.43. *Does knot Floer homology give a categorification of the Fox–Milnor condition?*

REMARKS.

(1) Fox and Milnor [**FM66**] prove that if K is slice in B^4, then the Alexander polynomial factors as
$$\Delta_K(t) = f(t)f(t^{-1}).$$
Given that the Euler characteristic of knot Floer homology is the Alexander polynomial, one might expect that the knot Floer homology of a slice knot K factors as
$$\widehat{HFK}(K) \cong V \otimes V^*$$
where V is a bigraded vector space, and V^* denotes its dual. This is false, since the knot Floer homology of the Kinoshita-Terasaka knot has rank 33. One precise conjecture would be that there is a natural spectral sequence from such a vector space of the form $V \otimes V^*$ to $\widehat{HFK}(K)$.

(2) Note that Gilmer [**Gil84**] proved that if there is a ribbon concordance from K_1 to K_0, then
$$\Delta_{K_1}(t) = \Delta_{K_0}(t) \cdot f(t)f(t^{-1}).$$
One could ask if this relation, or similar relations, also admit meaningful lifts to knot Floer homology.

(3) One could similarly ask for categorifications of restrictions on the Alexander polynomial coming from symmetries on the three-manifold or the knot. For example, the Alexander polynomial of the preimage $\widetilde{K}$ of a knot K in its n-fold branched cover $\Sigma_n(K)$ is
$$\Delta_{\widetilde{K}}(t) \doteq \prod_{i=1}^{n} \Delta_K(\xi_n^i t^{\frac{1}{n}})$$
where ξ_n is a primitive nth root of unity. Similarly, let K be a q-periodic knot with quotient $\overline{K}$, and let λ be the linking number of $\overline{K}$ with the

axis of periodicity $\overline{A}$. Let $\overline{L} = \overline{K} \cup \overline{A}$. Murasugi [**Mur71**] shows that the Alexander polynomial of K is

$$\Delta_K(t) \doteq \Delta_{\overline{K}}(t) \prod_{i=1}^{q} \Delta_{\overline{L}}(t, \xi_q^i)$$

implying in particular that $\Delta_{\overline{K}}(t)$ divides $\Delta_K(t)$ and that

$$\Delta_K(t) \doteq (1 + t + \dots t^{\lambda-1})^{q-1} \Delta_{\overline{K}}(t) \bmod q$$

where the equivalences above are up to a factor of $\pm t^{\pm i}$. Work of Hendricks shows that knot Floer homology recovers this last condition [**Hen15**], but a precise categorification of the first two remains unclear.

Scribed by: J. Hom, K. Hendricks, I. Zemke

PROBLEM 1.44.

(a) *If $K \in \mathcal{F}_n$ for all n, is K topologically slice?*
(b) *If $K \in \mathcal{T}_n$ for all n, is K smoothly slice?*

REMARKS.

(1) Let $\{\mathcal{F}_n\}_{n \in \frac{1}{2}\mathbb{N}}$ denote the solvable filtration of $\mathcal{C}$ due to Cochran-Orr-Teichner [**COT03**], and let $\{\mathcal{T}_n\}_{n \in \mathbb{N}}$ denote the bipolar filtration of $\mathcal{T}$ due to Cochran-Harvey-Horn [**CHH13**].
(2) It is known that $\mathcal{T} \subseteq \bigcap_n \mathcal{F}_n$. The solvable filtration subsumes several topological concordance invariants, e.g. a knot K is in $\mathcal{F}_0$ if and only if $\mathrm{Arf}(K) = 0$; it is in $\mathcal{F}_{0.5}$ if and only if it is algebraically slice; and if K is in $\mathcal{F}_{1.5}$ then all its Casson-Gordon sliceness obstructions vanish [**COT03**]. Likewise, the bipolar filtration [**CHH13**] subsumes many smooth concordance invariants, e.g. if a knot $K \in \mathcal{T}_0$ then $\tau(K) = \upsilon(K) = \Upsilon(K) = \nu^+(K) = 0$, referring to invariants from Heegaard Floer homology. It is not known whether $s(K) = 0$, referring to the s-invariant from Khovanov homology. However $s^{\#}(K) = 0$ [**KM13b**] (see also [**Gon21**]) and $\widetilde{s}(K) = 0$ [**DIS$^+$25**].

Scribed by: A. Ray

PROBLEM 1.45.

(a) *For arbitrary $n \geqslant 2.5$ and $g > 1$, does there exist a knot in $\mathcal{F}_n$ with topological slice genus at least g?*
(b) *For arbitrary $n \geqslant 0$ and $g > 1$, does there exist a knot in $\mathcal{T}_n$ with smooth slice genus at least g?*
(c) *Could the $L^{(2)}$-signature invariants used to obstruct membership in deeper levels of the solvable/bipolar filtration be used to give bounds on the slice genus?*

REMARKS.

(1) The problem originates from Cha [**Cha08**, Remark 5.6].
(2) The filtrations $\mathcal{F}_n$ and $\mathcal{T}_n$ of the smooth concordance group $\mathcal{C}$ and the subgroup $\mathcal{T}$ of topologically slice knots, respectively, are defined in Problem 1.44.
(3) For (a), an affirmative answer for the $n = 0, 1$ cases can be shown using Levine-Tristram and Casson–Gordon signatures respectively. Similar, an affirmative answer to the $n = 2$ case was established by Cha–Miller–Powell in [**CMP21**]. Unlike (a), question (b) is open even in the case $n = 0$.
(4) A knot is closer to being slice if it has low slice genus. Similarly it is closer to being slice if it lies deeper in the solvable/bipolar filtrations. Therefore the question is asking whether these two methods of approximating sliceness are related.
(5) We note that there are many examples of highly solvable/bipolar knots that are not known to have low slice genera–that is, the issue in part (c) is not a lack of potential examples but rather a lack of obstructions.

Proposed for K3 by: J. Cha
Scribed by: A. Ray, D. Ruberman

PROBLEM 1.46.

(a) *Determine the topological slice genera of torus knots. In particular, does the topological slice genus of a torus knot equal half the absolute value of its maximal Levine–Tristram signature?*
(b) *Is a two-bridge knot topologically slice if and only if it is ribbon, if and only if it is topologically homotopy-ribbon?*
(c) *Are the Casson–Gordon ribbon and topological sliceness obstructions* [**CG86**] *complete for two-bridge knots?*
(d) *Can the difference between the smooth and topological slice genus of a two-bridge knot be arbitrarily large?*

REMARKS.

(1) Most known methods to bound the topological slice genus from above are based on Freedman–Quinn's theorem that knots with Alexander polynomial 1 are topologically slice [**FQ90, BKK$^+$21**]; the exception to this is [**FT05**].
(2) The smooth slice genera of torus knots were determined by Kronheimer and Mrowka using gauge theory: $g_4(T(p,q)) = (p-1)(q-1)/2$ [**KM94**]. The first person to show that the topological slice genus of torus knots is typically less than their 3-genus was Rudolph [**Rud84**]. The topological slice genus $g_4^{top}(T(p,q))$ is less than the smooth slice genus $g_4(T(p,q))$ unless $T(p,q)$ has maximal signature, and in general $g_4^{top}(K) < pq/3 + p\log_2 q + q\log_2 p$ [**BFLL18, McC21**]. (See [**BBL20**] for improved bounds in the case of $p = 3$.) All of the known lower bounds on $g_4^{top}(T(p,q))$ derive from the Seifert form of $T(p,q)$, and the largest of these is half of the maximal Levine-Tristram signature (see also [**Tay79**]).

(3) The first part of the question is raised as a conjecture in [**FM16**, Conjecture 1]. The question whether the Casson-Gordon ribbon obstruction is complete is raised as a conjecture in [**EL09**, Conjecture 4.1].

Casson and Gordon introduced obstructions to a knot K being ribbon and to being topologically slice in [**CG86**]. They are defined in terms of signatures associated with n-fold cyclic branched covering spaces of K for a prime power n. When $K = K(p,q)$ is a two-bridge knot and $n = 2$, the ribbon obstruction is conjectured to be complete, but the topological slice obstruction is known not to be [**EL09, Mil18**]. The two obstructions coincide when p is a prime power and $n = 2$. Hence, if the ribbon obstruction is complete, then the answer to the first part of the question is affirmative when p is a prime power. The ribbon obstruction actually obstructs a knot being topologically *homotopy-ribbon*, meaning that the map induced by inclusion of the knot complement into the complement of a disk induces a surjection on fundamental groups. See Section 1.7 for more discussion of notions of ribbonness.

Casson and Gordon also identified a family $\mathcal{R}$ of ribbon two-bridge knots. Lisca used an obstruction based on Donaldson's diagonalization theorem to prove that $K(p,q)$ is smoothly slice iff $K(p,q) \in \mathcal{R}$ [**Lis07**]. It follows that $\mathcal{R}$ accounts for all ribbon two-bridge knots.

The Casson-Gordon ribbon obstruction is expressed in terms of a count of lattice points in triangular regions in the plane, while Lisca's smooth sliceness obstruction is expressed in terms of high-dimensional lattice embeddings. It is possible that the two obstructions are equivalent for number theoretic reasons. If so, it would follow that the Casson-Gordon ribbon obstruction is complete.

Lastly, it is possible that $K(p,q) \in \mathcal{R}$ if and only if $K(p,q)$ passes the topological slice obstructions for all prime powers n. If so, this would imply a positive solution to the stated question.

(4) The final question is raised by Feller and McCoy in [**FM16**, Question 3]. In contrast with the conjectured answer to the second question, they give examples of 2-bridge knots K for which $g_4^{top}(K) = 1$ and $g_4(K) = 2$.

Scribed by: J. Greene, D. Ruberman

PROBLEM 1.47. *Given a smooth knot $K \subset S^3$, determine its* nonorientable 4-genus γ_4, *i.e. the minimal first Betti number for all compact nonorientable smooth surfaces F properly embedded in B^4 with boundary K. This question may be developed in several directions:*

(1) *(Allen's Geography Question* [**All23**]*) For a given knot $K \subset S^3$, find the pairs $(e,b) \in \mathbb{Z}^2$ realized by the normal Euler number and first Betti number of a nonorientable bounding surface F.*

(2) *By analogy with the Milnor Conjecture for the orientable 4-genus of torus knots, what is the nonorientable 4-genus of a torus knot $T_{p,q}$?*

REMARKS. The nonorientable 4-genus was first defined by Murakami and Yasuhara [**MY00**] in 2000, though earlier results had been obtained by Viro [**Vir75**] and by Yasuhara [**Yas96**]. These authors, along with Gilmer and Livingston

[**GL11**], obstructed the existence of Möbius strips and punctured Klein bottles with boundary K using the classical signature and the Arf invariant of K.

The first examples of knots with $\gamma_4(K) > 3$ were produced by Batson, who used an elegant construction and the Heegaard Floer d invariant to prove that $\gamma_4(T_{2k,2k-1}) = k - 1$ [**Bat14**]. Lobb subsequently showed that Batson's construction does not always yield a minimal-genus nonorientable surface for torus knots [**Lob19**]. Allen added context by asking how the nonorientable 4-genus interacts with the normal Euler number of the bounding surfaces for torus (and other) knots [**All23**]. See [**BKST25, JVC21, OSS17a, Sab23**], among others, for further explorations of the nonorientable 4-genus for torus knots. In [**All23**], Allen made three conjectures (Conjectures 1.6, 1.7, and 1.8); Sato [**Sat24a**] has given counterexamples to Conjectures 1.6 and 1.8.

There are now many knots for which the nonorientable four-genus is known, including all knots with at most 10 crossings [**Gha22**]. There are also currently a variety of lower bounds for γ_4 using modern knot invariants. Many of these are what Sato [**Sat24b**] terms "unoriented slice-torus invariants", including the υ invariant from Heegaard Floer theory [**OSS17a**], the t invariant from Khovanov homology [**Bal20**], the $h_{\mathbb{Z}}$ invariant from instanton homology [**DS24a**]. Additional information about the nonorientable four-genus can be obtained by considering versions of Seiberg-Witten Floer homology of the double branched cover that take into account the involution coming from the covering transformation; see [**BH24b, KMT23b**].

Scribed by: J. Sabloff

Problem 1.48.

(a) *Suppose K and $K\#J$ are (smoothly) doubly slice knots. Must J be a (smoothly) doubly slice knot?*
(b) *Does there exist a knot that is smoothly slice, topologically doubly slice, and* not *smoothly doubly slice but such that $K\#K$ is smoothly doubly slice?*
(c) *Let M and N be closed 3-manifolds such that M and $M\#N$ embed (smoothly) in S^4. Must N embed (smoothly) in S^4?*

Remarks.

(1) A knot $K \subset S^3$ is called (smoothly) *doubly slice* if there is a (smoothly) unknotted 2-sphere $\mathcal{U} \subset S^4$ such that $K = S^3 \cap \mathcal{U}$.
(2) It is natural to try to equip the set of knots with the relation in which $K_0 \sim K_1$ if $K_0\#\overline{K}_1$ is doubly slice. However, to prove this is an equivalence relation, one must provide an affirmative answer to part (a) of the problem. For more details, see [**Mei15**] and [**Sto78**]. This question is also interesting in the topological category: Replace "smooth" with "topologically locally flat" above. Question (a) is originally due to Stoltzfus [**Sto78**].
(3) One motivation for Question (a) is that it would allow for the formulation of the Grothendieck group of knots modulo the relation discussed above. In the absence of a positive answer to Question (a), there is an alternative approach. Two knots K_0 and K_1 are called (smoothly) *doubly concordant*

if there exist (smoothly) doubly slice knots J_0 and J_1 such that $K_0 \# J_0 = K_1 \# J_1$. Let $\mathcal{C}_{\mathcal{D}}$ denote the group (under connected sum) of knots modulo this equivalence relation, which is called the *(smooth) double concordance group.* An affirmative answer to Question (a) would mean the double concordance group is the same as the aforementioned Grothendieck group. A natural problem is to exhibit nontrivial elements in these groups, as we now outline.

There is a topological version $\mathcal{C}_{\mathcal{D}}^{\text{top}}$ of this group obtained by replacing "smooth" with "topologically locally flat" above. There are epimorphisms from $\mathcal{C}_{\mathcal{D}}$ to $\mathcal{C}_{\mathcal{D}}^{\text{top}}$ and to the smooth concordance group $\mathcal{C}$. Let $\mathcal{K}$ denote the intersection of the kernels of these two homomorphisms. It is known that $\mathcal{K}$ contains an infinitely generated subgroup S with generators of order at least 3 [**Mei15**].

CONJECTURE. *These generators of S have infinite order.*

(4) There are many techniques for producing and studying 2-torsion in the smooth concordance group $\mathcal{C}$; see [**HKL16a**] for a nice overview. However, these techniques seem difficult to apply to $\mathcal{C}_{\mathcal{D}}$, which motivates Question (b).
(5) If K is doubly slice, then any cyclic cover of S^3 branched along K embeds (smoothly) in S^4. This suggests Question (c), which asks about embedding 3-manifolds in S^4. The connection between doubly slice knots, embedding problems and homology cobordisms was observed by Gilmer and Livingston [**GL83**]. See also Problem 4.25.

Proposed for K3 and scribed by: J. Meier

PROBLEM 1.49.

(a) *What is the structure of the equivariant concordance groups?*
(b) *Is the strongly negative amphichiral concordance group abelian?*
(c) *For any type of knot involution, is it possible to exhibit K_1 and K_2 that are equivariantly concordant but not standardly equivariantly concordant?*
(d) *Does there exist a freely periodic slice knot that is not equivariantly slice?*
(e) *(Boyle-Rouse) Is every periodic or freely periodic L-space knot either a torus knot or an iterated torus knot?*

REMARKS.

(1) Knots may have various sorts of symmetries; an equivariant concordance is a usual concordance that admits the same sort of symmetry. In the topological case, we require that the concordance be locally flat, and that any group action be locally linear. There are several variations discussed in this problem, each with its own particular set of concerns.
(2) Equivariant concordance groups can be defined for the classes of strongly invertible and strongly negative amphichiral knots. The strongly invertible concordance group has been studied by several authors: for instance, there exists a homomorphism to the Laurent polynomial ring $\mathbb{Z}[t, t^{-1}]$ [**Sak86**]. These groups are not *a priori* abelian due to (possible) dependence on the choice of the connected sum point, which must be taken to be

one of the two fixed points of the symmetry on the knot. In the strongly invertible case, it is known that the group is not abelian [**DP23b**], but in the amphichiral case this is not known.

Understanding strongly negative amphichiral knots is especially interesting, since the branched covers of such knots may provide examples of 2-torsion in the integer homology cobordism group. Examples of strongly negative amphichiral knots with determinant one were exhibited by Boyle and Chen in [**BC24b**].

(3) In part (c), we say that two knots K_1 and K_2 (equipped with the same type of involution) are *standardly* equivariantly concordant if they are connected by a smooth annular cobordism in $S^3 \times I$ that is invariant under an involution on $S^3 \times I$ that restricts to a standard involution on each $S^3 \times \{t\}$. One can also consider more general equivariant cobordisms, in which one requires that the annular cobordism be fixed by *some* involution on $S^3 \times I$ which extends the actions on $S^3 \times \{0\}$ and $S^3 \times \{1\}$. It is not known in any of the possible cases whether these notions coincide. The question may be asked in either the smooth or the topological category.

In the case of 2-periodic and strongly invertible knots, it is known that there exists a nonstandard involution on $S^3 \times I$ that restricts to the standard involution on $S^3 \times \{0\}$ and $S^3 \times \{1\}$. To construct one, start with an involution on S^4 with fixed-point set a knotted S^2 [**Sum75**]; puncturing such an example gives a nonstandard involution on $S^3 \times I$ which has fixed-point set S^1 on each of the two ends. It is not known whether an interesting such involution exists in the freely periodic or strongly amphichiral cases.

(4) Question (d) appears as Question 1 in [**BM23**]. Manolescu and Willis [**MW25**] gave examples of freely 2-periodic knots that are concordant but not (standardly) equivariantly concordant.

(5) Evidence for Question (e) comes from classic work of Murasugi [**Mur71**] and Hartley [**Har81**], who established factorization results for the Alexander polynomials of periodic or freely periodic knots. We also have several well-known restrictions on the Alexander polynomials of L-space knots; see [**OS05b**] or [**HW18**]. As discussed by Boyle and Rouse [**BR23b**], the only known examples of Alexander polynomials that simultaneously satisfy all of these conditions are products of cyclotomic polynomials. On the other hand, a conjecture of Li and Ni states that the only L-space knots with Alexander polynomials which are products of cyclotomic polynomials are torus knots or iterated torus knots [**LN15**]. Based on this, Boyle and Rouse [**BR23b**] conjectured a positive answer to Question (e).

Scribed by: I. Dai, K. Hendricks

1.7. Ribbon knots and ribbon concordance

A knot in S^3 is *ribbon* if it bounds a disk D in B^4 that can be isotoped to have no local maxima with respect to the radial Morse function on B^4. Equivalently, a knot is ribbon if it bounds an immersed disk in S^3 with only ribbon singularities. More generally, K_1 is *ribbon concordant* to K_0 if there is a cylinder $C \subset S^3 \times [0,1]$

such that $C \cap (S^3 \times \{i\}) = K_i$ for $i = 0, 1$ and C can be isotoped to have no local maxima with respect to the Morse function on $S^3 \times [0, 1]$ given by projection onto the interval factor.

PROBLEM 1.50.

(a) *Is every slice knot a ribbon knot?*
(b) *Is every slice link ribbon?*
(c) *Suppose K is a knot with smooth four-genus $g_4(K) = g$. Does K bound a smooth, genus g, ribbon surface F in B^4?*

REMARKS.

(1) The Slice-Ribbon Conjecture (a) was first posed as a question by Fox in 1961 [**Fox62**, Problem 25], and was discussed in [**Kir97**, Problem 4.22]. Very limited progress has been made on this problem, though the conjecture has been verified for 2-bridge knots by Lisca [**Lis07**] and most 3-stranded pretzel knots by Greene and Jabuka [**GJ11**] and by Lecuona [**Lec15**].
(2) There are other notions of "ribbonness" that can be considered: A knot $K \subset S^3$ is called *homotopy-ribbon* if there is a slice disk D for K with a surjection $\pi_1(S^3 \backslash K) \twoheadrightarrow \pi_1(B^4 \backslash D)$ and is called *handle-ribbon* if there is a slice disk D for K such that $B^4 \backslash \nu(D)$ admits a handle-decomposition with only 0-, 1-, and 2-handles; see [**Gor81, LM15, MZ23**].
(3) The implications

$$\text{ribbon} \Rightarrow \text{handle ribbon} \Rightarrow \text{homotopy-ribbon} \Rightarrow \text{slice}$$

are immediate, but all three converse implications are open.
(4) For many examples of handle ribbon knots and links that are not known to be ribbon, as well as connections to the Generalized Property R Conjecture, see [**GST10, MZ22**] and Problem 1.10.
(5) Miyazaki gave many examples of non-simple knots that are not homotopically ribbon, such as certain linear combinations of algebraic knots and, famously, the $(2, 1)$-cable of the figure-8 knot [**Miy94**]. Some of these examples are now known to be non-slice [**HKL12, DKM$^+$24, ACM$^+$26**]. Baker draws connections between the Slice-Ribbon Conjecture for fibered knots and contact topology in [**Bak16**]. See Problem 1.54 for more questions about ribbon concordances.
(6) In part (b), we say that an n-component link $L \subset S^3$ is called (strongly) *slice* if there is a disjoint collection $D \subset B^4$ of n disks with $\partial D = L$; the notions of ribbon, handle-ribbon, and homotopy-ribbon have obvious analogs for links.
(7) Many examples of handle ribbon links that are not known to be ribbon are constructed in [**GST10**] and [**MZ22, MZ25**].
(8) Eisermann [**Eis09**] gave an obstruction for a link (with at least two components) to be ribbon, in terms of the Jones polynomial. It is not clear whether his obstruction vanishes when the link is handle ribbon (or homotopy-ribbon, or slice) so, in principle, it could be used to show that a handle ribbon link is not ribbon.
(9) The question in part (c) generalizes the Slice-Ribbon Conjecture, and so it may be easier to find a counterexample. If $g_4(K) = g(K)$, then the

push-in of a minimal genus Seifert surface is a ribbon surface realizing the minimum possible value of $g_4(K)$, so the question has an affirmative answer in this case.

(10) Finally, the question of whether slice implies homotopy-ribbon can be phrased in the more general setting of knots in homology 3–spheres bounding disks in contractible 4–manifolds. For more details, see [**Kir97**, Problem 4.22] and [**CG83a**].

Scribed by: J. Meier

PROBLEM 1.51. *Does every ribbon knot arise as a symmetric union?*

REMARKS. Symmetric unions were introduced by Kinoshita and Terasaka to construct knots with Alexander polynomial 1 (by taking a symmetric union of an unknot diagram) [**KT57**]. It is straightforward to see that every symmetric union is a ribbon knot. A census shows that most small ribbon knots do arise as symmetric unions [**Lam21**], but there are candidates for ribbon knots that are not symmetric unions, such as $3_1\#8_{10}$ and $3_1\#8_{11}$. Other potential counterexamples obtained as certain satellites of knots of the form $K\#-K$ are given by Aceto [**Ace14**]. These knots have ribbon number two, but the symmetric ribbon number is shown to be arbitrarily large, if not infinite.

Scribed by: R. Lipshitz

PROBLEM 1.52. *Given K in S^3, is there an algorithm to detect if K is slice? Ribbon?*

REMARKS.

(1) This is Problem 1.34 from [**Kir97**], posed by A. Casson.
(2) There is still no algorithm, but there have been some interesting computer-aided searches: Owens and Swenton search for ribbon bands for alternating knots [**OS24**] and Gukov, Halverson, Manolescu, and Ruehle apply machine learning algorithms to the problem [**GHMR23**].
(3) Casson and Long gave an algorithm to determine if a surface automorphism compresses [**CL85**]. In conjunction with the main results of [**CG83a**], this algorithm can be used to determine whether a given fibered knot bounds a fibered, homotopy-ribbon disk in a homotopy four-ball; see also [**Lon86**]. Dunfield and Gong [**DG25**] have announced the results of a search for ribbon knots, slice knots, and concordances among the knots with at most 19 crossings.

Scribed by: J. Meier

PROBLEM 1.53.

(a) *Which knot properties are hereditary under ribbon concordance? Is the property of being alternating hereditary under ribbon concordance?*

(b) *Which knot properties imply minimality under ribbon concordance? Are positive knots minimal under ribbon concordance?*
(c) *Which knot invariants are monotone under ribbon concordance? Is crossing number, bridge number, or braid index monotone under ribbon concordance?*

REMARKS.

(1) A *knot property* is a subset $\mathcal{P}$ of the set of knots. It is *hereditary* under a directed relation $\mathcal{R}$ on the set of knots if $K_1 \in \mathcal{P}$, $K_0 \mathcal{R} K_1 \implies K_0 \in \mathcal{P}$.
(2) Using techniques of Silver and the solution of Rapoport's conjecture given by Kochloukova, Miyazaki showed that the property of being fibered is hereditary under ribbon concordance [**Koc06, Miy18, Sil92**] in the sense that if K_1 is ribbon concordant to K_0 and K_1 is fibered, then so is K_0; see also work of Zemke [**Zem19a**]. It is conjectured that the property of being alternating is hereditary under ribbon concordance [**GO23**, Conjecture 9].
(3) A knot K_1 is *minimal* under $\mathcal{R}$ if $K_0 \mathcal{R} K_1 \implies K_1 = K_0$. It is conjectured that positive knots are minimal under ribbon concordance [**BG24**, Conjecture 1.6]. Baker and Stoimenow have asked related questions [**Bak16, Sto15**]. See also Problem 1.54.
(4) Most knot invariants support natural directed relations $\preceq$ on their codomains: numerical (real-valued) invariants with respect to the standard ordering $\leqslant$; polynomial invariants with respect to divisibility; group-valued invariants with respect to injections; etc. A knot invariant ι is *monotone* under a directed relation $\mathcal{R}$ if $K_0 \mathcal{R} K_1 \implies \iota(K_0) \preceq \iota(K_1)$.

    Several knot invariants are known to be monotone under ribbon concordance. For example, if K_1 is ribbon concordant to K_0, the Alexander polynomial of K_0 divides that of K_1 [**Gil84, FP20**]. Zemke showed that a ribbon concordance from K_1 to K_0 induces a bigraded injection from the knot Floer homology of K_0 to K_1, with the consequence that the Seifert genus of K_0 is less than or equal to that of K_1 [**Zem19a**]. It is known that if K_1 has bridge number 2, then K_0 has bridge number at most 2 [**GO23**, Paragraph after Conjecture 9].
(5) Gordon's pioneering paper on ribbon concordance contains several fascinating problems related to the ones raised here [**Gor81**].

Scribed by: J. Greene

PROBLEM 1.54. *This problem is concerned with the restriction of the partial ordering $\geqslant$ coming from ribbon concordance to the concordance class $[K]$ of a knot K.*

(a) *What can be said about the order type of $[K]$?*
(b) *Does it depend on $[K]$? Must it contain a (unique) minimal element?*
(c) *(Gordon) Does there exist a concordance class of knots that contains two minimal elements with respect to ribbon concordance?*

REMARKS.

(1) The ribbon concordance relation was defined by Gordon in [**Gor81**], who proves that some classes $[K]$ contain minimal elements, and asked Question (c).

(2) The ribbon concordance relation was conjectured to be a partial ordering by Gordon. This was proved by Agol [**Ago22**]. Notice that for any knots K and L, the map from $[K]$ to $[L]$ defined for $J \in [K]$ by

$$J \mapsto J \# -K \# L$$

is an order preserving injection.

(3) The Slice-Ribbon Conjecture can be formulated to say the unknot U is the unique minimal element of $[U]$; see Problem 1.50.

See Problem 1.56 for a Heegaard Floer theory perspective.

Proposed for K3 by: C. Livingston
Scribed by: J. Meier

PROBLEM 1.55. *Suppose that C is a ribbon concordance from a fibered knot K_1 to a fibered knot K_0.*

(a) *Does the capped-off monodromy of K_1 (i.e. extended over a disk) extend to a compression body whose lower boundary gives the capped-off monodromy of K_0?*
(b) *Is C fibered by compression bodies?*
(c) *Suppose a fibered knot in a homology 3-sphere bounds a homotopy-ribbon disk D in a contractible 4-manifold V. Is $V \backslash \nu(D)$ fibered by handlebodies?*

REMARKS.

(1) Part (a) holds when K_0 is the unknot by the main theorem of [**CG83a**].
(2) Miller answers part (b) in the affirmative under the hypothesis that the ribbon concordance is defined by a collection of ribbon bands that are transverse to the fibration of the exterior of K_1 [**Mil21**, Theorem 1.9].
(3) Part (c) was posed by Casson and Gordon in the Remark before Corollary 5.4 of [**CG83a**].

Proposed for K3 by: A. Zupan
Scribed by: J. Meier

PROBLEM 1.56 (Hom). *If K_0 and K_1 are ribbon concordant and $\widehat{HFK}(K_0) \cong \widehat{HFK}(K_1)$, are K_0 and K_1 isotopic?*

REMARKS.

(1) Recall that Zemke [**Zem19a**] proved that a ribbon concordance induces a split injection on knot Floer homology. Gordon asked the following [**Gor81**, Question 6.2]: If $K_1 \geqslant K_2 \geqslant \ldots$, does there exist some m such that $K_n = K_m$ for all $n \geqslant m$? An affirmative answer to Problem 1.56 would answer Gordon's question in the affirmative. One large class of concordant knots with the same knot Floer homology is given in [**Wan22**, Theorem 1.2] (see also [**HW18**, Theorem 1]), but none of these are ribbon concordant to any other, by [**Wan22**, Theorem 1.8] and [**LZ19**].

(2) The same question could be asked with Khovanov homology $\overline{Kh}$ replacing knot Floer Homology $\widehat{HFK}$, or indeed other homology theories as treated in [**Kan22a**].

Proposed for K3 and scribed by: J. Hom

1.8. Knot concordance and 4-manifolds

Given a knot K, the 0-*trace* X_K is obtained by attaching a 0-framed 2-handle to B^4 along K and smoothing corners. The 0-trace is homotopy equivalent to S^2 and so $H_2(X_K;\mathbb{Z}) \cong \mathbb{Z}$. We say a knot is smoothly 0-*shake slice* if either generator of $H_2(X_K;\mathbb{Z})$ is represented by a smoothly embedded sphere.

More generally, define the smooth 0-*shake genus* of K to be the least genus of a smoothly embedded, closed, oriented surface representing either generator of $H_2(X_K;\mathbb{Z})$. Similarly, we can define topological 0-shake slice knots and 0-shake genus by considering locally flat embedded surfaces.

In either the smooth or topological setting, slice knots are 0-shake slice.

PROBLEM 1.57.

(a) *In either the smooth or topological settings, are* 0-*shake slice knots slice?*
(b) *Does there exist a knot K whose topological 0-shake slice genus is strictly less than the topological slice genus?*

REMARKS.

(1) The smooth analogue of the second question was [**Kir97**, Problem 1.41], which was answered in the negative by Piccirillo in [**Pic19**]. Her construction uses knots with diffeomorphic 0-traces whose smooth slice genera are distinct. A negative answer to part (a) would imply the existence of a knot that is slice in an integer homology ball but not in B^4; see Problem 1.60.
(2) One may also ask analogous questions for n-traces of knots, where some negative answers are known. For more on this see [**Akb77, Lic79, Akb93, AJOT13**].

Scribed by: A. Ray

PROBLEM 1.58. *What concordance information about a knot K is contained in its 0-trace $X_0(K)$ and in its* 0-*surgery $S^3_0(K)$? Specifically,*

(a) *Suppose K and K' have homeomorphic 0-traces.*
  (i) *Are K and K' topologically concordant?*
  (ii) *What if one just assumes K and K' have homeomorphic 0-surgeries?*
  (iii) *What if one just assumes K and K' have 0-surgeries that are topologically homology cobordant, preserving the homology class of a meridian.*
(b) *Suppose K and K' have homeomorphic 0-surgeries, where the image of the meridian of K is freely homotopic to the meridian of K'. Are K and K' smoothly concordant?*

REMARKS.

(1) This problem originates in a conjecture of Akbulut and Kirby [**Kir97**, Problem 1.19] that knots with the same 0-surgery are concordant. It was motivated, at least in part, by the fact [**KM78**] that it holds when one of the knots is unknotted. (The result of [**KM78**] predates the proof of Property R, which is of course much stronger.) Although the original conjecture is false, there are several variations, as one can ask about the homeomorphism or diffeomorphism type of $X_0(K)$ and about smooth or topological concordance.

(2) All of the questions in part (a) have negative answers if one is looking for smooth concordances; see [**CFHH13**], [**Yas17**], [**MP18**]. With reference to Question (a)(iii) and part (b), note that a concordance would preserve the free homotopy class (and hence homology class) of a meridian.

(3) Part (b) has a negative answer without the assumption on the meridian, see [**Yas17**]. There are no examples in the literature of homeomorphisms, satisfying the assumption on the meridian, between 0-surgeries on distinct knots.

Scribed by: L. Piccirillo

PROBLEM 1.59. *Let $K \subset S^3$ be a slice knot.*

(a) *Determine the set $\mathcal{R}(K)$ of ribbon disks bounded by K modulo isotopy.*

(b) *Determine the set $\mathcal{D}(K)$ of slice disks bounded by K modulo isotopy and local knotting.*

(c) *Determine the set $\mathcal{HR}(K)$ of homotopy-ribbon disks bounded by K modulo isotopy.*

(d) *If K is fibered, determine the set $\mathcal{FHR}(K)$ of fibered, homotopy-ribbon disks bounded by K in some homotopy 4-ball, modulo isotopy.*

REMARKS.

(1) There is no knot for which part (a) has been solved. A knot with $|\mathcal{R}(K)| = 0$ would be a counter-example to the Slice-Ribbon Conjecture. It is unknown whether or not $|\mathcal{R}(U)| = 1$ for the unknot U, though Scharlemann has shown that any non-standard ribbon disk for the unknot must have at least three minima [**Sch85a**]. It is known that $\mathcal{R}(Q_{p,2})$ is infinite [**MZ25**], where $Q_{p,q} = T_{p,q} \# \overline{T}_{p,q}$ is the connected sum of a torus knot with its mirror.

(2) There are many knots that bound isotopic ribbon disks that are distinct when considered up to isotopy rel. boundary: Any knot of the form $J \# -J$ bounds infinitely many isotopic ribbon disks that are pairwise non-isotopic rel. boundary, as follows from Zeeman's twist-spinning construction [**Zee65**]; see, for example, [**MZ25**, Proposition 2.2]. Additionally, many simple knots such as the stevedore knot 6_1 and pretzel knots $P(p, -p, p)$ bound pairs of such disks. The problems of studying disks modulo isotopy and modulo isotopy rel. boundary are closely related via an understanding of the symmetries of the boundary knot; see the proofs of [**JZ20b**, Lemma 2.2], [**MZ25**, Theorem 1.5], and [**MM25b**, Theorem 4.2].

(3) The set $\mathcal{D}(K)$ has recently been studied using modern invariants [**JZ20b**], classical invariants [**MP19**], and geometric techniques [**MM25b**]. It is immediate that $|\mathcal{D}(U)| = 1$ for U the unknot, but $\mathcal{D}(K)$ is not known for any nontrivial knot K.
(4) Consider the map $\iota_K \colon \mathcal{HR}(K) \to \mathcal{D}(K)$. Is there a knot K such that this map is injective, surjective, or has finite image? See [**MM25b**, Section 1.1]. There is no knot for which part (c) has been solved.
(5) Given the apparent difficulty in solving the above problems, it is reasonable to look for more tractable alternatives in which there is more structure, such as part (d).

Casson and Gordon found a beautiful connection between fibered, homotopy-ribbon disks bounded by a fibered knot K and handlebody extensions of the closed monodromy of K [**CG83a**]. Here, the extra structure yields more positive results. First, $|\mathcal{FHR}(U)| = 1$. Second, for all coprime $p > q > 1$, the set $\mathcal{FHR}(Q_{p,q})$ has been explicitly determined; the set $\mathcal{FHR}_\partial(Q_{p,q})$ of fibered, homotopy-ribbon disks modulo isotopy rel. boundary is in explicit bijection with $\{c/d \in \mathbb{Q} : c \text{ even}\}$ [**MZ25**].

Proposed for K3 and scribed by: J. Meier

PROBLEM 1.60. *Is there a knot in S^3 that is not smoothly slice in B^4 but is smoothly slice in an integer homology ball? What about a $\mathbb{Z}/2$-homology ball?*

REMARKS.

(1) As discussed in [**FGMW10**], one strategy to disprove the smooth Poincaré conjecture is to find a homotopy four-sphere X and a knot K in S^3 that is smoothly slice in $X - B^4$ but not in B^4; see also Problem 4.1. It is worth remarking that just about all known slice obstructions necessarily vanish for a knot that is slice in an integer homology ball. One invariant that can potentially make this distinction is the s-invariant from Khovanov homology [**Ras10**]. As a warm-up, it would be interesting to find a knot K with $s(K) \neq 0$ that is smoothly slice in a rational homology ball.
(2) Fintushel-Stern showed the figure-eight knot is smoothly slice in a rational homology ball with $\pi_1 = \mathbb{Z}/2$ [**FS84**], but is not even topologically slice in B^4 by the Fox-Milnor condition. For all known examples of knots that are slice in a rational ball but not B^4, the rational ball has 2-torsion in H_1. Kawauchi shows that all strongly negative-amphichiral knots are rationally slice [**Kaw09**] and Levine showed they are slice in the same rational homology ball Z_0 [**Lev23**]; in fact, he asks if it is possible that all rationally slice knots are slice in boundary sums of Z_0.

Scribed by: T. Lidman

PROBLEM 1.61. *A knot in S^3 bounds a topological disk in B^4 by coning (not necessarily locally flat); this problem asks about topological disks that a knot in a homology sphere Y might bound in a homology or homotopy ball bounded by Y.*

(a) *Given a locally flat knot in an integer homology 3-sphere, is it concordant to a knot in S^3, via a locally flat annulus in a topological integer homology cobordism? Variation: Can the homology cobordism be taken to be simply connected?*
(b) *Given an arbitrary knot in an integer homology 3-sphere Y, does it bound an embedded disc (not necessarily locally flat) in Δ, the Freedman filling of Y? Variation: Is there such a disk in a homology ball bounded by Y?*
(c) *Is there a link in an integer homology 3-sphere, each of whose components is independently concordant to a knot in S^3, that is not itself concordant to a link in S^3.*
(d) *Suppose that $\mathbb{Z}_n$ acts semi-freely on the homology sphere Y, with fixed point set a knot K. Does the action of $\mathbb{Z}_n$ extend semi-freely over Δ?*

REMARKS.

(1) By the proof of [**Fre82**, Theorem 1.4′] every integer homology 3-sphere bounds a compact, contractible 4-manifold Δ, which we call the *Freedman filling* of Y. It is known to be unique up to homeomorphism, by Freedman's h-cobordism theorem [**Fre82**, Theorems 1.3 and 1.6].
(2) The smooth analogue of (a) was disproved by A. Levine [**Lev16**], answering Problem 1.31 in [**Kir97**]; see also [**HLL22b, Zho21**]. An affirmative answer to (a) gives an affirmative answer to (b) for locally flat knots, via an embedding with an isolated, piecewise linear singularity obtained by coning off the knot in S^3. Note that topological embeddings need not be approximable by piecewise linear or locally flat embeddings. Such a wild topological embedding of a disk was constructed by Giffen using a shift-spinning construction in unpublished work, explained in [**DV09**, Section 6.6]. Recent work of Davis [**Dav20, Dav25**] gives evidence in support of an affirmative answer to (a), in terms of the solvable filtration of the knot concordance group [**COT03**].
(3) Part (d) is related to questions about extending topological group actions, without any requirement of local linearity. Recall [**FQ90**] that a free action of a cyclic group on a homology sphere Y extends to an action on the Freedman filling Δ of Y with one fixed point; the action is not necessarily locally linear at that point. Suppose instead that $\mathbb{Z}_n$ acts semi-freely on Y, with fixed point set a knot K. If n is a prime, say p, then the fixed point set of the extension would be a $\mathbb{Z}_p$-homology disk with boundary equal to K. So in some weak sense, K would be slice in Δ. This problem is rather different in character from the smooth version, discussed in problem 3.75.

Proposed for K3 and scribed by: Ray, Ruberman

PROBLEM 1.62.

(a) *Are all good boundary links topologically slice? Freely topologically slice?*
(b) *A special case of interest: Is the Whitehead double of the Borromean rings (with any choice of clasps) topologically slice?*

REMARKS.

(1) A *boundary link* (L, ϕ) is an n-component link $L \subseteq S^3$ together with a surjection $\phi \colon \pi_1(S^3 \backslash L) \twoheadrightarrow F_n$, sending a set of meridians of L to a free generating set of the free group on n generators. A link L admits such a homomorphism ϕ if and only if the components of L bound pairwise disjoint Seifert surfaces in S^3 [**Smy66**] (see also [**CS80**, Proposition 1.1]). Following [**FQ90**, Section 11.7C], a boundary link (L, ϕ) is said to be a *good boundary link* if the kernel of ϕ is a perfect group. Note that a 1-component good boundary link is a knot with Alexander polynomial one. Whitehead doubles of links with vanishing pairwise linking numbers are good boundary links.

(2) Part (a) is posed in [**Kir97**, Problem 1.36], with a slightly different definition of good boundary links than appears here. The first half of that problem asked whether Alexander polynomial one knots are topologically slice. The answer to that question is now known to be yes [**GT04**], [**FQ90**, Theorem 11.7B], [**Fre84**, Theorem 7] (see also [**BPR21**, Theorem 1.14] and [**PR21**, Section 21.6.3]).

(3) If the surgery sequence is exact (see Problem 4.46), then a link in S^3 is freely slice if and only if it is a good boundary link [**FQ90**, Corollary 11.7C]. More surprisingly, the topological surgery sequence in dimension four is defined and exact if and only if all good boundary links are freely slice [**FQ90**, Corollary 12.3C] (see also [**KOPR21a**, Section 23.2.1]).

(4) [**CKP20**, Corollary 2.2] gives a characterization of good boundary links in terms of Seifert matrices. This is very useful in confirming whether a given boundary link is good. For example, this gives an easy method to verify that Whitehead doubles of links with vanishing pairwise linking numbers are good boundary links.

(5) Several families of good boundary links have been shown to be freely topologically slice. The most general result is due to Cha, Kim, and Powell [**CKP20**, Theorem A], generalizing previous work focusing on Whitehead doubles, e.g. in [**FT95b**].

(6) Not all good boundary links are smoothly slice, even with more than one component. For example, Levine [**Lev12**] showed that the Whitehead double of the Borromean rings, with all positive clasps, is not smoothly slice. The question of whether the Whitehead double of the Borromean rings is topologically slice appeared as Problem 4.46 in [**Kir97**], and we restate it here in part (b) as a special case of the problem.

QUESTION. Is the Whitehead double of the Borromean rings (with any choice of clasps) topologically slice?

Proposed for K3 by: M. Freedman
Scribed by: A. Ray

1.9. Legendrian knots and Lagrangian surfaces

The next suite of problems focuses on Legendrian and transverse links in $\mathbb{R}^3$ and other contact 3-manifolds. A link is *Legendrian* if it is tangent to a specified

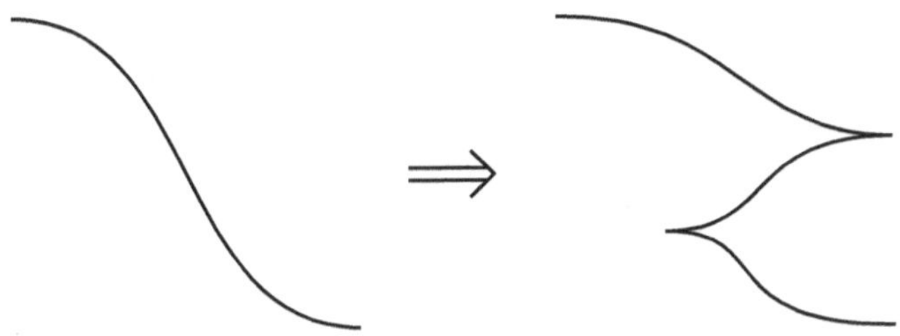

FIGURE 1. Adding a zig-zag

contact structure in a three-manifold. For links in the standard contact structure on $\mathbb{R}^3$ (denoted $(\mathbb{R}^3, \xi_{std})$) this simply means the tangents are in the kernel of the 1-form $\alpha = dz - ydx$. See Section 3.5 for more background and problems on contact structures. Legendrians are considered up to isotopy through Legendrians. Classically, there are two integer invariants of a Legendrian knot, the Thurston–Bennequin number tb and rotation number rot (for nullhomologous knots), defined in terms of how the contact planes twist while traveling along the knot. In $(\mathbb{R}^3, \xi_{std})$ or any other tight contact structure, these invariants satisfy the Bennequin inequality which says that for any Legendrian knot $tb(K) + |rot(K)| \leqslant 2g(K) - 1$ [**Eli92**]; in the case of the standard contact structure on $\mathbb{R}^3$ or S^3, this can be strengthened to the slice-Bennequin inequality where $g(K)$ is replaced by $g_4(K)$ [**Rud95**]. A Legendrian can be *stabilized* by adding a zig-zag in the xz-projection as in Figure 1; this operation lowers tb by 1 and either raises or lowers rot depending on orientations. There are other more subtle invariants of Legendrian knots, usually defined in terms of some version of Floer homology.

By analogy to studying surfaces in B^4 bounding knots in S^3, if (Y, ξ) is a contact 3-manifold that is filled by a symplectic 4-manifold (W, ω), such as a Stein filling, then we can ask if a Legendrian $K \subset (Y, \xi)$ bounds a Lagrangian surface Σ in W. Often we require that $\omega = d\lambda$ with $\lambda|_\Sigma$ also exact, i.e. Σ is an *exact* Lagrangian. There is a similar notion of Lagrangian cobordisms between Legendrian knots in symplectic cobordisms between contact 3-manifolds, and in the symplectization of $(\mathbb{R}^3, \xi_{std})$ these are often *decomposable*, which means they are built out of certain elementary pieces, similar to ribbon surfaces in B^4.

Any Legendrian knot can be perturbed to a knot that is transverse to the contact structure, called a *transverse* knot. For transverse knots, there is only one classical invariant, the self-linking number sl, which equals $tb + rot$ if the knot is a push-off of a Legendrian. In the presence of a symplectic filling, we are interested in when the transverse knot is filled by a symplectic surface. For an introduction to Legendrian and transverse knot theory see [**Etn05, Gei08**].

PROBLEM 1.63. *Is there a knot type with Legendrian representatives that do not destabilize but have arbitrarily negative Thurston–Bennequin number?*

REMARKS.

(1) This is Etnyre-Ng [**EN03**, Question 44]; see also [**CGH09**, Question 61].
(2) It is known by work of Etnyre and Honda that for any N there is a knot type having peaks of height N in its Legendrian mountain range [**EH03**]. For any choice of knot type and Thurston-Bennequin number,

there are finitely many distinct Legendrian representatives, so this question is equivalent to asking if there is a knot type with infinitely many non-destabilizable representatives. This has been proved for many knot types, including the unknot [**EF09**], torus knots [**EH01a**], cables of torus knots [**ELT12, LaF10**], and twist knots [**ENV13**].

(3) The finiteness of non-destabilizable Legendrian representatives problem is equivalent to whether there are finitely many non-destabilizable grid diagrams for a given knot [**DP13**]. It is proved in [**Dyn06**] that any non-minimal rectangular diagram of the unknot can be destabilized. This gives an algorithm to detect the unknot using monotonic simplification. A first step towards generalizing this result to arbitrary topological knots is to answer the question above.

Proposed for K3 by: L. Ng
Scribed by: K. Hendricks

PROBLEM 1.64.

(a) *Let $L \subset (S^3, \xi_{std})$ be a transverse link such that the branched double cover $(\Sigma_2(L), \xi_L)$ is Stein fillable. Is L transversely isotopic to the closure of a quasipositive braid?*

(b) *Is the same true if $(\Sigma_n(L), \xi_L)$ is Stein fillable for some $n > 2$?*

(c) *Does Stein fillability of the branched cover imply that the slice-Bennequin inequality must be sharp for the given link?*

REMARKS.

(1) Here, (S^3, ξ_{std}) denotes the standard contact structure on S^3. If $L \subset (S^3, \xi_{std})$ is a transverse link, then the cyclic n-fold branched cover $\Sigma_n(L)$ has an induced contact structure ξ_L, $n \geqslant 2$.

(2) See Problems 1.69 and 1.70 for definitions, background on quasipositive braids, and related questions.

(3) A closed braid is quasipositive if and only if it can be represented as a transverse intersection $S^3 \cap \mathcal{V}$ for some smooth, complex curve $\mathcal{V} \subset \mathbb{C}^2$ [**Rud83, BO01**].Thus if L is isotopic to the closure of a quasipositive braid, then $(\Sigma_n(L), \xi_L)$ is Stein fillable, because it bounds the Stein manifold obtained as the cover of B^4 branched over the complex curve $\mathcal{V} \cap B^4$. The question asks whether fillability of branched covers characterizes quasipositive braids. For more results on the branched covers of transverse links, see [**Pla06**]. More discussion of relations between different monoids related to quasipositivity and contact geometry is given in [**EVHM15**].

Proposed for K3 and scribed by: M. Hedden

PROBLEM 1.65. *Decomposable Lagrangian cobordisms between Legendrian knots or links in $\mathbb{R}^3$ are compositions of certain simple pieces admitting diagrammatic descriptions* [**EHK16**]. *Such cobordisms necessarily have no index 2 critical points.*

Is every exact Lagrangian cobordism without index 2 critical points Lagrangian isotopic to a decomposable Lagrangian cobordism?

REMARKS.

(1) At the time of writing, a candidate for a non-decomposable cobordism is given in [**CNS16**, Conjecture 3.3]. This can be built using another elementary construction, Guadagni's tangle move, see [**BLL**$^+$**21**, **GSY22**], which could in principle produce additional non-decomposable cobordisms.

(2) The hypotheses on index 2 critical points is necessary for a variety of reasons. For example, Lagrangian caps (see, e.g., [**Lin16**]) are, by definition, not decomposable; for decomposability, the positive boundary of the cobordism must be nonempty or else index 2 critical points are necessary. In the recent preprint [**DRG24**], an example of a non-decomposable Lagrangian concordance between knots is given. However, the authors there show that the concordance must contain index 2 critical points.

(3) Conway, Etnyre, and Tosun prove that decomposable Lagrangians are regular [**CET21**], which means there is a Weinstein structure on the cobordism where the Liouville vector field is tangent to the cobordism. Using this, [**EGL20**] implies the exterior of a decomposable Lagrangian disk necessarily has a Stein structure, and hence the exterior can be built out of particularly constrained elementary pieces, akin to the decomposable Lagrangian itself. (It is worth pointing out that the non-decomposable examples of [**DRG24**] above are not regular either.)

(4) Here is a variant of the problem: Is every filling (exact Lagrangian cobordism with empty negative boundary) Lagrangian isotopic to a decomposable Lagrangian cobordism? This question is sometimes stated more generally for exact Lagrangian cobordisms with negative boundary given by any (possibly empty) non-destabilizable Lagrangian link; the non-destabilizability condition rules out the presence of a Lagrangian cap as in [**Lin16**]. The possibly non-decomposable candidate from [**CNS16**], which is a concordance with a Legendrian trefoil as its negative boundary, can be glued to a filling of the trefoil to produce a potential counterexample to this version of the problem.

Scribed by: T. Lidman, R. Lipshitz, L. Ng

PROBLEM 1.66. *For Legendrian links $\Lambda_1, \Lambda_2 \subset (\mathbb{R}^3, \xi_{\text{std}})$, write $\Lambda_1 \preceq \Lambda_2$ if there is an exact Lagrangian cobordism in the symplectization $\mathbb{R} \times \mathbb{R}^3$ from Λ_1 to Λ_2. Is $\preceq$ a partial order?*

REMARKS.

(1) One can ask an analogous question in any contact 3-manifold, as well as in higher dimensions.

(2) The fact that $\preceq$ is not symmetric was first shown by Chantraine [**Cha15**].

(3) As for ribbon concordance in the smooth setting, many invariants of Legendrian knots are monotone under $\preceq$. For example, the behavior of rulings [**CNS16**], contact homology [**Pan17a**] and the contact invariant in monopole Floer homology [**BS18a**] support this conjecture.

(4) One can ask an analogous question about symplectic concordances between transverse knots.

Scribed by: R. Lipshitz, L. Ng

PROBLEM 1.67. *Given a Legendrian link in the standard contact $\mathbb{R}^3$ besides the standard unknot or Hopf link, classify its exact Lagrangian fillings up to compactly supported Hamiltonian isotopy.*

REMARKS. Eliashberg-Polterovich [**EP96**] showed that the unknot has a single filling. A recent preprint of Thomson claims that the Hopf link has exactly 2 fillings [**Tho25**]. These are the only links with known classification. The right-handed trefoil famously has 5 fillings [**EHK16**], distinguished through augmentations of the Legendrian contact homology. Casals and Gao showed there are examples of links with infinitely many non-isotopic fillings [**CG22**]. Casals has a more general conjectured classification for some large families of links in [**Cas22**, Conjecture 5.1]. A special case would be that the $T(2, n)$ torus link has exactly C_n fillings up to isotopy through exact Lagrangians, where C_n is the nth Catalan number, generalizing the count for the unknot and Hopf link. (See the work of Ekholm-Honda-Kalman for earlier questions on classifying Lagrangian fillings by augmentations of Legendrian contact homology [**EHK16**], including for the $T(2, n)$ torus links.) A lower bound of C_n in this case is due to Pan [**Pan17b**] and independently Shende-Treumann-Williams-Zaslow [**STWZ19**].

Scribed by: L. Ng

PROBLEM 1.68. *Determine the smooth knot types that have Legendrian representatives with orientable exact Lagrangian fillings.*

REMARKS.

(1) This problem appears in [**HS15**] together with a conjecture that such fillings exist if and only if the knot type K is quasi-positive and $\overline{tb}(K) = -\deg_a P_K - 1$, where $\overline{tb}(K)$ denotes the maximal Thurston-Bennequin number of a Legendrian representative of K and $P_K(a, z)$ the HOMFLYPT polynomial. These conditions are necessary for existence of a filling (see [**HS15**] and references therein) as is the sharpness of the slice-Bennequin inequality, $\overline{tb}(K) = 2g_4(K) - 1$, as shown in [**Cha10**, Theorem 1.4].
(2) For alternating knots [**CNS16**], an answer is that a filling exists if and only if the knot type is positive. In general, constructions of fillings exist for all positive knots [**HS15**] and all almost positive knots with diagrams of Type II [**Tag19**]. A variant of the question allows for nonorientable Lagrangian fillings; see [**CCPR$^+$24**].

(3) For comparison, the analogous question involving transverse knots has a succinct answer: Transverse knots with symplectic fillings exist within a smooth knot type if and only if the knot type is quasi-positive; see [**Rud83, BO01**].

Proposed for K3 and scribed by: D. Rutherford

PROBLEM 1.69. *Let $L \subset (S^3, \xi_{std})$ be a transverse link with*

$$sl_\Sigma(L) = -\chi(\Sigma),$$

for some Seifert surface Σ. Must L be strongly quasipositive?

REMARKS.

(1) The self-linking number of an oriented link transverse to the standard contact structure ξ_{std} on S^3 satisfies the well-known Bennequin bound [**Ben83, Etn05**]:

$$sl_\Sigma(L) \leqslant -\chi(\Sigma),$$

where Σ is any Seifert surface for L. This problem therefore attempts to characterize when this bound is sharp.

(2) Strongly quasipositive links are those that possess a *quasipositive Seifert surface*, constructed from parallel disks by attaching bands with positive half-twists [**Rud92**]. Such a surface is properly isotopic to the intersection of the 4-ball with a non-singular complex curve [**Rud83**]. It is known that if a link is strongly quasipositive, or bounds a complex curve of Euler characteristic equal to that of some Seifert surface for the link, then the Bennequin bound is sharp (see [**Shu07**, Proposition 1.G] or [**Hed10**, Proof of Theorem 1.5] for a proof).

(3) In light of the above remark, one can ask if sharpness of the Bennequin bound implies Σ is quasipositive. If so, then it follows that that *any* minimal genus Seifert surface for L is quasipositive. This is because the self-linking number depends only on the relative homology class of Σ, hence the Bennequin bound will be sharp for any other minimal genus Seifert surface Σ'.

(4) The question in this problem has an affirmative answer for fibered links [**Hed10**], links represented by minimal-index braids with fractional Dehn twist coefficient > 1 [**IK19**], 3-braids [**IK19**], positive knots [**Rud99**], almost positive knots [**FLL23**], and knots whose canonical and Seifert genus agree [**FLL23**]. Hedden and Tovstopyat-Nelip have announced an affirmative answer for links with two dimensional top group of Heegaard knot Floer homology.

(5) These notions generalize to arbitrary contact 3-manifolds, with the Bennequin bound generalized by Eliashberg to tight contact 3-manifolds [**Eli92**] and strong quasipositivity generalized via braids in open books [**BEH$^+$15, IK19, Hay21b**]. Geometrically, the class of quasipositive Seifert surfaces is the same as the class of *ribbons of Legendrian graphs* [**BI09, Hay22**]. One can therefore ask the following more general question: Suppose there exists a transverse link $L \subset (Y, \xi)$ for which

$$sl_{\xi,\Sigma}(Y, L) = -\chi(\Sigma),$$

for some Seifert surface Σ. Is Σ isotopic to the Legendrian ribbon of a Legendrian graph $\Gamma \subset (Y, \xi)$? This generalized question has an affirmative answer for overtwisted contact structures [**BCV09**].

Proposed for K3 and scribed by: M. Hedden

PROBLEM 1.70.

(a) *Let $L \subset (S^3, \xi_{std})$ be a transverse link with*

$$sl_\Sigma(L) = -\chi(\Sigma),$$

for some smooth surface $\Sigma \subset B^4$ bounded by L. Is L quasipositive?

(b) *Is the class of strongly quasipositive links equal to the class of quasipositive links whose Seifert and smooth slice genera are equal?*

REMARKS.

(1) Quasipositivity implies that L is isotopic to the transverse intersection of a complex curve in $\mathbb{C}^2$ with the unit 3-sphere [**Rud83**], and any link arising from such an intersection is quasipositive [**BO01**]. It is not true, however, that any surface in B^4 bounding L realizing equality in the slice-Bennequin bound is isotopic to a subsurface of a complex curve. Indeed, a complex curve is ribbon (in the sense that the radius function on the 4-ball, restricted to the curve, can be assumed to be Morse without critical points of index 2). On the other hand, tubing a complex curve to a smoothly knotted 2-sphere in its complement will produce a surface that is not ribbon, yet will still realize equality in the slice-Bennequin bound. One *can* ask, however, whether a ribbon surface realizing equality in the slice-Bennequin bound is properly isotopic to a piece of complex curve.

(2) Since quasipositive links have transverse representatives for which the slice-Bennequin bound is sharp (again, see [**Shu07**, Proposition 1.G] or [**Hed10**, Proof of Theorem 1.5]) we observe that Problem 1.69 implies an affirmative answer to part (b).

(3) Part (b) is asking whether there exists a link L bounding a complex curve whose Euler characteristic is the same as that of some Seifert surface for L, but for which no complex curve bounded by L is isotopic to a Seifert surface. The existence of such a link would answer part (b) in the negative, from which it would follow that Problem 1.69 is false. Note that part (b) has an affirmative answer for fibered links, by [**Hed10**]. Since complex curves are ribbon and Seifert surfaces are ribbon-immersed without singularities, this problem can be thought of as asking about the minimum number of ribbon singularities required across complex curves bounded by L.

Proposed for K3 and scribed by: M. Hedden

1.10. Physical knot theory

While knot theory was originally developed as a model for studying physical phenomena, classical knot theory elides essential features that would arise naturally in a real-life study of knots, such as the thickness of a rope. Physical knot theory studies a broader range of models for knots that are relevant to more physical settings. The first four problems of this section pertain to other geometric models for knots in physical knot theory. The last problem is about knot energies, which is motivated by how a knot would behave dynamically if it were electrically charged.

PROBLEM 1.71. *Does a Gordian unknot exist?*

REMARKS. A *physical link* is a collection of disjoint embedded loops of given finite lengths in a configuration with *thickness* at least 1. Thickness has many equivalent definitions [**GM99**, **CKS02**], but an appealing one is the same as Federer's definition [**Fed59**] of *reach*: the infimal r so that every point in $\mathbb{R}^3$ within r of the curve has a unique nearest neighbor on the curve. Alternatively, for smooth knots, thickness 1 means that a radius-1 tubular neighborhood is also embedded [**Sim02**]. A *physical isotopy* preserves the length of each loop while respecting this thickness constraint. The thickness lower bound yields an upper bound 1 for the curvature of a physical link, implying a least-length configuration exists in each physical isotopy class [**CKS02**]. By a *Gordian pair* we mean physical link configurations that are isotopic but not physically isotopic; see [**Nab95**, **LN21**] which provide Gordian pairs of higher dimensional knots, the latter giving an explicit construction of a Gordian unknot in S^4. Gordian pairs with two or more components [**CH15**, **KK23**] are known — even for unlinks [**AH25**, **KK25**] — but it remains unknown for knots, and especially for the unknot.

Scribed by: J. Hass, R. Kusner

PROBLEM 1.72 (The equilateral stuck unknots conjecture.). *Are there equilateral embedded polygons that are unknotted yet cannot be unknotted through polygons preserving edge lengths? Such an unknot is said to be* stuck.

REMARKS.

(1) The celebrated Carpenter's Rule problem asks whether an embedded polygon in the plane may be deformed to a convex configuration through embedded polygons with the same edge lengths. Solutions in the affirmative were given more or less simultaneously by Streinu [**Str00**] and Connelly, Demaine, and Rote [**CDR03**].
(2) The natural extension of this theorem to PL knots in codimension two asks whether polygons of knot type K may be deformed into one another through embedded polygons of the same edge lengths. For arbitrary edge lengths, the answer is known to be 'no': there are stuck unknots with as few as six edges [**CKS98**, **Tou01**, **AET04**] (see Figure 2) which cannot be convexified through embedded configurations with the same edge lengths.
(3) Interestingly, all of the known examples of stuck unknots have very different edge lengths, which they seem to require in an essential way. This leads to the question: are there *equilateral* stuck unknots?

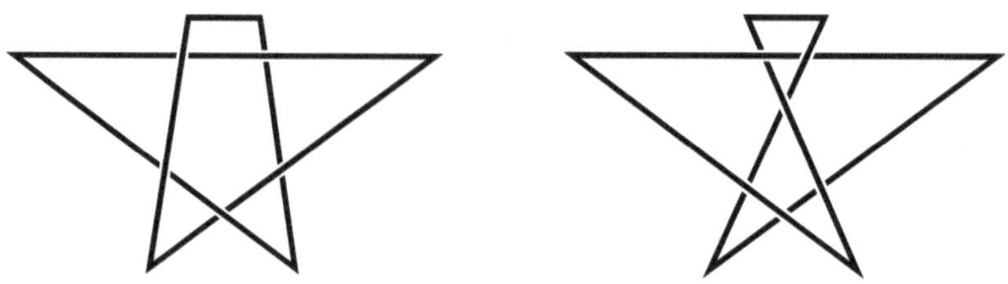

FIGURE 2. Stuck unknots

There is some reason to believe that there are not. Khoi [**Kho05**] proved that the symplectic volume of the space of equilateral n-gons in $\mathbb{R}^3$ is the largest symplectic volume of any space of n-gons with the same total length. That is, in this sense, equilateral polygons are the most flexible polygons. On the other hand, it is quite difficult to imagine an algorithm for unfolding an equilateral unknot of n edges.

Proposed for K3 by: J. Cantarella
Scribed by: D. Ruberman

PROBLEM 1.73 (The 15 pearls conjecture). *Is the pearl number of the trefoil equal to* 15*?*

REMARKS.

(1) A *pearl necklace* with n pearls is an embedding of an equilateral n-gon in $\mathbb{R}^3$ so that each vertex is surrounded by a sphere with diameter equal to the edgelength and the interiors of all the spheres are disjoint. This is a common off-lattice model of a self-avoiding polygon, usually called the *hard sphere polymer* (see [**SJ23**]). The *pearl number* of a knot K is the minimum number of pearls in any necklace whose polygon has knot type K.
(2) A self-avoiding polygon on a cubic lattice is a common model for random knots with self-interactions. The "minimum step numbers" are known for self-avoiding polygons on the cubic lattice for a number of knot types by a combination of geometric arguments and computer enumeration [**SIA$^+$09**]. For instance, the minimal step number of the trefoil on the simple cubic lattice is 24 [**Dia93**].
(3) Very little is known about the pearl number. Oshiro and Maehara [**MO99**] give a construction for a 15-pearl trefoil and conjecture that this configuration is minimal (so the pearl number of the trefoil is 15). Interestingly, their configuration is similar to the 15 step embedding of the trefoil on the *face-centered* cubic lattice (shown in Figure 3) found by Rechnitzer and Janse van Rensburg [**RR11**], which those authors conjecture to be minimal. By putting balls of radius 1/2 on the vertices of a realization on the cubic lattice one obtains a hard sphere polymer. Thus the pearl number is bounded above by the length of any lattice representation. Conversely, the pearl number is an upper bound on the equilateral stick number of any knot. The difference between the three integer invariants

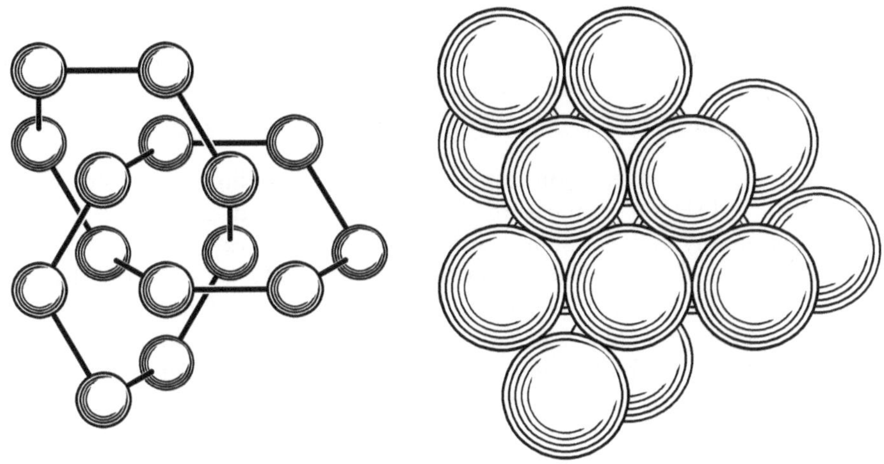

FIGURE 3. Face-centered cubic lattice embedding of trefoil

should be quite large for most knot types as can be seen for the trefoil with an equilateral stick number of six [**Jin97**].

(4) In [**Mae07**], Maehara proves that the pearl number of the trefoil is at least 11.

Proposed for K3 by: J. Cantarella
Scribed by: J. Cantarella, D. Ruberman

PROBLEM 1.74. *How does ropelength behave under connected sum of knots? Here are two conjectures, the second a weakening of the first.*

(a) *For any knot or link types K_1 and K_2,*

$$\mathrm{Rop}(K_1 \# K_2) \leqslant \mathrm{Rop}(K_1) + \mathrm{Rop}(K_2) - (4\pi - 4).$$

(b) *For any knot or link types K_1 and K_2,*

$$Rop(K_1 \# K_2) \leqslant Rop(K_1) + Rop(K_2) - c,$$

where $c > 0$ is some constant independent of knot type.

REMARKS.

(1) The (minimum) ropelength $\mathrm{Rop}(K)$ of a knot is defined to be the smallest quotient of length and thickness (defined in Problem 1.71) among all rectifiable curves in $\mathbb{R}^3$ realizing the knot type. It is known that ropelength minimizers exist in every knot and link type [**GMSvdM02**, **GL03**, **CKS02**], but their shapes (and rope-lengths) are known only in a few special cases of links with planar components. In all other case we only have numerical results approximating the rope-length.

(2) The ropelength minimizing Hopf link 2^2_1 is a pair of circles of equal radius in orthogonal planes, passing through each other's centers. Ropelength is scale-invariant, so we can assume that the circles have radius 2. Their thickness is then 1 and $\mathrm{Rop}(2^2_1) = 8\pi$. The ropelength-minimizing composite link $2^2_1 \# 2^2_1$ (a chain of three linked circles) is composed of two round

circles in the same plane, each passing through a center of the semicircular arc of a stadium curve in the orthogonal plane. This has ropelength 8π (for the circles) plus $4\pi + 4$ (for the stadium curve), so

$$\mathrm{Rop}(2^2_1 \# 2^2_1) = 12\pi + 4 = \mathrm{Rop}(2^2_1) + \mathrm{Rop}(2^2_1) - (4\pi - 4)$$

and we have saved some rope by splicing these links together. Katritch et al. [**KOP+97**] conjectured in 1997 that we can save at least as much rope on any connect sum; leading to Conjecture (a).

(3) The work of Diao [**Dia24**] on alternating knots uses the braid index as a lower bound on ropelength. The braid index of a connected sum is $b(K_1 \# K_2) = b(K_1) + b(K_2) - 1$ for two alternating knots K_1 and K_2. Thus the lower bound on ropelength of a connected sum of the two knots $b(K_1 \# K_2)$ will be slightly less than the sum of the lower bounds of the two knots K_1 and K_2.

Hence it is worth considering the weaker version, Conjecture (b), with an undetermined constant.

(4) In general, it would be important to have any theoretical lower bound on ropelength for knot types that approximates the numerical results. This has been done for the trefoil, but the techniques used do not generalize to other knot types, for which there remains a wide gap between the numerical results and the theoretical lower bounds.

Proposed for K3 by: J. Cantarella
Scribed by: D. Ruberman, R. Sazdanović

PROBLEM 1.75.

(a) *Find some knot energy on the space of smoothly embedded unknotted circles in $\mathbb{S}^3$ for which all unknotted critical points are great circles.*

(b) *Define a gradient flow for this knot energy which yields a deformation retract of the space of unknots to the subspace of great circles.*

REMARKS.

(1) These questions were posed by Freedman, He, and Wang [**FHW94**]. The motivation is one of the many equivalent formulations of the Smale conjecture mentioned in the Appendix to Hatcher's 1983 paper (which actually proves this conjecture). Variation (7) states [**Hat83**, p. 606] that "the space of smoothly embedded unknotted circles in $\mathbb{S}^3$ deformation retracts onto the space of great circles in $\mathbb{S}^3$, (i.e., $O(4)/O(2) \times O(2)$)." Of course, it would be of some interest to construct a more or less explicit example of such a retract.

(2) Beginning in the 1980s, several repulsive functionals, so-called knot energies, have been defined in the pursuit of disentangling complicated knotted curves and deforming them into 'simpler' curves within the same knot class; see, e.g., the survey by Strzelecki et al. [**SSvdM13**] and references therein. The most prominent example is the Möbius energy introduced by O'Hara [**O'H91**] in 1991.

(3) As pointed out by Freedman, He, and Wang [**FHW94**], it is tempting to conjecture that a suitable gradient flow for a knot energy actually defines

a retraction as stated by Hatcher, i.e., it will deform any curve from the unknot class to a round circle. Of course, this can only work if, except for the circles, there are no critical points within the unknot class; cf. Budney's post [**Bud16**]. Currently, we do not know whether any of the many smooth knot energies that have been proposed so far enjoys this property.

Scribed by: S. Blatt, H. Schumacher

1.11. Algorithmic and combinatorial questions

In spite of many powerful tools discussed above that can detect the unknot or bound the unknotting number (e.g. connections with 3-manifolds, knot homology theories, quantum invariants), these are often computationally difficult. Currently, we do not have an algorithm to compute unknotting number (Problem 1.80) or a fast algorithm to determine if a given knot is unknotted (Problem 1.76). The problems in this section ask for algorithms or other means of studying the computational difficulty associated with transforming one knot diagram to another.

As general background, the problem of recognizing an object X, such as the unknot, is in NP if there is a polynomial time algorithm that can certify that an object Y is indeed equivalent to X. On the other hand, a recognition problem is in co-NP if there is a polynomial time algorithm that can verify that Y is *not* equivalent to X. A problem is *NP-hard* if the existence of a polynomial time algorithm that solves the problem can be used to solve all other NP problems in polynomial time, i.e. solving it is equivalent to solving the most computationally difficult of all NP problems.

For a couple of other problems that ask for algorithms in knot theory see Problems 1.52 and 1.63. See also Section 3.2 for more discussion of computational complexity and a selection of algorithmic questions in 3-manifold topology.

PROBLEM 1.76.

(a) *Is there an algorithm to detect the unknot that runs in polynomial time (as a function of the number of crossings in an input diagram)?*
(b) *What is the structure of the unknot Reidemeister graph?*

REMARKS.

(1) There are now many algorithms to detect the unknot. The first was due to Haken [**Hak61**] and used normal surfaces. Hass, Lagarias and Pippenger [**HLP99**] refined his methods to prove that unknot recognition lies in NP. Dynnikov [**Dyn06**] used grid diagrams to find a very different algorithm. He showed that any grid diagram for the unknot can be reduced to the trivial diagram through a sequence of moves, none of which increases the grid number.
(2) There are now many invariants that detect the unknot, including Khovanov homology [**KM11**] and Heegaard Floer homology [**OS04b**]. However, none of these invariants seem to be computable in polynomial time.

A further significant result was proved by Kuperberg [**Kup14**] who showed that unknot recognition lies in co-NP, assuming the Generalized Riemann Hypothesis. In other words, if a knot is nontrivial, then there is an efficient way of certifying this. This result was proved unconditionally, removing the GRH assumption, by Lackenby [**Lac21a**] using sutured manifold hierarchies.

(3) The unknot Reidemeister graph in part (b) is formed by assigning a vertex to each diagram of the unknot and connecting two vertices by an edge if there is a Reidemeister move connecting them; it is locally finite. There are pairs of vertices representing two diagrams with n crossings such that the shortest path connecting them has length at least $n^2/25$ [**HN10**]. No two vertices representing diagrams with n crossings have distance greater than $2(236n)^{11}$ [**Lac15**].

This problem is complicated by the existence of "hard" unknot diagrams [**BCL**$^+$**24, PZ16**], which have the property that any sequence of Reidemeister moves taking them to the trivial diagram must go via diagrams with higher crossing number. The question may be more tractable if different sets of basic moves are allowed in addition to or in place of the standard Reidemeister moves; in the related setting of grid diagrams, monotonic simplification has been established [**Dyn06**].

Scribed by: M. Lackenby

PROBLEM 1.77. *How many Reidemeister moves are required to relate two diagrams of a knot (as a function of their numbers of crossings)?*

REMARKS.

(1) One reason why this question is interesting is that a computable upper bound leads to an algorithm to decide whether two knots are equivalent.

(2) Lackenby [**Lac15**] proved that, for a diagram of the unknot with c crossings, there is a sequence of Reidemeister moves taking it to the trivial diagram with length at most $(236c)^{11}$. It is not inconceivable that there is a polynomial upper bound that applies to every knot type. However, the best known upper bound on the number of Reidemeister moves required to relate two diagrams of a knot with c_1 and c_2 crossings is due to Coward and Lackenby [**CL14**]:

$$2^{2^{\cdot^{\cdot^{\cdot^{c_1+c_2}}}}}$$

where height of the tower is $k^{c_1+c_2}$ and where $k = 10^{1000000}$.

(3) A nontrivial lower bound was established by Hass and Nowik [**HN10**], who proved that for each natural number n there exists a diagram of the unknot with $7n-1$ crossings and where the number of Reidemeister moves required to take it to the trivial diagram is at least $2n^2 + 3n - 2$.

Scribed by: M. Lackenby

PROBLEM 1.78. *Let D be any diagram of the unknot with n crossings. Let $h(D)$ be the smallest number such that some series of Reidemeister moves that transforms D to a 0-crossing diagram has the property that all intermediate diagrams have at most $h(D)$ crossings. Define $h(n)$ to be the maximum of $h(D)$ over all n-crossing diagrams of the unknot. A diagram represents a* hard unknot *if $h(D) > n$. What is $h(n)$?*

REMARKS. It is known [**BCL$^+$24**] that $h(n) \geqslant n + 3$ for certain values of n.

Scribed by: J. Hass

PROBLEM 1.79. *Are there additional moves that, when added to the three Reidemeister moves, allow for strict monotonic descent in the crossing number of an unknot diagram?*

REMARKS.

(1) Additional moves allow increased efficiency in diagram simplification. Several attempts at finding such moves have been given, but none are known to suffice [**PZ16**]. Work of Dynnikov gives a monotonic, but not strictly monotonic, process [**Dyn03**].
(2) A strictly monotonic process would lead to a polynomial time algorithm for unknot detection; see Problem 1.76.
(3) The collection of additional moves can grow with the number of crossings, e.g. moving an arc across a twist region of a diagram. However the set of additional moves cannot be arbitrary; moves consisting of any sequence of Reidemeister moves would allow any knot equivalence to be carried out in one step. A reasonable restriction is to allow moves that reduce the crossing number and can be found in polynomial time (polynomial in the crossing number).

Scribed by: J. Hass

PROBLEM 1.80. *Is unknotting number computable? Is there even an algorithm to decide whether a knot has unknotting number one?*

REMARKS.

(1) Many elementary knot invariants, such as the crossing number, are known to be computable. However, it seems extremely challenging to determine whether the unknotting number is computable.
(2) Recall that the *unknotting number* of a knot is the minimum number of crossing changes required to unknot the knot, where the minimum is taken over all diagrams of the knot. One can also consider the *unlinking number* of a link, which is the minimal number of crossing changes required to turn it into an unlink. Closely related is the *splitting number*, which is the minimal number of crossing changes required to turn into a split link.
(3) Some lower bounds on the computational complexity of determining unlinking number are known: it was shown by de Mesmay, Rieck, Sedgwick and Tancer [**dMRST21**] and Koenig and Tsvietkova [**KT21**] that determining the unlinking number of a link is NP-hard.

(4) The second question in the problem seems more tractable, since many structural results are known about knots with unknotting number one [**ST89**], [**Sch85b**]. The analogous problem of determining whether a link has unlinking number one was solved for a large class of links by Lackenby [**Lac21b**].
(5) This problem was also stated in [**Lac17a**].

Proposed for K3 and scribed by: M. Lackenby

1.12. General knot theory

In this final section, we present a number of problems of interest that do not fit cleanly into any of the preceding sections.

PROBLEM 1.81.
(a) *Are all knots trivial?*
(b) *Conjecture: The Bing sling is knotted.*

REMARKS.
(1) For this problem, a knot is a topological embedding (a homeomorphism onto its image) of a circle in S^3. A knot is trivial if it is isotopic, i.e. connected by an arc of embeddings, to the unknot. A smooth knot in a string is trivial in this sense by simply pulling the string tight so that the knot becomes a point. A similar argument works for locally flat embeddings: the problem is only interesting for wild embeddings.
(2) The Bing Sling [**Bin56, DV09**] (see also [**Nan18, amd22**] and [**Shi73**]) is drawn in Figure 4.
(3) Brin [**Bri83**] shows that there are knots that at each point are locally equivalent to the Bing sling but are non-ambiently isotopic to the unknot.

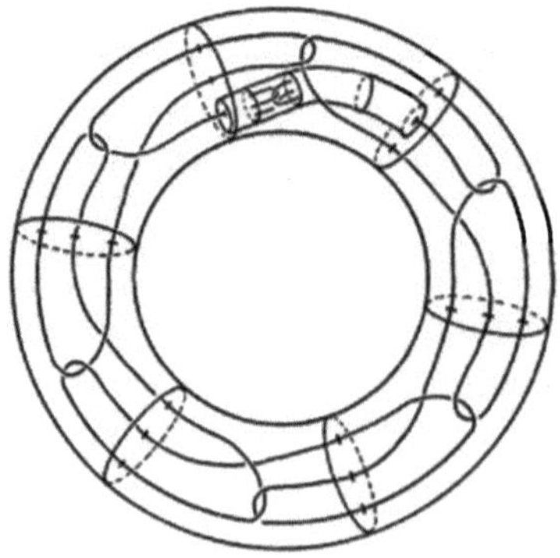

FIGURE 4. The Bing Sling

Proposed for K3 by: M. Freedman
Scribed by: D. Ruberman

PROBLEM 1.82.

(a) *What is a positive knot?*
(b) *Describe a simple set of moves to convert between two positive diagrams of the same knot or link.*

REMARKS.

(1) Positive knots and alternating knots are both defined in diagrammatic terms, but somewhat orthogonally to one another. The corresponding problems for alternating knots have been settled.
(2) The first problem is meant to evoke Ralph Fox's question, 'What is an alternating knot?' [**Lic97**, p.32]. It seeks a characterization of the class of positive knots in terms intrinsic to the knot exterior, e.g. a condition on the knot group or the existence of a special kind of geometric structure or spanning surface. It echoes a related question of Rudolph, who proved that positive links are strongly quasipositive. He asked, as a kind of converse to his result: "Can positive links be characterized as strongly quasipositive links that satisfy some extra geometric conditions?" [**Rud99**, Question, p.556]. Baader made progress on Rudolph's question by showing that a knot is positive if and only if it is strongly quasipositive and homogeneous [**Baa05**]. In light of this, an answer to Problem 1 could be obtained by giving a geometric characterization of homogeneity. Note that Fox's question was addressed in related papers by Greene and by Howie [**Gre17, How17**], which led to new applications for alternating knots.
(3) The second problem is meant to evoke the Tait flyping conjecture for alternating links, which was proven by Menasco and Thistlethwaite [**MT93**]. The quest for a set of moves satisfying Problem (b) is complicated by the fact that a knot can have reduced positive diagrams with different crossing numbers.

Proposed for K3 and scribed by: J. Greene

PROBLEM 1.83. *Determine the algebraic structure of the concordance group $\mathcal{O}$ of open strings.*

(a) *Is it abelian?*
(b) *Does it contain torsion?*

REMARKS.

(1) Virtual strings, also known as flat virtual knots [**Kau99**], were introduced by Turaev [**Tur04**] as homotopy classes of immersions of S^1 in compact oriented surfaces, up to stabilization of the surface.
(2) Two virtual strings in Σ_1 and Σ_2 are said to be *concordant* if they cobound an immersed annulus in an oriented 3-manifold M with boundary $\partial M = \Sigma_1 \cup -\Sigma_2$. A virtual string α in Σ is said to be *slice* if there exists an oriented 3-manifold M and immersed disk D in M such that $\partial M = \Sigma$ and $\partial D = \alpha$. The first example of an immersed curve in a surface that does not bound an immersed disk in any 3-manifold is due to Carter [**Car91**].

(3) Concordance classes of open strings form a group $\mathcal{O}$, which is known to be infinitely generated [**Tur04**]. Turaev also introduced the related notion of algebraic concordance for virtual strings, and Jie Chen [**Che23**, Example 3.19] has found examples of flat knots that are algebraically slice but not slice. There is a surjection to $\mathcal{O}$ from the concordance group of virtual knots (see the following Problem 1.84), which was shown to be non-abelian by Chrisman [**Chr22**].

Proposed for K3 by: H. Boden
Scribed by: D. Ruberman

PROBLEM 1.84. *For a classical knot, does its slice genus as a virtual knot agree with its slice genus as a classical knot?*

REMARKS.

(1) Virtual knots were originally defined combinatorially by Kauffman [**Kau99**]. For the purposes of this problem a virtual knot K may be regarded as an embedding of S^1 into $\Sigma \times [0,1]$, where Σ is a compact oriented surface. The *virtual slice genus* of K is the minimal genus, taken over all surfaces F and all 3-manifolds M, such that F embeds into $M \times [0,1]$ with $\partial F = K$. Here M is a compact oriented 3-manifold with $\partial M = \Sigma$. A virtual knot is said to be virtually slice if it bounds a disk in $M \times [0,1]$ for some 3-manifold M.
(2) This question is due to Dye-Kaestner-Kauffman [**DKK17**, §6], who gave an affirmative answer any knot for which the slice genus is determined by the Rasmussen invariant, including torus knots. Their results support the conjecture that for classical knots, their slice genus as virtual knots agrees with their slice genus as classical knots. It is known that a classical knot is virtually slice if and only if it is slice [**BN17**], thus the conjecture is true for knots with slice genus one.
(3) This problem has two versions, one for the smooth slice genus and the other for the topological (locally flat) slice genus.

Proposed for K3 by: H. Boden
Scribed by: D. Ruberman

PROBLEM 1.85. *Let K be a hyperbolic knot in S^3 and $\chi(K)$ the space of conjugacy classes of $PSL_2(\mathbb{C})$ representations of $\pi_1(S^3 - K)$. There is a distinguished component of $\chi(K)$ that contains the discrete faithful representation coming from the finite volume hyperbolic structure on the complement of K. Does this component always contain an arc of representations to $SO(3)$?*

REMARKS.

(1) Chinburg-Reid-Stover conjectured [**CRS22**, Conjecture 1.9] that there always is such an arc of representations. A positive answer would in fact prove that the knot groups of hyperbolic knots always have faithful $SO(3)$ representations.

(2) It is explained in [**Ago22**] that an affirmative answer to this question would prove there are no infinite descending chains of hyperbolic knots in the ribbon concordance partial order [**Gor81**, Question 6.2]. See Problem 1.56 for a knot Floer homology perspective on Gordon's question.

Proposed for K3 and scribed by: T. Lidman

PROBLEM 1.86.

(a) *Every connected cubic (i.e. trivalent) graph has freeness index at least 2.*
(b) *Every graph has freeness index at least two.*
(c) *There is a graph with freeness index two.*

REMARKS.

(1) A knot in the 3-sphere is the trivial knot if and only if its complement is a solid torus. One can extend the notion of triviality from knots to embedded (finite) graphs by considering a graph embedded in the 3-sphere to be "unknotted" if its complement is a connected sum of handlebodies. Every connected graph has an embedding into S^3 with handlebody complement; in fact, using induction and 2-handle addition, one can show that every graph has an unknotted embedding with the additional feature that every subgraph obtained by erasing the interior of a single edge is also unknotted [**Tho22**].
(2) Conway and Gordon showed that some graphs (for example, the complete graph with 7 vertices, K_7) have *some* knotted subgraphs no matter how they are embedded into S^3 [**CG83b**]. Planar graphs, by contrast, can be embedded in S^3 such that every subgraph is unknotted; so can some non-planar graphs, such as K_5.
(3) One can define an integer invariant of a finite (connected) graph Γ, the freeness index, which measures the extent to which Γ can be embedded in the 3-sphere such that it and its subgraphs are unknotted.

DEFINITION. An embedding f of Γ into S^3 such that the complement of Γ' is unknotted for every subgraph of Γ' of Γ obtained by deleting up to k edges from Γ is called *k-free*. The *freeness index of* Γ is the maximum of this integer over all embeddings f.

(4) In [**Tho22**] it is shown that every graph has freeness index at least one, while the graph K_6 has freeness index 8 and the Petersen graph has freeness index 4. [**Tho22**].
(5) We can relate the freeness index to the long-standing orientable cycle double cover conjecture (OCDCC) from combinatorics ([**Sze73**, **Sey80**]). The OCDCC for (bridgeless) cubic graphs implies the OCDCC for all graphs. For bridgeless cubic graphs Γ, the OCDCC is equivalent to the statement that Γ has an embedding into a closed orientable surface F such that every complementary region is a disk and the closure of every complementary region is also a disk (this is a *strong embedding* of Γ into F). In order to make a connection with the freeness index, one can exploit the fact that once a graph is embedded into a closed orientable surface F, F can then

be embedded into S^3 as a Heegaard surface. If both the embedding of the graph into the surface, and of the surface into the 3-sphere, are sufficiently "nice", one can conclude that the induced embedding of the graph into the 3-sphere satisfies some strong freeness conditions. This can be used to show that a cubic graph satisfying the OCDCC has freeness index at least 2 [**Tho22**].

(6) Note that a connected bridgeless cubic graph with freeness index 1 would provide a counter-example to the OCDCC.

Proposed for K3 and scribed by: A. Thompson

PROBLEM 1.87. *Is every fibered link in S^3 realized as the link of an isolated singular point of a polynomial map $\mathbb{R}^4 \to \mathbb{R}^2$?*

REMARKS.

(1) This problem is due to Benedetti–Shiota [**BS98**], and originates in the celebrated Milnor book [**Mil68b**].

(2) Let $f : \mathbb{R}^4 \to \mathbb{R}^2$ be a polynomial map with $f(0) = 0$ and set $V = f^{-1}(0)$. Assume that $0 \in \mathbb{R}^4$ is an isolated point of the intersection of V and the set of singular points of f. Such a point is often called a *weakly isolated singularity.*

As instances of Milnor's cone structure and fibration theorems [**Mil68b**], we have:

(i) There exists a small positive real number ε_0 such that for every $0 < \varepsilon < \varepsilon_0$, the sphere S^3_ε in $\mathbb{R}^4$ with radius ε centered at 0 is transverse to the surface V, and the link type L of $L_\varepsilon = (1/\varepsilon)(S^3_\varepsilon \cap V) \subset S^3$ does not depend on the choice of ε.

(ii) Under the stronger assumption that $0 \in \mathbb{R}^4$ is an isolated singular point of the map f, the link $L \subset S^3$ is fibered.

(3) In [**Mil68b**], one is mainly concerned with the links of (the realifications of) complex polynomial maps $\mathbb{C}^2 \to \mathbb{C}$ at an isolated singular point; in such a case, a weakly isolated singularity in (1) above is equivalent to a usual isolated singularity as in (2), and we obtain a distinguished and well understood family of fibered links (see [**Neu03, Web08**]). In the general case, there are no evident obstructions against a positive answer to the problem. As remarked in [**Mil68b**, Page 84], the first example of a fibered knot to be considered not belonging to that family for complex polynomials, was the figure-eight knot. Later, an explicit realization of the figure-eight knot as the link of a real isolated singularity was obtained in [**Per82**]. Further partial positive results have been worked out in [**Loo71, Rud87, Pic05, Bod19, Bod20, Bod23, AdSSQ24**].

(4) It is known [**AK81**] that *every* link L in S^3 can be realized as the link of a polynomial map as in (i) above, only requiring that 0 is an isolated point of the intersection of V and the set of singular points of f.

Proposed for K3 and scribed by: O. Saeki

PROBLEM 1.88. *Are there infinitely many congruence arithmetic links in the 3-sphere?*

REMARKS.

(1) This problem relates to the more general question posed by Thurston to find the topological and geometric properties of quotients of $\mathbb{H}^3$ by arithmetic subgroups of $PSL(2,\mathbb{C})$ [**Thu82**].
(2) A link complement is *arithmetic* if it is of the form $\mathbb{H}^3/\Gamma$ for some discrete subgroup $\Gamma \leqslant \mathrm{PSL}(2,\mathbb{C})$ with the property that Γ is commensurable with $\mathrm{PSL}(2, O_d)$, where O_d the ring of integers in $\mathbb{Q}(\sqrt{-d})$. Famously, Reid [**Rei91**] proved that the figure-eight knot is the only arithmetic knot in the 3-sphere. However, there are infinitely many arithmetic links (see for example the discussion in [**BR23a**]).
(3) The question is concerned with a particular type of arithmetic link called congruence links, which are defined as follows. A *principal congruence subgroup* Γ is equal to the kernel of
$$\mathrm{PSL}(2, O_d) \to \mathrm{PSL}(2, O_d/I)$$
for some ideal I in O_d. More generally, Γ is called *congruence* if it contains a principal congruence subgroup. When $\mathbb{H}^3/\Gamma$ is a link complement, the link is said to be *principal congruence* and *congruence* respectively.
(4) In [**BGR19**], Baker, Goerner and Reid gave a complete classification of principal congruence arithmetic links in the 3-sphere. In particular, there are finitely many such links. (See [**BGR22**] for all known link diagrams of principal congruence links.) The question asks whether this remains true for congruence arithmetic links.
(5) Much information is known about arithmetic links in the 3-sphere and in particular about congruence links. For example, any arithmetic link in the 3-sphere commensurable with $\mathrm{PSL}(2, O_d)$ must have
$$d \in \{1, 2, 3, 5, 6, 7, 11, 15, 19, 23, 31, 39, 47, 71\}.$$

Proposed for K3 by: A. Reid
Scribed by: M. Lackenby

PROBLEM 1.89.

(a) *Fix a long link L. What is the homotopy type of the embedding space of links isotopic to L?*
(b) *Fix a link L in a 3-manifold M. What is the homotopy type of the embedding space of links in M isotopic to L?*
(c) *Given a link L obtained by infecting a link L_1 by another link L_2, describe the homotopy type of component of L in the space of links in terms of the components of L_1 and L_2.*

REMARKS.

(1) A *long link* is an embedding of a disjoint union of lines that agree with a fixed collection of affine lines outside of a box. The answer to the first question, due to Hatcher and Budney, is known for long knots in the sense

that the knot spaces are shown to be homotopy equivalent to standard combinations of standard spaces; see Remark (3) below.

(2) Spaces of long links have applications in 4-manifold theory as well as high-dimensional manifold theory. For example, the recent work of Budney and Gabai [**BG19**] on the homotopy-type of $\mathrm{Diff}(S^1 \times D^{n-1})$ for $n \geqslant 4$ uses the low-dimensional homotopy groups of the space of long embeddings $\mathrm{Emb}(I, S^1 \times D^{n-1})$. This space can be thought of as the homotopy fiber of the forgetful map $\mathrm{Emb}(I \sqcup D^{n-2}, D^n) \to \mathrm{Emb}(D^{n-2}, D^n)$, i.e. it is the space of 2-component 'long links' where one component is trivial.

(3) A major step in the description of long knot spaces was Hatcher's proof of the Smale conjecture, which is equivalent [**Hat83**, Appendix] to the statement that $\mathrm{Diff}(D^3; \partial D^3)$ is contractible, which in turn implies that the component of the long unknot is contractible. Combined with work of Hatcher [**Hat76**] and Ivanov [**Iva76**] on diffeomorphisms of 3-manifolds, it also implies that each component in the space of long knots is a $K(\pi, 1)$ space [**Bud07**]. Hatcher [**Hat02**] determined the homotopy type of a long torus knot and, building on his work with McCullough [**HM97a**], of a long hyperbolic knot.

Budney [**Bud10**] determined the homotopy type of the component of any long knot K in $\mathrm{Emb}(\mathbb{R}, \mathbb{R}^3)$ in terms of the satellite decomposition of K, i.e. in terms of splicing operations, thus answering part (a) above when L has one component. The homotopy type is roughly a twisted product of factors of S^1, $S^1 \times S^1$, and configuration spaces of points in the plane, with one factor for each knot or link appearing in the satellite decomposition.

(4) Havens and Koytcheff [**HK21**] generalized Hatcher and Budney's results to the space of closed links in S^3 modulo rotations, whose components are $K(\pi, 1)$ spaces for irreducible links. They did not completely determine the homotopy types of spaces of split links because $\mathrm{Diff}(M)$ for a reducible 3-manifold M is not completely understood. However, Boyd and Bregman [**BB25**] obtained more detailed information on spaces of split links, especially their fundamental groups, using semi-simplicial spaces that model them. Kosanović [**Kos24**] related Vassiliev invariants to the Goodwillie–Weiss embedding calculus in the setting of a long knot K in any compact oriented 3-manifold M with boundary.

(5) In part (c), infection generalizes splicing from long knots to string (a.k.a. long) links. Budney [**Bud07**,**Bud12**] developed the operations of connect-sum and splicing at the space level, parameterizing them by operads. For both operations, he obtained space-level decomposition results, which for connect-sum he further developed at the homological level with F. Cohen [**BC09**].

Burke and Koytcheff [**BK15b**] generalized Budney's splicing operad to a (colored) operad for string link infection. They obtained a partial decomposition result for 2-component string links, building upon a result on isotopy classes in their joint work with Blair [**BBK15**]. This decomposition was completed for 2-component string links by Batelier and Ducoulombier [**BD23**]. It is essentially in terms of the stacking of string links rather than the more general infection operation. An answer to part (c) would be related to a decomposition of a 3-manifold analogous to the JSJ decomposition but with tori replaced by surfaces of higher genus.

Proposed for K3 and scribed by: R. Koytcheff

CHAPTER 2

Surfaces

Chapter 2 Editors: Carolyn Abbott, İnanç Baykur, Thomas Koberda, Priyam Patel, and Nick Salter

Dimension 2 occupies an unusual place in low-dimensional topology in that surfaces are completely classified (even in the exotic realm of infinite type lately so in vogue). But just as a course in linear algebra is not over once one learns that any two vector spaces of the same dimension over the same field are isomorphic, the story does not stop here—as in linear algebra, the heart of the subject is the study of *maps* between surfaces. A central character in 2-dimensional topology is, therefore, the *mapping class group* of a surface S, which is the group of isotopy classes of diffeomorphisms of S.

In the fifty-plus years that the mapping class group has been intensively studied following a prehistoric period spearheaded by Nielsen and Dehn in the earlier part of the twentieth century, enormous progress has been made in understanding its algebraic structure and its connections with other parts of the mathematical landscape. There are connections to geometry and topology in other dimensions, of course, but also algebraic geometry (by virtue of the deep connections with the theory of Riemann surfaces), complex analysis, symplectic geometry, geometric group theory, dynamics, Lie theory, and more. There are some long-standing structural conjectures left over from a previous generation of study, most notably the Ivanov conjecture (Problem 2.1) and the congruence subgroup problem (Problem 2.2). But the subject is continually refreshing itself by finding points of contact with other fields, and some of the deepest and most exciting unsolved problems concern how the mapping class group makes its presence felt in other areas.

For this reason, perhaps more so than in other chapters, *topology* has been interpreted quite liberally in the current chapter, to encompass many—although by no means all, or even a representative sampling of—points of contact with other fields. Even with this philosophy, there are too many sins of omission for us to enumerate here, but perhaps most notably, questions delving more deeply into Teichmüller theory or other quintessentially *geometric* topics have not been included in this volume. On the other hand, questions that were at least conceived to be in closer contact with other areas of low-dimensional topology were visibly favored in our list.

Below is a quick overview of how the subsections are organized:

Section 2.1 concerns the algebraic structure of the mapping class group. Many of the "greatest hits" (*Ivanov conjecture*, *congruence subgroup*, linearity, finite presentability of Torelli) can be found here.

Section 2.2 concerns the braid group, especially in its incarnation as the mapping class group of a punctured disk. The braid group reappears in Section 2.6, as well, in the context of geometric group theory.

Section 2.3 treats the subject of the topology of curves on surfaces. This includes topological questions about special configurations and behavior under lifting, as well as properties of Jordan curves (*peg problems*).

Section 2.4 deals with connections between the mapping class group and 3-manifolds. Given the close alliance between the study of 3-manifolds and hyperbolic geometry, we have also included some problems about hyperbolic surfaces, and the behavior of curves on such surfaces.

Section 2.5 is largely concerned with connections between the mapping class group and 4-manifolds, by way of the theory of Lefschetz fibrations and surface bundles. These topics have significant interaction with algebraic geometry, and we have also included here a well-known question originally posed by Kontsevich–Zorich concerning the topological structure of *strata of Abelian differentials*, which asks about a point of interaction between mapping class groups and algebraic geometry.

Section 2.6 moves away from more traditionally topological themes and into connections between braid and mapping class groups and other topics in geometric group theory. The literature here is vast and we have taken a somewhat narrow focus, mostly concerning the interaction of the mapping class group with various flavors of Artin groups.

Section 2.7 moves even further afield, into the realm of dynamics. Questions about homeomorphism and diffeomorphism groups of surfaces are posed here. Aspects of surface dynamics bleed over into logic (e.g., the first-order theory of groups), and a detour is taken to pose some questions of a logical flavor.

Section 2.8 concerns the mapping class groups of infinite-type surfaces. This includes many questions of a foundational nature in the sub-area.

Finally, Section 2.9 poses further questions originating in Floer homology and quantum topology about mapping class groups.

Basic notions: Throughout this chapter, all surfaces are oriented. The surface of genus g with b boundary components and n marked points or punctures will be denoted $S_{g,n}^b$. If b and/or n is zero, it will be omitted, so that S_g denotes the closed oriented surface of genus g. A surface S is of *finite type* if it is homeomorphic to some $S_{g,n}^b$ and is of *infinite type* otherwise. By default, a surface S will be assumed to be of finite type unless explicitly stated otherwise.

The *mapping class group* of S is the group of isotopy classes of orientation-preserving diffeomorphisms of S that preserve the boundary pointwise and preserve the set of marked points. It will be denoted $\mathrm{Mod}(S)$.

We will assume a familiarity with rudiments of the theory of mapping class groups, in particular the notion of a *Dehn twist* T_c about a simple closed curve $c \subset S$, and the *Nielsen–Thurston classification*. Let us briefly recall the latter. A mapping class $f \in \mathrm{Mod}(S)$ is *periodic* if f has finite order, and f is *reducible* if there is a finite set $\{c_1, \dots, c_k\}$ of essential simple closed curves on S admitting pairwise-disjoint representatives, such that f fixes $\{c_1, \dots, c_k\}$ as a set. If f is neither periodic nor reducible, it is *pseudo-Anosov*. The modern study of pseudo-Anosov mapping classes was initiated by Thurston; they are now known to have a remarkably rich and rigid structure.

As in any branch of low-dimensional topology, the study of embedded submanifolds of lower dimensions is a topic of central importance. Here, this amounts to the study of simple closed curves on surfaces. The set of isotopy classes of simple closed curves on S can be endowed with the structure of a simplicial complex, where pairwise non-isotopic curves $\{c_1, \ldots, c_k\}$ determine a $(k-1)$–simplex whenever they can be realized by pairwise-disjoint representatives. The resulting complex $\mathcal{C}(S)$ is called the *curve complex* of S.

We refer the reader to [**FM12**] for more background and a detailed treatment of these topics.

2.1. The algebraic structure of the mapping class group and its subgroups

In this section, we gather problems about the algebraic structure of the mapping class group, including long-standing open problems such as the Ivanov conjecture and the congruence subgroup problem.

PROBLEM 2.1 (Ivanov conjecture). *Let S be an orientable surface of finite type with genus at least three. If $G \leqslant \mathrm{Mod}(S)$ is a subgroup of finite index, does G have finite abelianization?*

REMARKS.

(1) The Ivanov conjecture is a weaker version of the question of whether or not the mapping class group has property (T). (The unpublished preprint [**And07**] announced a negative answer to the latter, which does not resolve the Ivanov conjecture in any case). It was recently shown that automorphism groups of most free groups do have property (T) and hence every finite index subgroup has finite abelianization: this was resolved for rank 5 by Kaluba–Nowak–Ozawa [**KNO19**] and for rank $\geqslant 6$ by [**KKN21a**]. A resolution for rank 4 was announced by Nitsche [**Nit22**].

(2) The Ivanov conjecture has an equivalent reformulation in terms of mapping class group actions on the homology of finite covers of surfaces [**PW13**]. The corresponding statement about these actions has become known as the *Putman–Wieland conjecture*. Some progress on the Putman–Wieland conjecture has been made [**LL23b, LL23a, MT24**].

(3) For braid groups and mapping class groups of surfaces of genus two or lower, virtual surjections to the integers exist by virtue of the existence of such maps for braid groups.

Scribed by: T. Koberda

PROBLEM 2.2 (Congruence subgroup problem). *Does every finite-index subgroup of the mapping class group of S contain a congruence subgroup?*

REMARKS.

(1) Let S be a finite-type surface, and let $K \leqslant \pi_1(S)$ be a finite-index characteristic (i.e., $\mathrm{Aut}(\pi_1(S))$-invariant) subgroup. A *congruence subgroup* of the mapping class group of S is the kernel of the natural map from the mapping class group of S to $\mathrm{Out}(\pi_1(S)/K)$. This definition is analogous to congruence subgroups of integral special linear groups.

(2) The congruence subgroup problem is known to have a positive answer in certain low-genus cases. For punctured spheres, this was established by Diaz–Donagi–Harbater [**DDH89**], with subsequent alternative proofs given by Asada [**Asa01**], Thurston, and McReynolds [**McR12**]. The case of genus one was originally addressed by Asada [**Asa01**], with further work done by Kent [**Ken16**] and Bux–Ershov–Rapinchuk [**BER11**].

(3) A positive resolution to the congruence subgroup problem has applications in anabelian geometry [**Mar19**].

Scribed by: T. Koberda

PROBLEM 2.3. *Is the mapping class group of a surface of finite type linear?*

REMARKS.

(1) Recall that a group is *linear* if it can be embedded as a subgroup of $\mathrm{GL}_n(\mathbb{C})$ for some n.

(2) Following the linearity of the braid group (see Remark 4 below), there are some low-complexity cases for which linearity is known to hold. This is true, in particular, for mapping class groups of punctured spheres and hyperelliptic mapping class groups, including the full mapping class group in genus 2 [**Kor00, BB01**]. In addition, if one can prove that a non-closed mapping class group is nonlinear (say of genus g with one puncture or boundary component) then all mapping class groups of genus $h > g$ have nonlinear mapping class groups. Unsurprisingly, there are examples of infinite-type surfaces with nonlinear mapping class groups [**APV21**].

(3) The analogous question of linearity for $\mathrm{Aut}(F_n)$, the automorphism group of a free group F_n of rank n, has been resolved in the negative by Formanek–Procesi [**FP92**]. They construct certain "poison subgroups" of $\mathrm{Aut}(F_n)$ that are known to be nonlinear. Subsequently, such poison subgroups were shown *not* to exist in mapping class groups by Brendle–Hamidi-Tehrani [**BHT01**].

(4) On the other hand, braid groups are known to be linear, due to the work of Bigelow and Krammer [**Big01, Kra02**]. The *Lawrence–Krammer–Bigelow* representations used to certify this are constructed as the action on the homology of a certain 4-manifold given as an infinite abelian cover of the configuration space of two points in the n-punctured disk. It is tempting to speculate about carrying out an analogous construction in higher genus, but various complications arise; ultimately, the natural homology

representations of braid groups are simply symmetric groups, and the Lawrence–Krammer–Bigelow representation arises as a deformation of a symmetric group representation [**Jon87**]. The corresponding homology actions of mapping class groups of surfaces in positive genus have infinite image.

(5) One possible approach to establishing nonlinearity would be to make effective the residual finiteness of mapping class groups. In particular, superpolynomial residual finiteness growth would imply nonlinearity by work of Bou-Rabee–McReynolds [**BRM15**].
(6) There is a categorified faithful, finite-dimensional, linear representation with mapping classes acting by functors; see [**LOT13**].
(7) Certain constraints on the properties of a faithful representation have been established. In [**Kor23**], Korkmaz shows that for $g \geqslant 3$, any faithful representation must have dimension $n > 3g - 3$. For any finite-dimensional linear representation of a mapping class group in genus $g \geqslant 3$, Bridson [**Bri10**] proved that Dehn twists are necessarily sent to quasi-unipotent matrices (i.e., ones whose eigenvalues are roots of unity). From this one can easily show that mapping class groups have no faithful linear representations in positive characteristic. For more recent further developments in this direction, see [**AS16, But19, KLS19**].
(8) For a much more extensive discussion and bibliography, see the article of Margalit [**Mar19**].

Scribed by: T. Koberda

PROBLEM 2.4. *Let S_1 and S_2 be orientable surfaces of finite type. Under what conditions do injective maps from (finite-index subgroups of) the mapping class group of S_1 to the mapping class group of S_2 necessarily arise from "manipulations" (e.g., inclusions, (branched) coverings, etc.) taking S_1 to S_2?*

REMARKS.

(1) It is known that all such injections arise from inclusions of surfaces provided that the genus of S_2 is not much larger than the genus of S_1 [**AS12**]. For braid groups, a detailed study of injective maps in a certain range was announced by Chen–Kordek–Margalit [**CKM19**]; see also Problem 2.7. Both of these results pertain only to injections at the level of the full mapping class group; the case of arbitrary finite-index subgroups seems much harder.
(2) There are many natural variations on this problem, replacing mapping class groups by curve graphs and related graphs associated to surfaces. The literature on this subject is vast, with two central results being [**Iva97**] and [**BM19**]. It has been announced that a (model-theoretic) interpretation between curve graphs of surfaces induces a virtual injection between mapping class group [**DKdlNG20**].

Scribed by: T. Koberda

PROBLEM 2.5. *For $g \geqslant 3$, determine a finite presentation for the Torelli group $\mathcal{I}_g$, or show that no finite presentation exists.*

REMARKS.

(1) The *Torelli group* $\mathcal{I}_g \leqslant \mathrm{Mod}(S_g)$ is the subgroup of $\mathrm{Mod}(S_g)$ consisting of mapping classes acting trivially on $H_1(S_g;\mathbb{Z})$. Work of Johnson [**Joh83**] shows that $\mathcal{I}_g$ is finitely generated for $g \geqslant 3$.

(2) This is [**Kir78**, Problem 2.9]. The question was raised by Magnus, and described by Birman in [**Bir71**].

(3) For $g = 1$, the Torelli group is trivial, and for $g = 2$, Mess showed it is a free group of infinite rank [**Mes92**].

(4) For any finitely presented group G, the second homology $H_2(G;\mathbb{Q})$ is finite dimensional, though the converse does not hold. Recent work of Minahan–Putman [**MP25**] announced that $H_2(\mathcal{I}_g;\mathbb{Q})$ is finite dimensional for $g \geqslant 6$. It is still open whether or not $H_2(\mathcal{I}_g;\mathbb{Z})$ is finitely generated for large g.

Scribed by: N. Salter

PROBLEM 2.6. *Give a classification or enumeration of the finite-index subgroups of* $\mathrm{Mod}(S_g)$ *that are generated by Dehn twists, Dehn multitwists, or powers thereof. Which of these subgroups are "geometrically meaningful", in the sense that they are given as the stabilizers of some kind of geometric or topological structure on the surface? Conversely, when is a geometrically meaningful subgroup (not necessarily of finite index) generated by powers of Dehn multitwists?*

REMARKS.

(1) The classical Dehn–Lickorish theorem asserts that the mapping class group of a closed orientable surface is generated by a finite collection of Dehn twists. Any finite-index subgroup contains *some power* of every Dehn twist, but a general finite-index subgroup need not contain any single twist T_c at all, nor is a general finite-index subgroup generated by the powers of Dehn twists that it does contain. Indeed, work of Funar [**Fun99**] and Masbaum [**Mas99**] shows that for general values of r, the subgroup of $\mathrm{Mod}(\Sigma_g)$ generated by r^{th} powers of Dehn twists is of infinite index.

(2) Salter and Calderon–Salter [**Sal19, CS21**] show that one such example is the class of *r-spin mapping class groups*, i.e., the stabilizers of a chosen r-spin structure on S_g, for $g \geqslant 5$. In [**CS23**], they proved that the *framed mapping class group* for a framing (i.e., the stabilizer of the isotopy class of that framing) is likewise generated by Dehn twists; these subgroups are of infinite index. Salter–Sane [**SS23**] announced a generating set for the stabilizer of a mod-2 homology class $[x]$ consisting of Dehn twists together with a single square-twist T_c^2.

(3) The terms "geometrically meaningful" and "geometric or topological structure" in the problem statement are intentionally vague, but the examples discussed in the remarks above are meant to be representative. To disambiguate, we do *not* (necessarily) mean "geometric structure" in the Thurstonian sense of a (G, X)-structure.

(4) One motivation for this question comes from the theory of (achiral) Lefschetz pencils, which are encoded in the mapping class group as sequences of Dehn twists whose product is the boundary multi-twist. When these Dehn twists lie in some "geometrically meaningful" subgroup, one can ask if this endows the ambient 4–manifold with an additional structure. One instance of this is the case of $r = 2$ (i.e., classical) spin mapping class groups, where the question of whether a symplectic 4-manifold is also spin is related to whether it admits a symplectic Lefschetz pencil whose twists lie in a spin mapping class group [**Sti01**, **BHM23**, **BH24c**, **AB23**].

Proposed for K3 and scribed by: N. Salter

2.2. Braid groups

The *braid group on n strands*, written B_n, can be defined in several equivalent ways (e.g., as the fundamental group of a configuration space), but most fittingly here, as the mapping class group of a disk with n punctures:

$$B_n := \operatorname{Mod}(\Sigma^1_{0,n}).$$

PROBLEM 2.7. *Classify the homomorphisms from the braid group B_n on n strands to the braid group B_m on m strands, where $n, m \in \mathbb{N}$ are arbitrary.*

REMARKS.

(1) There are several works in this direction including those of Artin [**Art47**], Lin [**Lin04**], Dyer–Grossman [**DG81**], Bell–Margalit [**BM06**], Castel [**Cas16**] and, finally, Chen–Kordek–Margalit [**CKM19**], where the most general result known to date was announced (the complete classification for $m \leqslant 2n$ and $n \geqslant 5$).

(2) There is a more precise formulation of this question. We view the group B_m as the mapping class group $\operatorname{Mod}(S^1_{0,m})$ of the m-punctured disk and denote by $\mathcal{C}(S^1_{0,m})$ the curve complex of $S^1_{0,m}$. We say that a homomorphism $\varphi\colon B_n \to B_m = \operatorname{Mod}(S^1_{0,m})$ is *reducible* if there exists a nonempty simplex $\mathcal{A}$ of $\mathcal{C}(S^1_{0,m})$ such that $\varphi(\beta)(\mathcal{A}) = \mathcal{A}$ for all $\beta \in B_n$. That is, every element of $\varphi(B_n)$ is reducible, and moreover preserves the same set of curves. Then we have the following conjecture, which can be compared with the Nielsen–Thurston classification.

CONJECTURE. *Let $n, m \in \mathbb{N}$ with $5 \leqslant n \leqslant m$, and let $\varphi\colon B_n \to B_m$ be a homomorphism. Then we have one of the following three possibilities.*
(a) *The image of φ is a cyclic group.*
(b) *$n = m$, and φ is an injective homomorphism as described in* [**BM06**].
(c) *φ is reducible.*

(3) In light of the identification of B_m with the mapping class group $\mathrm{Mod}(S^1_{0,n})$, this question admits a natural generalization to the problem of classifying all homomorphisms between mapping class groups of surfaces; see Problem 2.4. For further discussion, see the survey article [**AS16**].

Scribed by: T. Koberda, N. Salter

PROBLEM 2.8. *Fix distinct trivial tangles τ_1, τ_2 for which $\tau_1 \cup \overline{\tau_2}$ is the unknot. Describe the intersection of the associated wicket groups $W_n(\tau_1) \cap W_n(\tau_2) \leqslant B_{2n}$. Is this group finitely generated? Finitely presented?*

REMARKS.

(1) A *tangle* on n strands is a proper embedding of n disjoint arcs into the upper half-space in $\mathbb{R}^3$, considered up to isotopy rel boundary. Tangles τ_1, τ_2 based at the same set of $2n$ points in $\mathbb{R}^2 \subset \mathbb{R}^3$ can be combined into the link $\tau_1 \cup \overline{\tau_2}$, where $\overline{\tau}_2$ denotes the reflection of τ_2 through the xy-plane. A *trivial tangle* is any tangle that can be isotoped so that each strand has a single local maximum with respect to the standard height function on $\mathbb{R}^3$. There is a natural transitive action of the braid group B_{2n} on the set of trivial tangles based at a fixed set of $2n$ points, and the stabilizer of a fixed trivial tangle τ is known as a *wicket group*, written W_n or $W_n(\tau)$ when we wish to emphasize the particular trivial tangle.

(2) The questions in the problem originate from the works of S. Hirose, D. Iguchi, E. Kin, and Y. Koda that are discussed below.

(3) Note that $W_n(\tau_1)$ and $W_n(\tau_2)$ are conjugate in B_{2n}. Brendle–Hatcher [**BH13**] determined a finite presentation for W_n. This problem has a natural generalization in which one allows $\tau_1 \cup \overline{\tau}_2$ to be an unlink of c components for $c \geqslant 1$ [**HIKK22**, Question 2.10]. The group $W_n(\tau_1) \cap W_n(\tau_2)$ could be called the *Goeritz group* of the bridge splitting $\tau_1 \cup \overline{\tau}_2$ of the unlink. Hirose–Iguchi–Kin–Koda defined and studied this group as the *hyperelliptic Goeritz group*, proving that this group is finitely presented when the Heegaard splitting is a 3-bridge splitting of a 2-bridge link [**HIKK22**].

If the distance of an n-bridge decomposition of a link in S^3 with $n \geqslant 3$ is at least 5, then its Goeritz group is a finite group [**IK20**, Theorem 0.1].

(4) There is a connection with the study of knotted surfaces through the theory of bridge trisections [**MZ17a**]: Any smooth surface-link in $\mathbb{R}^4$ can be represented by a tri-plane diagram in which the first two tangle diagrams are arranged to have no crossings. Given this, the Goeritz group of the bridge splitting of the unlink given by the first two tangles is precisely the set of braids that act on the tri-plane diagram without changing the first two tangles. Thus, an understanding of this group could be applied to the problem of finding equivalences between bridge trisected surface-links.

Proposed for K3 and scribed by: J. Meier, N. Salter

PROBLEM 2.9. *Is there a nice presentation of the n-stranded braid group whose generating set is the set of all positive elementary braid half-twists?*

REMARKS.

(1) The adjective "nice" here should at least mean that all relations consist of equating two words in the positive generators of the same length, and perhaps also that there be a finite number of local pictures describing the required relations.
(2) The positive monoid in the standard Artin presentation consists of all positive braids. In the Birman–Ko–Lee presentation [**BKL98**], the positive monoid is the strongly quasipositive braids. The positive monoid for the desired presentation would correspond to the quasipositive braids and would hopefully provide interesting algorithmic tools similar to Garside [**Gar69**] for solving questions regarding quasipositivity. Similar infinite presentations hold for the mapping class groups of surfaces (Gervais [**Ger96**]) as well as the Torelli group (Putman [**Put07**]).

Proposed for K3 by: D. Calegari
Scribed by: J. Van Horn-Morris

2.3. Curves on surfaces

This section consists of problems regarding curves on surfaces, including topological questions about special configurations and behavior under lifting.

PROBLEM 2.10.

(a) *Is there an efficient algorithm to compute distances in the curve complex of a surface? The input to the algorithm should be the surface and the curves. One may ask similar questions for related complexes, such as the arc complex of a surface with boundary.*
(b) *Is there an algorithm (efficient or otherwise) to compute the distance between quasi-convex subsets of the curve complex? This question is particularly interesting in the case where the quasi-convex subsets are the disk sets of handlebodies.*

REMARKS.

(1) The curve complex has proved to be particularly important in Kleinian group theory and in the study of the mapping class groups of surfaces [**Min06**]. One notable feature of $\mathcal{C}(S)$, proved by Masur and Minsky [**MM99**], is that it is Gromov hyperbolic.
(2) The complex is not locally finite, and so there is no naive way of computing distances. However, distances in $\mathcal{C}(S)$ are known to be computable, by work of Leasure [**Lea02**], Shackleton [**Sha12**], Webb [**Web15**], Birman, Margalit, and Menasco [**BMM16**] and Watanabe [**Wat16**]. Bell and Webb [**BW16a**] announced an algorithm with running time that is bounded above by a polynomial function of the logarithm of the *weight* of

the two curves, given as normal curves in some triangulation of S. However, the running time of their algorithm grows very rapidly as the surface S is allowed to vary. The aim of Problem (a) is to find such an algorithm that runs in polynomial time even as the surface varies.

Baroni [**Bar25b**] has announced an efficient algorithm to compute distances up to a bounded multiplicative error, the bound depending only on S. He used this to find an efficient algorithm to determine whether a given surface automorphism is pseudo-Anosov, reducible, or periodic.

(3) A related problem is as follows. *Given a reducible element of the mapping class group, can an invariant multi-curve be found efficiently (in polynomial time as function of the 'size' of the mapping class and also in the Euler characteristic of S)?*

(4) When S is the boundary of a handlebody H, the associated *disk set* consists of the subcomplex of $\mathcal{C}(S)$ induced by the curves bounding disks in H. This was proved to be quasi-convex in $\mathcal{C}(S)$ by Masur and Minsky [**MM04**]. When S is a Heegaard surface for a closed 3-manifold, it bounds a handlebody on each side and so there are two associated disk sets. The distance between them in $\mathcal{C}(S)$ is the *Hempel distance* of the splitting [**Hem01**]. This is not known to be computable. However, it was proved by Masur and Schleimer [**MS13**] that it is computable up to a bounded additive error. The Hempel distance of a Heegaard splitting encodes important information. For example, the Hempel distance is zero if and only if the splitting is reducible, and the Hempel distance is one if and only if the splitting is weakly reducible but irreducible.

Proposed for K3 and scribed by: M. Lackenby

PROBLEM 2.11. *Find precise estimates for both the* extremal *and* average *behavior of the simple lifting degree of curves, in terms of combinatorial (e.g., intersection number) and/or geometric (e.g., geodesic length, word length) data.*

REMARKS.

(1) A result of Peter Scott [**Sco78**] tells us that for any closed geodesic γ on a hyperbolic surface S, there exists a finite degree cover $S' \to S$ such that γ lifts to S' (i.e., choosing basepoints, $\gamma \in \pi_1(S') \leqslant \pi_1(S)$) and is a simple closed curve in S'. The *simple lifting degree* of γ, which we will write as $\deg(\gamma)$, is defined to be the minimal degree of such a covering S' to which γ lifts simply.

(2) To precisely formulate the relationship between simple lifting degree and combinatorial/geometric complexity, define the functions

$$f_S(n) = \max\{\deg(\gamma) : i(\gamma, \gamma) \leqslant n\}$$

and, for a choice of hyperbolic metric ρ on S,

$$f_\rho(L) = \max\{\deg(\gamma) : \ell_\rho(\gamma) \leqslant L\}.$$

The extremal behavior (i.e., upper bounds) of these functions has been studied in work of Patel [**Pat14**], Gaster [**Gas16**], and Aougab–Gaster–Patel–Sapir [**AGPS17**]. A lower bound for average behavior of $\deg(\gamma)$ was obtained by Aougab–Gaster [**AG22**], but work of Sisto–Taylor [**ST19**] implies that some mechanism other than what Gaster finds in the extremal case is responsible for the average behavior. What is this mechanism?

Progress in the study of $f_S(n)$ has been obtained by Arenas and Neumann-Coto [**ANC20**]. They establish the inequality $f_S(n) \leqslant 5(n+1)$; note that this is independent of the surface S. They conjecture that the tighter bound $f_S(n) \leqslant 2(n+1)$ should hold. At the time of this writing, no examples exist with $f_S(n) - n$ more than 2.

(3) Variations of these questions can be posed by, e.g., restricting to the class of regular covers or to families of curves intersecting on a surface S. These have been largely unexplored.

Proposed for K3 by: T. Aougab, P. Patel
Scribed by: P. Patel, N. Salter

PROBLEM 2.12.

(a) *What is the maximum number of systoles on a closed, hyperbolic surface of genus g?*
(b) *What is the maximum cardinality of a set of pairwise non-isotopic, essential, simple closed curves on a closed surface of genus g with the property that no two curves in the set intersect in more than one point?*

REMARKS.

(1) A *systole* on a closed hyperbolic surface is a closed geodesic of shortest length. A systole is necessarily an essential, simple closed curve. On a closed surface, any two systoles intersect in at most one point (in the presence of punctures, systoles may intersect twice; see [**FP15**]). Hence the first problem arises as a special case of the second. This quantity is sometimes called the "kissing number"; kissing numbers of lattices have been studied in a variety of contexts.
(2) The first problem was first seriously studied by Schmutz Schaller [**Sch96**]. The answer is known to lie between $C_1(\epsilon) \cdot g^{4/3-\epsilon}$ and $C_2 \cdot g^2/\log(g)$. Here $C_1(\epsilon)$ is a function of an arbitrary value $\epsilon > 0$ and C_2 is an absolute constant. The lower bound is a construction of Schmutz Schaller using arithmetic surfaces [**SS97**]. The upper bound is due to Parlier using combinatorial and hyperbolic techniques [**Par13b**]. Schmutz Schaller conjectured that the lower bound is optimal, in that the maximum cardinality cannot exceed $C_3 \cdot g^{4/3}$ for some absolute constant C_3 [**SS97**, Conjecture].
(3) The second problem was raised independently by Juvan, Malnic, and Mohar [**JMM96**] and by Farb and Leininger, as recorded in [**MRT14**, Question 1]. The answer is known to lie between $C_1 \cdot g^2$ and $C_2 \cdot g^2 \log g$. The lower bound is a construction due to Malestein, Rivin, and Theran [**MRT14**]. The upper bound is due to Greene, using probabilistic combinatorics and geometric techniques [**Gre19**]. The lower bound is conjectured to be optimal, i.e., there exists an upper bound of the form $C_3 \cdot g^2$ to match it.

Przytycki solved the corresponding problem for arcs: if S is a connected, orientable, punctured surface of finite type, then the maximum number of pairwise non-isotopic, essential, simple arcs that limit to the punctures is $(|\chi(S)|+1)(|\chi(S)|+2)$ [**Prz15**]. His methods involve hyperbolic geometry.

Replacing "one" by k in the problem, the answer is known to lie between $C_1(k) \cdot g^{k+1}$ and $C_2(k) \cdot g^{k+1} \log g$, for some functions $C_1(k)$ and $C_2(k)$ independent of g. Once more, the lower bound is conjectured to be optimal, i.e., there exists an upper bound of the form $C_3(k) \cdot g^{k+1}$ to match it [**Gre19**, Conjecture 1].

Proposed for K3 and scribed by: J. Greene

PROBLEM 2.13. *Let S be a surface, and let Γ_1, Γ_2 be isotopy classes of embedded graphs in S. Determine when Γ_1 and Γ_2 are related by a sequence of surgeries.*

REMARKS.

(1) A *surgery* of an embedded graph $\Gamma \subset S$ is defined as follows. Suppose some component of $S \backslash \Gamma$ is a disk D. Let x and y be oriented edges in Γ bounding D. There is a unique isotopy class $\lambda_{x,y}$ of arc connecting x to y inside D. A regular neighborhood $N_{x,y}$ of $\lambda_{x,y}$ is a rectangle with two opposite sides comprising subarcs of x and y, and remaining sides running parallel to $\lambda_{x,y}$. The *surgery* of Γ along $\lambda_{x,y}$ is the graph Γ' obtained from Γ by deleting the edges of $N_{x,y}$ lying on x, y and replacing them with the two edges parallel to $\lambda_{x,y}$. Note that surgery can change the topology of the complement; there is a simple combinatorial criterion to determine when this does or does not happen [**San21**].

(2) The number of vertices and edges are manifestly preserved under surgery, but the isomorphism type of the graph can change. Sane has investigated the "surgery graph" that tracks when two embedded graphs are related by a surgery and has studied connectivity properties of this graph in a number of cases, discussed below.

When every vertex of Γ has even valence, Γ determines a class in $H_1(S; \mathbb{Z}/2\mathbb{Z})$. This homology class is readily seen to be invariant under surgery. The case where every vertex has valence 4 and the complement is a single disk was studied by Salter–Sane [**SS23**], who announced that in this case, graphs Γ and Γ' are related via a sequence of surgeries if and only if their mod-2 homology classes are equal. In the case where every vertex has valence 3 and the complement is a single disk, Sane announced in [**San20**] that any two such graphs on the same surface can be connected via a sequence of surgeries.

(3) This problem is more topological (and less purely combinatorial) than it might appear to be at first glance. It is closely connected to the action of the mapping class group on embedded graphs, and to the general problem of understanding subgroups of the mapping class group. A surgery equivalence class determines a "stabilizer subgroup" of the mapping class group of S, consisting of all mapping classes f for which $f(\Gamma)$ can be taken to Γ via surgery. The result of Salter–Sane mentioned above can

be rephrased as the assertion that the stabilizer subgroup for the class of graphs they consider is the stabilizer of a mod-2 homology class. Other classes of graphs appear to have smaller stabilizer subgroups that are not so easily characterized. Given the enduring mystery of classifying subgroups of the mapping class group (especially those of finite index), a primary motivation for understanding graph surgery is that it may lead to novel examples and phenomena in this arena.

(4) The literature mentioned above restricts attention to the *unicellular* case, when $S\backslash\Gamma$ is a single disk (note that these authors further restrict to surgeries that preserve this property). One reason to be particularly interested in this setting is that the enumeration of such "unicellular graphs" is a key ingredient in Harer and Zagier's computation of the Euler characteristic of the moduli space of Riemann surfaces [**HZ86**].

Proposed for K3 and scribed by: N. Salter

PROBLEM 2.14.

(a) *Does every Jordan curve in the Euclidean plane contain the vertices of a square?*

(b) *Does every Jordan curve in the Euclidean plane contain the vertices of the affine image of a regular hexagon?*

(c) *Let $\gamma \subset \mathbb{C}$ be a smooth Jordan curve, and let $S \subset \mathbb{C}$ consist of $2n \geqslant 4$ concyclic points. Does there exist a non-constant polynomial $p(z) \in \mathbb{C}[z]$ of degree $< n$ such that $p(S) \subset \gamma$?*

(d) *Suppose that h is a continuous, real-valued function defined on the unit sphere $S^2 \subset \mathbb{R}^3$ and that points $x_1, \ldots, x_4 \in S^2$ form the vertices of a square. Is there an isometry $\rho \in SO(3)$ such that h is constant on $\rho(x_1), \ldots, \rho(x_4)$?*

REMARKS.

(1) The first problem is known as the Square Peg Problem. It was posed by Otto Toeplitz in 1911 [**Toe11**]. It is solved affirmatively for several classes of curves, including C^1 curves (see [**Mat14**]).

More generally, a curve *inscribes* an n-gon (and the n-gon *inscribes in* the curve) if there exists an orientation-preserving similar copy of the n-gon whose vertices lie on the curve. Every Jordan curve inscribes every triangle (a noncolinear 3-gon) [**Nie92**], but no two dissimilar ellipses inscribe the same pentagon. Vaughn proved that every Jordan curve inscribes a rectangle (of some uncontrolled aspect ratio) [**Mey81**]. Greene and Lobb proved the optimal result for smooth curves: every smooth Jordan curve inscribes every cyclic quadrilateral (i.e., a quadrilateral whose vertices lie on a circle) [**GL23**]. By contrast, Pak observed that the quadrilaterals that inscribe in all triangles are the isosceles trapezoids [**Pak08**]. It is possible that the widest class of quadrilaterals that inscribe in all Jordan curves are the isosceles trapezoids.

(2) The second problem was raised as a conjecture by Grünbaum [**Gru72**, Conjecture 4.3]. It is known to hold if the curve is convex [**Gru72**, p.84]. It is known that every C^1 Jordan curve either contains the vertices of

the affine image of a regular hexagon or six colinear points that are the limit of the vertices of a sequence of affine images of regular hexagons [**Vv11**, Theorem 7]. However, the problem is open in general, even for smooth curves.

(3) The third problem was raised as a conjecture by Greene and Lobb [**GL24**, Conjecture 1.1]. The case $n = 2$ is true and is equivalent to the result about cyclic quadrilaterals noted above in the discussion of the first problem. The case $n = 3$, and some special instances when $n \geqslant 4$, was announced in [**GL24**]. These results all involve symplectic geometry.

(4) The fourth problem is known as the Table Problem, and it is related to the Knaster problem (which has been solved in the negative). It is posed as a conjecture in [**Mat14**, Conjecture 13]. The case in which $x_1, \ldots, x_4$ are equally spaced around a great circle was solved affirmatively by Dyson [**Dys51**]. The case in which h is an even function was announced by Naseri Sadr, who also announced a proof of a variation for Riemannian surfaces [**NS24**]. Call a set of four points $x_1, \ldots, x_4 \in S^2$ *balancing* if for every h there exists a ρ as in the problem statement. The Table Problem therefore asserts that the vertices of a square are a balancing set. The case of the standard height function shows that a balancing set of points must be concyclic. On the other hand, there exist concyclic $x_1, \ldots, x_4$ that are not a balancing set [**Kar13**]. Livesay proved that the vertices of a rectangle contained in a great circle form a balancing set [**Liv54**]. It is possible that the vertices of any rectangle on S^2 form a balancing set. A proof was claimed in [**Gri91**], but that paper was later invalidated [**Mat14**].

Remarks on these and related problems appear in Klee and Wagon's book [**KW91**] and Matschke's survey article [**Mat14**].

Proposed for K3 and scribed by: J. Greene

2.4. Connections with 3-manifolds and hyperbolic geometry

Surface topology interacts with hyperbolic geometry in both dimensions 2 and 3. In dimension 2, one equips the surface S itself with a complete hyperbolic metric. Such a metric is not unique; the *Teichmüller space* Teich(S) of S parametrizes complete hyperbolic metrics on S.

The connection between surface topology and hyperbolic 3-manifolds arises via Thurston's hyperbolization theorem (see [**Ota01**]). Given a diffeomorphism f of S, define the *mapping torus* M_f as

$$M_f := S \times [0,1]/(f(x),0) \sim (x,1).$$

Note that M_f is *fibered*: it has the structure of a fiber bundle with base S^1 and fiber S. Conversely, any fibered 3-manifold arises via the mapping torus construction (though the mapping class f may not be unique—see [**Thu86**]). The *hyperbolization theorem* asserts that the interior of M_f admits a complete hyperbolic metric if and only if f is pseudo-Anosov.

PROBLEM 2.15. *Is the genus g Goeritz group $\mathcal{G}_g$ finitely generated when $g \geqslant 4$? If so, find a set of generators.*

REMARKS.

(1) The *Goeritz group* $\mathcal{G}_g$ is the group of isotopy classes of orientation-preserving homeomorphisms of S^3 that preserve an unknotted genus g handlebody, that is, one side of a genus g Heegaard splitting. It is a rough analogue for S^3 of the braid group for S^2. The notion can be extended to Heegaard splittings of other 3-manifolds, as well [**JM13**].

(2) Goeritz [**Goe33**] exhibited a complete set of three generators for $\mathcal{G}_2$ and Akbas [**Akb08**] found associated relators, providing a complete presentation of $\mathcal{G}_2$. Powell [**Pow80**] believed he had found a complete set of 5 generators for $\mathcal{G}_g$, for any g, but there was a serious gap in the proof (see footnote of [**Sch04**]). This is now known as the *Powell Conjecture.* In [**Sch20**] it is shown that one of Powell's proposed five generators is redundant.

Freedman and Scharlemann [**FS18**] announced a proof of the Powell Conjecture for $g = 3$. Their methods break down for higher genus, and it is suspected that for $g \geqslant 4$ the group $\mathcal{G}_g$ is not even finitely generated. Freedman [**Fre22**] suggests a method by which this might be proven.

(3) In [**Sch24**], Scharlemann generalizes Powell's proposed generators and shows that this generalized set suffices. The set itself is infinite, but Scharlemann has shown that it can be used to demonstrate that the Powell Conjecture is stably true [**Sch25**]. That is, any element of $\mathcal{G}_g$, when viewed as acting on the genus $g + 1$ unknotted handlebody obtained by adding a single genus 1 summand to the original genus g handlebody, is a consequence of the 5 Powell elements of $\mathcal{G}_{g+1}$.

(4) In another approach, Zupan [**Zup20**] identifies a certain graph $\mathcal{R}_g$ in the curve complex of a genus g Heegaard surface and shows that, in general, the genus k Powell conjecture is true for all $k \leqslant g$ if and only if $\mathcal{R}_k$ is connected for all $k \leqslant g$.

Proposed for K3 and scribed by: M. Scharlemann

PROBLEM 2.16. *If two hyperbolic surfaces have the same unmarked simple length spectra (i.e., the same multiset of lengths that correspond to simple closed curves), are they isometric?*

REMARKS.

(1) Let M be a closed Riemannian manifold with negative sectional curvature, and let $\mathcal{C}$ denote the set of free homotopy classes of closed curves in M. In each homotopy class, there is a unique closed geodesic. This defines a *marked length spectrum* function $\ell\colon \mathcal{C} \to \mathbb{R}_{>0}$ that assigns to each class g the length $\ell(g)$ of this closed geodesic. Burns and Katok asked whether the function ℓ determines M up to isometry [**BK85**]. While this question is open in general, the answer is yes in the case of surfaces; this is called *marked length spectrum rigidity.* Fricke and Klein showed this

holds for closed hyperbolic surfaces, in which case there are $6g-5$ geodesic curves whose lengths determine the surface us to isometry [**FK65**]. Otal and, independently, Croke showed that surfaces with variable negative curvature are also marked length spectrum rigid [**Ota90, Cro90**]; see [**CFF92, HP97**] for other classes of metrics with respect to which surfaces are marked length spectrum rigid.

In this problem, the length spectrum is unmarked, in the sense that we no longer consider the function ℓ but only the image of ℓ as a multiset in $\mathbb{R}_{>0}$; that is, we simply have a list of the lengths of closed geodesics on the surface, counted with multiplicity. Moreover, we consider only lengths corresponding to simple closed curves.

If we drop the assumption that curves we consider are simple, then the answer is no; that is, hyperbolic surfaces do not enjoy unmarked length spectrum rigidity. Vignéras constructed the first examples of isospectral but non-isometric closed surfaces [**Vig80**], and Sunada gave a general construction of such surfaces [**Sun85**]. However, the length spectrum does determine the isometry class of the surface in some low-complexity cases, such as a one-holed torus with one boundary component [**Haa85, BS88b**].

On the other hand, the question of simple length spectrum rigidity is still open. It is known that the examples of isospectral but non-isometric surfaces constructed by Sunada can be distinguished by their simple length spectra [**Mau13**].

(2) One can ask a family of analogous questions for each natural number: suppose two hyperbolic surfaces have the same unmarked "n-spectra", which is the super-set of lengths corresponding to all curves with at most n self-intersections. Are they isometric? Above, "superset of lengths" means the sets of lengths counting multiplicity.
(3) Baik–Choi–Kim [**BCK21**] recently announced that there is an open dense set in Teichmüller space where the unmarked simple length spectrum is indeed rigid. The introduction to this paper contains a thorough discussion of the history and context of this problem.
(4) See also Problem 3.17 for a discussion of length spectrum regidity for higher-dimensional hyperbolic manifolds.

Proposed for K3 by: T. Aougab
Scribed by: C. Abbott, P. Patel

PROBLEM 2.17. *Suppose that c is a geodesic current on a hyperbolic surface, and suppose that, on the space of hyperbolic metrics on the surface, c has the same length function as a closed geodesic. Must c be a convex combination of closed curves?*

REMARKS.

(1) For an introduction to the theory of geodesic currents in general, see the foundational work of Bonahon [**Bon86, Bon88**], as well as the contemporary book of Erlandsson–Souto [**ES22**].
(2) As an example, suppose that c_1 and c_2 are so-called "length-twins," i.e., non-homotopic closed curves that have the same length in any hyperbolic

metric. Then the geodesic current $\frac{1}{2}c_1 + \frac{1}{2}c_2$ is not a closed curve, but it will also have the same length as c_1 and c_2. It *is* a convex combination of c_1 and c_2; thus, the question asks whether there exists some more exotic type of current (for example one whose support in $\mathbb{H}^2$ is not a discrete set of geodesics) that can also have the same length as either c_1 or c_2.

Proposed for K3 and scribed by: T. Aougab, P. Patel

PROBLEM 2.18. *What is the best lower bound on the volume of a fibered hyperbolic 3-manifold one can give in terms of the translation length of the monodromy with respect to various metric spaces, such as Teichmüller space with the Weil–Peterson metric or the pants complex?*

REMARKS.

(1) The monodromy of a hyperbolic 3–manifold that fibers over the circle is a pseudo-Anosov homeomorphism, which acts with positive translation length on several well-studied spaces, including Teichmüller space with either the Teichmüller or Weil-Peterson metric and the pants complex.

Brock [**Bro03**] shows that for a finite-type surface S and a pseudo-Anosov homeomorphism f of S, the hyperbolic volume of the mapping torus M_f is comparable to the translation length $\tau_{WP}(f)$ of f acting on the Teichmüller space of S with the Weil-Peterson metric. In particular, he shows that there exist constants κ_1 and κ_2, depending on the topology of S, such that

$$\kappa_2 \tau_{WP}(f) \leqslant \mathrm{Vol}(M_f) \leqslant \kappa_1 \tau_{WP}(f). \tag{3}$$

He also proves the analogous inequality with the translation length of f acting on the pants complex.

The heart of the question, therefore, lies in making the constant κ_2 (or the analogous constant for other metric spaces) effective by describing precisely how it depends on the genus g of S.

(2) Kojima–McShane [**KM18**] and Brock–Bromberg [**BB16**] give effective *upper* bounds on the volume in terms of the translation length of the monodromy with respect to the Teichmüller and Weil-Peterson metrics on Teichmüller space, respectively. A related question is to find an effective upper bound on $\mathrm{Vol}(M_f)$ in terms of the translation length with respect to the pants complex.

A lower bound was obtained by Bridgeman–Brock–Bromberg for the Weil-Peterson translation length in the case the 3–manifold is relatively acylindrical [**BBB23**], but the bound is not effective. In the unpublished work of Aougab–Taylor–Webb, they show that when considering the translation length of f on the pants complex, the constant κ_2 decays like $\frac{1}{g!}$. No effective lower bounds are known.

(3) There is no such lower bound in terms of the translation length of the monodromy with respect to the Teichmüller metric on Teichmüller space. McMullen gives an explicit counterexample in which the Teichmüller translation lengths of the monodromies f_n tend to infinity but the mapping tori M_{f_n} have bounded volume [**McM14**].

(4) The analogue of the upper bound of the inequality (3) was proved for irreducible end-periodic homeomorphisms of infinite-type surfaces by Field–Kim–Leininger–Loving [**FKLL23**], and the analogue of the lower bound was recently announced by Field–Kent–Leininger–Loving [**FKLL25**]. They give a universal constant for κ_1 and show that the constant κ_2 depends on the topology of the finite-type subsurface on which the irreducible end-periodic is acting "interestingly" (not just by translation). It would interesting to make their lower bound effective and compare it to the description of κ_2 one might achieve in the finite-type case.

(5) This problem falls into the larger category of questions asking what hyperbolic-geometric data about a fibered 3-manifold one can obtain from the surface and homeomorphism data used to construct it. For instance, in any of the cases mentioned above, is it possible to gather data about the short geodesics in the 3-manifold M_f from information about S and f?

Proposed for K3 by: T. Aougab
Scribed by: P. Patel, C. Abbott

PROBLEM 2.19. *Which mapping classes give rise to arithmetic hyperbolic 3-manifolds as their mapping tori?*

REMARKS.

(1) Roughly speaking, arithmetic hyperbolic manifolds are a special class of hyperbolic manifolds constructed by "number-theoretic" methods. They enjoy many special properties, e.g., possessing a very rich group of commensurations (isometries of finite covers). For a general introduction, see [**MR03**]; also see Problem 3.10.

(2) This is likely to be a deep and difficult question, since Thurston's proof of the existence of the hyperbolic structure on M_f is non-constructive. It would be useful to amass some systematic data on the monodromies of known fibered arithmetic 3-manifolds. Are the elements that appear distinctive or exceptional from the point of view of the mapping class group?

As alluded to above, arithmetic hyperbolic manifolds are distinguished in the class of all hyperbolic manifolds by the property that their fundamental groups have infinite index in their commensurators (indeed, the commensurator is *dense* in $\mathrm{PSL}_2(\mathbb{C})$). One place to start would be to better understand how the commensurator of the mapping torus is encoded in the mapping class itself.

(3) In the world of knots, there is an analogous question which, remarkably, has been resolved. Alan Reid [**Rei91**] has shown that the only knot $K \subset S^3$ for which the complement $S^3 \backslash K$ is arithmetic hyperbolic is the figure-8 knot. The figure-8 knot is fibered with fiber a once-punctured torus. More generally, work of Bowditch-Maclachlan-Reid [**BMR95**] classifies the monodromies of all arithmetic hyperbolic once-punctured torus bundles over S^1. In spite of this, it is not clear what special properties these elements possess that single them out from other elements of $\mathrm{SL}_2(\mathbb{Z})$.

(4) The paper [**BMR95**] also proves some other results of relevance. For example, it shows that there are at most finitely many cyclic commensurability classes of arithmetic hyperbolic surface bundles with surface of fixed non-compact topological type.

(5) Associated to any hyperbolic manifold is an invariant called the *invariant trace field*, the field generated by the traces of squares of elements of the fundamental group, viewed via the hyperbolic structure as isometries in $\mathrm{PSL}_2(\mathbb{C})$. Arithmetic hyperbolic 3-manifolds are known to be characterizable in terms of properties of their traces (see [**MR03**]). A more general problem is to determine the invariant trace field associated to any pseudo-Anosov element.

Proposed for K3 and scribed by: N. Salter

2.5. Surface fibrations and connections with symplectic and algebraic geometry

A genus–g *Lefschetz fibration* on a closed 4–manifold X is a smooth map $f\colon X \to B$, where B is a closed orientable surface, such that f is

- a submersion except at finitely many critical points $\{p_i\} \subset X$, and
- modeled around each p_i in complex coordinates as $f(z_1, z_2) = z_1 z_2$

where a *regular fiber* is S_g. The local monodromy around each p_i is a right-handed Dehn twist, and the corresponding singular fiber, which is a nodal curve, is obtained by crushing the core curve of the associated Dehn twist to a point.

A Lefschetz fibration (with sections) can be encoded by a *monodromy factorization.* A monodromy factorization is an expression of the form

$$T_{c_1} \dots T_{c_\ell}[A_1, B_1] \dots [A_h, B_h] = T_{\delta_1} \dots T_{\delta_b}$$

in $\mathrm{Mod}(S_g^b)$, where $\{c_i\}$ are essential simple closed curves and $\{\delta_i\}$ are parallel to distinct boundary components on S_g^b. Here T_c denotes a right-handed Dehn twist along $c \subset S_g^b$ and $[x, y] := xyx^{-1}y^{-1}$ is a commutator. When $b = 0$, this is a monodromy factorization of a genus–g Lefschetz fibration over $B = S_h$ with ℓ nodes obtained by crushing $\{c_i\}$ on Σ_g. It is the monodromy factorization of a *surface bundle over a surface* if, in addition, one has $\ell = 0$. When $b > 0$ and $h = 0$, it prescribes a monodromy factorization of a genus–g *Lefschetz pencil* with b base points and ℓ nodes.

PROBLEM 2.20. *Can one detect holomorphicity from a monodromy factorization of a Lefschetz pencil, fibration, or a surface bundle over a surface? What are the special properties of monodromy factorizations of holomorphic fibrations?*

REMARKS.

(1) Unless $\ell = g - 1 = 0$, any fibration corresponding to a monodromy factorization as above can be made symplectic and J-holomorphic for an almost complex structure J compatible with the symplectic form. However, J is not necessarily integrable.

(2) Let $f\colon X \to S_h$ be a relatively minimal genus-g Lefschetz fibration. Clearly, for it to be *holomorphic*, the total space X should have the homotopy type of a compact complex surface, e.g., the fundamental group one reads from the monodromy factorization should be a Kähler group, the odd Betti numbers should be even, and such.

Set $\chi_f := \frac{1}{4}(\chi(X) + \sigma(X)) - (g-1)(h-1)$ and $K_f^2 := c_1^2(X) - 8(g-1)(h-1)$. A few other known constraints are:
- Beauville's inequality [**Bea79**]: $\chi_f \geqslant 0$;
- Arakelov's inequality [**Ara71**]: $K_f^2 \geqslant 0$; and
- Xiao's slope inequality [**Xia87**]: $(4 - 4/g)\chi_f \leqslant K_f^2 \leqslant 12\chi_f$.

Further constraints on χ_f and K_f^2 are available for holomorphic fibrations of special types (hyperelliptic, non-hyperelliptic, trigonal, non-trigonal, bielliptic, etc.); see e.g., [**AK02**].

(3) There is an additional constraint formulated in terms of the Nielsen-Thurston classification. A subgroup $\Gamma < \mathrm{Mod}(S_g)$ is *reducible* if there is a multi-curve $C \subset S_g$ with only essential components fixed up to homotopy by each element of Γ. The *monodromy group* Γ of a nontrivial holomorphic surface bundle over $B \backslash f(\mathrm{Crit}(f))$ must be *infinite* and *irreducible* [**Shi97**].

Proposed for K3 and scribed by: I. Baykur

Problem 2.21. *What is the minimum number, $m_{g,b}$, of right-handed Dehn twists along essential curves into which the boundary multi-twist, $\Delta := T_{\delta_1} \cdots T_{\delta_b}$, can be factorized in $\mathrm{Mod}(S_g^b)$? Find factorizations realizing the minimum.*

Remarks.

(1) A factorization $T_{c_1} \cdots T_{c_\ell} = \Delta$, where no c_i is homotopic to a boundary component or point, is often called a *positive factorization* of Δ. A positive factorization of the boundary multi-twist $\Delta = T_{\delta_1} \cdots T_{\delta_b}$ in $\mathrm{Mod}(S_g^b)$, where each δ_j is parallel to a distinct boundary component of S_g^b and $b > 0$, corresponds to a genus-g Lefschetz pencil with b base points and ℓ nodes.

With these in mind, the problem amounts to determining the smallest second Betti number b_2 of genus-g Lefschetz pencils/fibrations with b base points. It is motivated by a central problem regarding the existence of small symplectic and exotic 4-manifolds. See, e.g., [**BK17, Bay22, BH24c**] for some successful implementations of this approach to 4-dimensional exotica.

(2) The pairs $g, b \in \mathbb{N}$ for which there is no positive factorization of the boundary multi-twist are determined in [**BMVHM17**]; namely, when $g = 1$ and $b > 9$, or $g \geqslant 2$ and $b > 4g + 4$. In all other cases, the minimum is known to exist. The only cases when the problem has been completely settled are $g = 1$ [**Kas77, Moi77b**] and $g = 2$ [**BK17**], for any $b \geqslant 0$.

For $g \geqslant 3$, there are reasonable upper bounds for every genus; the positive factorizations discovered in [**Mat96, Cad98, Kor01, Ham17**] imply that $m_{g,b} \leqslant 2g + 4$ when g is even and $b \leqslant 4$, and $m_{g,b} \leqslant 2g + 10$ when g is odd and $b \leqslant 8$. These bounds are known to not be sharp, at least in low genera. For example, $m_{g,b} = 7$ when $g = 2$ and $b \leqslant 3$ [**BK17, Bay22**], whereas $m_{g,b} \leqslant 12$ when $g = 3$ and $b \leqslant 4$ [**Smi01b, Bay22, HH18a**]. For large $b \leqslant 4g + 4$, another upper bound is $m_{g,b} \leqslant 8g + 4$, which is realized by genus-g pencils on $S^2 \times S^2$ [**SS94, Tan12**].

There are also lower bounds available in every genus, derived from Seiberg-Witten theory of symplectic 4-manifolds, roughly implying that $m_{g,b} \geqslant g$ [**Li00, Sti99, BK03**]—which is unlikely to be sharp in *any* genus.

(3) There are several variations of this problem studied in the literature.

The minimum number of Dehn twists in positive factorizations of any $h \geqslant 1$ commutators in $\mathrm{Mod}(S_g)$ (which correspond to genus-g Lefschetz fibrations over S_h) is almost completely determined in [**KO01, Mon12, Ham14, SY17, BH23**]; the only open cases are when $g \geqslant 3$ and $h = 1$.

When $b > 0$, one can also inquire about the *maximum* number of Dehn twists, related to the uniform topology of Stein fillings of contact 3-manifolds. This problem was completely answered in [**BVHM15, BVHM16, BMVHM17**], establishing in particular that in many cases one may have arbitrarily long positive factorizations of Δ in $\mathrm{Mod}(S_g^b)$.

In some cases, one can obtain more leverage on the stated problem by imposing additional conditions, say by assuming that the factorization lies in a smaller mapping class group (e.g., [**Alt20**]) or by fixing some topological invariants of the corresponding Lefschetz pencils (e.g., [**Sti02**]).

Proposed for K3 and scribed by: I. Baykur

PROBLEM 2.22. *Does every Lefschetz fibration over the 2–sphere admit a section?*

REMARKS.

(1) Here we assume the fibration has critical points; otherwise, there exist T^2–bundles over S^2 without sections.

Lefschetz fibrations that arise from blowing up the base locus of Lefschetz pencils naturally admit sections—namely, the exceptional divisors. However, if say the fibration is constructed through branched covers, or abstractly via a group-theoretic factorization, it is not *a priori* clear whether a section exists.

Equivalently, one can ask if every positive Dehn twist factorization of the identity in $\mathrm{Mod}(S_g)$ lifts to $\mathrm{Mod}(S_g, p)$, the mapping class group of the (fiber) surface fixing a point $p \in \Sigma_g$.

(2) This question has been raised repeatedly in the literature; e.g. [**Smi01a, Aur05, Ona10, Gom25**], and Problems 2.8, 4.1, 7.1 and 8.1 in [**Kor06**], [**Sti15**], [**KS09**], and [**End21**] respectively.

(3) More generally we can ask about the existence of *multisections* (in the sense of [**DS03, BH16b**]):

QUESTION. Does every Lefschetz fibration over the 2–sphere admit a multisection? How about a 2-section?

Similarly, one can reformulate the existence of an n–section in terms of the existence of a lift of a positive Dehn twist factorization of the identity in $\mathrm{Mod}(S_g)$ to a positive factorization also involving positive arc twists in $\mathrm{Mod}(S_g, \{p_1, \ldots, p_n\})$, the mapping class group of S_g fixing the *set* of n distinct points $p_1, \ldots, p_n \in S_g$ [**BH16b, BH16a**].

(4) It is worth noting that for *achiral* Lefschetz fibrations, where the Dehn twists in the factorization can be a mix of right-handed and left-handed twists, there are counterexamples to the question. For instance, there is a genus–1 achiral Lefschetz fibration on S^4 [**Mat82**], which, for homological reasons, cannot admit a section. (In contrast, *every* genus–1 Lefschetz fibration admits a section.) More examples are given in a recent preprint of Gompf [**Gom25**].

(5) An ad hoc method for finding sections of a Lefschetz fibration with an explicit monodromy factorization is to reverse-engineer the monodromy factorization using elementary relations (among Dehn twists) in the mapping class group (e.g., [**KO08, Ham17, Tan12**]).

A potential counterexample to the existence of a section, where this ad hoc method essentially fails, is the (spin, signature-zero) genus–9 Lefschetz fibration of Baykur–Hamada in [**BH24c**].

Proposed for K3 and scribed by: I. Baykur

PROBLEM 2.23. *Does there exist a surface bundle over a surface that admits a complete hyperbolic metric, or, more generally, a complete metric of variable negative curvature?*

REMARKS.

(1) Thurston's hyperbolization theorem provides for an enormous wealth of 3-manifolds that are surface bundles over S^1, and that moreover admit complete hyperbolic metrics. Moving up a dimension, the same phenomenon becomes quite mysterious. As in the 3-manifold case, it is *necessary* that the monodromy of every nontrivial loop in the base be pseudo-Anosov. A construction of such *purely pseudo-Anosov surface subgroups* of mapping class groups has been announced in very recent work of Kent–Leininger [**KL24**], resolving a major open question in its own right. It is not yet clear whether every purely pseudo-Anosov subgroup induces a word-hyperbolic surface-by-surface group extension, which is a necessary condition for the total space to admit a metric of negative curvature, and indeed, it is presently not known if the Kent-Leininger examples are word-hyperbolic. A weak version of Problem 2.23 is then simply to furnish examples of surface bundles over surfaces with word-hyperbolic fundamental group.

(2) Bowditch [**Bow09**] shows that for fixed base and fiber genus, there are only finitely many possible examples of hyperbolic surface bundles over a surface.

(3) As detailed in [**Rei06**], the existence of a hyperbolic surface bundle over a surface would contradict the long-standing conjecture of LeBrun [**LeB02**] that the Seiberg-Witten invariants of a closed hyperbolic 4-manifold must vanish (see Problem 4.95).

(4) Work of Kapovich [**Kap98**] shows that surface bundles over surfaces can never admit complex hyperbolic metrics.
(5) For an extensive account of the questions raised here, see [**KL24**, Section 1].

Scribed by: N. Salter

PROBLEM 2.24. *Does there exist a complex surface X that admits three or more non-isomorphic structures as a surface bundle over a surface?*

REMARKS.

(1) The analogous situation in the case of surface bundles over the circle is organized by the Thurston norm [**Thu86**]: if a surface bundle over the circle admits two different fibrations, then it admits infinitely many. By contrast, work of F.E.A. Johnson [**Joh99**] shows that a given 4-manifold admits only finitely many structures as a surface bundle over a surface with base and fiber both of genus $\geqslant 2$.
(2) Atiyah [**Ati69**] and Kodaira [**Kod67**] constructed complex surfaces X that are surface bundles over surfaces in two distinct ways. These are constructed as branched covers of products of algebraic curves; the two bundle structures arise from the projections onto either factor. Certain families of these examples admit exactly two fiberings by work of Chen [**Che18**], Salter–Tshishiku [**ST20**], and Landesman–Litt–Sawin [**LLS25**]. The problem formulated here was also raised by Catanese [**Cat17**].
(3) In the smooth category, Salter [**Sal15**] gave a method of constructing surface bundles over surfaces (with base and fiber each of genus at least 2) with an arbitrary finite number of fiberings. However, these are known not to admit complex structures.

Proposed for K3 and scribed by: N. Salter

PROBLEM 2.25. *Consider surface bundles over surfaces where both fiber F and base B have genus $\geqslant 2$ and where $\pi_1(B)$ injects in the mapping class group of the fiber. Does such a bundle have a k–multisection for some $k > 0$ (i.e., a continuous choice of k everywhere distinct points)?*

REMARKS.

(1) This is Problem 2.17 in [**Kir97**] by G. Mess. In [**Mor89**], Morita gives a cohomological obstruction for a surface bundle to admit a section (i.e. the case $k = 1$). Using this obstruction, Hillman gave examples of surface bundles over surfaces (with both base and fiber genus $\geqslant 2$) that do not admit sections [**Hil15**]. Other examples of S_g-bundles over the 2-torus with no section were constructed by Li–Litt–Salter–Srinivasan [**LLSS23**]; these were constructed with applications to arithmetic geometry in mind. Both classes of examples *virtually* admit sections, that is, they admit sections after pulling back to some finite cover of the base.

(2) In spite of the similarity between this question and Problem 2.22, it is likely that substantially different techniques will be necessary. Section problems for surface bundles tend to be of a more group-theoretic flavor, boiling down to obstructing a lifting of the monodromy representation from an unpointed to a pointed mapping class group. Available techniques include cohomological considerations as well as approaches based on the Nielsen–Thurston classification. The multisection question for Lefschetz fibrations is more delicate, since here it is necessary to lift the Dehn twists in the monodromy factorization on $\mathrm{Mod}(S_g)$ to Dehn twists in $\mathrm{Mod}(S_g^b)$, not simply to arbitrary elements in $\mathrm{Mod}(S_g^b)$ projecting back to the original twists.

Proposed for K3 and scribed by: I. Baykur, N. Salter

PROBLEM 2.26 (Kontsevich–Zorich conjecture). *Understand the homotopy types of strata of abelian differentials. Which stratum-components are $K(\pi, 1)$ spaces? What are the fundamental groups? For which stratum-components $\mathcal{H}$ is some version of the monodromy map $\rho\colon \pi_1(\mathcal{H}) \to \mathrm{Mod}(S_g)$ injective? Do the answers to these questions change in a predictable way as one moves from the principal stratum down to the minimal?*

REMARKS.

(1) A *translation surface* is a surface T that admits an atlas of charts to $\mathbb{C}$ for which transition maps are translations ($z \mapsto z + c$). An *abelian differential* is a pair (X, ω) of a Riemann surface together with a holomorphic differential. It is a basic but profound fact that every translation surface corresponds to an abelian differential and vice-versa.

Fixing a genus, the space of all abelian differentials is identified with the *Hodge bundle* over the moduli space of Riemann surfaces of genus g, the vector bundle whose fiber over a Riemann surface X is the space $H^{1,0}(X)$ of all holomorphic 1-forms on X. This space is *stratified* according to the multiplicities of the zeroes. Every differential on a Riemann surface of genus g has $2g-2$ zeroes when counted with multiplicity; the generic case where all zeroes are simple (the partition 1^{2g-2}) is called the *principal stratum*, and the case where there is a single zero of multiplicity $2g-2$ is called the *minimal stratum*.

In general, the stratum $\mathcal{H}_\kappa$ associated to the partition $\kappa = k_1 + \cdots + k_n$ is a quasiprojective complex orbifold of complex dimension $2g + n - 1$, and therefore (after passing to a finite cover to resolve the orbifold issue) has the homotopy type of a finite CW complex. Strata are fundamental objects in the study of *Teichmüller dynamics*, where they host a dynamical system induced from a natural action of $\mathrm{GL}_2(\mathbb{R})$, but their topology is quite mysterious, despite their close relationship to the moduli space of Riemann surfaces and the mapping class group.

(2) In [**KZ03**], Kontsevich–Zorich compute π_0 for all strata. They find that each stratum has between between one and three connected components. For the partitions $2g-2$ and $g-1, g-1$, there are special "hyperelliptic" components that can be identified with configuration spaces of points in $\mathbb{C}$, resolving Problem 2.26 in this case.

In [**KZ97**], they pose a version of Problem 2.26, conjecturing that each stratum component should be a $K(\pi, 1)$ and that the fundamental groups should be some flavor of mapping class group.

(3) Work of Calderon–Salter [**CS23**] identifies the image of $\pi_1(\mathcal{H})$ in the mapping class group under the natural monodromy homomorphism for all non-hyperelliptic stratum components in genus $g \geqslant 5$. They find that these are "framed mapping class groups"—the stabilizer of the distinguished framing on the translation surface inherited from the standard framing on $\mathbb{C}$. This shows that whenever the monodromy homomorphism is *injective*, the fundamental group of the stratum component is indeed some flavor of mapping class group, as predicted by the conjecture.

The question of monodromy injectivity appears to be subtle, however. In the hyperelliptic setting, injectivity was established in the work of Kontsevich–Zorich. On the other hand, in genus 3, there is a stratum-component known as $\mathcal{H}_4^{odd}$ for which injectivity is known not to hold. Work of Looijenga–Mondello [**LM14**] computes the fundamental group to be $A(E_6)$ (the Artin group of type E_6) modulo its center. Under the monodromy map, the Artin generators are sent to Dehn twists about simple closed curves whose intersection pattern is given by the E_6 Dynkin diagram. Work of Wajnryb [**Waj99**] shows that this homomorphism $A(E_6) \to \mathrm{Mod}(\Sigma_3)$ has a non-central element in the kernel.

Proposed for K3 and scribed by: N. Salter

2.6. Surfaces and geometric group theory

This section concerns problems relating to connections between braid and mapping class groups and other topics in geometric group theory. While there are many such connections, the narrow focus here mostly concerns the interaction of mapping class groups and Artin groups.

An *Artin group* is a group with a finite generating set $g_1, \ldots, g_k$, where every relation is of the form $g_i g_j \cdots = g_j g_i \ldots$, such that each side of the relation has the same length ℓ_{ij} and alternates between g_i and g_j. For $\ell_{ij} = 2$ this asserts that g_i and g_j commute, and for $\ell_{ij} = 3$ this asserts that g_i, g_j satisfy the braid relation $g_i g_j g_i = g_j g_i g_j$. A *right-angled Artin group* is an Artin group where each pair of generators that satisfy some relation have $\ell_{ij} = 2$. Thus, every pair of generators either commute or satisfy no relation. A right-angled Artin group can be encoded by a graph, with vertices encoding generators and edges between every pair of commuting generators.

PROBLEM 2.27. *For $n \geqslant 4$, does the braid group B_n admit a finite-index subgroup that embeds in a right-angled Artin group?*

REMARKS.

(1) In CAT(0) geometry, a cube complex is called *special* if it satisfies certain conditions on its hyperplanes [**HW08**]. Groups acting by combinatorial isometries on special cube complexes are of particular interest in geometric group theory, and are closely related to subgroups of right-angled Artin groups. A group is said to be *virtually special* if it contains a finite-index subgroup isomorphic to the fundamental group of a special cube complex.
(2) By [**BKK16**], any group that embeds in a right-angled Artin group acts by C^∞-diffeomorphisms on $\mathbb{R}$. As braid groups do not admit such actions [**FF20**], they do not embed in right-angled Artin groups, so it is necessary to pass to proper subgroups.
(3) One could go further and ask whether braid groups are virtually special. This is strictly stronger than Question 2.27.
(4) In light of the resolution of the congruence subgroup problem for braid groups (see [**Mar19**] and the references therein), to resolve Problem 2.27 in the negative, it suffices to obtain a cofinal sequence of subgroups of B_n, none of which embed in a right-angled Artin group.

Scribed by: T. Koberda

PROBLEM 2.28. *Let Γ be a graph that is not a nontrivial join, and let $A(\Gamma)$ be the associated right-angled Artin group. Does there exist an injective map from $A(\Gamma)$ to the mapping class group of a surface such that every loxodromic element of $A(\Gamma)$ is sent to a pseudo-Anosov mapping class?*

REMARKS.

(1) There are many injective homomorphisms from right-angled Artin groups into mapping class groups [**CP01, CLM12, Kob12**], and general criteria are fairly well-understood [**KK14b**]. Right-angled Artin groups have a Nielsen–Thurston classification, not unlike mapping class groups [**BC12, KK14a**], wherein each element is either elliptic or loxodromic.
(2) The set of loxodromic elements of $A(\Gamma)$ is nonempty if and only if $A(\Gamma)$ does not split as a nontrivial direct product, which is true if and only if Γ does not split as a nontrivial join.
(3) By [**MT16, KMT17**], the restriction of any such injective map to a subgroup where every nontrivial element is loxodromic is necessarily a convex cocompact subgroup of the mapping class group of S. However, any such subgroup obtained in this way is free (see [**KK14a**]). In particular, this does not provide a strategy to construct convex cocompact surface subgroups as in Problem 2.23.
(4) In light of [**KMT17**], the problem can be reformulated entirely in terms of the combinatorial topology of the underlying surface. Let Γ be fixed, and let Λ be the opposite graph, i.e., the one obtained by reversing the adjacency relation in Γ. It suffices to find a surface S and π_1-injective,

pairwise non-nested subsurfaces $\{S_v\}_{v\in V(\Gamma)}$ indexed by the vertices of Γ such that the following hold:

(i) Each S_v supports a pseudo-Anosov mapping class (in the mapping class group of S_v). In particular, each S_v is non-annular.
(ii) Subsurfaces S_v and S_w intersect essentially if and only if v and w are not adjacent in Γ (or equivalently, adjacent in Λ).
(iii) Let $Y \subset \Lambda$ be a connected subgraph such that every vertex of Λ is adjacent to a vertex of Y. Then every essential, simple, nonperipheral curve on S essentially intersects S_v for some vertex v of Y.

The difficulty of this reformulation is the non-canonical nature of Y for a general graph Λ.

Scribed by: T. Koberda

PROBLEM 2.29. *Determine the Artin groups that can be embedded into a mapping class group.*

REMARKS.

(1) If the Artin group is simply laced and the homomorphism sends each standard generator to a Dehn twist, then there is an almost complete classification. It is known that the Artin groups of type A_n (see [**BH73**]), D_n (see [**PV96**]) and $\tilde{A}_n$ (see [**Ryf23**]) can be embedded into a mapping class group in this way. It is also known that an Artin group does not embed in this way if its Coxeter graph is connected and contains a full subgraph isomorphic to either $\tilde{D}_n$ with $n \geqslant 4$ (see [**Lab97**]), or E_6 (see [**Waj99**]). These leave a finite number of possible open cases.
(2) Say a homomorphism from an Artin group to a mapping class group is *geometric* if each standard generator is sent to a Dehn multi-twist. The question is very open in the case of non-geometric embeddings. There are by now various examples of non-geometric embeddings of Artin groups into mapping class groups; see e.g., [**CMM21**] for the case of right-angled Artin groups and [**Sze10, BT12**] for the case of braid groups.
(3) The question is also closely related to another folklore question on Artin groups: which Artin groups are *virtually special* (possessing a finite-index subgroup isomorphic to a subgroup of a right-angled Artin group). It is known that every right-angled Artin group embeds into a mapping class group, and so obstructions to embedding an Artin group into mapping class groups also obstruct embeddability into right-angled Artin groups. From this point of view it is also natural to consider a "virtual" version of the above question, namely to determine which Artin groups have some finite-index subgroup that embeds into a mapping class group.
(4) This problem is historically attributed to J. Crisp and L. Paris [**CP01**].

Scribed by: T. Koberda

PROBLEM 2.30. *Which right-angled Artin groups contain closed hyperbolic surface groups? Is there an algorithmic or graph-theoretic criterion to decide this?*

REMARKS. All but finitely many surface groups embed in right-angled Artin groups [**CW04**], and right-angled Artin groups on cycles of length at least five contain hyperbolic surface groups [**SDS89**]. Right-angled Artin groups on graphs up to eight vertices containing surface groups were classified by Crisp–Sageev–Sapir [**CSS08**]; they also posed the question of whether a right-angled Artin group contains a surface group if and only if it contains a one-ended word-hyperbolic group. See also [**Kim10**].

Scribed by: T. Koberda

PROBLEM 2.31. *Let S be a closed surface of genus at least 2. Show that the stable commutator length is rational on the commutator subgroup of $\pi_1(S)$.*

REMARKS.

(1) If G is a group and $g \in [G, G]$, the *commutator length* of g, denoted $\mathrm{cl}(g)$, is the least number of commutators in G whose product is g, and the *stable commutator length* of g, denoted $\mathrm{scl}(g)$, is the limit of $\mathrm{cl}(g^n)/n$ as $n \to \infty$.

(2) Suppose X is a space with $\pi_1(X) = G$, suppose L is an oriented circle, and suppose $f\colon L \to G$ takes L to the conjugacy class of $g \in [G, G]$. After replacing f by inclusion, consider the the affine subspace V of $H_2(X, L; \mathbb{R})$ that is mapped by the boundary map in homology to the fundamental class $[L]$ of L in $H_1(L; \mathbb{R})$. Then the stable commutator length of g is $1/4$ times the infimum of the (relative) Gromov norm on the subspace V.

 Taking L to be a 1-manifold and $f\colon L \to X$ to be a map for which $f_*[L] = 0$ in $H_1(X; \mathbb{R})$ extends the definition of stable commutator length to all homologically trivial formal chains (a purely group-theoretic definition is given in [**Cal09b**]).

 The conjecture above may be broadened as follows. Let G be an arbitrary word-hyperbolic group and $f\colon L \to X$ an arbitrary map from an oriented 1-manifold L to X with $f_*[L] = 0$ in $H_1(X; \mathbb{R})$. Here are a few related conjectures:

 (a) *the stable commutator length of L is rational;*
 (b) *the Gromov norm of any rational class in $H_2(X, L; \mathbb{Q})$ is rational;* and
 (c) *the Gromov norm of any rational class in $H_2(X; \mathbb{R})$ is rational.*

 For X a closed surface of genus at least 2, the answer to (c) is positive, and for every L the answer to (b) is positive for all rational $\alpha \in V$ mapping to the fundamental class of $[L]$ outside a compact interval.

(3) This is known to be true for elements represented by curves supported in an essential proper subsurface $S' \subset S$ (see [**Cal09b**]).

(4) If S is a compact oriented surface, possibly with boundary and every component having negative Euler characteristic, the Gromov norm of the fundamental class of S in $H_2(S, \partial S; \mathbb{R})$ is $-2\chi(S)$.

If $f\colon L \to X$ is as above, a map $F\colon S \to X$ *virtually bounds* f if there is a degree n (oriented) covering map $\pi\colon \partial S \to L$ for which $f\pi = F$ on ∂S (we allow the possibility that L is empty). Such a map is *extremal* if it is an isometry for the Gromov norm. Extremal maps are π_1-injective.

A further conjecture: *In each case of the conjecture in Remark (1), there is an extremal map of a surface virtually bounding* L *(or virtually representing a given class in* $H_2(X;\mathbb{Q})$ *if* L *is empty).*

This conjecture could be thought of as bearing on Gromov's question of whether every one-ended hyperbolic group contains a closed surface subgroup.

(5) The questions above have a positive answer for free groups. See e.g., [**Cal09b**].

Proposed for K3 and scribed by: D. Calegari

2.7. Homeomorphism groups and interactions with dynamics and logic

In this section, we consider connections between surfaces and dynamics, studying homeomorphism and diffeomorphism groups of surfaces. We also touch on connections with logic, in particular the first-order theory of the mapping class group and solutions to systems of equations.

PROBLEM 2.32. *Does every surface bundle over a surface admit a flat connection? What about surface bundles over* 3*-manifolds?*

REMARKS.

(1) An S_g-bundle $p\colon E \to B$ admits a *flat connection* if it admits a codimension 2 foliation everywhere transverse to the fibers. Equivalently, the bundle admits a flat connection if the monodromy map $\rho\colon \pi_1(B) \to \mathrm{Mod}(S_g)$ admits a lifting $\tilde{\rho}\colon \pi_1(B) \to \mathrm{Diff}(S_g)$.

(2) Work of Morita [**Mor87**] shows that the MMM classes $e_i \in H^{2i}(\mathrm{Mod}(S_g))$ obstruct the existence of a flat connection for $i \geqslant 3$. In practical terms, this means that there are surface bundles over manifolds of dimension at least 6 that admit no flat connection. Conversely, the first MMM class is known *not* to obstruct flatness; see [**KM05**]. New techniques, presumably of a more dynamical flavor, will be required to obstruct the existence of flat connections on surface bundles when the base is of lower dimension.

(3) This question can be asked for varying degrees of regularity, all the way from lifting to the group of homeomorphisms, up to real-analytic or volume-preserving diffeomorphisms.

In the volume-preserving case, techniques from symplectic geometry may be relevant. This discussion follows ideas outlined in [**KM05**]. Let ω be a volume form on S_g, and let $\mathrm{Diff}(S_g,\omega)$ denote the group of volume-preserving diffeomorphisms; the identity component is denoted $\mathrm{Diff}_0(S_g,\omega)$. An S_h-bundle over S_g is determined by a set $a_1, b_1, \ldots, a_h, b_h$ of mapping classes for which the surface relation $[a_1,b_1]\ldots[a_h,b_h]$ holds. From this point of view, a flat volume-preserving connection consists of a choice of lifts $\alpha_i, \beta_i \in \mathrm{Diff}(S_g,\omega)$ such that $[\alpha_1,\beta_1]\ldots[\alpha_h,\beta_h] = 1$.

One possible method for obstructing such a lift proceeds as follows. Beginning with a surface bundle specified by the mapping classes $a_1, b_1, \ldots, a_h, b_h$ as above, the set of *all* lifts α_i, β_i is a torsor for the group $\mathrm{Diff}_0(S_g, \omega)^{2h}$. There is a map (not a group homomorphism)

$$\mu\colon \mathrm{Diff}_0(S_g, \omega)^{2h} \to \mathrm{Diff}_0(S_g, \omega)$$

that sends a set of lifts $\alpha_1, \beta_1, \ldots, \alpha_h, \beta_h$ to the surface word $\prod_{i=1}^{h}[\alpha_i, \beta_i]$; the fiber $\mu^{-1}(\mathrm{id})$ then describes the set of flat volume-preserving connections on the bundle.

$\mathrm{Diff}_0(S_g, \omega)$ admits a *flux homomorphism* Flux: $\mathrm{Diff}_0(S_g, \omega) \to H^1(S_g; \mathbb{R})$ whose kernel is the group of *Hamiltonian diffeomorphisms* $\mathrm{Ham}(S_g, \omega)$. A first question is whether the flux homomorphism can serve as an obstruction to admitting a flat connection, although this seems unlikely to be the case.

Considering next the set of lifts for which $[\alpha_1, \beta_1] \ldots [\alpha_h, \beta_h]$ is Hamiltonian, one is led to wonder if the *Hofer metric* on $\mathrm{Ham}(S_g, \omega)$ could be of use in obstructing the existence of a flat connection. Perhaps it is possible to show that the image of μ must stay *far away* from 0 in the Hofer metric? (See [**MS17a**] for an introduction to the Hofer metric.)

Proposed for K3 and scribed by: N. Salter

PROBLEM 2.33. *Let S be a closed compact surface (without boundary). Is there a finitely generated, torsion-free group G such that G cannot act faithfully by homeomorphisms on S?*

REMARKS.

(1) There are non-compact surfaces for which every countable group acts by homeomorphisms [**APV21**].
(2) We pose the question for closed surfaces only, since the presence of boundary gives access to special techniques that may be able to resolve the question as stated without developing new methods for the closed case. On the other hand, the question is open even for the disk, relative to the boundary. Potential candidates include lattices in high rank Lie groups. For more regular actions (i.e., smooth actions) more is known; see [**BFH22**], for instance.
(3) The torsion-free hypothesis is necessary, since there are often a priori bounds on the size of torsion in surface homeomorphism groups.
(4) In the case of the plane, Le Roux [**LR11**] has studied actions of the fundamental group of the Klein bottle under the assumption that the standard generators act freely and preserving orientation. Among other results, he finds obstructions for the action of certain torsion-free groups by fixed-point free orientation-preserving homeomorphisms.
(5) A more general question would be to find an algebraic characterization of groups acting on S by homeomorphisms. Such characterizations are

known for groups acting on the interval and on the circle, via orderability. It may be difficult to find concise characterizations; see e.g., [**dlNG22**, **Ros13**, **Hyd19**].

Scribed by: T. Koberda

PROBLEM 2.34. *Let S be a compact surface. For $0 \leqslant r < s$, does there exist a nontrivial finitely generated subgroup $G_r \leqslant \mathrm{Diff}_0^r(S)$ such that every action of G_r on S by C^s diffeomorphisms is trivial? Can these examples be torsion-free?*

REMARKS.

(1) For $0 \leqslant r \leqslant \infty$, let $\mathrm{Diff}_0^r(M)$ denote the identity component of the group of C^r diffeomorphisms of a manifold M. For nonintegral r, we write $r = k + \epsilon$ for $k \in \mathbb{N}$ and $0 \leqslant \epsilon < 1$, and we consider C^k diffeomorphisms whose k^{th} order derivatives are ϵ–Hölder continuous.
(2) The condition that every higher-regularity action of G_r be *trivial* can be relaxed, for instance, to ask that every such action of G_r be non-faithful, or abelian.
(3) Problems of this nature have been extensively investigated in dimension one, where Kim and Koberda gave a positive answer to this question [**KK20**, **KK21**]; one obtains finitely generated groups with infinite simple commutator subgroups such that each smoother action has abelian image. No instances are known in dimension greater than one. Prior to this, Plante–Thurston showed that nilpotent groups of C^2-diffeomorphisms of the interval are abelian [**PT76**], but Farb–Franks showed that there can be non-abelian nilpotent actions of regularity C^1 [**FF03**]. More generally, Castro–Jorquera–Navas build nilpotent groups that can act at a certain regularity but whose higher-regularity actions are all abelian [**CJN14**]. Further examples of finitely generated groups acting with critical regularities on 1-manifolds are constructed in [**MW23**].
(4) In the case $r = 0$ and $s \geqslant 1$, it is possible to use *Thurston's Stability Theorem* [**Thu74b**] to exhibit particular groups of homeomorphisms that cannot act faithfully by diffeomorphisms. See, for instance, [**Cal06a**].

Proposed for K3 and scribed by: S. Kim, T. Koberda

PROBLEM 2.35. *Is the first-order theory of the mapping class group of a surface decidable?*

REMARKS.

(1) The *first order theory* of a group G in the language of group theory is the formal first-order language equipped with group multiplication and the identity element. A first-order theory is *decidable* if there is an algorithm (i.e., Turing machine) that decides which sentences are true in that theory and which are false.

(2) For homeomorphism groups of surfaces (and indeed all compact manifolds of positive dimension), the corresponding theory is not decidable because it interprets arithmetic [**KKdlNG25**].
(3) The theory of free groups is decidable; this is part of the resolution of the Tarski problem [**Sel02, KM06**].

Proposed for K3 and scribed by: T. Koberda

PROBLEM 2.36. *Are systems of equations over mapping class groups and braid groups decidable?*

REMARKS.

(1) A *system of equations over a group* G is a finite conjunction of formulae of the form
$$r(x_1, \ldots, x_n, a_1, \ldots, a_m) = 1$$
and
$$r(x_1, \ldots, x_n, a_1, \ldots, a_m) \neq 1,$$
where r is a word in a free group of rank $n + m$, where $(x_1, \ldots, x_n)$ is a tuple of variables, and where $(a_1, \ldots, a_m)$ (with $m \geqslant 0$) is a tuple of *parameters* in G. A *solution* to a system of equations is a tuple of elements of G such that when the elements are substituted for variables, one obtains a valid expression in G.
(2) Decidability of equations would consist of an algorithm to decide whether or not a particular system has a solution or not. Precisely, one might fix a finite generating set for G (if it exists) and use a Gödel numbering to encode equations with parameters as natural numbers. Then, one would require a Turing machine to take a natural number corresponding to an equation and decide if it admits a solution.
(3) One could allow variations on the problem; for instance, one could fix the underlying surface or allow it to vary, pass to finite index subgroups, or restrict the topological type of the underlying surface.
(4) Decidability of the theory of a group or class of groups has a long history, starting at least with the Tarski problem. The Tarski problem asked whether the nonabelian free groups are elementarily equivalent to each other and whether their first order theories are decidable; see [**Sel02, KM06**]. For general hyperbolic groups (possibly with torsion), Dahmani and Guirardel proved that equations are solvable; see [**DG10**]. Conjecturally, the full first order theory of a vast generalization of hyperbolic groups, namely hierarchically hyperbolic groups, should be decidable. A particularly interesting special case of such a general result would be the decidability of the first order theory of mapping class groups and braid groups, which in turn would give a positive answer to the proposed question.
(5) There is a vast literature on equations over groups, including finite groups and free groups, though for mapping class groups and braid groups the problem appears to be open. See [**Rom12**] for a survey.

(6) It seems likely that equations over homeomorphism groups of surfaces (with or without parameters) are undecidable.

Proposed for K3 and scribed by: T. Koberda

2.8. Infinite-type surfaces

We now delve into the world of *infinite-type surfaces*, that is, those that are not homeomorphic to any $S^b_{g,n}$. As in the finite-type case, infinite-type surfaces are completely classified, in this case by their genus (which may be infinite), the number of compact boundary components, and the pair (E, E^g), where E is the end space of the surface (which is necessarily a closed subspace of a Cantor set) and E^g is the subspace of E consisting of ends accumulated by genus; note that E and E^g are topological spaces, and a homeomorphism between surfaces induces a homeomorphism of pairs on (E, E^g).

In this section, we gather problems on the algebraic structure of their mapping class groups (for example, their subgroup structure), the geometric structure of their homeomorphisms (for example, the existence of quasiconformal representatives), and their dynamical structure (for example, understanding the properties of flows on translation surfaces). We also consider more traditional questions from geometric group theory about their mapping class groups, in particular considering their actions on graphs. The study of infinite-type surfaces from many of these perspectives is relatively recent, and so many of these questions are foundational.

PROBLEM 2.37. *Give a Nielsen–Thurston-type classification for the mapping class groups of infinite-type surfaces. In particular, which homeomorphisms are the appropriate generalization of pseudo-Anosov homeomorphisms in this setting?*

REMARKS.

(1) For finite-type surfaces S, there is a powerful classification due to Nielsen and Thurston of the mapping classes of S: every mapping class is periodic, reducible, or pseudo-Anosov. The most important aspect of this classification is that if a mapping class f is not periodic or reducible, it *must* be pseudo-Anosov, meaning that f is highly chaotic and that there exist two transverse measured foliations λ^+ and λ^- on S, where f stretches along λ^+ and contracts along λ^-. The exact analog of this classification for infinite-type surfaces is false. For instance, the homeomorphisms called handleshifts (introduced by Patel-Vlamis [**PV18**]) are neither periodic, reducible, nor pseudo-Anosov. The classification, therefore, needs to be modified in the infinite-type setting. First, the definition of a reducible mapping class should be generalized, but more importantly, the third category (the analog of pseudo-Anosov mapping classes) needs to be expanded and likely broken into subcategories.

(2) Some work has been announced on this problem by Bestvina, Fanoni and Tao [**BFT23**], who classify so-called "tame" homeomorphisms that satisfy an additional finiteness condition. Such homeomorphisms do not exhibit any pseudo-Anosov-like behavior.

(3) Irreducible end-periodic mapping classes should form an important class within the third category. For end-periodic homeomorphisms of infinite-type surfaces, Handel and Mosher outlined a theory paralleling that of pseudo-Anosov homeomorphisms in unpublished work from the 1980s, describing a pair of transverse geodesic laminations on the surface preserved by the end-periodic mapping class. This theory was further developed by Cantwell, Conlon, and Fenley [**CCF21**]. In addition to the above, irreducible end-periodic mapping classes share many properties with finite-type pseudo-Anosov mapping classes; for example, they give rise to hyperbolic mapping tori [**FKLL23**] and many strongly irreducible ones are loxodromic isometries of complexes of curves and arcs [**PT25**].

Proposed for K3 by: P. Patel
Scribed by: C. Abbott, P. Patel

PROBLEM 2.38. *Give an appropriate analogue of the curve graph for infinite-type surfaces, and characterize the surfaces for which no such graph exists.*

REMARKS.

(1) A motivating philosophy in geometric group theory is that one can study the algebra of a group G via the geometry of the spaces on which G acts and the dynamics of such actions. When G is the mapping class group of a finite-type surface, one of the most fruitful actions to study is that of the mapping class group of S on the *curve graph*, $\mathcal{C}(S)$, associated to S.

Part of the reason that this action is so dynamically rich is due to a famous result of Masur and Minsky [**MM00**], which shows that the curve graph is infinite-diameter and Gromov-hyperbolic: there exists $\delta > 0$ so that, for all geodesic triangles in $\mathcal{C}(S)$, each side of the triangle is contained in the δ-neighborhood of the other two. This means that $\mathcal{C}(S)$ shares properties with hyperbolic space so that some of the key tools in hyperbolic geometry can be used to study $\mathcal{C}(S)$ and groups acting on it. The curve graph has played a pivotal role not only in the study of mapping class groups, but also Teichmüller theory, Kleinian groups, and the moduli space of Riemann surfaces.

By contrast, for infinite-type surfaces the curve graph has finite diameter. This makes it much harder to study the above topics for infinite-type surfaces via the action of groups on this graph. An analogue of the curve graph would be an infinite-diameter hyperbolic graph associated to an infinite-type surface S on which the mapping class group of S acts coboundedly and by isometries.

(2) Vlamis showed that there are infinite-type surfaces whose mapping class groups do not act on any metric space with unbounded orbits [**Vla24b**]; additional examples were announced in [**Vla24a**]. Thus, it is not possible to find an analogue of the curve graph that can be associated to *any* infinite-type surface.

(3) Among infinite-type surfaces whose mapping class groups do admit unbounded actions on metric spaces, there are several associated graphs that have been fruitful to study thus far, for instance the ray graph (Calegari

[**Cal09a**]), the relative arc graph (Aramayona–Fossas–Parlier [**AFP17**]), the omnipresent arc graph (Fanoni–Ghaswala–McLeay [**FGM21**]), and the grand arc graph (Bar-Natan–Verberne [**BNV23**]). Unfortunately, this is a piecemeal approach, since these graphs are not defined for all infinite-type surfaces in this class, and may fail to be infinite-diameter and hyperbolic in general. In light of this, one wishes to first classify which infinite-type surfaces admit an analogue of the curve graph, and to then find a universal graph that can be associated to any such infinite-type surface that is both infinite-diameter and hyperbolic.

(4) This problem appeared on an AIM problem list for a workshop on surfaces of infinite type [**LPRT**] and was proposed for inclusion there by P. Patel and K. Mann.

Proposed for K3 by: P. Patel
Scribed by: C. Abbott, P. Patel

PROBLEM 2.39.

(a) *Does the mapping class group of an infinite-genus surface with no planar ends contain every countable group?*
(b) *Does the mapping class group of the one-ended infinite-genus surface have any proper finite-index subgroups?*
(c) *For an infinite-type surface, describe and characterize the subgroup of mapping classes with quasi-conformal representatives. In particular, when is it a normal subgroup of the mapping class group?*

REMARKS.

(1) The work of Aougab, Patel, and Vlamis [**APV21**] shows every mapping class group of an infinite-genus surface with no planar ends contains every finite group as a subgroup. Thus the problem is really about countably infinite groups. The same work of Aougab, Patel, and Vlamis also shows that the above question is true for infinite-genus surfaces with no planar ends and self-similar end space, e.g., the Loch Ness monster surface. To do this, they realize every countable group as the isometry group of a hyperbolic structure, and moreover, show that the above surfaces are the only ones in which this is possible. Therefore, to answer the above question, it is necessary to construct countably infinite subgroups of the mapping class group that are not subgroups of the isometry group of the surface.
(2) It was shown by G. Domat [**Dom22**] that the mapping class group of the one-ended infinite genus surface has nontrivial abelianization, and the abelianization has many copies of $\mathbb{Q}$ as direct summands. It follows that the mapping class group of the Loch Ness monster surface has many proper subgroups of countably infinite index. In the same work, Domat also shows that the abelianization does contain torsion subgroups, though they are not known to be direct summands.
(3) When S is a Riemann of finite type, it is a basic fact that every mapping class admits a representative as a quasi-conformal homeomorphism, but this is no longer true if S is of infinite type. In this setting, the group

of mapping classes with quasi-conformal representatives is sometimes referred to as the *Teichmüller modular group*; the reader should be aware that this term is sometimes used to refer to the entire mapping class group in the finite-type setting. While this usage is consistent, there is the possibility for confusion.

(4) This problem appeared on a problem list compiled at the 2021 Nearly Carbon Neutral Geometric Topology conference [**CPV21**].

Proposed for K3 by: P. Patel
Scribed by: P. Patel, C. Abbott

PROBLEM 2.40.

(a) *Let S be an infinite-type surface and ϕ a mapping class for which there is a (marked) conformal structure Σ on S with respect to which ϕ is realized as a quasi-conformal homeomorphism. Then for each quasi-conformal structure in the connected component of Teichmüller space containing Σ, there is a quasiconformal homeomorphism representing ϕ which minimizes the dilatation. If the infimum of the minimal dilatations for each structure in this component is achieved and is greater than 1, is the minimizing quasiconformal homeomorphism unique? What is the structure of such a minimizer?*

(b) *If ϕ is parabolic (see Remark* 1 *below) for T, must a sequence of conformal structures with dilatation converging to the infimum degenerate by pinching along some sequence of essential simple closed curves? Is this system unique? Can it depend on the component T?*

(c) *If ϕ fixes more than one component T, T' of* Teich(S)*, can the type be different for T and T'? In particular, can ϕ be parabolic for T and hyperbolic for T'?*

REMARKS.

(1) Let S be a surface of infinite type, and let ϕ be a mapping class for which there is some (marked) conformal structure Σ on S for which ϕ is realized by a quasiconformal homeomorphism $f\colon \Sigma \to \Sigma$. Then there is *some* representative homeomorphism which realizes the infimum of the dilatation. This follows from compactness (for any finite K) of the space of K-quasiconformal homeomorphisms of the unit disk, normalized to fix 3 points on the boundary (see, e.g., [**Ahl06**]).

We may then try to vary the conformal structure on S in the component $T \subset \text{Teich}(S)$ containing Σ, and for each point in T we get a minimum value for the dilatation of a homeomorphism representing ϕ, and thereby a function $K_T\colon T \to [1, \infty)$. Note K_T takes the value 1 if and only if ϕ is represented by an isometry for some conformal structure in the component T.

Say ϕ is *elliptic for T* if K_T achieves the value 1, *parabolic for T* if the infimum is not achieved, and *hyperbolic for T* if the infimum is achieved and is bigger than 1. Part (a) can be restated as follows: if ϕ is hyperbolic for T, is the minimizer unique? What is its structure?

(2) When ϕ is a mapping class on a surface S of *finite* type, there is a (marked) conformal structure on S with respect to which ϕ can be realized as a quasiconformal homeomorphism. Moreover, there is a unique quasiconformal homeomorphism representing ϕ that minimizes the dilatation; this is the *Teichmüller map*. Note that in the finite-type setting, Teichmüller space has a single connected component.

(3) There are elementary examples of topological surfaces of infinite type S and mapping classes ϕ that are not realized by quasiconformal maps for *any* choice of (marked) conformal structure on S. Perhaps the simplest example is: let S be the infinite ladder surface, which we think of as the infinite union of twice-punctured tori T_n indexed by integers $n \in \mathbb{Z}$ joined end to end, let α be any pseudo-Anosov diffeomorphism of a single twice-punctured torus, and let ϕ be the mapping class that does α^n on T_n. No matter what conformal structure we choose on S, the restriction of α^n to T_n has dilatation at least as big as $|n|$ times the minimum dilatation of α on a twice-punctured torus. This example is (highly) reducible; it would be nice to have a straightforward irreducible example.

(4) If S is a surface of infinite type, it is possible to define the Teichmüller space of S to be the space of marked conformal surfaces $f: S \to \Sigma$ up to equivalence; see the recently announced constructions of Basmajian–Chandran [**BC24a**] and Tappu [**Tap23**] for details. The connected components of this space will be contractible, but (with the hypothesis that S has infinite type) there will always be infinitely many components [**Bas97**, **BK08**]. Some mapping classes will fix no component (as in the previous remark), some will fix some components and not others, some will fix every component (e.g., those supported in compact subsurfaces). What are the possible orbit types? It was recently announced by Basmajian and Chandran that the subgroup that fixes every component is the compactly supported subgroup [**BC24a**].

(5) Recent work of Basmajian and Chandran [**BC24a**] announced that there exist mapping classes ϕ that are represented by quasiconformal homeomorphisms on some components of Teichmüller space but not others. In particular, the function K_T defined in Remark 1 above may not be well-defined for every component $T \subseteq \text{Teich}(S)$.

(6) A version of this problem appeared on a problem list compiled at the 2021 Nearly Carbon Neutral Geometric Topology conference [**CPV21**].

Proposed for K3 by: D. Calegari, P. Patel
Scribed by: C. Abbott, D. Calegari, P. Patel

PROBLEM 2.41. *Give a finite list of practically computable invariants of the mapping class group or pure mapping class group of an infinite-type surface S that determine the topology of S.*

REMARKS.

(1) Let S be a surface, possibly of infinite type. When $S = S^p_{g,b}$ is of finite type, the homeomorphism type of the surface is encoded in the isomorphism type of its mapping class group, with two pairs of exceptions for

small values of g, b, and p. Most cases can be distinguished by considering well-known group invariants: work of Birman–Lubotsky–McCarthy [**BLM$^+$83**] computes the algebraic rank, and Harer [**Har86**] computes the virtual cohomological dimension, both in terms of g and b. Combining these formulas, it is possible to determine the topology of the surface S, although note that this argument breaks down when either $g = 0$ or $b = 0$ (see [**RS11**, Remark A.2]).

(2) The classification of infinite-type surfaces tells us that an infinite-type surface is determined by its end space E, the closed subgroup of non-planar ends E^g of S, the genus, and the number of compact boundary components of S. However, the end space of an infinite-type surface can be any closed subset of a Cantor set, and in practice, it is hard to tell when two such subsets are homeomorphic. In addition, Bavard–Dowdall–Rafi [**BDR20**] prove algebraic rigidity for mapping class groups of infinite-type surfaces: if $\mathrm{Map}(S)$ is isomorphic to $\mathrm{Map}(S')$, then S and S' are homeomorphic. But again, big mapping class groups are uncountable groups and determining when two of these groups are isomorphic in general is difficult. Thus, we pose the question above.

(3) There has been one result by Aougab–Patel–Vlamis [**APV21**, Theorem 8.1] in this direction for n-ended orientable infinite-genus surfaces with no planar ends.

Proposed for K3 and scribed by: P. Patel

PROBLEM 2.42. *Is the geodesic flow in almost every direction on the Chamanara surface ergodic? What about on the translation surface considered by Bruin and Lukina, which is similar to the Chamanara surface?*

REMARKS.

(1) Translation surfaces have geodesic flows in every direction that preserve Lebesgue measure. Kerckhoff, Masur, and Smillie proved that for every finite-type translation surface, this is the only probability measure preserved by the flow (that is, the flow is uniquely ergodic) in almost every direction [**KMS86**]. As a corollary, they obtain that the flow in almost every direction is ergodic with respect to the Lebesgue measure. It is natural to ask to what extent this generalizes to infinite-type surfaces.

Fraczek and Ulcigrai produced many examples of infinite-area, infinite-type translation surfaces where the straight line flows are not ergodic in almost every direction [**FU14**]. For finite-area translation surfaces, however, this is widely open.

(2) The Chamanara surface, first described by Reza Chamanara in [**Cha04**], is one of the best known translation surfaces. Chamanara defines a family of translation surfaces with a parameter $\alpha \in [0, 1]$. For the case $\alpha = 1/2$, called the standard Chamanara surface, consider a square with edge length 1, and divide the top and bottom sides into two halves. Identify the top right half with the bottom left half via a translation, and then divide the unidentified top and bottom edges in half, identify the top right half with the bottom left half via a translation, and repeat. Do the same for the

left and right edges, always identifying the upper part of the right edge with the lower part of the left edge. Excluding the corners of the square and the points where we divided the edges, we obtain the Chamanara surface, a one-ended infinite-genus translation surface of finite area. See also [**DHV24**, Example 2.4.22] for another description of the surface. The Chamanara surface has a close relationship to baker's map and is sometimes called the baker's map surface; see [**CGL06**] for further details.

Bruin and Lukina consider a family of translation surfaces similar to the Chamanara surfaces [**BL23**], but their surfaces lack certain metric symmetries enjoyed by the Chamana surfaces, and so different techniques may be necessary to approach this problem.

(3) Similarly to proving that there exists a billiard in a polygon where the billiard flow is ergodic (in the 3-dimensional unit tangent bundle), one can use a Baire Category argument to show that there exists a finite-area, infinite-type surface where the flow in almost every direction is ergodic with respect to Lebesgue measure. Moreover, by work of Vorobets [**Vor97**], one can even find an explicit example, but these arguments do not seem applicable in the two situations above.

Proposed for K3 by: J. Chaika
Scribed by: C. Abbott, P. Patel

PROBLEM 2.43. *Let X be a compact, totally disconnected subset of $\mathbb{R}^2$ with $|X| \geqslant 2$, and let Γ_X denote the mapping class group of $\mathbb{R}^2 - X$.*

(a) *For $Y \subset X$, let $\Gamma_{X,Y}$ be the subgroup of Γ_X permuting Y. Classify invariants q that are* functorial.
(b) *If $X \subset \mathbb{R}^2$ is totally disconnected and α is a homeomorphism of $\mathbb{R}^2$ fixing X as a set, let $C_\alpha(X)$ denote the space of α-invariant closed subsets of X, in the Hausdorff topology. Let q be functorial as in part* (a). *Classify those q that are* continuous *or* semi-continuous.
(c) *What is the relationship between the translation length τ_X and topological entropy? More precisely, is there a positive constant C so that if α is a homeomorphism of $\mathbb{R}^2$ permuting X and g is the class of α in Γ_X, then the (topological) entropy of α is at least as big as $C \cdot \tau_X(g)$?*
(d) *Suppose X is a Cantor set in the plane, and let $P\Gamma_X$ denote the 'pure' subgroup of Γ_X fixing X pointwise. Let N be a subgroup of $P\Gamma_X$ that is normal in Γ_X, and for each integer n, let N_n be the subgroup of $P\Gamma_n$ that is the image of N restricted to any n-element subset of X. What possible sequences of subgroups N_n can arise in this way?*
(e) *For which $g \in \Gamma_X$ is there an embedding $X \to \mathbb{R}^2$ for which g is represented by a diffeomorphism α which is C^∞? Or C^k for fixed k?*

REMARKS.

(1) In Part (a), an invariant is *functorial* if for each X, there is a map $q_X \colon \Gamma_X \to \mathbb{R}$ so that the following holds. For each inclusion $Y \subset X$ and each $g \in \Gamma_{X,Y}$ with image $g_Y \in \Gamma_Y$ under the natural restriction map

$\Gamma_{X,Y} \to \Gamma_Y$, there is an inequality $q_X(g) \geqslant q_Y(g_Y)$. In part (b), an invariant is *(semi-)continuous* if the map $C_\alpha(X) \to \mathbb{R}$ taking Y to $q_Y([\alpha])$ is (semi-)continuous for all α, where $[\alpha]$ is the class of α in Γ_Y.

(2) The *ray graph* $\mathcal{R}_X$ is the graph whose vertices are isotopy classes of rays in $\mathbb{R}^2 - X$ from infinity to a point in X, and whose edges are pairs that may be realized disjointly. This graph is hyperbolic, connected, and infinite diameter, and its flag complex is contractible. The group Γ_X acts on it by isometries.

There is a cyclic order on the set of vertices of $\mathcal{R}_X$ according to the cyclic order on the geodesic representatives at infinity (with respect to any complete hyperbolic structure on $\mathbb{R}^2 - X$) which may be order completed to a circle S^1_X on which Γ_X acts by (orientation-preserving) homeomorphisms.

Numerical invariants of elements of Γ_X may be defined in terms of these actions, including

(a) rotation number $\mathrm{rot}_X(g) \in \mathbb{R}/\mathbb{Z}$ for $g \in \Gamma_X$ acting on S^1_X;
(b) translation length $\tau_X(g)$ for $g \in \Gamma_X$ acting on $\mathcal{R}_X$; and
(c) counting quasimorphisms $H_{X,\sigma}(g)$ for $g \in \Gamma_X$ and σ a Γ_X-orbit of path in $\mathcal{R}_X$.

Some of these invariants have a well-defined 'name' (rotation number, translation length) independent of X. Others depend on choices that are special to X.

Another way to express the fact that some invariants have a well-defined name is to say that they are functorial in the sense of part (a). For example, an inclusion $Y \to X$ induces a 1-Lipschitz map $\mathcal{R}_X \to \mathcal{R}_Y$ well defined up to bounded distance, and translation length τ_X in the graph $\mathcal{R}_X$ is functorial.

(3) There is a partial order $\prec$ due to Boyland [**Boy88**] on the set of all conjugacy classes in all (finite type) braid groups where $g \prec h$ if for every homeomorphism α of the disk fixing the boundary with a finite α-invariant set X such that α represents g relative to X, there is another α-invariant set Y such that α relative to Y represents h. We say that *g forces h*, and this partial order is called *braid forcing*. A numerical invariant q from (finite type) braid groups to $\mathbb{R}$ is *monotone* if it is monotone with respect to braid forcing partial order.

All monotone invariants arise as follows: let G denote the group of all homeomorphisms of the disk fixed on the boundary, and for each braid g, let G_g be the subset of G of homeomorphisms representing g relative to some finite invariant set. Then $g \prec h$ if and only if $G_g \subset G_h$. For any function $f\colon G \to \mathbb{R}$, we obtain a monotone invariant by $f(g) = \sup_{\alpha \in G_g} f(\alpha)$.

Braid forcing makes perfect sense for mapping class groups instead of braid groups, and one may extend this notion in the obvious way to a partial order on all conjugacy classes in all Γ_X simultaneously. Thus, another natural question is: *What is the relationship between (semi-)continuity and Boyland's braid forcing?*

(4) In [**CC22**] the individual subgroups N_n that can arise as in part (d) are identified; they satisfy the so-called 'inertia' condition.

(5) Let $\mathcal{R}_X$ be the ray graph and S^1_X the circle on which Γ_X acts by homeomorphisms as in Remark 2. An additional question is: *Can the (non)-realization property of part* (e) *be related to numerical properties of g that can be computed from the action on $\mathcal{R}_X$ or S^1_X?*

Proposed for K3 and scribed by: D. Calegari

PROBLEM 2.44. *Given an infinite-type surface S, which homeomorphisms $f\colon S \to S$ give rise to mapping tori M_f that admit a hyperbolic structure? For those which do admit a hyperbolic metric, is M_f homeomorphic to the interior of a compact hyperbolic manifold with totally geodesic boundary?*

REMARKS.

(1) Thurston's theorem on the hyperbolization of mapping tori states that for a finite-type surface S and a homeomorphism f of the surface S, the mapping torus

$$M_f = S \times [0,1]/(x,0) \sim (f(x),1)$$

is hyperbolic if and only if f is pseudo-Anosov. This question works towards an analogue of the theorem for infinite-type surfaces.

A first approach to this problem could be to give a list of criteria on f that ensures certain properties necessary for the mapping torus to be hyperbolic. For instance, to give criteria on f that ensure that M_f is atoroidal.

(2) It is known from work of Field–Kim–Leininger–Loving [**FKLL23**] that for a strongly irreducible end-periodic homeomorphism f of an infinite-type surface S, the mapping torus M_f is hyperbolic and homeomorphic to the interior of a compact hyperbolic manifold with totally geodesic boundary. This motivates the second part of the question above.

Proposed for K3 and scribed by: P. Patel

PROBLEM 2.45. *Compute the end-periodic cobordism group Δ_2^e of end-periodic automorphisms (diffeomorphisms or homeomorphisms) of surfaces.*

REMARKS.

(1) The (oriented) cobordism group of n-dimensional manifolds, denoted Δ_n, consists of equivalence classes of pairs (M^n, f) where M is a compact oriented manifold and f is a homeomorphism. The equivalence relation is given by $(M_0, f_0) \sim (M_1, f_1)$ if there is a compact cobordism between M_0 and M_1 with a homeomorphism extending f_0 and f_1. This group was computed for $n \geqslant 4$ by Kreck [**Kre84b**] and for $n = 3$ by Melvin [**Mel79**]. Bonahon [**Bon83**] (see also [**EE82**]) showed that $\Delta_2 \cong \mathbb{Z}^\infty \oplus \mathbb{Z}_2^\infty$. As in usual cobordism theories, compactness is used to ensure that one gets a nontrivial group.

(2) An end-periodic manifold M is a noncompact manifold with finitely many ends, such that each end ϵ has a neighborhood that is half of an infinite cyclic cover of a compact manifold X_ϵ. End-periodic manifolds behave in many respects (topological, geometric, and analytic) as if they were actually compact. An end periodic automorphism $f\colon M \to M$ is a homeomorphism or homeomorphism that is a covering translation over X_ϵ on some neighborhood of each end ϵ. End-periodic homeomorphisms arise naturally in the study of depth-one foliations [**CCF21**], with the behavior in the end corresponding to the limiting behavior of a non-compact leaf as it approaches a compact leaf. Cobordisms of end-periodic automorphisms are defined as in the compact case, with the extension required to be end-periodic, and one defines the end-periodic cobordism group $\Delta_n^{\mathfrak{e}}$ as above.

(3) This problem is meaningful in all dimensions; the calculation of the usual Δ_2 makes use of special techniques related to the Nielsen-Thurston classification of surface automorphisms. A similar end-periodic Nielsen-Thurston theory, originating in unpublished work of Handel-Miller, is developed in depth in [**CCF21**], and it seems reasonable to approach the computation of $\Delta_2^{\mathfrak{e}}$ using similar tools.

Proposed for K3 and scribed by: D. Ruberman

PROBLEM 2.46.

(a) *Which coarsely boundedly generated mapping class groups of infinite-type surfaces are hyperbolic?*

(b) *Consider the class of surfaces with $n \geqslant 2$ ends, all accumulated by genus. Are the mapping class groups of these surfaces quasi-isometric for different values of n?*

REMARKS.

(1) Big mapping class groups are not finitely, countably, or compactly generated. There is a generalization of compactness called coarse boundedness, defined and studied extensively by C. Rosendal [**Ros13, Ros22**] for Polish groups. It turns out that when the (pure) mapping class group of an infinite-type surface admits a coarsely bounded generating set, the quasi-isometry type of the group is well-defined. Mann–Rafi [**MR23**] and T. Hill [**Hil25**] give a classification of those surfaces whose mapping class groups and pure mapping class groups have this property, respectively.

This problem asks which coarsely boundedly generated mapping class groups of infinite-type surfaces are quasi-isometric to a hyperbolic metric space. In other words, for which infinite-type surfaces is the mapping class group with the word metric coming from a coarsely bounded generating set a hyperbolic metric space?

(2) The mapping class groups of some surfaces are not only coarsely boundedly generated, they are themselves coarsely bounded. In particular, they are finite diameter (when equipped with a word metric coming from a coarsely bounded generating set), and so are elementary hyperbolic.

Schaeffer-Cohen [**SC24**] showed that the mapping class group of a plane minus a Cantor set is quasi-isometric to the loop graph, which is an infinite diameter hyperbolic graph associated to the surface. This is the only known example of a mapping class group of an infinite-type surface that is non-elementary hyperbolic.

(3) Problem 2.46.(b) is a special case of the broad (and possibly unanswerable) problem of determining the quasi-isometry type of coarsely boundedly generated mapping class groups of infinite-type surfaces.

Proposed for K3 by: C. Abbott, K. Mann
Scribed by: C. Abbott, P. Patel

2.9. Mapping class groups in other contexts: Floer homology and quantum topology

This section considers questions originating in Floer homology, symplectic topology, and quantum topology, including the AMU Conjecture and a version of the Volume Conjecture for surface diffeomorphisms.

PROBLEM 2.47.

(a) *Given a mapping class ψ of a based surface S, there is an induced endomorphism of the symmetric product $Sym^i(S)$ and hence an endofunctor ψ_* of the partially wrapped Fukaya category $\mathcal{F}(S)$ as constructed, say, by Auroux* [**Aur10**]. *Compute the categorical entropy of ψ_*, in the sense of* [**DHKK14**], *for ψ pseudo-Anosov.*

(b) *Given a fibered knot K with pseudo-Anosov monodromy, what is*

$$\limsup_{n\to\infty}\left(\dim \widehat{HFK}(\Sigma^n(K), \widetilde{K}; i)\right)^{1/n},$$

where $\widetilde{K}$ is the branch locus in the n-fold cyclic branched cover $\Sigma^n(K)$, and i denotes the Alexander grading (in $\mathbb{Z}$) with respect to the fiber surface. Here, $\widehat{HFK}$ denotes the knot Floer homology of K [**OS04c**, **Ras03**].

(c) *With notation as in the previous question, what is*

$$\limsup_{n\to\infty}\left(\dim \widehat{HF}(\Sigma^n(K))\right)^{1/n}$$

for the Heegaard Floer invariant $\widehat{HF}$ [**OS04e**] *of the closed 3-manifold $\Sigma^n(K)$?*

(d) *Given a closed surface S and a pseudo-Anosov automorphism ψ, let M_ψ denote the mapping torus of ψ, and $\widehat{HF}(M_\psi; i)$ the summand of $\widehat{HF}(M_\psi)$ spanned by the* spin^c*-structures $\mathfrak{s}$ with $\langle c_1(\mathfrak{s}), [F]\rangle = i$. What is*

$$\limsup_{n\to\infty}\left(\dim \widehat{HF}(M_{\psi^n}, i)\right)^{1/n}?$$

What about

$$\limsup_{n\to\infty} \left(\dim_{\mathbb{F}_2} HF^+(M_{\psi^n}, i)\right)^{1/n}$$

for $i \neq 0$ (or for twisted HF^+ for $i = 0$)?

REMARKS.

(1) In (a), the categorical entropy is equal to

$$\limsup_{n\to\infty} \left(\dim H_*(\widehat{CFDA}(\psi^n, i-g))\right)^{1/n}$$

where $\widehat{CFDA}(\psi)$ is the bordered Floer bimodule [**LOT15**] and g is the genus of S. In the case $i = 1$, this entropy is equal to the dilatation $\lambda(\psi)$ [**LOT13, DHKK14**]. For (d), see [**OS04e**] for the definition of HF^+, which is a Floer homology relative to a divisor; $\widehat{HF}$ is Floer homology in the complement of that divisor.

(2) Properties of bordered Floer homology imply that the categorical entropy is an upper bound for the answers to the remaining questions (perhaps up to a constant factor) [**LOT15**]. Also, the growth rate of $\bigoplus_i \widehat{HFK}(\Sigma^n(K), \widetilde{K}, i)$ is an upper bound for the growth rate of $\widehat{HF}(\Sigma^n(K))$.

(3) The growth rate of $|H_1(\Sigma^n(K))|$ is a lower bound for the growth rate of $\widehat{HF}(\Sigma^n(K))$. The growth rate of $|H_1(\Sigma^n(K))|$ is the Mahler measure of the Alexander polynomial [**SW02**] when the first homology is finite. In particular, in cases where all branched covers are L-spaces (see, e.g., [**BBG19a, BBG19b**]), the growth rate is given by this Mahler measure.

(4) Some results about the growth rate of Heegaard Floer homology of branched covers were proved in [**HM18**], though these results are about linear, not exponential, growth.

(5) In (d), the reason for the restriction to HF^+ with $i \neq 0$ is that HF^+ is finitely generated over $\mathbb{F}[U]$ but not over $\mathbb{F}$ for $i = 0$.

(6) Some further observations on these questions can be found in [**Cor18**].

(7) One can ask similar questions about the growth of knot invariants like knot Floer homology, Khovanov homology, or symplectic Khovanov homology when one inserts a high power of a braid.

Scribed by: R. Lipshitz

PROBLEM 2.48. *The mapping class group of a closed, orientable, genus g surface S acts by symplectomorphisms on the symmetric product* $\mathrm{Sym}^g(S)$. *In particular, it acts smoothly on* $\mathrm{Sym}^g(S)$.

(a) *What is the kernel of the map from the mapping class group of S to the smooth mapping class group of* $\mathrm{Sym}^g(S)$? *What about in the based case?*

(b) *Given disjoint simple closed curves $\alpha_1, \ldots, \alpha_g$ in S, there is a corresponding torus T_α in* $\mathrm{Sym}^g(S)$: *the image of $\alpha_1 \times \cdots \times \alpha_g \subset S^g$. These are the tori that appear in Heegaard Floer theory. When do $\alpha_1, \ldots, \alpha_g$ and*

$\beta_1, \ldots, \beta_g$ give smoothly isotopic tori? When are they Lagrangian isotopic for an appropriate choice of symplectic form?

REMARKS.

(1) The question relates to whether Heegaard Floer homology is truly a symplectic invariant, or should be determined by the smooth isotopy classes of the Heegaard tori. (Examples of the difference between smooth and symplectic topology in other settings have received substantial interest, though many such examples are now known; [**Sei97**] is particularly relevant.)

(2) The fact that the mapping class group acts symplectically on $\mathrm{Sym}^g(S)$, for appropriate symplectic forms, was shown by Perutz [**Per08a**]. (The obvious map, where a diffeomorphism ϕ of the surface S sends a point $\{x_1, \ldots, x_g\} \in \mathrm{Sym}^g(S)$ to $\{\phi(x_1), \ldots, \phi(x_g)\}$ gives a homeomorphism but is not typically smooth; an exception is if ϕ is holomorphic with respect to some complex structure on S and the smooth structure on $\mathrm{Sym}^g(S)$ is induced from this complex structure, in which case the obvious induced map of $\mathrm{Sym}^g(S)$ is also holomorphic.) Clarkson used Heegaard Floer homology to show that the map from the based mapping class group to the group of symplectomorphisms of $\mathrm{Sym}^g(S\backslash\{z\})$ modulo the Hamiltonian diffeomorphisms is injective [**Cla17**]. (This uses the fact that, for appropriate choices of symplectic forms [**Per08b**], the Heegaard tori are exact [**Hen12, HLL22a**], and so Heegaard Floer homology agrees with the usual Lagrangian intersection Floer homology.)

(3) The fundamental group of $\mathrm{Sym}^g(S)$ (for $g > 1$) is isomorphic to $H_1(\Sigma)$. In particular, the Torelli group of S acts trivially on $\pi_1(\mathrm{Sym}^g(S))$. (It also acts trivially on the homology of $\mathrm{Sym}^g(S)$.) Also, by Perutz's work, if α_1' is obtained from α_1 by a handleslide, then $\alpha_1 \times \cdots \times \alpha_g$ and $\alpha_1' \times \cdots \times \alpha_g$ are smoothly isotopic (and, in fact, Hamiltonian isotopic for appropriate symplectic forms) [**Per08b**].

(4) In the case $g = 2$, one might be able to use the Abel-Jacobi map (which in this case presents $\mathrm{Sym}^2(S)$ as a blow-up of the Jacobian torus) to show that the Torelli group acts trivially, since it acts trivially on the Jacobian torus $H^1(S;\mathbb{R})/H^1(\Sigma;\mathbb{Z})$. Perhaps this suggests that the Torelli group or the Johnson kernel acts trivially in the higher-genus case as well.

Proposed for K3 and scribed by: R. Lipshitz, T. Perutz

PROBLEM 2.49 (AMU conjecture). *Let S be a surface with negative Euler characteristic. If $\phi \in \mathrm{Mod}(S)$ acts by a pseudo-Anosov on some subsurface (including S itself), show that there exists $p_0(\phi)$ such that the Witten–Reshetikhin–Turaev representation $\rho_p(\phi)$ has infinite order for $p \geqslant p_0(\phi)$.*

REMARKS.

(1) Given a compact Lie group G, the *Witten–Reshetikhin–Turaev G-quantum representations*, introduced by Witten [**Wit89**] and rigorously defined by Reshetikhin and Turaev [**RT91**], are families of representations of $\mathrm{Mod}(S)$. The most studied case is when $G = SU(2)$ or $G = SO(3)$,

in which case these representations are indexed by an integer $p \geqslant 3$. More precisely we have:

$$\rho_p\colon \operatorname{Mod}(S) \to PGL(d_p(S), \mathbb{C}),$$

where $d_p(S)$ is given by the famous Verlinde formula and is a polynomial in p of degree $3g-3$ (where g is the genus of S). The representation ρ_p sends a Dehn twist to a torsion element, and when S has negative Euler characteristic, $\rho_p(\operatorname{Mod}(S))$ is infinite for p big enough. Moreover, Andersen (see [**And06**]) proved that if $\phi \in \operatorname{Mod}(S)$ is non-central, then $\rho_p(\phi)$ is nontrivial for p sufficiently large. This property is called *asymptotic faithfulness.*

(2) Geometric properties of these representations are quite mysterious. For instance, it is not known if the Nielsen-Thurston classification can be detected by quantum representations. In [**AMU06**], Andersen, Masbaum, and Ueno studied the case of the four-holed sphere. They proved that if ϕ is pseudo-Anosov in the mapping class group of the four-holed sphere, then $\rho_p(\phi)$ has infinite order for p big enough. In the same paper, they stated the above problem as a conjecture; it is now known as the *AMU conjecture.*

(3) If $\phi \in \operatorname{Mod}(S)$ does not act as a pseudo-Anosov on a subsurface, then up to some power, it is the product of powers of commuting Dehn twists, and so $\rho_p(\phi)$ has finite order for all p. Also, as the quantum representations enjoy some nice "splitting" properties regarding subsurfaces, it is enough to prove this conjecture for pseudo-Anosov elements.

(4) The AMU conjecture is only known for two surfaces: the four-holed sphere (proved in [**AMU06**]) and the one-holed torus (see [**San12**]). In general, finding pseudo-Anosov elements satisfying the conjecture is already a difficult task. In [**EJ16**], it was proved that a certain class of "homological" pseudo-Anosov elements on holed spheres satisfy the AMU conjecture. Examples of point pushing mapping classes satisfying the AMU conjecture have been studied in [**KS16**] and [**MS21**]. See also [**DK22**] where examples were found using exponential growth of certain quantum invariants; this paper shows that the Volume Conjecture (see Problem 1.27) implies the AMU conjecture.

Scribed by: R. Santharoubane

PROBLEM 2.50 (Volume conjecture for surface diffeomorphisms). *Let S be a closed oriented surface, let $q = e^{\frac{2\pi i}{n}}$ be a root of unity, and let $\mathcal{K}^q(S)$ be the Kauffman bracket skein algebra. Let $\varphi\colon S \to S$ be a pseudo-Anosov diffeomorphism with mapping torus M_φ, and let $r\colon \pi_1(S) \to \mathrm{SL}_2(\mathbb{C})$ be a smooth point in the character variety that is φ-invariant. Finally, let L be the intertwiner realizing an isomorphism between the representations of $\mathcal{K}^q(S)$ corresponding to $[r]$ and $[r \circ \phi_*]$, normalized so that* $\det(L) = 1$. *Show that*

$$\lim_{n\to\infty} \frac{1}{n} \log|\operatorname{Trace} L| = \frac{1}{4\pi} \operatorname{vol}_{\mathrm{hyp}} M_\varphi.$$

REMARKS.

(1) For a closed oriented surface S and a root of unity q, results of [**BW17, FKBL19, GJS25, KK22, FKBL25**] more or less establish a one-to-one correspondence between irreducible representations of the Kauffman bracket skein algebra $\mathcal{K}^q(S)$ and smooth points in the $\mathrm{SL}_2(\mathbb{C})$ character variety of S. Thus in the above setup, the representations constructed from $[r]$ and $[r \circ \phi_*]$ must be isomorphic by some map L called the *intertwiner*. After normalizing L so that $\det(L) = 1$, $|\mathrm{Trace} L|$ is dependent only on the choice of ϕ, r, q, and the puncture invariants.

(2) This conjecture appears in [**BWY21**] and is a toy version of the Kashaev Volume Conjecture [**Kas97**] (see Problem 1.27), as revisited by Baseilhac-Benedetti [**BB04**]. In preprints, the conjecture has been claimed for some examples [**BWY22, Pan24, Wan25**] and for torus knots. There also exist versions of this conjecture for punctured surfaces, involving puncture invariants on the quantum side and 3-manifolds obtained from M_φ by Dehn filling on the hyperbolic side [**PW24**].

Proposed for K3 and scribed by: F. Bonahon, H. Wong

CHAPTER 3

3-manifolds

Chapter 3 Editors: John Baldwin, Robert Lipshitz, and Marc Lackenby

Perhaps more than any other part of low-dimensional topology, the landscape of 3-manifold topology has changed since the publication of **K2**. On the one hand, several longstanding conjectures central to the field have been resolved—most notably, the Poincaré Conjecture and Thurston's Geometrization Conjecture (Problems 3.1 and 3.45 of **K2**, proved by Perelman, building on work of of Hamilton), but also Thurston's Virtual Fiberedness Conjecture (Problem 3.51 of **K2**, proved by Agol, building on work of Wise and others), the Property P Conjecture (Problem 1.15 of **K2**, proved by Kronheimer-Mrowka), and others. On the other hand, both Floer theory and contact topology, which played only a small role in 3-manifold topology at the time of **K2**, have developed into central techniques in the field, leading to their own questions. Computability and computational complexity have also come to be seen as more central topics in the field, and development in Floer theory and 4-manifold topology have led to renewed interest in connections between 3- and 4-manifolds.

We have (somewhat artificially) divided the problems in this chapter into different themes:

- Section 3.1 collects problems related to hyperbolic geometry. Hyperbolic geometry remains central to the area, and this section has more problems than any other.
- Section 3.2 gives a few questions related to algorithmic topology.
- Section 3.3 collects a few questions related to Heegaard splittings.
- Section 3.4 has questions about the fundamental groups of 3-manifolds that do not fit naturally with other topics. (Questions about fundamental groups of hyperbolic 3-manifolds are mostly in Section 3.1.)
- Section 3.5 has a range of questions about contact 3-manifolds.
- Section 3.6 has questions about Floer theory invariants.
- Section 3.7 has questions about other 3-manifold invariants.
- Section 3.8 has questions about 3-manifolds as the boundaries of 4-manifolds.

Each section begins with some context and a few definitions that are used in it.

3.1. Hyperbolic geometry

A 3-manifold M is *hyperbolic* if it admits a (complete) Riemannian metric of constant sectional curvature -1. Equivalently, M is the quotient of hyperbolic 3-space by a discrete group of isometries acting properly discontinuously. By Mostow Rigidity, finite-volume hyperbolic 3-manifolds are determined, up to isometry, by their fundamental groups [**Mos68, Mar74, Pra73**]. In particular, given a (finite volume) hyperbolic 3-manifold M, any geometric invariant of M, such as its volume or the lengths of its closed geodesics, are homeomorphism invariants of M and, in fact, isomorphism invariants of $\pi_1(M)$.

By Thurston's Geometrization Conjecture [**Thu82**] (proved by Perelman [**Per02, Per03b, Per03a**]), every 3-manifold can be decomposed into geometric pieces, with hyperbolic pieces in some sense the most common and complicated. So, since the 1970s, hyperbolic geometry has played a central role in 3-manifold topology.

Problems 3.1–3.4 focus on hyperbolic metrics and their volumes. Problems 3.5–3.9 are about covers and fundamental groups of hyperbolic manifolds. A relevant notion for several of these is the *profinite completion* of a group G, which is the inverse limit of its finite quotients (with the inverse limit topology). Two finitely generated groups have isomorphic (as topological groups) profinite completions if and only if their sets of finite quotients are equal [**DFPR82**].

Problem 3.10 is about a special class of hyperbolic manifolds, the arithmetic ones. Arithmetic 3-manifolds also appear in Problems 3.2, 3.6, 3.17, 3.35, and 3.77.

Problems 3.11–3.17 are about submanifolds and geometric invariants of hyperbolic manifolds. For several of these, it is helpful to recall that a *π_1-injective surface* in a manifold M is a continuous map f from a closed, connected surface Σ to M so that the induced map $f_* : \pi_1(\Sigma) \to \pi_1(M)$ is injective. Often, one seeks such surfaces with additional properties, such as being an immersion or an embedding.

Problems 3.18–3.21 concern foliations, laminations, flows, and related topics. Problems 3.22–3.26 are about hyperbolic groups and their kin.

Problems 3.29 and 3.30 in Section 3.2, 3.35 in Section 3.3, 3.40 in Section 3.5, 3.63 in Section 3.7, and 3.77 in Section 3.8 are also about hyperbolic manifolds. (Of course, hyperbolic manifolds are an important class to consider in many of the other problems, as well.)

PROBLEM 3.1. *Classify the smallest volume hyperbolic 3-manifolds of various types. In particular:*

(a) *Determine the nonorientable closed hyperbolic 3-manifolds of least volume.*
(b) *Determine the n-cusped hyperbolic 3-manifolds of least volume for each $n \geqslant 3$.*
(c) *Determine the smallest volume hyperbolic 3-manifolds with n orientable cusps for each n.*
(d) *Determine the n-cusped orientable hyperbolic 3-manifolds of least volume for each n.*

REMARKS.

(1) Parts (a)–(c) appeared in [**Kir97**, Problem 3.60].
(2) There has been a lot of progress in determining the hyperbolic 3-manifolds of low volume [**GHM$^+$21**]. In particular, the orientable hyperbolic manifold of smallest volume was shown to be the Weeks manifold by Gabai, Meyerhoff, and Milley [**GMM09, Mil09**]; its volume is approximately 0.9427. Extensive evidence from SnapPy indicates that the hyperbolic 3-manifold with smallest volume is orientable (hence is the Weeks manifold). Closed nonorientable 3-manifolds appear to have larger volumes,

and so the currently available techniques do not readily apply to them, which suggests why part (a) remains open.

(3) The two smallest orientable hyperbolic 3-orbifolds were identified by Gehring, Marshall, and Martin [**GM09**], [**MM12**], as was the smallest volume nonorientable 3-orbifold.

(4) The n-cusped manifolds with smallest volume are known for $n = 1$ and 2; in both cases, these manifolds are nonorientable. For $n = 1$, Adams [**Ada87**] showed that the Gieseking manifold is the unique smallest manifold, with volume $v_3 \simeq 1.0149$, the volume of a regular ideal tetrahedron. For $n = 2$, Adams [**Ada88**] proved that the least volume is $2v_3$, and showed that there is a unique 2-cusped manifold with this volume.

(5) The smallest volume orientable hyperbolic 3-manifolds with n cusps are known for $n = 1$, 2 and 4. For $n = 1$, this is due to Cao and Meyerhoff [**CM01**], who showed that there are two manifolds with smallest volume $2v_3$, the figure-eight knot complement and the figure-eight knot sister. For $n = 2$, Agol [**Ago10**] proved that there are two orientable 2-cusped manifolds with smallest volume, the Whitehead link complement and the $(-2, 3, 8)$ pretzel link complement, both of which have volume v_8, which is the volume of a regular ideal octahedron, approximately 3.6638. For $n = 4$, Yoshida [**Yos13**] proved that the link 8^4_2 is the unique orientable 4-cusped manifold with smallest volume $2v_8 \simeq 7.3276$. For $n = 3$, the minimal volume is conjectured to be realized by the 3-chain link complement, which has volume approximately 5.33 (see [**Zha23**] for some results in this direction).

(6) Agol [**Ago10**] has conjectured that for $n \leqslant 10$, the minimal volume of an orientable n-cusped hyperbolic manifold is realized by the minimally twisted n-chain link complement. However, for larger values of n ($11 \leqslant n \leqslant 25$ and $n \geqslant 60$), it has been shown [**KPR12**] that these manifolds do not have minimal volume, as the $(n-1)$-fold cyclic cover over one component of the Whitehead link has smaller volume.

Scribed by: M. Lackenby

PROBLEM 3.2. *Show that the volumes of hyperbolic 3-manifolds are not all rationally related.*

REMARKS.

(1) This was proposed by Thurston in [**Thu82**]. One could interpret "rationally related" to mean the volumes are all rational multiples of each other, or the weaker claim that they span a finite-dimensional $\mathbb{Q}$-vector space. (The expectation is that neither holds.)

(2) By Borel [**Bor81**] there is a real number v_k such that the volume of an arithmetic hyperbolic 3-manifold with invariant trace-field k is an integral multiple of v_k. More generally, for any number field k, there are real numbers $v_1, \ldots, v_n$ such that for any (possibly non-arithmetic) hyperbolic 3-manifold with invariant trace-field k, its volume is an integral linear combination of $v_1, \ldots, v_n$.

(3) A 250+ year old problem asks whether Catalan's constant $\mathcal{G}$ is rational. Agol [**Ago10**] proved that the minimal volume 2-cusped hyperbolic 3-manifolds has volume equal to $4\mathcal{G}$, the volume of the regular ideal octahedron.

(4) Recently, F. Calegari, Dimitrov, and Tang [**CDT24**] established the irrationality of $L(2, \chi_{-3})$, which is closely related to the volume of a regular ideal 3-simplex, and hence the volume of the figure-eight knot complement.

(5) There is an analogous open question concerning the rationality of the Chern–Simons invariant. One can find an extensive discussion of this and related questions about the Chern–Simons invariant in [**Kir97**, Problems 3.62 and 3.63]; see also Problem 3.64.

(6) The volume of an ideal tetrahedron with interior angles α, β and γ is $\Lambda(\alpha)+\Lambda(\beta)+\Lambda(\gamma)$, where Λ is the Lobachevsky function. Milnor [**Mil82**] has a conjecture that specifies precisely the rational relations between values of Λ.

(7) For a more in-depth discussion of this and related problems, see [**GMM10**, Problem 10.34] and [**Neu98**].

Scribed by: M. Lackenby

PROBLEM 3.3. *Does every cusped hyperbolic 3-manifold have a geometric ideal triangulation?*

REMARKS.

(1) A *geometric ideal tetrahedron* is the convex hull of any four points on the 3-sphere at infinity of $\mathbb{H}^3$ that do not all lie on a circle. Topologically, it is a tetrahedron with its vertices removed. A *geometric ideal triangulation* of a hyperbolic 3-manifold M is an expression of M as a union of geometric ideal tetrahedra glued along their faces. Geometric ideal triangulations are useful, for example when studying hyperbolic Dehn surgery [**BP92**]. The question is whether they always exist. This was probably first asked in print by Yoshida [**Yos96**]; see also [**WYY96**], [**PP00**].

(2) Epstein and Penner [**EP88**] showed that any cusped hyperbolic 3-manifold can be constructed by gluing geometric ideal polyhedra along their faces. Furthermore, it is known that any geometric ideal polyhedron can be subdivided into geometric ideal tetrahedra. However, it is not known if this can be done compatibly across the manifold.

(3) Luo, Schleimer, and Tillmann [**LST08**] showed that the question has a positive answer virtually, in the sense that any cusped hyperbolic 3-manifold is finitely covered by a manifold admitting a geometric ideal triangulation. On the other hand, Choi [**Cho04**] gave an example of an incomplete hyperbolic structure on the figure-eight knot complement that does not admit a geometric ideal triangulation.

Scribed by: M. Lackenby

PROBLEM 3.4 (Chen–Yang Volume Conjecture).

(a) *Prove that, for any hyperbolic* 3*-manifold* M,

$$\lim_{\text{odd}\, r\to\infty} \frac{1}{r}\log(TV(M;e^{2\pi i/r})) = \frac{1}{2\pi}\mathrm{Vol}(M).$$

(b) *Prove that, for any closed, oriented, hyperbolic* 3*-manifold* M,

$$\lim_{\text{odd}\, r\to\infty} \frac{1}{r}\log(\mathrm{WRT}(M;e^{2\pi i/r})) = \frac{1}{4\pi}(\mathrm{Vol}(M) - i\mathrm{CS}(M)).$$

REMARKS.

(1) Given a root of unity q so that q^2 is a primitive root of unity of odd order, $TV(M;q)$ denotes the corresponding Turaev–Viro invariant [**TV92**]. Similarly, $\mathrm{WRT}(M;q)$ denotes the Witten–Reshetikhin–Turaev invariant [**RT91**]. For a hyperbolic 3-manifold M, $\mathrm{Vol}(M)$ denotes the volume of M and $\mathrm{CS}(M)$ denotes the Chern–Simons invariant of M.
(2) Both forms of the conjecture were formulated by Q. Chen and T. Yang, in [**CY18**].
(3) In Part (a), M can be closed, cusped, or have totally geodesic boundary.
(4) Part (a) has been verified in many cases [**DKM25**], including for large families of 3-manifolds with cusps. A related version concerns the Witten–Reshetikhin–Turaev invariant $\mathrm{WRT}(M;q)$ where M is a 3-manifold and q is a primitive root of unity of odd order r.
(5) For a generalization of Part (b) to relative WRT invariants, see [**WY23**].

Scribed by: H. Wong

PROBLEM 3.5.

(a) *Do there exist closed non-Haken hyperbolic 3-manifolds with arbitrarily large injectivity radius?*
(b) *Does there exist a cofinal tower of regular covers of closed hyperbolic 3-manifolds where all of the manifolds in the tower are non-Haken?*
(c) *Is there a tower of hyperbolic rational homology spheres for which all fundamental groups are not left-orderable, the manifolds contain no co-orientable taut foliations, or the manifolds are L-spaces? (The three properties listed are connected via the L-space conjecture. See Problem* 3.48*)*

REMARKS.

(1) A *cofinal tower of regular covers* is a sequence of covers

$$\cdots \to M_2 \to M_1 \to M$$

such that $\bigcap \pi_1(M_i) = 1$ and $\pi_1(M_i)$ is a normal subgroup of $\pi_1(M)$ for all i.
(2) Question (a) is due to Cooper and appeared as [**Kir97**, Problem 3.58].
(3) Clearly a negative answer to part (a) implies a negative answer to (b), since the injectivity radii of the manifolds in a tower tend to infinity. Any hyperbolic 3-manifold has a cofinal tower of regular covers, and so a negative answer to (a) or (b) would give an alternative proof of the Virtual Haken Conjecture, proved by Agol [**Ago13**].

(4) F. Calegari and Dunfield [**CD06**] constructed a sequence of closed, hyperbolic rational homology 3-spheres for which their injectivity radii tend to infinity. One of the examples discussed in their paper was a tower that covers the Weeks manifold. This tower is a candidate for part (b).
(5) The manifolds in F. Calegari–Dunfield's towers are rational homology 3-spheres but not integral homology 3-spheres. Indeed, it is expected (see Problem 3.6) that there do not exist cofinal towers of regular covers consisting of hyperbolic integral homology 3-spheres.
(6) D. Calegari and Dunfield [**CD03**] showed that the fundamental group of the Weeks manifold is not left-orderable. Are the fundamental groups of the tower constructed by F. Calegari–Dunfield also not left-orderable?

Proposed for K3 by: A. Reid, P. Patel
Scribed by: M. Lackenby

PROBLEM 3.6. *Given a cofinal tower of covers* $M \leftarrow M_1 \leftarrow M_2 \leftarrow \cdots$, *is it true that the torsion subgroups* $\mathrm{Tor}(M_n)$ *of* $H_1(M_n, \mathbb{Z})$ *and the hyperbolic volumes* $\mathrm{vol}(M_n)$ *satisfy*

$$\lim_{n\to\infty} \frac{\log|\mathrm{Tor}(M_n)|}{\mathrm{vol}(M_n)} = \frac{1}{6\pi}\,?$$

REMARKS.

(1) A tower of finite covers $M \leftarrow M_1 \leftarrow M_2 \leftarrow \cdots$ is called *cofinal* if the corresponding fundamental groups satisfy $\bigcap \pi_1(M_n) = \{1\}$.
(2) Bergeron and Venkatesh conjectured that the above limit holds whenever M is a closed arithmetic 3–manifold and $\{M_n\}$ is a cofinal tower of congruence covers [**BV13**]. See also Lê for a closely related formulation [**LÔ9**]. In the non-arithmetic setting, Brock and Dunfield conjectured that the above limit holds for any cofinal tower of regular covers $\{M_n\}$ such that $b_1(M_n) = 0$ for all n [**BD15**, Conjecture 1.13].
(3) Brock and Dunfield have also compiled extensive computational data that supports both the Bergeron–Venkatesh conjecture and their extension to non-arithmetic manifolds [**BD15**, Section 4]. See also work of Şengün [**Ş11, Ş12**], where computations are given that motivated the conjecture of Bergeron and Venkatesh. The computational data suggests that the above limit may fail to hold when M is non-arithmetic and the covers M_n are allowed to have positive Betti numbers. In [**BcV16**], a more extensive conjectural framework is developed and more data is given, that might explain the distinction between the arithmetic and non-arithmetic cases.

Scribed by: D. Futer, M. Lackenby

PROBLEM 3.7. *Does every finite-volume hyperbolic 3-manifold admit a finite-sheeted cover fibering over the circle with orientable pseudo-Anosov monodromy?*

REMARKS.

(1) Agol, building on work of Wise and Kahn–Markovic, proved that every finite-volume hyperbolic 3-manifold admits a finite-sheeted covering that fibers over the circle [**Ago08, Ago13, Wis21, KM12**]. The monodromy of the fibration is pseudo-Anosov. It is said to be *orientable pseudo-Anosov* if the stable and unstable laminations are transversely orientable. An obstruction to a pseudo-Anosov being orientable is the existence of a singularity of odd order. Indeed, in the situation where there is no odd order singularity, one may pass to a finite cover of the manifold where the monodromy is orientable pseudo-Anosov [**McM13**].

(2) One rationale for this question is that orientable pseudo-Anosovs have many nice properties. For example, the dilatation of an orientable pseudo-Anosov is the root with largest modulus of the Alexander polynomial of the fibered manifold [**Thu22**]. This may be useful for understanding profinite rigidity, since Alexander polynomials can be used to analyze the profinite properties of 3-manifolds (for example [**Liu23**]); see also Problem 3.8.

Scribed by: I. Agol, M. Lackenby

PROBLEM 3.8. *If M_1 and M_2 are finite-volume hyperbolic 3-manifolds whose fundamental groups have isomorphic profinite completions, must M_1 and M_2 be isometric?*

REMARKS.

(1) The problem was stated by Reid in [**Rei15**, Question 9].

(2) Apart from its intrinsic interest, a positive answer to this problem would have some useful consequences. For example, it would provide a new solution to the homeomorphism problem for finite-volume hyperbolic 3-manifolds, since one could distinguish non-homeomorphic manifolds by finding a finite quotient of one fundamental group that is not a finite quotient of the other one.

(3) It was shown by Liu [**Liu23**] that only finitely many finite-volume hyperbolic 3-manifolds have fundamental groups that can share the same profinite completion. Furthermore, the profinite completion of $\pi_1(M)$ is known to contain quite a lot of information about M. For example, it controls not only $H_1(M)$ but also the first homology of any finite covering space. Jaikin–Zapirain [**JZ20a**] also proved that the profinite completion determines whether M fibers over the circle. However, it is not currently known whether the volume of a hyperbolic 3-manifold M is determined by the profinite completion of M.

(4) Examples of distinct compact 3-manifolds whose fundamental groups have isomorphic profinite completions were provided by Funar [**Fun13**] (torus bundles with sol geometry) and Hempel [**Hem14**] (Seifert fibered spaces). So it is reasonable to restrict this problem to hyperbolic 3-manifolds.

(5) Bridson, McReynolds, Reid, and Spitler [**BMRS20**] gave some interesting examples of hyperbolic 3-orbifolds M that are profinitely rigid in a much stronger sense. In their examples, whenever $\pi_1(M)$ has profinite completion equal to the profinite completion of some finitely generated residually finite group G, then $\pi_1(M)$ must be isomorphic to G.

Scribed by: M. Lackenby

PROBLEM 3.9. *Is being Haken a profinite invariant amongst 3-manifolds? That is, if M_1 and M_2 are 3-manifolds so that $\pi_1(M_1)$ and $\pi_1(M_2)$ have isomorphic profinite completions, and M_1 is Haken, must M_2 also be Haken?*

REMARKS.

(1) The key case to consider is if M is a hyperbolic Haken rational homology sphere and N is a closed hyperbolic 3-manifold. A positive answer to Problem 3.8 implies a positive answer to this problem.
(2) Note that, for general finitely presented residually finite groups, being a free product with amalgamation is not a profinite property [**CWLRS25**].

Scribed by: M. Lackenby, P. Patel, A. Reid

PROBLEM 3.10.

(a) *Are there infinitely many commensurability classes of arithmetic rational homology 3-spheres?*
(b) *Are there infinitely many arithmetic integral homology 3-spheres?*

REMARKS.

(1) Recall that two hyperbolic manifolds are *commensurable* if they have a common finite cover. Two lattices Γ_1 and Γ_2 in a Lie group G are *commensurable* if there is some $g \in G$ such that $\Gamma_1 \cap g\Gamma_2 g^{-1}$ has finite index in Γ_1 and in $g\Gamma_2 g^{-1}$.
(2) The general construction of arithmetic lattices is as follows. Start with a connected semisimple Lie group G with trivial center and no compact factor, and a semisimple algebraic subgroup H of $GL(n, \mathbb{R})$ defined by some polynomial equations with integer coefficients. Suppose that there is surjective homomorphism ϕ from the identity component H^o of H to G with compact kernel. Then any lattice in G that is commensurable with $\phi(H^o \cap GL(n, \mathbb{Z}))$ is *arithmetic*. When G is $SO(3,1)$, then $\mathbb{H}^3/\Gamma$ is an *arithmetic hyperbolic 3-orbifold*, and when Γ is also torsion-free, $\mathbb{H}^3/\Gamma$ is an *arithmetic hyperbolic 3-manifold*. Arithmetic hyperbolic 3-manifolds and 3-orbifolds can alternatively be defined in terms of orders in quaternion algebras. See [**MR03**] for this definition, as well as a comprehensive introduction to the subject.
(3) It is known that there are infinitely many arithmetic rational homology 3-spheres, by constructions of F. Calegari–Dunfield [**CD06**] and Boston–Ellenberg [**BE06**]. However, their constructions do not immediately give infinitely many commensurability classes.

(4) It appears that there are only two known arithmetic integral homology 3-spheres. These are the 1/2 Dehn filling on the knot 5_2 and the 3-fold cyclic branched cover of the $(-2, 3, 7)$-Pretzel knot.
(5) See [**Rei07**] for further discussion of this problem, as well as a potential route to find infinitely many arithmetic hyperbolic rational homology 3-spheres up to commensurability.
(6) When a hyperbolic 3-manifold M has cusps, it is never a rational homology 3-sphere, but in this case one can consider the homomorphism $H^1(M) \to H^1(\partial M)$ induced by inclusion. The kernel of this homomorphism is the *cuspidal cohomology* of M. It is known that there are only finitely many commensurability classes of cusped arithmetic hyperbolic 3-manifolds that contain a manifold with trivial cuspidal cohomology [**MR03**, Theorem 9.3.6]. This was used by Reid [**Rei91**] to show that the figure-eight knot is the unique arithmetic hyperbolic knot complement.
(7) An analogous question in dimension two is whether there are infinitely many commensurability classes of genus 0 hyperbolic arithmetic 2-orbifolds. This was proved to be false in [**LMR06**].

Proposed for K3 by: A. Reid
Scribed by: T. Lidman, M. Lackenby

PROBLEM 3.11. *Does ever hyperbolic knot in the 3-sphere have meridian length at most 4?*

REMARKS.

(1) Every hyperbolic knot has a maximal cusp, the boundary of which is a Euclidean torus. The *length* of a slope on the knot exterior is the length of a geodesic representative of this slope in the Euclidean torus. This is a quantity that appears frequently in the theory of Dehn surgery. It was shown by Agol [**Ago00**] and Lackenby [**Lac00**] that when M is filled along a slope with length more than 6, then the resulting manifold is hyperbolic. Hence, the meridian of any knot in the 3-sphere is known to have length at most 6. The question is whether this upper bound can be improved to 4.
(2) The problem is interesting for several reasons. First, it would show that the complements of knots in the 3-sphere have distinctive geometry. Secondly, it would also lead to improvements to many results about surgery on knots in the 3-sphere.
(3) Examples of hyperbolic knots in the 3-sphere with meridian lengths tending to 4 from below were given by Agol [**Ago00**], who asked whether it is possible to find hyperbolic knots with meridian length greater than 4. Further examples were given by Purcell [**Pur08**]. The question is known to have a positive answer for 2-bridge knots [**Ada96**] and alternating knots [**ACF$^+$06**].

(4) Note that, in this problem, it is necessary to restrict to knots rather than links. Görner [**GÏ5**] has examples of hyperbolic links in the 3-sphere where the shortest length of any slope on any boundary component is $\sqrt{21} \simeq 4.582$. It would be interesting to know whether $\sqrt{21}$ is optimal here.

Scribed by: M. Lackenby

PROBLEM 3.12.

(a) *Considering all closed, orientable, π_1-injective surfaces (possibly non-embedded) in all closed hyperbolic 3-manifolds, what is the infimum of the areas of all these surfaces?*

(b) *One can ask the same question, but considering just closed, orientable, π_1-injective surfaces of a fixed genus g.*

(c) *One can also consider only closed, orientable, embedded π_1-injective surfaces.*

(d) *Alternatively, one can consider all closed, orientable, essential surfaces in all hyperbolic 3-manifolds M, allowing M not to be closed.*

REMARKS.

(1) A totally geodesic genus-two surface in a hyperbolic 3-manifold has area 4π. A universal lower bound for orientable surfaces is 2π [**Has95**], based on unpublished work of Uhlenbeck. Hence, the answer to part (a) lies somewhere between 2π and 4π.

(2) A lower bound for the area of a closed nonorientable essential surface is π, since such a surface is double covered by an orientable one.

(3) Any closed orientable π_1-injective surface is homotopic to a least area surface S, which is immersed [**SY82, SU82**]. Its sectional curvature κ is then everywhere at most -1. Gauss-Bonnet gives that the integral of κ over the surface is $2\pi\chi(S)$. On the other hand, the area is the integral of 1 over the surface. Hence, the more negative that κ is, on average over S, the smaller the area of S.

Proposed for K3 by: J. Hass
Scribed by: J. Hass, M. Lackenby

PROBLEM 3.13. *Does every closed hyperbolic 3-manifold admit an immersed π_1-injective surface with only double points? More precisely, if M is a closed, connected, hyperbolic 3-manifold, is there a closed, connected surface Σ (other than a 2-sphere) and an immersion $f\colon \Sigma \to M$ such that $f_*\colon \pi_1(\Sigma) \to \pi_1(M)$ is injective and $|f^{-1}(\{x\})| \leqslant 2$ for all $x \in M$?*

REMARKS.

(1) This is true for Haken 3-manifolds, which by definition have embedded π_1-injective surfaces, and for small Seifert-fibered spaces with infinite fundamental group, which can easily be seen to have immersed essential tori with only double points; hence the restriction to hyperbolic manifolds (by the Geometrization Theorem).

(2) This is true for finite volume orientable connected non-compact hyperbolic 3-manifolds and for most Dehn fillings by [**CL01**], hence should be true for all but finitely many closed, hyperbolic 3-manifolds of bounded volume. The immersed surfaces produced in closed hyperbolic 3-manifolds by [**KM12**] will usually have triple points of intersection.

Scribed by: I. Agol, M. Lackenby

PROBLEM 3.14. *Can a hyperbolic knot complement in the 3-sphere contain a closed, embedded totally geodesic surface?*

REMARKS.

(1) In 1990, Menasco and Reid conjectured that the answer is no [**MR92**]. There is evidence for and against this conjecture. For example, as evidence that it may be false, there are embedded totally geodesic spanning surfaces [**AS05**]. There are closed, embedded, totally geodesic surfaces in hyperbolic link complements in the 3-sphere with just two link components [**Lei06**]. For any $\epsilon > 0$, there are closed, embedded surfaces in knot complements with principal curvatures no more than ϵ [**Lei06**]. There are hyperbolic knots in rational homology 3-spheres with closed, embedded, totally geodesic surfaces in their complement [**DeB06**].

(2) In favor of Menasco–Reid's conjecture, many known families of knots have been proved not to admit a closed, embedded, totally geodesic surface, such as alternating knots [**MR92**], knots with tunnel number 1 [**MR92**], Montesinos knots [**Oer84**], 3-bridge knots and double torus knots [**IO00**].

(3) Related questions for other types of surfaces are also open. For example, can a hyperbolic knot complement in the 3-sphere contain a totally geodesic separating surface? What about one with meridional boundary components?

Scribed by: J. Purcell, M. Lackenby

PROBLEM 3.15. *Let M be a closed hyperbolic 3-manifold with positive first Betti number.*

(a) *Which elements of $H^2(M;\mathbb{R})$ are realized as the Euler classes of taut foliations?*
(b) *Which elements of $H^2(M;\mathbb{R})$ are realized as the Euler classes of tight contact structures?*
(c) *How about universally tight contact structures?*
(d) *The question can also be asked for pseudo-Anosov flows...*
(e) *... and for quasi-geodesic flows.*
(f) *How about Euler classes of representations into* $\mathrm{Homeo}^+(S^1)$*?*

REMARKS.

(1) For any compact, orientable 3-manifold M, Thurston [**Thu86**] defined a pseudo-norm on the second homology groups $H_2(M;\mathbb{R})$ and $H_2(M,\partial M;\mathbb{R})$, now called the *Thurston norm.* The Thurston norm induces *dual norms* on $H^2(M;\mathbb{R})$ and $H^2(M,\partial M;\mathbb{R})$. The unit ball of the Thurston dual norm is a convex polyhedron with integral vertices [**Thu86**].

(2) Thurston showed that the Euler class $e(\mathcal{F})$ of any taut foliation $\mathcal{F}$ has dual norm at most one, and if $\mathcal{F}$ has any compact leaf, then the dual norm of $e(\mathcal{F})$ is equal to one [**Thu86**]. Conversely, he conjectured that any integral class in $H^2(M;\mathbb{R})$ of dual norm equal to one is the Euler class of a taut foliation on M [**Thu86**, page 129, Conjecture 3]. Thurston was aware that $e(\mathcal{F})$ satisfies the *parity condition* meaning that for every closed orientable surface S, the pairing $\langle e(\mathcal{F}),[S]\rangle$ has the same parity as $\chi(S)$. Gabai proved that every vertex of the dual unit ball is realized as the Euler class of a taut foliation [**Gab97, Yaz20**], but Gabai and Yazdi constructed counterexamples to Thurston's conjecture [**GY20, Yaz20**]. It is therefore not clear which elements of $H^2(M;\mathbb{R})$ are realized as the Euler classes of taut foliations, which is the first of the above questions.

(3) Tight contact structures also have an Euler class. Eliashberg proved that the Euler class of a tight contact structure has dual norm at most one [**Eli92**]. Every C^0 taut foliation (other than the product foliation on $S^2\times S^1$) can be C^0-approximated by a universally tight contact structure, by Eliashberg–Thurston [**ET98**], Kazez–Roberts [**KR17**], and Bowden [**Bow16**]. So, for any closed, orientable, irreducible 3-manifold M, the set of Euler classes of universally tight contact structures on M includes Euler classes of taut foliations. It is possible that every integral class in $H^2(M;\mathbb{R})$ of dual norm one and satisfying the parity condition is realized as the Euler class of a tight contact structure on M. Sivek and Yazdi [**SY23**] have shown that, in the counterexamples given by Gabai and Yazdi, the classes in $H^2(M;\mathbb{R})$ that were proved not to be realized by taut foliations *are* realized by tight contract structures.

(4) Pseudo-Anosov flows and quasi-geodesic flows also admit Euler classes, so one can ask the analogous question for them. Similarly, a representation $\pi_1(M)\to \mathrm{Homeo}^+(S^1)$ defines a circle bundle over M, and hence an Euler class. This satisfies the Milnor–Wood inequality [**Mil58, Woo71**], and hence has dual norm at most one. It does not necessarily satisfy the parity condition. It is possible that every integral class in $H^2(M;\mathbb{R})$ of dual norm one is realized by a representation to $\mathrm{Homeo}^+(S^1)$.

Proposed for K3 by: M. Yazdi
Scribed by: M. Yazdi, M. Lackenby

PROBLEM 3.16. *Does every finite-volume hyperbolic 3-manifold contain infinitely many simple closed geodesics?*

REMARKS. It seems reasonable to suppose that, in some sense, a typical closed geodesic in a finite-volume hyperbolic 3-manifold does not self-intersect, and so one would expect the question to have a positive answer. Kuhlmann [**Kuh06**] proved that the answer is positive for cusped hyperbolic 3-manifolds. Moreover, she gave geometric conditions that imply that a closed hyperbolic 3-manifold contains infinitely many simple closed geodesics. Among the first 200 closed hyperbolic 3-manifolds in the census, she was able to verify these conditions for 178 of them [**Kuh08**].

It is straightforward to prove that every closed hyperbolic 3-manifold contains at least one simple geodesic, since a shortest geodesic is always simple.

Proposed for K3 by: A. Reid
Scribed by: M. Lackenby

PROBLEM 3.17. *Let M_1 and M_2 be finite-volume hyperbolic n–manifolds. If the length spectra of M_1 and M_2 coincide, must the two manifolds be commensurable?*

REMARKS.

(1) For a hyperbolic manifold M, the *length spectrum* $\mathcal{L}(M)$ is the ordered tuple of all lengths of closed geodesics in M, listed with multiplicity. The length spectrum is closely related to the eigenvalue spectrum $\mathcal{E}(M)$ of the Laplace–Beltrami operator. Specifically, $\mathcal{L}(M)$ determines $\mathcal{E}(M)$, and $\mathcal{E}(M)$ determines $L(M)$, the set of all lengths of closed geodesics, stripped of multiplicity information.

(2) In all dimensions $n \geqslant 2$, there are well-known constructions of *isospectral twins*: pairs of hyperbolic manifolds M_1 and M_2 for which the length spectra coincide, but that are not isometric. Vignéras constructed the first isospectral twins, using arithmetic methods [**Vig80**]. Subsequently, Sunada described a flexible group-theoretic construction in which the M_i are covers of a fixed manifold or orbifold [**Sun85**]. In all known constructions of isospectral twins, the manifolds M_1 and M_2 are *commensurable*, meaning they share a common finite-sheeted cover. This prompted Reid to pose Problem 3.17 [**Rei14**, Question 2.1].

(3) The question in this problem is known to have a positive answer for arithmetic manifolds in dimensions $n \not\equiv 1 \mod 4$. See Reid [**Rei92**]; Chinburg–Hamilton–Long–Reid [**CHLR08**]; and Prasad–Rapinchuk [**PR09, PR15**]. Outside the arithmetic setting, Futer and Millichap have constructed incommensurable pairs M_1 and M_2 that share an arbitrarily large finite portion of their length spectra, up to a cutoff that grows linearly with volume [**FM17**].

Proposed for K3 by: A. Reid
Scribed by: D. Futer

PROBLEM 3.18. *Is there a closed hyperbolic 3-manifold that is foliated with minimal leaves?*

REMARKS. This problem was raised by Thurston and Uhlenbeck around 1980; see [**Uhl83**, Page 154]. Each leaf of such a foliation is an area minimizing surface, that is, any compact subsurface of a leaf minimizes area among all homologous surfaces with the same boundary. Taut foliations can always be made minimal in some (but perhaps not the hyperbolic) Riemannian metric [**Sul79a**, **Has86**]. Hass and Thurston, at the 1984 Durham Symposium on Kleinian groups, 3-Manifolds, and Hyperbolic Geometry, gave examples of foliations in a hyperbolic manifold that cannot be made minimal [**Has15**]. See also [**HW19**]. Additional geometric conditions on a foliation can rule out the possibility of each leaf being minimal [**WW20**].

There is no local obstruction, since there are many foliations of $\mathbb{H}^3$ by minimal planes.

Scribed by: J. Hass

PROBLEM 3.19.

(a) *Does every closed hyperbolic 3-manifold have a nowhere zero vector field whose lift to the universal cover has proper flow lines?*
(b) *Can one ensure that the leaf space of the foliation by flow lines in the universal cover is Hausdorff?*
(c) *If so, the leaf space in the universal cover is homeomorphic to a plane and one can then ask: can the induced action of $\pi_1(M)$ on a plane be compactified to an action on the closed disk?*

REMARKS.

(1) The basic form of this question is due to W. Thurston.
(2) If the answer to part (c) is "yes," it seems likely one can show the action on the boundary must be faithful, in which case there are non-examples, since the fundamental group of the Weeks manifold is known (by Calegari–Dunfield [**CD03**]) not to act faithfully on a circle.
(3) Quasigeodesic flows are proper in this sense. However it is known that there are examples of hyperbolic 3-manifolds that do not admit quasi-geodesic flows, since such flows give rise to a faithful action on a universal circle, as in the previous remark, by [**Cal06b**].

Proposed for K3 and scribed by: D. Calegari

PROBLEM 3.20. *Let M be a closed hyperbolic 3-manifold with a faithful homomorphism $\rho\colon \pi_1(M) \to \mathrm{Homeo}_+(\mathbb{R})$. Prove that M supports a dual transversely orientable, essential lamination λ.*

REMARKS.

(1) Let ρ be a faithful homomorphism as above. Suppose λ is a lamination on M, and let $\tilde{\lambda}$ be its lift to the universal cover. Then λ is said to be *dual* to ρ if there is a map from the leaf space of $\tilde{\lambda}$ to $\mathbb{R}$ that is equivariant with respect to the actions of $\pi_1(M)$. It is well known that ρ gives rise to a dual lamination that is transversely orientable; the problem asks whether there is one that is also essential.

(2) There is a faithful homomorphism ρ as above if and only if $\pi_1(M)$ is left-orderable, by [**Con59**] and [**CR16**]. In this case, the L-space Conjecture (Problem 3.48) asserts that M supports a transversely orientable taut foliation. By contrast, this problem asks merely for a transversely orientable essential lamination, but one that is also dual to ρ.
(3) Morgan and Shalen [**MS88**] nearly solved the analogous problem where $\pi_1(M)$ acts isometrically on an $\mathbb{R}$ tree. They showed that M supports an incompressible measured lamination, but they did not show that this lamination is dual to ρ in the corresponding sense.
(4) There is a faithful homomorphism $\rho\colon \pi_1(M) \to \mathrm{Homeo}_+(\mathbb{R})$ if and only if there is one with nontrivial image, by [**BRW05**].

Proposed for K3 by: D. Gabai
Scribed by: D. Gabai, M. Lackenby

PROBLEM 3.21. *In this problem, all 3-manifolds are orientable, while all flows are considered up to orbit equivalence and are assumed to be transitive.*

(a) *Are there only finitely many Anosov flows on a given closed 3-manifold?*
(b) *Are there only finitely many pseudo-Anosov flows on a given closed 3-manifold?*
(c) *Are there only finitely many quasigeodesic pseudo-Anosov flows on a given closed 3-manifold?*
(d) *Are there only finitely many pseudo-Anosov flows without perfect fits on a given closed 3-manifold?*
(e) *Are there only finitely many veering triangulations on a given finite volume cusped hyperbolic 3-manifold?*

REMARKS.

(1) Two flows on a 3-manifold are *orbit equivalent* if there is an orientation-preserving homeomorphism of the manifold sending orbits of one flow to those of the other. A flow is *transitive* if it has a dense orbit; every pseudo-Anosov flow on a closed, atoroidal (in particular, hyperbolic) 3-manifold is automatically transitive [**Mos92**].
(2) Note that (an affirmative answer to) (b) implies (c) implies [**Fen16**] (d), and also that (b) implies (a).
(3) Barthelme, Mann, and Bowden have made partial progress on (a), showing that there are finitely many contact Anosov flows on a closed 3-manifold [**BM24b**]. Towards (b), Zung has shown that there are only finitely many transitive pseudo-Anosov flows with *positive Birkhoff sections* on a given rational homology 3-sphere [**Zun25**]; see also [**BSZ25**].
(4) A *veering triangulation* on a 3-manifold is an ideal triangulation with a particular combinatorial structure. These were first introduced by Agol on punctured mapping tori of pseudo-Anosov surface homeomorphisms [**Ago11**]. Agol and Gueritaud later showed more generally that if a 3-manifold admits a pseudo-Anosov flow without perfect fits, then there is a canonical veering triangulation in the complement of the singular orbits (see [**LMT23**]). Frankel, Schleimer, and Segerman have outlined an

inverse to this construction, which should provide a one-to-one correspondence between veering triangulations and pseudo-Anosov flows without perfect fits; see [**FSS22**], [**SS24a**]. Tsang independently proved such a correspondence in his thesis [**Tsa23**] via a different approach based on work in [**AT24**].

(5) Given the correspondence between pseudo-Anosov flows and veering triangulations, one can show that an affirmative answer to (e) would imply an affirmative answer to (d).

Proposed for K3 by: I. Agol
Scribed by: J. Baldwin, M. Lackenby, J. Zung

PROBLEM 3.22. *Let $G = \pi_1(M)$ be the fundamental group of a finite volume hyperbolic 3–manifold M. What is the regularity of the smoothest (virtual) action of G on S^1? On the interval?*

REMARKS.

(1) Not all hyperbolic 3–manifold groups can act faithfully on the interval, as this is equivalent to left orderability. Calegari–Dunfield showed that the Weeks manifold group cannot act faithfully on the interval [**CD03**], and having finite first homology is an obstruction for a finitely generated group to act by diffeomorphisms on the interval by Thurston Stability [**Thu74b**].

(2) Every hyperbolic 3–manifold group has a finite index subgroup that embeds into a right-angled Artin group and therefore admits a faithful C^1 action on the circle by [**FF03, Jor12**]. It is unclear whether higher levels of regularity can be achieved, or whether the existence of such an action persists in the ambient supergroup; moreover, it is interesting to consider whether or not the C^1 actions of finite index subgroups are conjugate to higher regularity actions, or if there are other actions of higher regularity not arising from right-angled Artin groups.

(3) One can ask similar questions about other geometric 3–manifold groups, or indeed general 3–manifold groups. In some cases, the answers are known or partially known (e.g. Euclidean manifold groups virtually admit C^∞ actions whereas nilmanifold fundamental groups cannot admit faithful C^2 actions).

(4) Some nonexistence results are known for actions of fibered 3–manifold groups which lie in a sufficiently small neighborhood of the identity; [**BKKT20**].

(5) Actions of fundamental groups on the interval and the circle are closely related to further topological structure that a 3–manifold may have, such as certain taut foliations, Anosov flows, and so on [**CR16**]. Are the actions arising from these kinds of structures conjugate to smoother actions?

Proposed for K3 and scribed by: T. Koberda

PROBLEM 3.23. *What is the Margulis constant in dimension 3? Is it realized uniquely by the Weeks manifold W, where $\mu(W) = 0.77442\ldots$?*

REMARKS.

(1) The fundamental *Margulis lemma* in dimension 3 asserts that there is an $\epsilon > 0$ so that for any discrete subgroup Γ of $\text{Isom}(\mathbb{H}^3)$, if $f, g \in \Gamma$ and each move some point $x \in \mathbb{H}^3$ less than ϵ, then the subgroup generated by f and g has an abelian subgroup of finite index. If $\Gamma = \pi_1(N)$ for N a complete, orientable hyperbolic 3-manifold, then $\langle f, g\rangle$ is abelian. The Margulis lemma is the foundation of the thick-thin decomposition of hyperbolic 3-manifolds; see [**Thu97**, §5]. For a given Γ, call the supremum $\mu(\Gamma)$ of such ϵ the *Margulis number* of Γ. For a class of discrete subgroups of $\text{Isom}(\mathbb{H}^3)$ one can define the *Margulis constant* as the infimum of the $\mu(\Gamma)$ over all groups Γ in that class. The problem asks for the *Margulis constant* μ_3, the infimum over the fundamental groups of all complete, finite volume, orientable hyperbolic 3-manifolds.

(2) There are many partial results in this direction, e.g. [**Mey87**], [**Sha11**], [**Sha13**], [**CS92**], [**CS12b**], [**FPS22**].

Scribed by: D. Gabai

PROBLEM 3.24.

(a) (*Cannon Conjecture*) *If G is a finitely presented, Gromov hyperbolic group with space at infinity equal to the 2-sphere, must G be a cocompact Kleinian group?*

(b) *More generally, if G is a Gromov hyperbolic group whose boundary is planar, is G virtually a convex cocompact Kleinian group?*

REMARKS.

(1) The first of these questions is Cannon's conjecture [**Can94, CS98**].

(2) Recall that a Gromov hyperbolic group G has a space at infinity $\partial_\infty G$, which is the space of geodesic rays based at some basepoint, up to a suitable notion of equivalence. A group is *Kleinian* if it acts properly discontinuously on $\mathbb{H}^3$. Cocompact Kleinian groups have, as a finite-index subgroup, the fundamental group of a closed hyperbolic 3-manifold.

(3) It was shown by Bestvina and Mess [**BM91**] that a Gromov hyperbolic group is a PD3-group if and only if it is torsion free and has space at infinity the 2-sphere. Hence, Cannon's conjecture is equivalent to Problem 3.38 restricted to PD3-groups that are Gromov hyperbolic.

(4) Any group G acts on itself by isometries, and in the case of Gromov hyperbolic groups, this extends to an action of $\partial_\infty G$. So, in the setting of this problem, G comes equipped with an action on the 2-sphere, and the question is whether one can identify the 2-sphere with the Riemann sphere in such a way that the action is by Möbius transformations. A good deal of work has been done on the analytic quality of the action of G on $\partial_\infty G$ (see, for example, [**BK05, BK02b, BK02a**]).

(5) Markovic has shown that Cannon's conjecture holds for a group G if it has, in a certain sense, enough surface subgroups [**Mar13**]. More precisely, the conjecture is true if every pair of points in $\partial_\infty G$ are separated by the circle at infinity of a quasi-convex surface subgroup. By work of Kahn and Markovic [**KM12**], Kleinian groups always satisfy this hypothesis.

(6) A higher-dimensional version of Cannon's conjecture is known to be true, by work of Bartels, Lück and Weinberger [**BLW10**]: if G is a Gromov hyperbolic group whose boundary is homeomorphic to S^{n-1} with $n \geqslant 7$, then G is the fundamental group of an aspherical closed n-dimensional manifold.

(7) Question (b) in the case where the boundary is homeomorphic to the limit set of a convex cocompact Kleinian group was asked by Walsh in [**DHM15**]. When ∂G is homeomorphic to S^1 the answer to (b) is yes; this is due to Tukia [**Tuk88**], Gabai [**Gab92**] and Casson–Jungreis [**CJ94**]. When ∂G contains no Sierpinski carpet and G has no 2-torsion, the answer to (b) is again yes, due to Haissinsky [**Haï15**]. Question (b) in the case where ∂G is a Sierpinski carpet is a conjecture of Kapovich–Kleiner [**KK00**].

Scribed by: M. Lackenby

PROBLEM 3.25 (Bending Conjecture).

(a) *Is a quasi-Fuchsian group determined by the hyperbolic metric on the boundary of its convex core?*

(b) *Is a quasi-Fuchsian group with parabolics, other than a Fuchsian group, determined by the bending measure on the boundary of its convex core?*

REMARKS. Let S be a closed surface (possibly with punctures) and let $\Gamma = \pi_1(S)$. Consider the space QH(S) of quasi-Fuchsian representations, which corresponds to quasi-conformal deformations of Fuchsian representations, that is to the interior points of the space AH(S) of hyperbolic structures on $S \times \mathbb{R}$ (such that the punctures of S are mapped to parabolic elements). Quasi-Fuchsian groups are important examples of Kleinian groups with infinite covolume.

Let $\rho \in \mathrm{QH}(S)$ and let $\Gamma_\rho = \rho(\Gamma)$. The manifold $M_\rho := \mathbb{H}^3/\rho(\Gamma)$ is homeomorphic to $S \times \mathbb{R}$. The limit set Λ_ρ is a Jordan curve and its complement $\Omega_\rho = \mathbb{C}\mathrm{P}^1 \backslash \Lambda_\rho$ is a disjoint union of two topological disks: $\Omega_\rho = \Omega_\rho^+ \sqcup \Omega_\rho^-$. Bers' Simultaneous Uniformization Theorem proves that QH(S) can be parametrized by the conformal structures of Ω_ρ/Γ_ρ, that is by two copies of Teichmüller spaces $\mathcal{T}(S)$.

The manifold M_ρ has a convex core $C(\rho)$, which corresponds to the smallest convex subset of M_ρ such that the inclusion is an homotopy equivalence. This convex core can be defined as the quotient by ρ of the convex hull of the limit set of ρ. In particular, provided ρ is not Fuchsian, $C(\rho)$ is homeomorphic to $S \times [-1, 1]$. The boundary of $C(\rho)$ is the disjoint union of two embedded pleated surfaces, which are surfaces totally geodesic almost everywhere, and bent along a measured geodesic lamination. In particular, from the pleated surfaces $\partial C(\rho)^+$ and $\partial C(\rho)^-$ one can consider the induced metric (that is, a point in $\mathcal{T}(S)$) and the amount of bending, which is quantified by the bending laminations (that is, a point in the space $\mathcal{ML}(S)$

of measured laminations on S). This defines two maps:

$$\mu\colon \mathrm{QH}(S) \to \mathcal{T}(S) \times \mathcal{T}(S),$$

and

$$\beta\colon \mathrm{QH}(S)\backslash\mathcal{F}(S) \to \mathcal{ML}(S) \times \mathcal{ML}(S),$$

where $\mathcal{F}(S)$ is the space of Fuchsian representations. W. Thurston conjectured that the maps μ and β are homeomorphisms onto their images. As steps in proving Thurston's conjecture, Sullivan [**Sul85**] proved that the map μ is surjective, and Bonahon and Otal [**BO04**] described the image of the map β, which consists of pairs of laminations on S that fill and such that the measure on each isolated leaf is less than π. Dular and Schlenker [**DS24b**] in 2024 showed the injectivity of the map β when S has no punctures. Unfortunately, their proof does not show that quasi-Fuchsian manifolds are infinitesimally rigid with respect to the measured bending lamination on the boundary of their convex core. As Bonahon [**Bon96**] proved, this infinitesimal rigidity is equivalent to the infinitesimal rigidity with respect to the induced metric on the boundary of the convex core.

We can then state the following still open questions. The first two questions are (a) and (b) above:

(i) When S is closed, is the map μ a homeomorphism?
(ii) When S has punctures, are the maps μ and β homeomorphisms?
(iii) Are the maps μ and β infinitesimally rigid?

Partial results for the map β are due to Series [**Ser06**] in the case of the once-punctured torus $S_{1,1}$, Bonahon [**Bon05**] in a neighborhood of the Fuchsian locus and Hodgson-Kerckhoff [**HK98**] for quasi-Fuchsian manifolds bent along rational laminations. (Gueritaud [**Gué09**] gave an alternative and independent proof of Series' result.)

Anti-de Sitter geometry is a Lorentzian analogue of hyperbolic geometry and the space of globally hyperbolic maximal compact (GHMC) AdS manifolds on $S\times\mathbb{R}$ has many similarities with the space of quasi-Fuchsian manifolds discussed above. In particular, the conjectures discussed above have an analogue also in this context. In [**BDMS21**] you can see a discussion for the universal version of the same problem both in hyperbolic and anti-de Sitter space.

Scribed by: S. Maloni

PROBLEM 3.26. *Let M be a finite volume hyperbolic 3-manifold, and let M' be a minimal index finite cover of M such that $\pi_1(M')$ embeds in a right-angled Artin group. Can one obtain good control over the complexity of the smallest right-angled Artin group containing $\pi_1(M')$ from the data of M?*

REMARKS.

(1) Work of Agol and Wise [**Ago13, Wis21**] established that there is always such a finite cover M'.
(2) The complexity of a right-angled Artin group could be interpreted in many different ways, including:
    (i) Number of vertices in the defining graph;
    (ii) Cohomological dimension;

(iii) Chromatic number of the defining graph;
(iv) Diameter of the defining graph.

(3) This question is related to, but different from questions such as finding the minimal index of a subgroup contained in a right-angled Artin group, the minimal index Haken cover, etc.

Proposed for K3 and scribed by: T. Koberda

3.2. Algorithmic 3-manifold topology

Computations have long played a central role in 3-manifold topology, and so the questions of what information can be computed and how fast computations can be done have been of substantial interest. In this section, we present six problems along these lines. Problems 3.27 and 3.30 are about computational complexity. Problems 3.31 and 3.32 are decidability questions, asking if there are any algorithms to determine orderability of π_1 or isotopy of surfaces. Problems 3.28 and 3.29 are not phrased as computational complexity problems, but ask for effective versions of relating triangulations by Pachner moves and the virtual b_1 conjecture. Some problems in other sections, including (parts of) Problems 3.44, 3.56, 3.60, and 3.63, are also decidability questions.

Recall from Section 1.11 that a decision problem (a problem with a yes-or-no answer) is in NP if, whenever there is a positive answer, there is a certificate that proves this that can be verified in polynomial time. A decision problem is in co-NP if its negation is in NP, that is, if there is a polynomial time certificate for a negative answer. When the required output is more complicated than a yes/no answer (for example, when the output is a hyperbolic structure), then this is a *function problem.* When the answer to a function problem can be certified in polynomial time, the problem lies in the class FNP.

PROBLEM 3.27.

(a) *What is the computational complexity of the homeomorphism problem for compact, orientable 3-manifolds?*
(b) *Is there a polynomial-time algorithm to recognize the 3-sphere?*
(c) *What is the computational complexity of the recognition problem for other 3-manifolds?*

REMARKS.

(1) The *homeomorphism problem* for 3-manifolds is as follows. One is given triangulations of two 3-manifolds, M and N, and the problem asks whether N is homeomorphic to M.
(2) For a fixed 3-manifold M, its *recognition problem* is as follows. One is given a triangulation of a 3-manifold N, and the problem asks whether N is homeomorphic to M. It is a restricted version of the homeomorphism problem, where one of the manifolds is fixed.
(3) The homeomorphism problem for compact orientable 3-manifolds has been solved [**Kup19a**], [**SS14**]. However, there is currently a very large gap between known upper bounds on its complexity and known lower bounds.

If we are given two 3-manifolds triangulated with t_1 and t_2 tetrahedra, the fastest known running time for an algorithm to determine whether they are homeomorphic is

$$2^{2^{\cdots^{t_1+t_2}}}$$

where the tower of exponentials has some fixed (but currently unknown) height [**Kup19a**]. An explicit bound was given by Scull [**Scu21**] in the case where the manifolds are known to be hyperbolic (but the hyperbolic structure is not part of the input).

(4) The best known lower bound on the computational complexity of the homeomorphism problem is rather weak. Lackenby [**Lac17b**] showed the homeomorphism problem is at least as hard as the graph isomorphism problem. The graph isomorphism problem is widely believed not be solvable in polynomial time (partly because of [**Sch88**]). However, it is solvable in quasi-polynomial time [**Bab16**].
(5) The recognition problem is significantly more tractable than the homeomorphism problem. For example, in the case of the 3-sphere, the recognition problem is known to lie in NP [**Sch11**], [**Iva08**], and co-NP, assuming the Generalized Riemann Hypothesis, [**Zen18**]. The first solution to 3-sphere recognition was given by Rubinstein [**Rub95, Rub97**] and Thompson [**Tho94**].
(6) There are now many manifolds for which the recognition problem is known to be in NP, including the solid torus [**Iva08**], a genus g handlebody [**Iva08**], and the product of a compact orientable surface and an interval [**Lac21a**]. None of these problems is known to be solvable in polynomial time, however.
(7) An efficient solution to 3-sphere recognition would give an efficient solution to the unknot recognition problem (Problem 1.76). This is because one could perform ± 1 Dehn surgery on the knot and, by work of Gordon and Luecke [**GL89**], the result is the 3-sphere if and only if the given knot is the unknot. (This assumes that the knot is given by means of a diagram, rather than via a triangulation of its exterior, so that it is possible to build a triangulation of the filled-in manifold efficiently.)

Scribed by: M. Lackenby

PROBLEM 3.28. *How many Pachner moves are needed to pass between two triangulations of a compact 3-manifold?*

REMARKS.

(1) Pachner [**Pac91**] proved that any two combinatorial triangulations of a compact piecewise-linear n-manifold differ by a finite sequence of modifications that are now called *Pachner moves*. In the case of closed n-manifolds, these take the following form: pick a subcomplex of the triangulation that is combinatorially isomorphic to a union B of facets in the boundary of an $(n+1)$-simplex Δ^{n+1}, and replace the copy of B by $\mathrm{cl}(\partial\Delta^{n+1}\backslash B)$. When $n = 3$, there are four types of such move: a $(1,4)$-move, which replaces a single tetrahedron with four tetrahedra arranged around a new vertex;

a $(4,1)$-move, which is the reverse of this procedure; a $(2,3)$-move, which replaces two tetrahedra that share a common facet with three tetrahedra arranged around a new edge; and a $(3,2)$-move, which is the reverse of the above move. For n-manifolds with boundary, there is a further type of Pachner move, which involves attaching an n-simplex onto a union of facets in the boundary of the manifold, or the reverse of this.

(2) One can ask this question for general compact 3-manifolds, or for a fixed 3-manifold, or for a class of 3-manifolds (such as all Seifert fibered spaces).

(3) The question can be formulated in terms of the Pachner function, which is defined as follows. For a compact 3-manifold M and positive integers t_1 and t_2, consider all pairs of triangulations of M with t_1 and t_2 tetrahedra and let $P_M(t_1, t_2)$ be the maximal number of moves needed to pass between any two such triangulations. Let $P(t_1, t_2)$ be the maximum of $P_M(t_1, t_2)$ over all compact 3-manifolds M. The question asks how $P(t_1, t_2)$ grows as a function of t_1 and t_2.

(4) Any computable upper bound on P leads to a solution to the homeomorphism problem for compact 3-manifolds, as follows. Given two triangulations of two 3-manifolds M_1 and M_2 with t_1 and t_2 tetrahedra, the task is to decide whether they are homeomorphic. Pachner's theorem [**Pac91**] states that if they are homeomorphic, then the triangulations differ by a sequence of Pachner moves, with length at most $P(t_1, t_2)$. Thus if P has a computable upper bound, then one need only try all sequences of Pachner moves with at most this length, starting from one triangulation. If none of the resulting triangulations is combinatorially isomorphic to the second triangulation, then one can deduce that the manifolds are not homeomorphic.

(5) Mijatovic [**Mij05a**] proved that for all knot exteriors M, $P_M(t_1, t_2)$ is at most

$$2^{2^{\cdots^{t_1+t_2}}}$$

where the height of the tower is $c^{t_1} + c^{t_2}$ and where $c = 2^{200}$. The same bound holds for all 'fibre-free' Haken 3-manifolds [**Mij05b**]. When M is the 3-sphere, Mijatovic [**Mij03**] and King [**Kin01**] found upper bounds on P_M of exponential type (more precisely, of the form $k^{{t_1}^2} + k^{{t_2}^2}$ for some constant k). However, an explicit bound is not known for P in general, or even for P_M for a general hyperbolic 3-manifold M.

(6) Very few nontrivial lower bounds on P are known (for example [**Bur11**] and [**JRST20**, Remark 15]).

Scribed by: M. Lackenby

PROBLEM 3.29.

(a) *Given a closed hyperbolic 3-manifold M, can one find an explicit bound on the degree of a finite cover $\widetilde{M}$ having $b_1(\widetilde{M}) > 0$?*

(b) *For example, is there such a finite cover with degree at most a polynomial function of the number of tetrahedra in some triangulation of M?*

(c) *Similarly, can one find an explicit upper bound on the degree of a finite cover $\widetilde{M}'$ that fibers over the circle?*

REMARKS.

(1) Agol and Wise proved that M has a finite cover $\widetilde{M}$ with $b_1(\widetilde{M}) > 0$, and indeed a finite cover $\widetilde{M}'$ that fibers over the circle [**Ago13, Wis21**]. However, the proof is non-constructive, and it seems hard to extract an upper bound on the covering degree.
(2) Given M, one can compute the degrees of covers $\widetilde{M}$ and $\widetilde{M}'$ as above, since one can start building all the finite covers of M, ordered by covering degree, and for each one, one can determine whether $b_1 > 0$ or whether it fibers over the circle.
(3) Agol showed that an irreducible 3-manifold whose fundamental group satisfies a certain group-theoretic property called RFRS (residually finite $\mathbb{Q}$-solvable) is virtually fibered [**Ago08**]. One approach to the above problem would be to find an explicit finite cover with RFRS fundamental group, at which point it seems more likely that one could get good bounds needed for a fibered cover. Currently, explicit towers of RFRS covers are only known in a few specific examples, including for certain Bianchi groups due to Agol–Stover [**AS23**].

Scribed by: M. Lackenby

PROBLEM 3.30.

(a) *What is the computational complexity of determining whether a compact 3-manifold admits a hyperbolic structure?*
(b) *If a compact 3-manifold does admit a hyperbolic structure, what is the computational complexity of finding it?*
(c) *In particular, do these problems lie in the complexity classes NP and FNP respectively?*

REMARKS.

(1) The most efficient known algorithm to find a hyperbolic structure on a 3-manifold if one exists is due to Scull [**Scu21**], based on work of Kuperberg [**Kup19a**]. If the 3-manifold is given as a triangulation with t tetrahedra, it runs in time

$$2^{2^{t^{O(t)}}}.$$

However, in practice, SnapPy [**CDGW**] is good at finding hyperbolic structures if they exist, and hence, it seems quite plausible that hyperbolicity can be efficiently certified, as the question proposes.
(2) The problem of deciding whether a 3-manifold is hyperbolic, without actually finding a hyperbolic structure, may be much easier. To achieve this, one might establish the existence of an alternative structure that is equivalent to hyperbolicity. For example, one might show that its fundamental group is infinite and has no $\mathbb{Z} \times \mathbb{Z}$ subgroup. Alternatively, one might give a metric triangulation that is CAT(-1). Another possibility is to exhibit

a finite cover that fibers over the circle with pseudo-Anosov monodromy, which exists by work of Agol [**Ago13**]. However, none of these approaches seems straightforward.

(3) The problem of deciding hyperbolicity is known to be in the complexity class co-NP assuming the Generalized Riemann Hypothesis, by work of Hass and Kuperberg [**HK12a**]. In the case of manifolds with nonempty toroidal boundary, this was established by Haraway and Hoffman [**HIH22**], without requiring GRH. Hence, assuming GRH, the problem of deciding whether a compact 3-manifold is hyperbolic runs in exponential time.

Scribed by: M. Lackenby

PROBLEM 3.31. *Suppose M is a closed 3-manifold.*

(a) *Can one decide if the fundamental group of M is left-orderable?*
(b) *What is the complexity of a certificate of left-orderability, or of non-left-orderability?*

REMARKS.

(1) A nontrivial group G is *left-orderable* if it admits a left-invariant total order.
(2) There is no algorithm to determine in general, from a finite presentation of a group G, whether G is left-orderable.
(3) It is known that $\pi_1(M)$ is left-orderable if and only if there is a surjective homomorphism from $\pi_1(M)$ to a nontrivial left-orderable group. Thus, one may ask: among all such $\rho : \pi_1(M) \to G$, what is the least complexity of a machine that will recognize exactly the elements $a \in \pi_1(M)$ (expressed as the canonical representative in some regular prefix-closed bijective language of geodesics if one exists, say) for which $\rho(a) > \mathrm{id}$? For example, if $b_1(M) > 0$ one may take $G = \mathbb{Z}$ and $\rho : \pi_1(M) \to G$ any nontrivial element of $H^1(M)$, and then the recognition may be done with a counter automaton (but no simpler machine for typical M, e.g. M hyperbolic).
(4) One may also ask for the least complexity of a left order on $\pi_1(M)$ itself. This might be considerably more complicated than the least complexity of a left order on an infinite quotient.
(5) The L-space Conjecture (Problem 3.48) asserts that for an irreducible 3-manifold M the following conditions are equivalent:
    (i) M has left-orderable fundamental group;
    (ii) M admits a coorientable taut foliation;
    (iii) M is not an L-space.

    The second property is known to be algorithmically decidable for atoroidal manifolds by Agol–Li [**AL03**], who proved more generally that whether a 3-manifold has a coorientable Reebless foliation is decidable; the third is algorithmically decidable for the elementary reason that Heegaard Floer homology can be calculated. Thus, if the property of having a left-orderable group is not decidable, the L-space Conjecture is false.

(6) This problem is similar to Question 8.7 in [**Cal03**].

Proposed for K3 and scribed by: D. Calegari

PROBLEM 3.32. *Is there an algorithm to determine whether two closed, embedded surfaces in* $\mathbb{R}^3$ *are isotopic?*

REMARKS.

(1) The problem of deciding whether two *incompressible* surfaces in a compact orientable irreducible 3-manifold are isotopic is solvable, using work of Waldhausen [**Wal68b**, Proposition 5.4] (see also [**Bar25a**, Theorem 1.2]). However, surfaces in $\mathbb{R}^3$ (other than 2-spheres) are compressible.
(2) All spheres embedded in $\mathbb{R}^3$ are isotopic by the Schoenflies theorem.
(3) Any torus embedded in $\mathbb{R}^3$ is the boundary of a regular neighbourhood of a knot, and therefore the isotopy problem for tori is basically the same as the equivalence problem for knots, which has been solved by Haken [**Hak68**], Hemion [**Hem79**] and Matveev [**Mat07**].
(4) The case of genus two surfaces was recently solved by Baroni [**Bar25a**]. The challenge for higher genus surfaces is as follows. Given two closed connected surfaces embedded in $\mathbb{R}^3$, one must decide whether their exteriors are homeomorphic. If they are not, then the surfaces are not isotopic. However, if their exteriors are homeomorphic, then this is not enough to deduce that the surfaces are isotopic; one must check that there is a homeomorphism compatible with the gluing maps along the surfaces. As the exteriors might have infinite mapping class group, it is not clear that this can be checked algorithmically.

Scribed by: M. Lackenby

3.3. Heegaard splittings

A *Heegaard splitting* of a 3-manifold Y is a decomposition of Y as a union of two handlebodies, glued along their boundaries. Here are three longstanding problems about Heegaard splittings.

PROBLEM 3.33. *Let* M *and* N *be closed orientable 3-manifolds. Prove that if there is a degree-1 map* $f\colon M \to N$ *then* $g(M) \geqslant g(N)$*, where* $g(M)$ *is the Heegaard genus of* M.

REMARKS.

(1) This is a long-standing conjecture, dating back to Haken and Waldhausen [**Hak66, Wal70**]. The conjecture implies the Poincaré Conjecture: suppose a closed 3-manifold N is homotopy equivalent to S^3. Since a homotopy equivalence is a degree-1 map, the conjecture implies that $0 = g(S^3) \geqslant g(N)$ and therefore that N is S^3.

(2) A general approach to this conjecture was discussed in [**RW92, Li22**]. However, not much progress has been made. Let $M = W \cup_T V$ be an amalgamation of two manifolds W and V along an incompressible torus, with W being the exterior of a knot in a homology 3-sphere. In this case, there is a canonical degree-1 map that pinches W into a solid torus. It is proved in [**Li22**] that the Heegaard genus of the resulting 3-manifold is at most as large as the Heegaard genus of M, verifying the conjecture in a special case.

Proposed for K3 and scribed by: T. Li

PROBLEM 3.34. *Do any two genus-g Heegaard splittings of a closed, orientable 3-manifold M become equivalent after at most g stabilizations?*

REMARKS. The answer "yes" was conjectured by W. Thurston. This has been studied extensively and upper bounds on the number of necessary stabilizations are known [**Joh95**], [**RS96**], but the conjectured optimal bound of g is open. It is known that g stabilizations may be needed [**HTT09**]; see also [**Kir97**, Problem 3.89].

Scribed by: A. Thompson

PROBLEM 3.35. *Given a compact manifold M, let $r(M)$ denote the rank of its fundamental group and $g(M)$ denote its Heegaard genus.*

(a) *Does every closed orientable hyperbolic 3-manifold M with $r(M) = 2$ satisfy $g(M) = 2$?*
(b) *Is there a non-Haken manifold M such that $r(M) < g(M)$?*
(c) *Is there a constant $k > 0$ such that $g(M) \leqslant kr(M)$ for every closed orientable 3-manifold M?*
(d) *If $\{M_i\}$ is the set of congruence covers of an arithmetic hyperbolic 3-manifold M, is the infimum of $r(M_i)/\mathrm{vol}(M_i)$ zero or positive?*
(e) *Is there a knot $K \subset S^3$ such that $r(\overline{S^3 \backslash \nu(K)}) < g(\overline{S^3 \backslash \nu(K)})$?*

REMARKS.

(1) For a closed orientable 3-manifold M, $r(M)$ is always at most $g(M)$. There are examples of Seifert fibered spaces where this inequality is strict [**BZ84**], and indeed the difference between Heegaard genus and rank can be arbitrarily large for graph manifolds [**SW07**]. Problem 3.92 asks for further examples of this phenomenon.
(2) Li [**Li13**] was the first to find hyperbolic 3-manifolds where rank does not equal Heegaard genus. In fact, he showed that the difference between Heegaard genus and rank can be arbitrarily large. But all of the examples in [**Li13**] are Haken, hence part (b). However, it is not known whether the ratio of Heegaard genus to rank can be arbitrarily large, which is the content of part (c). This is Question 7 in Section 8 of [**Li13**]. Moreover, it is possible that in the case of hyperbolic 3-manifolds with small rank, the Heegaard genus has to be equal to the rank. Part (a), which was asked by Li in [**Li13**], raises this in the case of rank 2 fundamental group.

(3) The congruence covers of an arithmetic hyperbolic 3-manifold were shown by Lackenby [**Lac06**] to have the property that the ratio of their Heegaard genus to their volume is bounded away from zero. So a positive answer to part (c) would imply that the infimum in part (d) is positive. Moreover, by work of Abért and Nikolov [**AN12**], if the infimum in (d) is positive, this would yield a counterexample to the Fixed Price Conjecture [**Gab00**], an outstanding question in measured group theory.

Proposed for K3 by: T. Li
Scribed by: T. Li, M. Lackenby

3.4. Other problems on π_1

Problems about the fundamental groups of hyperbolic manifolds are in Section 3.1, and problems relating the fundamental group with Floer-theoretic invariants (through the L-space Conjecture or representation varieties) are in Section 3.6. In this section are four problems related to the fundamental group that do not fit neatly in either category.

PROBLEM 3.36. *(Simple loop conjecture) Let $f\colon F \to M$ be a 2-sided immersion of a surface into a 3-manifold such that $f_*\colon \pi_1(F) \to \pi_1(M)$ is not injective. Must some simple closed curve lie in the kernel of f_*?*

REMARKS. This is [**Kir97**, Problem 3.96], proposed by J. Hass. When a simple closed curve lies in the kernel of f_*, (F, f) is compressible to an immersion of a simpler surface. When f is an embedding, the conjecture is true, by the Loop Theorem. It is also true when M is an I-bundle over a closed surface [**Gab85**], when M is Seifert fibered [**Has87**], when M is a graph manifold [**RW98**] and when M is a sol manifold [**Zem16**].

Proposed for K3 by: J. Hass
Scribed by: M. Lackenby

PROBLEM 3.37.

(a) *Is every finitely generated 3-manifold group linear (over some field with characteristic zero)?*
(b) *If so, can one bound the dimension of a faithful linear representation?*

REMARKS.

(1) Every finitely generated 3-manifold group is the fundamental group of a compact 3-manifold (possibly with boundary) [**Sco73**]. Any compact 3-manifold is finitely covered by a compact, orientable 3-manifold. Such a cover has linear fundamental group if and only if the original manifold does. A free product of finitely many linear groups is linear. Hence, the problem reduces to the case of compact, orientable irreducible 3-manifolds.

(2) Compact, orientable manifolds are known to satisfy the Geometrization Conjecture, and hence admit decompositions into geometric pieces. The fundamental group of any hyperbolic 3-manifold admits a faithful representation into SO(3, 1), and hence is linear. Seifert fibered spaces are known to have linear fundamental groups [**AFW15**], although it does not appear to be known whether there is a universal upper bound on the dimension of such a representation. Moreover, it is not known how to obtain a faithful linear representation of the fundamental group of a 3-manifold from faithful representations of its JSJ pieces.

(3) Any closed, orientable 3-manifold with a hyperbolic piece in its JSJ decomposition is virtually special [**PW18**], by work of Przytycki–Wise and Agol–Wise [**Ago13, Wis21**], and hence its fundamental group virtually embeds in $\mathrm{SL}(m, \mathbb{Z})$ for some integer m. Similarly, any compact, orientable, aspherical 3-manifold with nonempty boundary has virtually special fundamental group [**PW18**] (see also [**AFW15**]). So, the problem of linearity is unresolved only for closed graph manifolds.

(4) In Problem 3.33 of the Kirby problem list, W. Thurston asked whether every finitely generated 3-manifold group admits a faithful representation into $\mathrm{GL}(4, \mathbb{R})$. By the above discussion, this is true for hyperbolic 3-manifolds. However, Button [**But14**] showed that the fundamental groups of some graph manifolds do not admit faithful representations into $\mathrm{GL}(4, k)$ for any field k.

(5) Apart from the intrinsic interest in this problem, linearity has various consequences. For example, finitely generated linear groups are residually finite and Hopfian. These properties are already known to hold for finitely generated 3-manifold groups [**Hem87**]. Even having a nonabelian (but possibly non-faithful) linear representation can be useful. Zentner [**Zen18**] proved that the fundamental group of every homology 3-sphere other than the 3-sphere has such a representation into $\mathrm{SL}(2, \mathbb{C})$, and used this to provide a new algorithm to recognize the 3-sphere. Representations into SU(2) are a particularly active area of research because of their relationship with instanton homology (see, e.g., [**KM04**]). See also Problem 3.52.

Scribed by: M. Lackenby

PROBLEM 3.38. *Is every PD3-group the fundamental group of a closed, aspherical 3-manifold?*

REMARKS.

(1) This is [**Kir97**, Problem 3.77(A)], due to G. P. Scott, and is very closely related to the Cannon Conjecture (Problem 3.24). The more general question whether every PDn-group is the fundamental group of a closed, aspherical n-manifold was raised by Wall [**Wal79**, Problem G2].

(2) A *Poincaré duality group G of dimension n* (a *PDn-group*) is a finitely presented group whose homology and cohomology satisfy Poincaré duality over $\mathbb{Z}[G]$ with a fundamental class in dimension n. The only known

finitely presented PDn-groups are fundamental groups of closed, aspherical n-manifolds, and every PD2-group is the fundamental group of a closed, aspherical surface [**EM80, EL83**]. However, there are PDn-groups that are not finitely presentable, for every $n \geqslant 4$ [**Dav98**].

(3) Despite its long heritage, there has been little progress on this problem. However, the theory of PD3-groups has some parallels with the theory of 3-manifold groups. Every PD3-group admits an essentially unique decomposition into 'irreducibles', and there is also a good analogue of a JSJ decomposition for PD3-groups [**KR88a, KR88b, KR89**]. The question has been resolved positively for PD3-groups with subgroup structure corresponding to mapping tori or Seifert fibred 3-manifolds. See [**Hil20**] for an extensive survey.

Scribed by: M. Lackenby

PROBLEM 3.39.

(a) *Is every finitely generated perfect group the normal closure of a single element?*
(b) *Is there an integral homology sphere whose fundamental group has weight bigger than one?*
(c) *For any n, is there an integral homology sphere whose fundamental group has weight bigger than n?*
(d) *Specifically, are the weights of the fundamental groups of Seifert fibered homology spheres $\Sigma(a_1, \ldots, a_n)$ unbounded?*

REMARKS.

(1) Part (a) is a longstanding question due to Wiegold; see [**MK14**, Problem 5.52]. Note added in proof. Chen and Lodha [**CL25**] have claimed a counterexample to the Wiegold problem.
(2) The *weight* of a group is the minimum number of elements needed to normally generate it. So, part (a) asks whether every finitely generated perfect group has weight 1.
(3) Since the fundamental group of a link complement is normally generated by a collection of meridians, the weight of the fundamental group of a 3-manifold obtained by surgery on a link is bounded above by the number of components. Thus, the weight provides a bound for the *surgery number* of a 3-manifold, and group-theoretic progress on the Wiegold question would likely be useful in tackling the similarly difficult problem of finding homology spheres with large surgery number (see Question 1.15). Conversely, topological methods for bounding surgery numbers could perhaps be useful in solving the Wiegold question.
(4) Dave Auckly suggests the following argument. Suppose Σ is a homology sphere with $\pi_1(\Sigma)$ of weight n. Then by choosing n embedded loops in $I \times \Sigma$ that normally generate the fundamental group, and performing surgery on these elements, we obtain a simply-connected cobordism from

Σ to itself whose intersection form agrees with $\#^n S^2 \times S^2$. If one can obstruct the existence of such a cobordism, it follows that $\pi_1(\Sigma)$ has weight larger than n.

Scribed by: M. Hedden

3.5. Contact topology

A *contact structure* on a closed 3-manifold Y is a 2-plane distribution ξ that is nowhere integrable in the following sense: locally, $\xi = \ker(\alpha)$ for some 1-form α so that $\alpha \wedge d\alpha$ is nonvanishing. Typically, we assume that ξ is coorientable, in which case the word "local" is not needed in the previous sentence; a 1-form α on Y so that $\alpha \wedge d\alpha$ is nonvanishing is called a *contact form*. A (coorientable) contact structure induces an orientation on Y, and if Y was oriented, we typically require that that orientation agrees with the chosen orientation on Y. Contact manifolds (Y_1, ξ_1) and (Y_2, ξ_2) are *contactomorphic* if there is a diffeomorphism $\phi : Y_1 \to Y_2$ so that $\phi^*\xi_2 = \xi_1$.

A contact form α induces a vector field R on Y, characterized by $\alpha(R) = 1$ and $\iota_R d\alpha = 0$. The vector field R is called the *Reeb vector field* of α, and its periodic orbits are *Reeb orbits*. Given a point p on a Reeb orbit, flowing along R induces a map $\xi_p \to \xi_p$, the *Poincaré return map*. A Reeb orbit is called *elliptic* if the eigenvalues of the restriction of the linearized return map to the contact structure have unit length; otherwise, the orbit is called *hyperbolic*.

An *overtwisted disk* in a contact manifold (Y, ξ) is a disk $D \subset Y$ so that the interior of D is transverse to ξ except at one point, and the boundary of D is tangent to ξ. A contact structure is *overtwisted* if it contains an overtwisted disk, and *tight* otherwise. Eliashberg showed that overtwisted contact structures are flexible (or satisfy an h-principle) [**Eli89**]. So, most of contact topology focuses on tight contact structures.

The *Giroux torsion* of a contact 3-manifold (M, ξ) is the supremum of the integers $n \geqslant 1$ for which there is a contact embedding of

$$(T^2 \times [0,1], \ker(\cos(2\pi n z)dx - \sin(2\pi n z)dy))$$

into (M, ξ), where (x, y, z) are the standard coordinates on $T^2 \times [0,1] = \mathbb{R}^2/\mathbb{Z}^2 \times [0,1]$. We say the Giroux torsion of (M, ξ) is 0 if no such embedding exists. Overtwisted contact manifolds have infinite Giroux torsion.

A contact manifold is *virtually overtwisted* if some finite cover is overtwisted, and *universally tight* if every cover (or, equivalently, if its universal cover) is tight.

Contact manifolds often appear as the boundaries of symplectic manifolds. (The boundary of a symplectic manifold need not be contact.) A symplectic 4-manifold (W, ω) with convex contact boundary (Y, ξ) is a *filling* of (Y, ξ). Symplectically fillable contact 3-manifolds are tight [**Gro85, Eli90**]. The problems mention various kinds of fillings, including Stein fillings, strong symplectic fillings, weak symplectic fillings; see, for instance, [**Etn04a**].

Contact structures on 3-manifolds can be represented combinatorially by *open books*, another word for (and perspective on) fibered links [**Win73, ET98**]. The data of an (abstract) open book consists of a compact, oriented surface S with boundary (the *page*), and a self-diffeomorphism φ of S (the *monodromy*) that restricts to the identity on ∂S. The *Giroux correspondence* asserts that open books up to *positive stabilization* are in bijection with contact 3-manifolds up to contactomorphism [**Gir02**]; see also [**BHH24, LV24**].

Problems 3.40–3.42 (and in some sense 3.43) concern the existence of tight contact structures and their geometry. Problems 3.44 and 3.45 focus on open books. Problems 3.46 and 3.47 are about the dynamics of Reeb vector fields.

PROBLEM 3.40. *Does every closed, orientable, hyperbolic 3-manifold admit a tight contact structure?*

REMARKS.

(1) There exist closed, orientable contact 3-manifolds without tight contact structures, such as the connected sum $P\#\overline{P}$ of the Poincaré homology sphere with its orientation reversed [**EH01b**]. On the other hand, if M is orientable and irreducible, and Σ is an oriented surface in M representing a nontrivial second homology class and realizing the minimal genus among such representatives, then Gabai [**Gab83**] showed that M has a taut foliation in which Σ is a leaf. So, in light of Eliashberg–Thurston's theorem [**ET98**], which produces a symplectically fillable contact structure from a taut foliation (see also [**KR17, Bow16**]), any orientable, irreducible 3-manifold with nontrivial second homology admits a tight contact structure. Moreover, Honda–Kazez–Matić [**HKM02**], and independently Colin [**Col01**], proved that toroidal manifolds admit tight contact structures. Lisca–Stipsicz's work [**LS09**] then implies that every Seifert manifold admits a tight contact structure unless it is $(2n-1)$-surgery on the torus knot $T(2, 2n+1)$. The remaining geometric manifolds are hyperbolic.

(2) Kaloti–Tosun [**KT17**] and independently Li–Liu [**LL19**] proved that there are hyperbolic rational homology spheres without fillable contact structures; it is open whether these manifolds admit tight contact structures.

(3) There has been some success in linking the Riemannian geometry of a 3-manifold with its contact geometry [**EKM12, EKM16**], and similar techniques may provide an approach to this problem.

Proposed for K3 and scribed by: J. Baldwin

PROBLEM 3.41. *Is it true that for every knot $K \subset S^3$, there is an integer n_K such that $S^3_r(K)$ admits a tight contact structure for all $r \geqslant n_K$?*

REMARKS.

(1) This appears as Conjecture 4.5 (high surgery conjecture) in the notes for Stipsicz's 2010 ICM talk [**Sti10**], but Stipsicz reports that he learned of the question from Ko Honda.

(2) This question can be viewed as a step towards the following fundamental problem in 3-dimensional contact topology: Which closed oriented 3-manifolds admit tight contact structures? See also Problem 3.40 and its remarks for further context and progress on that question.

(3) There has been some partial progress towards an affirmative answer to this question. For example, the work of Golla [**Gol15**, Proposition 6.18] (see also Mark–Tosun [**MT18**]) implies an affirmative answer for L-space knots, for which one can take $n_K = 2g(K)$ (the point being that L-space knots are fibered and strongly quasipositive and thus have Legendrian representatives satisfying $tb + |rot| = 2g_4 - 1 = 2g - 1$). Further evidence comes from deep connections between tight contact structures and taut foliations (see the section introduction). Roberts [**Rob01a, Rob01b**] proved that for any fibered hyperbolic knot K and any slope $r \in (-1, \infty)$, either r-surgery on K or r-surgery on its mirror admits a taut foliation, and hence a tight contact structure, providing evidence for an affirmative answer to Problem 3.41.

(4) An interesting variation of this problem would be whether one can replace "tight" with "fillable."

Proposed for K3 by: D. Ruberman
Scribed by: B. Tosun

PROBLEM 3.42. *Does every tight contact* 3*-manifold have finite Giroux torsion?*

REMARKS.

(1) An affirmative answer to the question would show that a contact manifold is tight if and only if it has finite torsion.

(2) Towards an affirmative answer, Gay proved in [**Gay06**] that strongly symplectically fillable contact manifolds have Giroux torsion equal to 0. Moreover, Colin showed that every neighborhood of certain incompressible tori in a universally tight contact manifold has finite Giroux torsion [**Col01**] (see also [**HKM02**]).

Proposed for K3 and scribed by: J. Etnyre

PROBLEM 3.43. *Understand how various properties of contact structures behave under different kinds of symplectic cobordism. For instance:*

(a) *Is tightness preserved under exact symplectic cobordism?*

(b) *Is weak fillability preserved under weak symplectic cobordism?*

(c) *How do the support genus and Giroux torsion behave under Legendrian surgery?*

REMARKS.

(1) Wand showed that tightness is preserved by Legendrian surgery [**Wan15**]. It follows that tightness is preserved under Stein cobordism. Stein cobordisms are examples of exact symplectic cobordisms. It is known that tightness is not in general preserved under symplectic cobordisms that are not exact. For instance, a result of Gay shows that tightness is not preserved under strong symplectic cobordism [**Gay06**]; see also [**BE13**].

(2) Many forms of symplectic fillability—including strong, exact, and Stein fillability—are preserved under the same type of cobordisms. However, it is unknown whether weak symplectic fillability is preserved under weak symplectic cobordism.
(3) The behavior of Giroux torsion and the support genus of a contact structure under Legendrian surgery is completely open, though the expectation is that Giroux torsion is nonincreasing and the support genus is nondecreasing under Legendrian surgery.

Proposed for K3 and scribed by: J. Etnyre, O. Plamenevskaya

PROBLEM 3.44. *Is there an algorithm to decide, given an open book, whether the corresponding contact 3-manifold is tight or fillable?*

REMARKS.

(1) See [**Wan15**] for work of Wand related to this problem.
(2) A similar question is whether there is an algorithm to decide, given some presentation of a 3-manifold, whether it admits a tight or fillable contact structure.
(3) There are known criteria for tightness and Stein fillability. For example, Giroux proved [**Gir02**] that a contact 3-manifold is Stein fillable if and only if it has a compatible open book whose monodromy is a product of positive Dehn twists; see also Loi–Piergallini [**LP01**]. Similarly, Honda–Kazez–Matić proved [**HKM07**] that a contact 3-manifold is tight if and only if all of its compatible open books are right-veering.

Proposed for K3 and scribed by: J. Baldwin

PROBLEM 3.45.

(a) *Are there contact 3-manifolds with support genus greater than one?*
(b) *Are there contact 3-manifolds with arbitrarily large support genus?*

REMARKS.

(1) The *support genus* of a contact 3-manifold (Y, ξ) was defined by Etnyre–Ozbagci [**EO08**] to be the minimum of

$$\{g(S) \mid (S, \varphi) \text{ in an open book supporting } (Y, \xi)\}.$$

(2) Overtwisted contact 3-manifolds have support genus zero [**Etn04b**]. However, Etnyre proved [**Etn04b**] that there are contact 3-manifolds with nonzero support genus, and introduced the problem of finding the minimal genus of a supporting open book for a given contact structure. Ozsváth–Stipsicz–Szabó gave another proof of the same using Heegaard Floer homology [**OSS05**].

(3) For candidate contact structures with large support genus, one might consider connected sums of contact manifolds with support genus one. Another potential source of examples are contact structures compatible with open books (S, φ) where S is a surface of large genus and one boundary component, and φ is a composition of many Dehn twists along a curve parallel to ∂S.

Proposed for K3 and scribed by: J. Baldwin

PROBLEM 3.46. *Let λ be a contact form on a closed 3-manifold that is not a lens space. Must the associated Reeb flow have infinitely many simple periodic orbits?*

REMARKS.

(1) On lens spaces, there are examples of contact forms with exactly two simple Reeb orbits; see [**CGHHL23**, Ex. 1.1]. Conversely, it is shown in [**CGHHL23**] that if a closed three-manifold admits a contact form with exactly two simple Reeb orbits, then it must be a lens space.
(2) By [**CGH16**], two is the minimum number of simple Reeb orbits for any contact form on a closed three-manifold. In [**CGHP19**], it was shown that for a nondegenerate contact form on a closed three-manifold, such that the associated contact structure has torsion Chern class, there are either two or infinitely many simple Reeb orbits. Later, this result was generalized by removing either the Chern class assumption [**CDR22**], or the nondegeneracy assumption [**CGHHL24**]; however, removing both assumptions simultaneously remains open, and this is equivalent to Problem 3.46, by Remark (1).
(3) Once results on infinitely many orbits are established, a next natural step is to study the growth rate of the number of orbits in terms of the period.

Proposed for K3 and scribed by: D. Cristofaro-Gardiner

PROBLEM 3.47.

(a) *Does every Reeb flow on S^3, associated to a contact form giving the standard contact structure, have an elliptic periodic orbit?*
(b) *What about if one restricts to Reeb flows on S^3 arising from boundaries of strictly convex domains in $\mathbb{R}^4$?*

REMARKS.

(1) It is known that no flow on S^3 can be Anosov.
(2) For some dynamical implications of the existence of an elliptic orbit; see [**New77**].
(3) For more about what is known the problem, including in higher dimensions, see the summary in the introduction of [**AM17**].

(4) It is a longstanding problem whether every Riemannian metric on S^2 has an elliptic closed geodesic; an affirmative answer to Part (a) would imply this. For more about the history of this problem, see [**CM24**], which in particular proves the existence of an elliptic closed geodesic for C^2-generic Riemannian metrics on S^2.

Proposed for K3 and scribed by: D. Cristofaro-Gardiner

3.6. Floer homology of 3-manifolds

Floer homology invariants of 3-manifolds come in several variants: instanton Floer homology, a relative version of Donaldson invariants for 4-manifolds, originally formulated (for homology spheres) by Floer [**Flo88**]; monopole Floer homology, a relative version of the Seiberg-Witten invariants, constructed in full generality by Kronheimer-Mrowka [**KM07**]; Ozsváth-Szabó's Heegaard Floer homology [**OS04e**]; Hutchings's embedded contact homology [**Hut02, HT07, HT09**]; and others. Both Heegaard Floer homology and embedded contact homology are holomorphic curve analogues of monopole Floer homology, inspired by the Atiyah-Floer Conjecture [**Ati88a**] and Taubes's SW=Gr theorem [**Tau00**]. In particular, for 3-manifolds, the monopole, Heegaard Floer, and embedded contact invariants are known to be isomorphic [**Tau10a, Tau10b, Tau10c, Tau10d, Tau10e, CGH17a, CGH17b, CGH17b, KLT20c, KLT20d, KLT20d, KLT20a, KLT20b**]. (For cobordisms and 4-manifolds, the question of whether these invariants agree is open; see Problem 3.58.)

The relationship between these invariants and instanton Floer homology is largely open (see Problem 3.59). By contrast, a precise relationship between the Donaldson and Seiberg-Witten 4-manifold invariants was proposed by Witten, and is called Witten's Conjecture [**Wit94**]. See [**FL18**] for progress on that conjecture.

Heegaard Floer homology was extended to knots [**OS04c, Ras03**], links [**OS08**], and sutured manifolds [**Juh06**], and also gives invariants of a contact manifolds [**OS05a**]. Inspired by this, instanton and monopole Floer homology have also been extended to sutured manifolds [**KM10**].

The basic Heegaard Floer group is denoted $HF^-(Y)$, and is a module over $\mathbb{Z}[U]$; it is the homology of a chain complex $CF^-(Y)$. Other variants $HF^+(Y)$, $HF^\infty(Y)$, $HF_{\text{red}}(Y)$, and $\widehat{HF}(Y)$ are obtained by algebraic operations on $CF^-(Y)$ and taking homology; in particular, $\widehat{HF}(Y) = H_*(CF(Y)/U\,CF(Y))$. All of these group decompose as direct sums along spin^c-structures on Y. The signs in the differential on $CF^-(Y)$ depend on a choice of orientation system. There is a canonical orientation system [**OS04d**], and problems where the coefficients are not $\mathbb{F}_2$ intend the canonical orientation system be used.

For a rational homology sphere Y, the Euler characteristic of the simplest variant of Heegaard Floer homology, $\widehat{HF}(Y)$, is the number of elements of $H_1(Y)$. So, the rank of $\widehat{HF}(Y)$ is bounded below by $|H_1(Y)|$ (when finite). Manifolds for which the rank or dimension of $\widehat{HF}(Y)$ is equal to $|H_1(Y)|$ are called *L-spaces.* (In principle, this definition depends on the ground field or ring, but at present there is no known 3-manifold Y for which $\widehat{HF}(Y)$ has torsion; see Problem (a).) This class has played an important role in many applications of Floer theory to 3-manifolds. Problems 3.48–3.53 all relate to L-spaces.

For Y a rational homology sphere, Manolescu gave a stable homotopy refinement $SWF(Y)$ of the monopole Floer homology of Y [**Man03**]. That is $SWF(Y)$, the *Seiberg–Witten Floer spectrum*, is an S^1-equivariant (or, in fact, $Pin(2)$-equivariant [**Man16b**]) CW spectrum whose equivariant homology recovers monopole Floer homology (see [**LM18a**]). The construction of $SWF(Y)$ constitutes one of the first successful instances of the Floer homotopy program initiated by Floer [**Flo89**] and Cohen–Jones–Segal [**CJS95**]. The Seiberg–Witten Floer spectrum was instrumental in Manolescu's disproof of the high-dimensional triangulation conjecture [**Man16b**]. Several extensions of SWF beyond rational homology spheres have also been given [**KM14b, KLS18, SS25**].

PROBLEM 3.48 (L-space Conjecture). *For prime rational homology 3-spheres Y, are the following equivalent?*

(a) $\pi_1(Y)$ *is left-orderable.*
(b) Y *is not an L-space.*
(c) Y *admits a coorientable taut foliation.*

REMARKS.

(1) This problem is known as the *L-space Conjecture*. The question of whether (b) is equivalent to (c) was first raised by Ozsváth–Szabó following their proof (completed by Bowden and Kazez–Roberts) that (c) implies (b); see below. This equivalence was proved for Seifert fibered rational homology spheres with orientable base (i.e., S^2) by Lisca–Stipsicz [**LS07**]. Boyer–Rolfsen–Wiest had previously established the equivalence of (a) and (c) for the same class of manifolds [**BRW05**]. Together, these results implied the L-space Conjecture for Seifert fibered rational homology spheres with orientable base, and inspired the conjecture that (a) and (b) are equivalent in general, as put forth by Boyer–Gordon–Watson in [**BGW13**]; see also [**Pet10**]. The formulation of the L-space Conjecture as stated above first appears in Juhász's survey [**Juh15**].
(2) A nontrivial group G is *left-orderable* if it admits a left-invariant total order.
(3) The only direction that is known to hold in general is that if Y admits a coorientable taut foliation, then Y is not an L-space. This was proved by Ozsváth–Szabó for C^2-foliations in [**OS04b**]; see also [**KMOS07**]. Later results of Bowden [**Bow16**] and Kazez–Roberts [**KR17**] completed the proof for C^0-foliations. If Y is additionally an integer homology sphere, then having a coorientable taut foliation implies having a left-orderable fundamental group [**BB15, CD03**]. A notable partial converse is that if $\pi_1(Y)$ is left-orderable and Y has a genus 2 Heegaard splitting, then Y has a taut foliation [**Li24**].
(4) The conjecture has now been solved for large families of 3-manifolds, including all those with Seifert or Sol geometry [**BGW13**] and all graph manifolds [**BC17b, HRRW20, Ras17**]. Consequently, it remains to prove the conjecture for 3-manifolds admitting hyperbolic pieces in their JSJ decompositions.

(5) By [**BRW05**], if $f : Y \to Z$ is a nonzero degree map between prime rational homology spheres and $\pi_1(Z)$ is left-orderable, then so is $\pi_1(Y)$. Consequently, a proof of the L-space conjecture would imply that if Y is an L-space, then so is Z. See Problem 3.62.

Scribed by: T. Lidman

PROBLEM 3.49. *Are any of the three conditions in the L-space Conjecture equivalent, for all prime rational homology 3-spheres Y, to the condition that Y admits a contact structure which is filled by a symplectic 4-manifold with $b_2^+ > 0$.*

REMARKS. In proving that the existence of a coorientable taut foliation on Y implies that Y is not an L-space, Ozsváth–Szabó construct [**OS04b**] a weak symplectic filling of a particular contact structure on Y. This filling can be chosen to have $b_2^+ > 0$. Recall that for rational homology spheres, admitting a weak symplectic filling is equivalent to admitting a strong one [**OO05**].

Scribed by: T. Lidman

PROBLEM 3.50 (The Floer Poincaré Conjecture). *If Y is an integral homology sphere that is an L-space, show that Y is S^3 or the connected sum of some copies of the Poincaré sphere (with arbitrary orientations).*

REMARKS.

(1) This is in the spirit of [**Kir97**, Problem 3.106], which asks a similar question about instanton Floer homology.
(2) The statement is already known to be true for several classes of homology spheres: Dehn surgeries on knots in S^3 by [**OS11**]; Brieskorn spheres by [**OS03b**] in combination with [**Eft09**]; and manifolds with incompressible tori by [**Eft18**].
(3) In combination with the L-space Conjecture (Problem 3.48), a solution to the problem would provide an alternative route to the Poincaré Conjecture: if Y is a homotopy 3-sphere, then $\pi_1(Y)$ is not left-orderable, hence an L-space, hence S^3 or a connected sum of Poincaré spheres (but the latter possibility contradicts simple connectivity).
(4) One can phrase an analogous conjecture for framed instanton homology, which would say that if $I^\sharp(Y) \cong \mathbb{Z}$ then $Y = S^3$. If this were true then it would imply the Poincaré Conjecture immediately (without the L-space conjecture), because trivial π_1 implies $I^\sharp(Y) \cong \mathbb{Z}$. It would also imply a positive answer to Problem 3.52. Note that $I^\sharp$ is somewhat different from $\widehat{HF}$; for instance, Bhat has announced that $I^\sharp$ of the Poincaré sphere has 2-torsion [**Bha24**].

Scribed by: C. Manolescu

Problem 3.51. *Suppose Y is a rational homology 3-sphere such that every homomorphism $\pi_1(Y) \to \mathrm{SU}(2)$ has abelian image. Does it follow that*

$$\dim_{\mathbb{C}} I^{\sharp}(Y;\mathbb{C}) = |H_1(Y;\mathbb{Z})|,$$

i.e., that Y is an instanton L-space*?*

Remarks.

(1) The conclusion follows if the corresponding set of reducible flat connections is Morse–Bott nondegenerate for the Chern–Simons functional, which is equivalent to Y being cyclically finite [**BN90, BS18b**]. This holds automatically, for instance, if Y is a homology 3-sphere. The problem is whether the conclusion follows without assuming that Y is cyclically finite.
(2) If the L-space conjecture and the conjectured isomorphism between instanton homology and Heegaard Floer homology both hold (see Problems 3.59 and 3.48), then an affirmative answer to the question in the problem would imply the following: for Y an irreducible rational homology 3-sphere, if $\pi_1(Y)$ is left-orderable, then $\pi_1(Y)$ admits an irreducible SU(2)-representation. Proving the latter would be interesting in its own right.
(3) One can ask a similar question regarding instanton Floer homology and representations from $\pi_1(Y)$ to $\mathrm{SU}(N)$ for $N > 2$; see [**DIS24**] for recent work in the $N = 3$ case.

Proposed for K3 and scribed by: J. Baldwin

Problem 3.52.

(a) *Does every closed 3-manifold M besides the 3-sphere admit a nontrivial representation $\pi_1(M) \to \mathrm{SU}(2)$?*
(b) *For which M with nonabelian fundamental group does every homomorphism $\pi_1(M) \to \mathrm{SU}(2)$ have abelian image?*

Remarks.

(1) The question in (a) is [**Kir97**, Problem 3.105(A)].
(2) The answer to (a) is *yes* if $M \not\cong S^3$ is Seifert fibered [**FS90**] or toroidal [**LPCZ23**] (see also [**BS22**]), leaving only the hyperbolic case open. It is also *yes* whenever M is Dehn surgery on a knot in S^3 [**KM04**], or the branched double cover of a knot in S^3 [**CNS16, Zen17**] (using results of Kronheimer and Mrowka [**KM10**]), or the boundary of a Stein domain with nontrivial homology [**BS18b**]. These ultimately rely on nonvanishing results for some form of instanton Floer homology, or closely related techniques.
(3) An affirmative answer to (a) would follow from two widely believed conjectures, that (i) framed instanton homology is isomorphic to the "hat" version of Heegaard Floer homology over $\mathbb{Q}$; and (ii) the only irreducible integer homology sphere L-spaces are S^3 and the Poincaré homology sphere with either orientation; see 3.59 and 3.50 for discussions of (i) and (ii).

(4) The analogue of (a) for representations $\pi_1(M) \to \mathrm{SL}(2, \mathbb{C})$ has been answered affirmatively by Zentner [**Zen18**].
(5) Regarding the question in (b), there are many closed 3-manifolds M such that $\pi_1(M)$ is nonabelian but every representation $\pi_1(M) \to \mathrm{SU}(2)$ has abelian image. This includes infinitely many graph manifolds [**Mot88**], built by gluing together pairs of torus knot exteriors. The Seifert fibered examples are classified [**SZ22**], and only a handful of hyperbolic examples are known, including the manifold known as Vol3, as reported by Dunfield. Each of these examples has nontrivial first homology.
(6) A rational homology sphere for which every representation $\pi_1(M) \to \mathrm{SU}(2)$ has abelian image is generally (and perhaps always) an instanton L-space; see Problem 3.51.

Scribed by: S. Sivek, J. Baldwin

PROBLEM 3.53. *Are all strong L-spaces branched double covers of alternating links in S^3?*

REMARKS.
(1) A rational homology 3-sphere Y is an *L-space* if
$$\operatorname{rank} \widehat{HF}(Y) = |H_1(Y)|,$$
where $|H_1(Y)|$ denotes the number of elements in $H_1(Y)$. A *strong L-space* is a rational homology 3-sphere Y which admits a Heegaard diagram $\mathcal{H}$ so that
$$\operatorname{rank} \widehat{CF}(\mathcal{H}) = |H_1(Y)|.$$
Call such a Heegaard diagram a *strong Heegaard diagram.* If $\mathcal{H}$ is strong, then the differential on $\widehat{CF}(\mathcal{H})$ vanishes
(2) The notion of a strong L-space was studied by Levine and Lewallen [**LL12**]. They proved, for example, that the Poincaré homology 3-sphere is not a strong L-space (despite being an L-space). Greene [**Gre13c**] observed that the double branched cover of a non-split alternating link is a strong L-space. The question above was first asked by Greene and Levine in [**GL16**], who proved the case of L-spaces that admit genus 2 strong Heegaard diagrams.
(3) It is open whether being a strong L-space is equivalent to admitting a Heegaard diagram $\mathcal{H}$ for which the differential on $\widehat{CF}(\mathcal{H})$ vanishes.
(4) One motivation for studying strong L-spaces is their relation to the L-space conjecture (Problem 3.48). The main result of [**LL12**] is that strong L-spaces have non left-orderable fundamental group.
(5) Usui [**Usu12**] announced a proof that if Y has a genus 2 or 3 strong Heegaard diagram, then under mild hypothesis Y is the double branched cover of an alternating link, and also is a graph manifold.
(6) Any two basepoints z, w in a strong Heegaard diagram $\mathcal{H}$ specify a Floer minimal knot (compare Problem 1.20).
(7) There are analogues of strong L-spaces in the monopole and instanton Floer settings—manifolds that admit corresponding Floer complexes with

no irreducible generators. In the monopole setting these are the *minimal L-spaces* defined by Lin and Lipnowski [**LL22**]. In the instanton setting, these are $SU(2)$-*abelian manifolds* (manifolds for which all homomorphisms from the fundamental group to $SU(2)$ have abelian image). Such manifolds are generally instanton L-spaces; see Problem 3.51.

The relationships between these three classes of manifolds—strong L-spaces, minimal L-spaces, and $SU(2)$-abelian manifolds—are not completely understood. There are branched double covers of alternating knots that are not $SU(2)$-abelian [**Zen11**] and which are likely not minimal L-spaces [**LL22**]. The Weeks manifold is a minimal L-space but not $SU(2)$-abelian, and is not the branched double cover of an alternating knot. The latter suggests that the Weeks manifold is not a strong L-space, but this is open.

A strong L-space is an instanton L-space, by [**BLY23**].

Scribed by: J. Baldwin, I. Zemke

PROBLEM 3.54.

(a) *Is there a closed 3-manifold M whose Heegaard Floer homology $\widehat{HF}(M;\mathbb{Z})$ has torsion?*
(b) *Is there a rational homology 3-sphere M for which $\widehat{HF}(M;\mathbb{Z})$ has torsion?*
(c) *Is there a knot $K \subset S^3$ whose knot Floer homology $\widehat{HFK}(S^3,K;\mathbb{Z})$ has torsion?*

REMARKS.

(1) The Heegaard Floer homology of a closed 3-manifold M with integer coefficients depends on a choice of isomorphism class of coherent orientation systems (and similarly for the knot Floer homology of links). There are $2^{b_1(M)}$ such choices in general. The Heegaard Floer group referenced in part (a) is defined using the *canonical* isomorphism class of orientation systems described in [**OS04d**]. By contrast, $\widehat{HF}(S^1 \times S^2, \mathfrak{o};\mathbb{Z}) \cong \mathbb{Z}/2\mathbb{Z}$, with respect to the non-canonical orientation system $\mathfrak{o}$. For rational homology 3-spheres the canonical choice is the only choice.
(2) Jabuka–Mark [**JM08**] proved that there is 2-torsion in $HF^+(\Sigma_g \times S^1;\mathbb{Z})$ and $HF^\infty(\Sigma_g \times S^1;\mathbb{Z})$, where Σ_g is the genus-g surface and $g \geqslant 3$, but they showed there is no torsion in $\widehat{HF}(\Sigma_g \times S^1;\mathbb{Z})$ for any g.
(3) In a related direction, Li and Ye in [**LY24**], and Bhat in [**Bha24**] announced the existence of 2-torsion in the framed instanton Floer homology $I^\#(S^3_r(K);\mathbb{Z})$ for any nontrivial knot K with $r = 1, 1/2, 1/4$, and 2-torsion in the unreduced singular instanton knot homology $I^\#(S^3,K;\mathbb{Z})$ for many knots K.

Scribed by: D. Chen, J. Baldwin

PROBLEM 3.55.

(a) *For K a nontrivial knot in S^3, does $HFK^-(K)$ always admit an $\mathbb{F}_2$-summand, as an $\mathbb{F}_2[U]$-module?*
(b) *For Y a rational homology sphere, if $HF_{\text{red}}(Y)$ is nontrivial, does $HF_{\text{red}}(Y)$ always admit an $\mathbb{F}_2$-summand?*

REMARKS.

(1) The answer to Part (a) is "yes" for fibered knots [**BVV18**] and more generally for knots with $\widehat{HFK}(K, g)$ supported in a single $\mathbb{Z}/2\mathbb{Z}$-grading [**Ni22**]. See also [**HW18**].
(2) Regarding part (b), Lin [**Lin24**] proved that if a rational homology sphere Y admits a taut foliation, then $HF_{\text{red}}(Y)$ admits an $\mathbb{F}_2$-summand. In particular, if the taut foliation part of the L-space conjecture holds, then the answer to the second question is "yes"; and this question is implicitly raised in his paper.
(3) The proofs of [**BVV18**] and [**Ni22**] use geometric perspectives (i.e., essential surfaces in knot exteriors) to deduce the existence of an $\mathbb{F}_2$-summand, and [**Lin24**] uses taut foliations. Thus, it is natural to ask: does the generator of this summand hold special geometric meaning?
(4) If $HF_{\text{red}}(Y) = \mathbb{F}_2$, then [**HKL19**] implies the $\mathbb{F}_2$-summand is either in grading d or $d-1$, where d denotes the correction term of Y.
(5) Lastly, [**AB24a**, Question 1.10] asks the following refined version of the second question: If $HF_{\text{red}}(Y)$ contains an $\mathbb{F}_2[U]/U^k$-summand, does $HF_{\text{red}}(Y)$ also contain an $\mathbb{F}_2[U]/U^\ell$-summand for all $1 \leqslant \ell < k$? The authors claim that the answer is "yes" for surgery on a knot in S^3 and large surgery on links in S^3.

Proposed for K3 and scribed by: J. Hom

PROBLEM 3.56.

(a) *For Y^3 a rational homology sphere, is the Seiberg–Witten Floer spectrum $SWF(Y)$ always a wedge of spheres?*
(b) *Is every monopole Floer homology L-space a Seiberg–Witten Floer homotopy L-space?*
(c) *Find an example of two rational homology spheres with the same monopole Floer homology but with different Seiberg–Witten Floer spectra.*
(d) *Develop an algorithm for computing Seiberg–Witten Floer spectra.*

REMARKS.

(1) There are some explicit computations of $SWF(Y)$. In certain cases, the critical points and gradient trajectories of the Chern–Simons–Dirac functional can be identified explicitly by work of Mrowka, Ozsváth, and Yu [**MOY97**]; in simple situations these suffice to determine the Floer homotopy type. (This holds for elliptic manifolds and a few families of Brieskorn spheres [**Man14**].) In [**DSS23**], $SWF(Y)$ is calculated for a general class of plumbed manifolds, including (for example) all Seifert fibered rational homology spheres with base orbifold S^2.

(2) Problem 3.56 should be interpreted differently depending on whether $SWF(Y)$ is considered as a nonequivariant spectrum, an S^1-equivariant spectrum, or a Pin(2)-equivariant spectrum.

Part (a) may be asked in the nonequivariant setting or the S^1-equivariant setting. In the nonequivariant setting, the meaning of Problem 3.56 is clear. In the S^1-equivariant setting, the simplest possible form of $SWF(Y)$ is given by the wedge of a single S^1-representation sphere $S^{n\mathbb{C}}$ with a collection of (suspensions of) free S^1-spheres S_+^{2m+1}; the problem asks if every $SWF(Y)$ is of this form.

Currently, there are no known examples of rational homology spheres whose Seiberg–Witten Floer spectrum is not a wedge of spheres in the sense of (a). All examples in [**Man03, Man14, Man16b**] and [**DSS23**] are (up to suspension) homotopy equivalent to wedges of spheres, although in some cases this is not immediately obvious. For example, as discussed in [**Man14**, Section 5.2], the Seiberg–Witten Floer spectrum of $-\Sigma(2,3,11)$ is (up to suspension) a torus. Nonequivariantly, the suspension of a torus is homotopy equivalent to $S^2 \vee S^2 \vee S^3$: suspending the torus gives a CW complex with a single 0-cell, two 2-cells, and one 3-cell; the attaching map for the 3-cell is nullhomotopic. As an S^1-space, the suspension of the torus is homotopy equivalent to $S^{\mathbb{C}} \vee \Sigma^{\mathbb{R}^2}(S_+^1)$. One can check that (a) is closed under orientation reversal and connected sums.

Part (b) may be asked in either the S^1- or Pin(2)-equivariant settings; the nonequivariant case is trivial. We say Y is an *S^1-homotopy L-space* if the S^1-Seiberg–Witten-Floer spectrum is a suspension of S^0 in each *spin*^{c}-structure. If (in addition) the Pin(2)-homotopy type of $SWF(Y)$ is a suspension of S^0 in each self-conjugate *spin*^{c}-structure, we say that Y is a *Pin(2)-homotopy L-space.*

Parts (c) and (d) may be interpreted in the nonequivariant, S^1-equivariant, or Pin(2)-equivariant settings.

Scribed by: I. Dai

PROBLEM 3.57. *Construct an S^1- or* Pin(2)*-equivariant lattice homotopy type that computes the Seiberg–Witten Floer homotopy type.*

REMARKS. Lattice homology is an invariant for plumbed 3-manifolds introduced by Némethi [**Ném05, Ném08**], based on earlier work by Ozsváth and Szabó [**OS03b**]. Zemke has announced that lattice homology is isomorphic to Heegaard Floer homology whenever Y is the boundary of a plumbing tree [**Zem25**]. Since lattice homology is completely combinatorial, this provides an effective calculation of the Heegaard Floer homology of these manifolds. In addition to computational applications for 3-manifolds (see for example [**CK14, DM19, AKS20**]), further uses of this relation include a determination of the link Floer complex of an algebraic link from its multivariable Alexander polynomial [**BLZ24**].

Many constructions in Floer theory have analogously been developed for lattice homology. For instance, one can define knot lattice homology [**OSS14a**] and construct a surgery exact sequence [**Ném11, Gre13d**]. Nevertheless, the parallel between lattice homology and Floer theory is incomplete. For instance, cobordism maps for lattice homology have not yet been defined in the literature.

In [**DSS23**], it is indeed shown that in a restricted range of cases, Némethi's construction may be modified to define a lattice spectrum that is provably equal to the Seiberg–Witten Floer spectrum of Manolescu. However, it is unclear whether this is the correct construction to coincide with the Seiberg–Witten Floer spectrum in all cases. Such a result would allow for the possible computation of interesting Seiberg–Witten Floer spectra (compare Problem 3.56). Additional avenues of investigation include the calculation of Steenrod squares, or even equivariant cohomological operations, for such manifolds. These could lead to more refined numerical invariants (see for example [**LS14**] in the case of Khovanov homology) that capture more homology cobordism information for graph manifolds than current methods involving Floer homology.

Proposed for K3 and scribed by: I. Dai

PROBLEM 3.58. *Prove that there is an isomorphism relating Heegaard Floer homology and monopole Floer homology that commutes with the cobordism maps in the two settings. In particular, prove that the mixed invariant in Heegaard Floer homology is equal to the Seiberg–Witten invariant for closed 4-manifolds.*

REMARKS. As discussed in the section introduction, the Heegaard Floer homology of a closed 3-manifold is isomorphic to its monopole Floer homology, but it is not known whether these isomorphisms respect the TQFT-like structures of the theories.

Scribed by: J. Baldwin

PROBLEM 3.59. *Prove an isomorphism relating instanton Floer homology and Heegaard Floer homology.*

REMARKS.

(1) There is some context for this question in the section introduction.
(2) Kronheimer–Mrowka conjectured that sutured instanton homology is isomorphic to sutured Heegaard Floer homology with coefficients in $\mathbb{C}$,

$$SHI(M,\gamma;\mathbb{C}) \cong SFH(M,\gamma;\mathbb{C}) \tag{4}$$

for every balanced sutured manifold (M,γ) [**KM10**]. If true, this would imply that the non-equivariant versions of the closed 3-manifold invariants and knot invariants in the two theories are also isomorphic over $\mathbb{C}$, as these are special cases of the sutured invariants. In particular, it would show that the *framed instanton homology* of a closed 3-manifold Y is isomorphic to the *hat* version of its Heegaard Floer homology with coefficients in $\mathbb{C}$,

$$I^{\sharp}(Y;\mathbb{C}) \cong \widehat{HF}(Y;\mathbb{C}). \tag{5}$$

One can interpret the problem above as asking for a proof of (4).

(3) As evidence for the above conjecture, Baldwin–Li–Ye proved that the dimension of sutured instanton homology is bounded by the number of generators in the sutured Heegaard Floer chain complex; that is,

$$\dim_{\mathbb{C}} SHI(M,\gamma) \leqslant \dim_{\mathbb{C}} SFC(\mathcal{H}),$$

where $\mathcal{H}$ is any admissible Heegaard diagram for (M,γ) [**BLY23**]. Moreover, the conjectured isomorphism (5) is known to hold for several classes of manifolds, including all Seifert fibered rational homology spheres [**ABDS22**].

(4) More directly related to Witten's conjecture, one could also ask for a relationship between the equivariant instanton theory introduced by Austin–Braam [**AB95**] and studied extensively by Miller Eismeier [**ME23**] and Daemi–Miller Eismeier [**DME22**], and the equivariant (*plus* or *minus*) versions of Heegaard Floer homology.

Scribed by: J. Baldwin

PROBLEM 3.60. *Find an algorithm to compute instanton Floer homology of closed 3-manifolds and the Donaldson invariants of closed 4-manifolds.*

REMARKS. There are algorithms to compute the Heegaard Floer invariants of knots [**MOS09**], 3-manifolds [**SW10, MOT25**], cobordisms [**LMW08, MOT25**], and the Heegaard Floer mixed invariant of closed 4-manifolds [**MOT25**]. Per Problem 3.58 and Witten's Conjecture (see the section introduction), this should be closely related to the Donaldson invariant for manifolds of simple type.

Scribed by: J. Baldwin

PROBLEM 3.61. *Is the dimension of Heegaard Floer homology invariant under genus 2 mutation?*

REMARKS.

(1) Let Σ be a closed, genus 2 surface embedded in a 3-manifold M. The *genus 2 mutant* M_Σ is the 3-manifold obtained by cutting M along Σ and regluing by the hyperelliptic involution of Σ. This operation was introduced in [**Rub87**] as an analogue for 3-manifolds of Conway mutation for knots. If $K \subset M$ is a knot and Σ is disjoint from K, then its image $K_\Sigma \subset M_\Sigma$ is called the *genus 2 mutant* of K, and any Conway mutation can be achieved by a sequence of at most two genus 2 mutations.

(2) Several invariants, like hyperbolic volume [**Rub87**] and the Alexander polynomial [**CL99**] are preserved by genus 2 mutation; see [**DGST10**] for a survey.

(3) A broad interpretation of the problem is whether

$$\dim \widehat{SFH}(M,\gamma) = \dim \widehat{SFH}(M_\Sigma,\gamma)$$

over $\mathbb{Z}/2$, for any balanced sutured manifold (M,γ) and genus 2 surface $\Sigma \subset M$. If true, this would imply that the dimensions of the Heegaard

Floer invariants $\widehat{HF}$ and $\widehat{HFK}$ for closed 3-manifolds and knots are preserved by genus 2 mutation, as these are special cases of sutured Floer homology. It would also imply that the concordance invariant τ is preserved by genus 2 mutation, and hence Conway mutation, as in [**BS21**, Proposition 1.30]. The latter is related to Problem 1.53 of [**Kir97**], which asked whether Conway mutation preserves the concordance class of a knot, and to which the answer is "no" [**Kea89, KL99**].

(4) The spin^c-graded $\widehat{HF}$ is not preserved by genus 2 mutation [**Cla17**]. Neither is the bigraded, or even the δ-graded, $\widehat{HFK}$ [**MS15**], though δ-graded knot Floer homology is preserved by Conway mutation [**Zib23**].

(5) The dimension of $\widehat{HF}$ was shown to be invariant under genus-1 mutation in [**HRW22**].

(6) Floer's original instanton homology with $\mathbb{Z}/2$ coefficients is invariant under genus 2 mutation [**Rub99a**] (the paper claims the result over $\mathbb{Z}$ but there is a gap in the proof).

Proposed for K3 and scribed by: J. Baldwin

PROBLEM 3.62. *How do Floer homological invariants behave under maps of nonzero degree? For instance, let Y and Z be closed, oriented 3-manifolds, and suppose that there is a degree-1 map $f : Y \to Z$. Is it true that*

$$\operatorname{rk} HF(Y) \geqslant \operatorname{rk} HF(Z), \tag{6}$$

where HF is the $\widehat{HF}$ or HF_{red} version of Heegaard Floer homology?

REMARKS.

(1) This question is inspired by the fact that many 3-manifold invariants, from the first Betti number to simplicial volume, exhibit monotonic behaviors under maps of nonzero degree; see [**Wan02**]. Perhaps more to the point, the order of first homology is monotonic under degree-1 maps between rational homology 3-spheres, and $\widehat{HF}$ categorifies the order of H_1 for such manifolds.

(2) If M and N are 3-manifolds and d is a nonzero integer for which there exists a degree-d map from M to N, then M is said to *d-dominate* N.

(3) The inequality (6) is known to hold in the setting of $\widehat{HF}$ for Seifert fibered homology spheres [**KL15**], and when f is pinching a homology solid torus in Y to a solid torus [**HRW24**].

(4) If (6) holds for $\widehat{HF}$ or HF_{red}, then one could prove that homotopy 3-spheres are L-spaces without using the Poincaré Conjecture, as every homotopy 3-sphere is 1-dominated by S^3.

(5) If (6) holds for HF_{red}, then it would follow that if Y is an L-space that 1-dominates Z, then Z is also an L-space. This would also follow directly from the L-space Conjecture under the added assumption that Y and Z are prime, as discussed in Problem 3.48.

(6) Gadgil [**Gad07**] proved that Y 1-dominates Z if and only if Y embeds into

$$(Z \times [0,1]) \# n(\mathbb{C}P^2 \# \overline{\mathbb{C}P}^2)$$

as a hypersurface homologous to $Z \times \{0\}$ for some n. In the case that Y embeds in the same manner into $(Z \times [0,1])\#n\overline{\mathbb{CP}}^2$, the inequality (6) follows from standard properties of Floer homology.

(7) Gadgil [**Gad07**] also proved that Y 1-dominates Z if and only if Y can be obtained by surgery on a link $L \subset Z$ such that every component of L is nullhomotopic. A special case is when Y is obtained via p/q-surgery on a nullhomotopic knot $K \subset Z$. In this case, one expects the stronger inequality

$$\operatorname{rk} \widehat{HF}(Y) \geqslant p \cdot \operatorname{rk} \widehat{HF}(Z), \tag{7}$$

as conjectured by Ni.

(8) There are some known results for maps of degree greater than one. For example, if $f : Y \to Z$ is a p^n-sheeted regular covering map between rational homology 3-spheres, for p a prime, there is a Smith-type inequality relating the Heegaard and/or monopole Floer homologies of Y and Z with $\mathbb{Z}_p$ coefficients [**LT16, LM18b, Lar19**]; see [**HLL22a**] for similar results on branched coverings.

(9) It might also be interesting to study analogous questions for knot Floer homology, sutured Floer homology, or Khovanov homology. Suppose, for example, that K and L are knots in S^3 such that there is a proper degree-1 map from the exterior of K to that of L. It is known in this case that the Alexander polynomial of K divides that of L, and the genus behaves monotonically,

$$g(K) \geqslant g(L).$$

(One can find references for these and related facts in [**BBRW16**].) Ni has therefore asked whether knot Floer homology also behaves monotonically; i.e., does the rank inequality

$$\operatorname{rk} \widehat{HFK}(S^3, K) \geqslant \operatorname{rk} \widehat{HFK}(S^3, L)$$

hold? Is there a rank inequality in each Alexander grading?

Proposed for K3 and scribed by: T. Lidman, Y. Ni

3.7. Other 3-manifold invariants

This section collects problems about invariants of 3-manifolds not coming from hyperbolic geometry or Floer theory. Problem 3.63 is about the η-invariant from index theory, Problem 3.64 asks about rationality of the Chern–Simons invariant, Problems 3.65 and 3.66 concerns skein modules of 3-manifolds, Problem 3.67 requests a categorification of the Witten-Reshetikhin-Turaev invariants, and Problem 3.68 asks about Gukov–Pei–Putrov–Vafa's $\widehat{Z}$ invariants.

PROBLEM 3.63. *Give a method for computing the η invariant for the Dirac operator, $\eta_{Dirac}(Y, \mathfrak{s})$, associated to a spin structure $\mathfrak{s}$ on a hyperbolic 3-manifold Y.*

REMARKS.

(1) The Atiyah–Patodi–Singer η-invariant [**APS75a, Goe12**] is a real number defined in terms of the spectrum of a first-order self-adjoint elliptic differential operator A on a closed odd-dimensional oriented Riemannian manifold Y; denote it by $\eta_A(Y)$. When the operator $D = \partial/\partial t + A$ on $[0,1) \times Y^3$ extends to an elliptic operator on a manifold X with boundary Y, then $\eta_A(Y)$ serves as a boundary correction term for the Atiyah-Singer formula determining the index of D (with appropriate boundary conditions.) The main geometric examples are of 'Dirac type', where D is the signature operator, Dolbeault operator, or the Dirac operator associated to a spin or spin^c structure on X. In the last case the operator A is essentially a Dirac operator on Y. The η-invariant is not a topological invariant: for the signature and Dirac operators, it varies with the metric.

For hyperbolic manifolds, Mostow rigidity gives a canonical metric and so the η-invariant is a topological invariant. The special case when Y is hyperbolic and D is the signature operator has been extensively studied, and in fact can be computed using formulas of [**MN92, Ouy97, CGHN00**]. This computation been implemented numerically in the program Snap [**GN**]. However, it seems that there has been little research into the case of the Dirac operator $\eta_{\text{Dirac}}(Y, \mathfrak{s})$ on a spin hyperbolic 3-manifold $(Y, \mathfrak{s})$.

(2) The only known examples of spin hyperbolic manifolds for which $\eta_{\text{Dirac}}(Y, \mathfrak{s})$ is known are those that admit an orientation-reversing (and spin-structure preserving, in an appropriate sense) isometry. For these, it follows that $\eta_{\text{Dirac}}(Y, \mathfrak{s}) = 0$. One might start by trying to prove cut/paste formulas analogous to [**MR90a, MR90b**] or to understand the reduction of $\eta_{\text{Dirac}}(Y, \mathfrak{s})$ modulo integers (in a suitable normalization). This is related to a version of the Chern–Simons invariant; see [**APS76**].

(3) One can pose the same problem for manifolds with other geometries. Nicolaescu used adiabatic limit techniques to compute $\eta_{\text{Dirac}}(Y, \mathfrak{s})$ when Y is a circle bundle [**Nic99**] and more generally when Y is a Seifert-fibered 3-manifold with Seifert geometry [**Nic00**].

(4) A related, and perhaps easier, problem is to calculate the reduced η-invariant $\xi_\alpha(Y) \in \mathbb{R}$, defined in [**APS75b**], for a spin 3-manifold together with a unitary representation α. The metric dependence here is less impactful, as there can be at most integer jumps along a path of metrics.

Proposed for K3 and scribed by: D. Ruberman

PROBLEM 3.64. *Let A be a flat connection on the trivial* SU(2) *bundle on a closed three-manifold M. The Chern–Simons invariant* $\mathrm{CS}(M, A)$ *is defined to be*

$$\mathrm{CS}(M, A) = \frac{1}{4\pi^2} \int_M Tr(A \wedge dA + \frac{2}{3} A \wedge A \wedge A).$$

Is $\mathrm{CS}(M, A)$ *always rational?*

REMARKS.

(1) This was asked in [**Auc94**], and implicitly in [**KK90**]. It is related to Problems 3.62, 3.63, and 3.104 in [**Kir97**].
(2) Rationality holds for all Seifert fibered spaces and graph manifolds [**FS90, Fur90, Auc94**].
(3) One potential counterexample comes from $+1/2$-surgery on the mirror of 5_2. As described in [**NST24**], numerical calculations based on [**KK90**] suggest that a certain flat connection on this manifold has irrational Chern–Simons invariant. In addition to answering this problem, a proof of irrationality for this flat connection would imply that the integer homology cobordism group is not spanned by all graph manifold homology spheres. In general, a homology sphere with an irrational Chern–Simons invariant cannot admit a homology cobordism without 3-handles to a graph manifold.
(4) On the other hand, Auckly has proposed some strategies to answer this question in the affirmative (see [**Auc94**] and [**Kir97**, Problem 3.104]). For example, he notes that if two flat connections are bordant, then the Chern–Simons invariants agree, and asks if one can show that any flat connection is bordant to one on a connected sum of lens spaces. This would then show all Chern–Simons invariants are rational.
(5) It is natural to ask about rationality of Chern–Simons invariants for flat G-connections on all 3-manifolds, where $G = \mathrm{SU}(n)$ or indeed any compact Lie group. A starting point might be to show rationality for all G when M is a graph manifold. See [**Nis98**] for some calculations of $\mathrm{SU}(n)$ Chern–Simons invariants for Seifert fibered 3-manifolds.
(6) A hyperbolic 3-manifold M has two *a priori* different Chern–Simons invariants: $\mathrm{CS}(M)$ associated to its Levi-Civita connection, and the complex Chern–Simons invariant $\mathrm{CS}(M, A_\rho)$ associated to the flat $\mathrm{PSL}(2,\mathbb{C})$ connection coming from the discrete faithful representation $\rho\colon \pi_1(M) \to \mathrm{PSL}(2,\mathbb{C})$. These are related to the hyperbolic volume by Yoshida's formula [**Yos85, KK93**]

$$\mathrm{CS}(\rho) = \frac{1}{2}\,\mathrm{CS}(M) - \frac{i}{2\pi}\,\mathrm{vol}(M).$$

Hence, for hyperbolic manifolds the rationality question here is closely tied to the rationality question for volumes discussed in Problem 3.2.

Scribed by: T. Lidman, D. Ruberman

PROBLEM 3.65. *Let $\mathcal{S}_{2,\infty}(Y)$ denote the Kauffman bracket skein module of a closed, oriented 3-manifold Y; this is a module over $R = \mathbb{Z}[A, A^{-1}]$. For Y prime and connected, does $\mathcal{S}_{2,\infty}(Y)$ having rank one, in the sense that*

$$\mathcal{S}_{2,\infty}(Y) \otimes_R \mathbb{Q}(A) \cong \mathbb{Q}(A),$$

imply that Y is S^3 or $S^1 \times S^2$?

REMARKS.

(a) This problem appears as [**DKS25**, Question 10.3].
(b) Przytycki [**Prz00**] showed that $\mathcal{S}_{2,\infty}(Y) \otimes_R \mathbb{Q}(A)$ is multiplicative under connected sum. Hence, the set of (closed, connected) 3-manifolds that satisfy $\mathcal{S}_{2,\infty}(Y) \otimes_R \mathbb{Q}(A) \cong \mathbb{Q}(A)$ is closed under connected sum. Thus, the problem can be alternatively formulated as asking whether this set consists only of S^3 and $\#^k(S^1 \times S^2)$ for $k \geqslant 1$.
(c) The skein module is expected to be isomorphic to the degree 0 part of the framed $SL(2,\mathbb{C})$ Floer homology $HP^\bullet_\#$ from [**AM20**]; see Problem 3.66. Zentner [**Zen18**] proved that every nontrivial integral homology sphere has an irreducible representation to $SL(2,\mathbb{C})$; however, this does not immediately imply that $HP^\bullet_\#$ has rank greater than one.
(d) The answer to the problem is affirmative if we add the condition that $\mathcal{S}_{2,\infty}(Y) \otimes_R \mathbb{Q}[A,A^{-1}]$ is finitely generated over $\mathbb{Q}[A,A^{-1}]$. Indeed, in that case, the character variety $\mathcal{X}(Y)$ is finite, because the skein module at $A=-1$ gives the ring of functions on $\mathcal{X}(Y)$; see [**Bul97**], [**PS00b**]. A result of Detcherry, Kalfagianni, and Sikora [**DKS25**] implies that the rank of the skein module is greater or equal than the number of $SL(2,\mathbb{C})$ representations. Thus, there must be a unique representation to $SL(2,\mathbb{C})$, so Y is an integral homology sphere, and Zentner's theorem mentioned above implies that $Y = S^3$.
(e) The answer is also affirmative under the weaker condition that $\mathcal{S}_{2,\infty}(Y) \otimes_R \mathbb{Q}[A,A^{-1}]$ is *tame*, i.e., it is a direct sum of cyclic $\mathbb{Q}[A^{\pm 1}]$-modules and does not contain any $\mathbb{Q}[A,A^{-1}]/(\phi_{2N})$ as a submodule, for some N odd, where ϕ_{2N} is the $(2N)$-th cyclotomic polynomial. For then, the results of [**DKS25**] and [**GJS23**] imply that $\mathcal{X}(Y)$ is finite, and we conclude that $Y = S^3$, just as in (d).
(f) The manifold $Y = S^1 \times S^2$ has $b_1(Y) > 0$ and $\mathcal{S}_{2,\infty}(Y)$ of rank one; see [**HP95b**]. The skein module $\mathcal{S}_{2,\infty}(Y) \otimes_R \mathbb{Q}[A,A^{-1}]$ is not tame: by [**HP95b**], it contains $\mathbb{Q}[A^{\pm 1}]/(\phi_{2N})$ as a submodule, for all odd N.
(g) The skein module of a rational homology sphere with coefficients in $\mathbb{Q}(A)$ is at least 1-dimensional, by [**DKS25**].

Scribed by: R. Detcherry, C. Manolescu

PROBLEM 3.66. *Suppose that Y is a closed, oriented 3-manifold, and let $\mathcal{S}_{2,\infty}(Y)$ denote the Kauffman bracket skein module over $R = \mathbb{Z}[A,A^{-1}]$ as in Problem* 3.65.

(a) *Is the dimension*

$$\dim_{\mathbb{Q}(A)} \mathcal{S}_{2,\infty}(Y) \otimes_R \mathbb{Q}(A)$$

equal to the dimension of the degree 0 part of Abouzaid–Manolescu's sheaf-theoretic Floer homology $HP^\bullet_\#(Y)$ from [**AM20**]*?*
(b) *When is $\mathcal{S}_{2,\infty}(Y) \otimes_R \mathbb{Q}[A,A^{-1}]$ tame, as defined in Problem* 3.65*?*

REMARKS.

(1) An affirmative answer to the first question was conjectured in [**GJS23**], which also contains the proof that

$$\mathcal{S}_{2,\infty}(Y) \otimes_R \mathbb{Q}(A)$$

is finite dimensional over $\mathbb{Q}(q)$.

(2) When $\mathcal{S}_{2,\infty}(Y) \otimes_R \mathbb{Q}[A, A^{-1}]$ is tame, the dimension

$$\dim_{\mathbb{Q}(A)} \mathcal{S}_{2,\infty}(Y) \otimes_R \mathbb{Q}(A)$$

is bounded above by the dimension of the unreduced coordinate ring of the $\mathrm{SL}(2,\mathbb{C})$ character variety of Y, and is bounded below by the number points in this character variety [**DKS25**].

(3) It is conjectured in [**DKS25**] that if Y is a small 3-manifold, then

$$\mathcal{S}_{2,\infty}(Y) \otimes_R \mathbb{Q}[A, A^{-1}]$$

is finitely generated over $\mathbb{Q}[A, A^{-1}]$, and hence tame. Positive progress on computing $\mathcal{S}_{2,\infty}(Y)$ in the case of the Cartesian product of a circle with a surface has been made by Gilmer and Masbaum [**GM19**] and Detcherry and Wolff [**DW21**]. Moreover, the skein modules of 3-manifolds obtained by surgery on the figure-eight knot and on $(2,p)$ torus knots are shown to be tame [**DKS25**].

Scribed by: H. Wong

PROBLEM 3.67. *Categorify the Witten–Reshetikhin–Turaev invariants of 3-manifolds.*

REMARKS. The Witten–Reshetikhin–Turaev (WRT) invariants [**Wit89**, **RT91**] are maps from the set of roots of unity to $\mathbb{C}$. They are an analogue of the Jones polynomial for 3-manifolds. Khovanov [**Kho00**] constructed a categorification of the Jones polynomial for links in S^3, in the form of a bigraded abelian group whose graded Euler characteristic is the Jones polynomial. Since the WRT invariants are complex numbers, it is less clear what it means to categorify them. The first ideas in this direction appeared in the work of Crane and Frenkel [**CF94**]. For recent work on categorification at roots of unity, see [**Kho16**, **EQ16**, **QRSW21**].

Scribed by: C. Manolescu

PROBLEM 3.68.

(a) *Give a mathematical definition of the $\widehat{Z}$ invariants for all 3-manifolds.*
(b) *Categorify the $\widehat{Z}$ invariants.*

REMARKS.

(1) In [**GPV17**, **GPPV20**], Gukov, Pei, Putrov, and Vafa conjectured the existence of some power series with integral coefficients associated to 3-manifolds, denoted $\widehat{Z}_a(q)$. They converge in the unit disk $|q| < 1$, and their limits as q approach some roots of unity should recover the Witten–Reshetikhin–Turaev invariants.

(2) Rigorous definitions of the $\widehat{Z}$ invariants exist for plumbed 3-manifolds and certain surgeries on knots; see [**GPPV20, GM21, CGPS20**].
(3) Since the $\widehat{Z}$ invariants have integral coefficients, they are good candidates for categorification.
(4) There are many other interesting questions about $\widehat{Z}$, as well. For example, it is interesting to study its large N limit, which seems to be connected to enumerative geometry [**EGG$^+$22**].

Scribed by: C. Manolescu

3.8. Bounding 4-manifolds

Leading into Chapter 4, we conclude with questions about how 3-manifolds bound 4-manifolds. Rokhlin proved that every 3-manifold arises as the boundary of a smooth 4-manifold [**Roh51**], but if one restricts the topology of the 4-manifold or of the pair, the problem becomes subtle.

For example, given a ring R, let Θ^3_R denote the 3-dimensional *R-homology cobordism group*. That is, Θ^3_R is the quotient of the monoid of 3-manifolds with the same R-homology as S^3 (*R-homology 3-spheres*), with the operation of connected sum, by the relation that $Y_1 \sim Y_2$ if there is a smooth 4-dimensional bordism W from Y_1 to Y_2 so that $H_*(W;R) \cong H_*(S^3;R)$. The special case $\Theta^3 = \Theta^3_{\mathbb{Z}}$ has received particular interest, because of its sensitivity to smoothness in dimension 3 and relation to triangulations in higher dimensions. (Recent surveys of the topic include [**Man18, Şav24**].) Problems 3.69–3.72 concern Θ^3_R.

Problem 3.73 concerns a non-smooth version of Θ^3_R. Specifically, call two R-homology 3-spheres *topologically* R-homology cobordant if they co-bound a topological 4-manifold with the R-homology of $S^3 \times [0,1]$. Connected sum again makes this set of equivalence classes into a group, the *topological homology cobordism group*, which we denote Θ^{TOP}_R.

In a slightly different direction, Problems 3.74 and 3.75 ask about extending diffeomorphism or group actions from 3-manifolds to 4-manifolds they bound. Problem 3.77 asks about hyperbolic 3-manifolds that bound a hyperbolic 4-manifold. The section concludes with Problem 3.78, which would more naturally be in Section 3.7, but was placed here in the final production of the book.

PROBLEM 3.69.

(a) *What is the isomorphism type of* $\Theta^3_{\mathbb{Z}}$*?*
(b) *Does there exist a torsion element* $[Y]$ *in* $\Theta^3_{\mathbb{Z}}$*?*
(c) *Does there exist a torsion element* $[Y] \in \Theta^3_{\mathbb{Z}}$ *having Rokhlin invariant* $\mu([Y]) = 1$*?*
(d) *Do there exist infinitely-divisible elements in* $\Theta^3_{\mathbb{Z}}$*; that is, does there exist a homology 3-sphere* Y *such that for infinitely many* $n \in \mathbb{N}$ *there exists a* $[Z_n] \in \Theta^3_{\mathbb{Z}}$ *such that* $[Y] = n[Z_n]$*?*

REMARKS.

(1) The integer homology cobordism group $\Theta^3_{\mathbb{Z}}$ is known to be a countable abelian group that has a summand isomorphic to $\mathbb{Z}^\infty$ by work of [**DHST23**].

(2) In dimension $n = 1, 2$, the analogous group $\Theta^n_{\mathbb{Z}}$ vanishes. In dimension $n \geqslant 4$, the group $\Theta^n_{\mathbb{Z}}$ is isomorphic to the group of homotopy spheres up to h-cobordism [**GA70**], and therefore finite [**KM63**]. However, in dimensions $n \geqslant 4$, the analog of this group in the PL category vanishes; it is this latter group that arises as an obstruction in the study of triangulations [**GS80, Mat78**].

(3) Manolescu showed that there does not exist an element $[Y]$ of order two having Rokhlin invariant $\mu(Y) = 1$ [**Man16b**]. By previous work of Galewski–Stern [**GS80**] and Matumoto [**Mat78**], this was sufficient to disprove the outstanding cases of the Triangulation Conjecture, showing the existence of topological manifolds in every dimension greater than or equal to five that do not admit triangulations.

Since $Y\# -Y$ bounds a smooth homology 4-ball for all integer homology 3-spheres Y, to find a 2-torsion element, it suffices to exhibit a homology 3-sphere Y with an orientation-reversing diffeomorphism, such that $[Y] = [-Y] = -[Y]$ in homology cobordism, but $[Y] \neq [S^3]$. Noteworthy examples of 3-manifolds with such orientation-reversing diffeomorphisms include double branched covers of determinant 1, non-slice knots, such as those in [**BC24b**].

Various authors have given conditions that imply that a homology 3-sphere $[Y]$ must have infinite order in $\Theta^3_{\mathbb{Z}}$, for example [**LRS18**, Theorems C and D] or [**HHL21**, Theorem 1.13].

(4) By work of Galewski–Stern, a negative answer to part (c) would imply that a topological manifold M of dimension $\geqslant 5$ is triangulable if and only if a particular obstruction class in $H^5(M;\mathbb{Z})$ vanishes.

(5) If $\mu([Y]) = 1$, then $[Y]$ cannot be equal to $n[Z]$ for even n. Furthermore, any infinitely divisible $[Y]$ must necessarily have Heegaard Floer correction term $d([Y]) = 0$, and similarly for any other numerical invariant that is additive under connected sum. Any torsion element of a group is infinitely divisible, so in the presence of torsion part (d) might be modified to ask about the presence of infinitely divisible elements in the quotient of $\Theta^3_{\mathbb{Z}}$ by its torsion subgroup.

Scribed by: K. Hendricks and C. Manolescu

PROBLEM 3.70. *Is $\Theta^3_{\mathbb{Z}}$ generated by the classes of knot surgeries $\left[S^3_{1/n}(K)\right]$, where n ranges over all integers and K ranges over all knots in S^3?*

REMARKS. There are homology spheres that are not given by surgery on a knot in S^3 [**GL89, Auc97, HKL16b**]. Nozaki–Sato–Taniguchi used filtered instanton Floer homology to exhibit a homology three-sphere that is not even homology cobordant to any knot surgery [**NST24**]. Hendricks–Hom–Lidman showed that $\Theta^3_{\mathbb{Z}}$ is not generated by surgeries on knots of bounded genus [**HHL21**].

Scribed by: K. Hendricks

PROBLEM 3.71. *Is there a nontrivial element in the kernel of the natural map*

$$\Theta^3_{\mathbb{Z}} \to \Theta^3_{\mathbb{Z}/2\mathbb{Z}};$$

that is, does there exist an integer homology 3-sphere that does not bound a smooth $\mathbb{Z}$*-homology 4-ball but does bound a smooth* $\mathbb{Z}/2\mathbb{Z}$*-homology four-ball?*

REMARKS. There is a natural homomorphism from the concordance group $\mathcal{C}$ to $\Theta^3_{\mathbb{Z}/2\mathbb{Z}}$ via taking double branched covers. As discussed in Problem 3.72, $\Sigma(2,3,7)$ bounds a rational homology ball but, because its Rokhlin invariant is nontrivial, does not bound a $\mathbb{Z}/2\mathbb{Z}$ homology ball. See also [**BL02**] for some further discussion of $\mathbb{Z}/2\mathbb{Z}$-homology cobordism.

Scribed by: K. Hendricks

PROBLEM 3.72.

(a) *Does the kernel of the map* $\Theta^3_{\mathbb{Z}} \to \Theta^3_{\mathbb{Q}}$ *contain a subgroup that is isomorphic to* $\mathbb{Z}^\infty$*?*
(b) *If so, does it contain such a subgroup as a summand?*

REMARKS. Fintushel and Stern [**FS84**] showed that $\Sigma(2,3,7) = S^3_{+1}(4_1)$ bounds a smooth rational homology ball; as it has nontrivial Rokhlin invariant, it is a nontrivial element of the kernel of the map above. Indeed, as $\Sigma(2,3,7)$ has non-vanishing Neumann–Siebenmann invariant $\bar{\mu}$ [**Neu80, NR78**], it is of infinite order in the homology cobordism group, implying that the kernel is infinite. This was the only known example of a Brieskorn sphere bounding a rational homology ball but not an integer homology ball until infinite families of examples were given by Akbulut and Larson [**AL18**], followed by further examples from Şavk [**Şav20**] and Simone [**Sim21**]. It is unknown whether these families are linearly independent.

Scribed by: K. Hendricks

PROBLEM 3.73.

(a) *Calculate* $\Theta^{\mathrm{TOP}}_{\mathbb{Z}/p}$.
(b) *Calculate* $\Theta^{\mathrm{TOP}}_{\mathbb{Q}}$.
(c) *Is the linking form homomorphism* [lk] *injective?*

REMARKS.

(1) Freedman proved that any integral homology 3-sphere bounds a contractible topological 4-manifold, unique up to homeomorphism rel boundary [**FQ90**, Corollary 9.3C], [**Fre82**, Theorem 1.4′][**BKK**$^+$**21**, Chapter 21.3.2]. This implies the topological integer homology cobordism group is trivial, $\Theta^{\mathrm{TOP}}_{\mathbb{Z}} = 1$.
(2) Let $W(\mathbb{Q}/\mathbb{Z})$ denote the Witt group of non-singular bilinear forms over $\mathbb{Q}/\mathbb{Z}$. The $\mathbb{Q}/\mathbb{Z}$-valued linking form on $H_1(Y;\mathbb{Z})$ defines a homomorphism

$$[\mathrm{lk}] : \Theta^{\mathrm{TOP}}_{\mathbb{Q}} \to W(\mathbb{Q}/\mathbb{Z}).$$

See [**Bre97**, Chapter VI, Section 10] for a review of Poincaré–Lefschetz duality, and how one uses it to define the linking form, or [**CFH16**, Section 2] for a self-contained synopsis of the same. That the linking form defines a homomorphism to $W(\mathbb{Q}/\mathbb{Z})$ is an immediate consequence of the following fact: if a rational homology 3-sphere Y bounds a rational ball, then its first homology has a subgroup of order $\sqrt{|H_1(Y)|}$ on which the linking form vanishes. This is proved in [**CG86**, Theorem 1 and Theorem 2] (see also [**Gil82**, Lemma 1] for a self-contained treatment). One can easily show that this homomorphism is surjective; indeed, any form can be presented with an integral matrix, which can be used to build a 4-dimensional 2-handlebody whose boundary will be a 3-manifold representing the given form.

(3) Homology cobordism groups are closely connected to knot concordance, by way of Dehn surgery and cyclic branched covers. For instance, both constructions define maps from the knot concordance group to $\Theta_{\mathbb{Q}}$, with each factoring through $\Theta_{\mathbb{Z}/p}$ for some p depending on the surgery slope or degree of cyclic branched covering, respectively. (For the latter, one should restrict to p a prime power to ensure the branched covers are $\mathbb{Z}_p$-homology spheres and, for slice knots, bound $\mathbb{Z}_p$-homology balls [**CG78**].) In the case of branched coverings, these maps are actually homomorphisms. Both observations have been extensively used in the smooth setting. Understanding the problems above could have applications to the theory of topological concordance groups. For instance, if [lk] fails to be injective, and refined invariants of $\Theta_{\mathbb{Q}}^{\mathrm{TOP}}$ can be defined, then the latter could be used to study the topological concordance group through the surgery and branched covering maps.

Proposed for K3 and scribed by: M. Hedden

PROBLEM 3.74. *Let Y be a rational homology sphere and $f\colon Y \to Y$ be a self-diffeomorphism of Y. Suppose W is a 4-manifold with boundary Y such that f extends to a self-diffeomorphism of W.*

(a) *What constraints are there on the intersection form of W?*
(b) *Does there exist a pair (Y, f) such that Y bounds a homology ball but $f\colon Y \to Y$ does not extend over any definite manifold of either sign?*
(c) *Develop methods for constraining the intersection form of W that apply to indefinite, non-spin W.*

REMARKS.

(1) It is known that the 3-dimensional bordism-with-diffeomorphism group is trivial [**Mel79**]; that is, if Y is a 3-manifold equipped with a self-diffeomorphism f, then there always exists a smooth 4-manifold W over which f extends. It is thus natural to ask what homological restrictions occur for the class of such W. This is analogous to the fact that the usual 3-dimensional bordism group is trivial, but the set of intersection forms bounded by Y reflects the topology of Y.
(2) The extension question for diffeomorphisms is also motivated by the theory of corks [**Akb91**]. If $\pi_1(W) = 1$, then every self-homeomorphism of

Y extends as a self-homeomorphism of W by work of Freedman [**FQ90**]. In contrast, there are many examples in which a self-diffeomorphism of Y does not extend as a self-diffeomorphism over a particular contractible manifold bounded by Y, or even any homology ball with boundary Y. See for example [**Akb91, LRS23a, AKS20, DHM23**].

(3) While it is possible to obstruct the extension of f over definite manifolds of a fixed sign, it is unknown whether there are any pairs (Y, f) for which Y bounds a homology ball, but f does not extend over any definite manifold of either sign. Disregarding the self-diffeomorphism f, it has only recently been shown that there are integer homology spheres Y that bound no definite manifold of either sign; see [**NST24**]. If the action of the extension on the cohomology of W is constrained, then partial results have been obtained in [**ADMT23**] using the Chern–Simons filtration in instanton Floer homology.

Scribed by: I. Dai

PROBLEM 3.75. *Let Y be a rational homology 3-sphere equipped with an action of a cyclic group $\mathbb{Z}/p\mathbb{Z}$. Suppose W is a 4-manifold with boundary Y over which the action of $\mathbb{Z}/p\mathbb{Z}$ extends smoothly.*

(a) *What constraints are there on the intersection form of W?*
(b) *Does there exist an integer homology sphere Y bounding a homology ball such that Y admits a cyclic group action that does not extend over any definite manifold of either sign?*
(c) *Develop methods for constraining the intersection form of W that apply to indefinite, non-spin W.*

REMARKS. While this is similar to Problem 3.74, the motivation and techniques for the present question are slightly different. If $f\colon Y \to Y$ generates the action of a cyclic group, then obstructing the extension of f as a diffeomorphism clearly obstructs the extension of f as a group action. However, there are many interesting examples where the two notions differ. For instance, one may consider the standard S^1-action on a Brieskorn sphere $\Sigma(p, q, r)$. Any root of unity $\zeta \in S^1$ generates the action of a cyclic group on $\Sigma(p, q, r)$ whose generator f is isotopic to the identity (as a diffeomorphism). In [**AH16, AH21**], it is shown that this never extends as a smooth group action over any contractible manifold that $\Sigma(p, q, r)$ may bound, even though it is not hard to see that f extends as a diffeomorphism. See also [**Edm87, KL93, BH24a**].

Scribed by: I. Dai

PROBLEM 3.76. *What is the structure of the equivariant homology cobordism groups?*

REMARKS. There are several equivariant homology cobordism groups one can construct. One version is the $\mathbb{Z}/p\mathbb{Z}$-equivariant cobordism group as appropriate homology cobordism classes of pairs consisting of integer homology spheres with an orientation-preserving $\mathbb{Z}/p\mathbb{Z}$ action having nonempty fixed point set. One can also define more general groups that take into account a diffeomorphism on each homology sphere and have no restriction on the fixed-point set; this is done in [**DHM23**]. In various cases, one can produce interesting $\mathbb{Z}^\infty$-subgroups. For instance, it follows from [**DHM23**] that there exists a $\mathbb{Z}^\infty$-subgroup spanned by cork boundaries in the case $p = 2$; similar techniques can be attempted for other p. The structure of these groups is otherwise poorly understood.

It is well-known that there is a homomorphism from the smooth concordance group to the rational homology cobordism group given by taking branched covers. There is likewise a homomorphism from the strongly invertible concordance group to the $\mathbb{Z}/2\mathbb{Z}$-equivariant rational homology cobordism group; see [**AB24b**]. There is also a homomorphism from the usual concordance group to the equivariant cobordism group, by taking branched covers but remembering the branching action, which may be used to produce more refined sliceness or concordance obstructions; see e.g. [**BH24a**].

Scribed by: I. Dai, K. Hendricks

PROBLEM 3.77. *Does there exist a hyperbolic rational homology 3-sphere that is the totally geodesic boundary of a compact, orientable hyperbolic 4-manifold?*

REMARKS.

(1) When a compact hyperbolic 3-manifold is the totally geodesic boundary of a compact hyperbolic 4-manifold, it is said to *geometrically bound*. Questions about geometric boundaries of hyperbolic manifolds of dimension greater than 4 are discussed in Problem 4.126(b).
(2) A result of Ferrari–Kolpakov–Reid [**FKR23**] shows that there exist infinitely many arithmetic hyperbolic rational homology 3-spheres that geometrically bound hyperbolic 4-manifolds, but in their examples, all of the 4-manifolds are nonorientable. Prior to [**FKR23**], all constructed examples of hyperbolic 3-manifolds that geometrically bound had non-zero first Betti number.
(3) Most constructions of hyperbolic 3-manifolds that geometrically bound (including the ones in [**FKR23**]) are arithmetic of simplest type and bound hyperbolic 4-manifolds cut out from ones that are arithmetic of simplest type. It would be interesting to construct non-arithmetic examples.
(4) When a compact, orientable 3-manifold M geometrically bounds a compact, orientable, hyperbolic 4-manifold, this imposes restrictions on the η-invariant of M, due to [**LR00**]. For example, the Weeks manifold does not geometrically bound a compact orientable hyperbolic 4-manifold.

Scribed by: P. Patel, M. Lackenby

PROBLEM 3.78.

(a) *Is there a non-semisimple 3-TQFT whose mapping class group representation is faithful or has an element in its kernel?*

(b) *Define a 4-manifold invariant via non-semsimple categories that can distinguish 4-manifolds that are not distinguished by classical or gauge-theoretic invariants.*

(c) *Is there is a full 3-TQFT extension of the representations, constructed by Bonahon and Wong, of the Kauffman bracket skein algebra? In particular, do non-semisimple TQFTs recover the Bonahon–Wong representations?*

REMARKS. (1) In 1989, Atiyah [**Ati88b**] gave the definition of an n-dimensional topological quantum field theory (n-TQFT for short) that assigns objects in some category to n-manifolds and morphisms to cobordisms. Many 3-TQFTs have been constructed via semisimple modular categories; this problem is asking about applications of TQFTs constructed via non-semisimple categories that do not seem accessible to semisimple TQFTs.

(2) Non-semisimple 3-TQFTs are relevant to the linearity problem for the mapping class group of a surface; see Problem 2.3 for a discussion of classical approaches. For example, the 3-TQFTs of [**DRGG$^+$22**] lead to mapping class group representations with the property that the action of a Dehn twist has infinite order and thus is a candidate answer to part (a). By contrast, in the usual semisimple quantum mapping class group representations, all Dehn twists have finite order and the representations are not faithful.

(3) The underlying 4-manifold invariants of 4-TQFTs coming from semisimple modular categories, such as [**CKY97**, **CY93**, **BB18**], are conjecturally determined by the Euler characteristic, the signature and the fundamental group, see [**BB18**, Conjecture 8.1]. This conjecture is verified in many cases in [**Reu23**]. It is unknown whether the 4-TQFTs of [**CGHPM23**] or the invariants of [**BDR23a**, **BDR23b**] go beyond such classical invariants. Note that in the 3-dimensional setting, there are invariants of closed 3-manifolds underlying a 3-TQFT distinguishes diffeomorphism types of homotopically equivalent lens spaces which were not distinguished by semisimple quantum invariants; see [**CGPM14**].

(4) In [**BW16c**, **BW16b**, **BW17**, **BW19**] Bonahon and Wong constructed a family of finite-dimensional representations of the Kauffman bracket skein algebra of surface. Their construction is based on the theory of quantum Teichmüller space of Chekhov and Fock [**FC99**] and Kashaev [**Kas98**]. Recently, Frohman, Kania-Bartoszynska and Le showed these representation have a unicity condition, see [**FKBL19**]. Partial results in the direction are known: certain BW-representations arise from the non-semisimple TQFT of [**BCGPM16**], see [**KK22**]. If a TQFT extension of the link invariants of [**BGPMR20**] exists it should contain the general BW-representation

Proposed for K3 by: N. Geer
Scribed by: D. Ruberman

CHAPTER 4

4-manifolds

Chapter 4 Editors: İnanç Baykur, Ciprian Manolescu, Maggie Miller, and Mark Powell

Many central problems in dimension 4 are yet to be solved, including the following famous problems, which are worthy of special mention.

- The smooth 4-dimensional Poincaré conjecture.
- The Wall stabilization problem for homeomorphic 4-manifolds.
- The 11/8 conjecture on the existence of certain spin 4-manifolds.
- The Schoenflies conjecture on smooth embeddings of S^3 in S^4.
- The unknotting conjecture for smooth 2-knots in S^4.
- The topological surgery conjecture for 4-manifolds with arbitrary fundamental group.
- The question of whether all 2-links are slice.
- The existence of nontrivial mapping classes of S^4.
- The existence of a symplectic analogue of the Bogomolov–Miyaoka–Yau inequality.

The prevalence of fundamental mysteries is reflected in the length of the chapter. Each of the above famous problems mentioned above stands at the apex of a subfield of related problems. Here is a summary of the contents of this chapter.

(1) Section 4.1 presents problems on exotic smooth structures on 4-manifolds, generalizing the Poincaré conjecture. Other problems concern the effect of certain surgical constructions, such as knot surgery, Gluck twists, and cork twists, in altering the smooth structures, as well as stable equivalence of exotic pairs of orientable 4-manifolds.
(2) Section 4.2 concerns the existence of smooth 4-manifolds with prescribed properties, beginning with geography problems such as the 11/8 conjecture for spin 4-manifolds and questions on irreducible smooth structures, before looking at Wall realization, connect sums of star partners, and more.
(3) Section 4.3 lists questions on embeddings of 3-manifolds in S^4, headed by the smooth Schoenflies problem, before considering questions on embeddings of other 3-manifolds in the 4-sphere.
(4) Section 4.4 poses problems on surfaces in 4-manifolds, asking for conditions for isotopy, exotic surfaces, non-smoothable surfaces, and the possible algebraic topology of surface complements, as well as problems on concordance of surfaces, in particular whether 2-links are slice. There is also consideration of how surfaces can be used to detect exotic structures, and what properties spaces of embedded surfaces in 4-manifolds have.

(5) Section 4.5 is devoted to problems on the homeomorphism and homotopy classification of 4-manifolds, starting with the famous topological surgery conjecture asserting that all groups are good. However, the majority of the section focuses on the search for classifications of 4-manifolds with known good fundamental groups, including conditions for two 4-manifolds to be equivalent, and asking for realization results for the algebraic invariants used in the classifications.

(6) Section 4.6 contains problems on cobordisms, such as the 4-dimensional smooth s-cobordism question, a 4-dimensional version of Lickorish–Wallace, and codimension one embeddings of cobordisms.

(7) Section 4.7 asks about diffeomorphism invariants of 4-manifolds, invaluable tools in the search for exotic smooth structures, with problems concerning their relationships, limitations, and potential to distinguish distinct smooth structures.

(8) Section 4.8 considers symmetries of 4-manifolds. This refers to homeomorphisms and diffeomorphisms of 4-manifolds, mapping class groups, and group actions on 4-manifolds. A central question is the status of the smooth mapping class group of S^4. There are then questions on the smooth and topological mapping class groups, exotic mapping classes, and the behavior of such symmetries under stabilization, as well as on the structure of the higher homotopy groups of diffeomorphism and homeomorphism groups, and their properties purely as topological groups. The closely related topic of group actions on 4-manifolds also features, including exotic and non-smoothable group actions, circle actions, and involutions on S^4.

(9) Section 4.9 represents the extensive field of symplectic and complex structures on 4-manifolds, with many problems exploring the divide between the smooth, symplectic, and complex categories. Some central problems in this section concern the geography and botany of symplectic 4-manifolds and fibrations, the diversity of symplectic Calabi–Yau surfaces and fake projective planes, and the famous symplectic isotopy problem and its variations on the isotopy classification of symplectic surfaces and complex curves in $\mathbb{C}P^2$. Other questions address the challenges in distinguishing smooth (resp. symplectic) structures on homeomorphic (resp. diffeomorphic) complex algebraic surfaces, as well as similar problems concerning Weinstein and Stein surfaces.

(10) Section 4.10 lists problems on trisections of 4-manifolds. Some of these offer an alternative perspective on several of the hardest questions in 4-manifold topology.

(11) Section 4.11 collects questions about other structures on 4-manifolds, such as handle decompositions, CW complexes, achiral fibrations, and open books, as well as on geometric structures such as hyperbolic metrics, positive scalar curvature metrics, and Lipschitz structures.

Despite the importance of 4-dimensional reasoning for them, this chapter does not tend to contain problems on concordance of knots and links, which typically appear in Chapter 1, nor problems on homology cobordism of 3-manifolds, which typically appear in Chapter 3.

4.1. Exotic structures on 4-manifolds

This section presents problems on exotic smooth structures. A *smooth structure* on a manifold X is a maximal atlas of charts with smooth transition functions. Two smooth structures are diffeomorphic if there is a homeomorphism $f\colon X \to X$ that pulls back one smooth structure to the other. If a manifold is endowed with a smooth structure, for example if it is a well-known standard 4-manifold, then any other smooth structure that is not diffeomorphic to the original is called an *exotic* smooth structure. One can also consider smooth structures up to isotopy and up to concordance, but unless specified we will consider them up to diffeomorphism.

If X has nonempty boundary, then an exotic smooth structure with respect to the above definition is called *absolutely exotic*. If there is no homeomorphism f as above with $f|_{\partial X}$ the identity, then the smooth structure is called *relatively exotic*.

We describe the operation of *Gluck twisting* a 4-manifold along an embedded 2-sphere with trivial normal bundle. Given an embedding $S^2 \times D^2 \hookrightarrow M$ we can remove the interior and then reglue $S^2 \times D^2$ using the diffeomorphism $G\colon S^2 \times S^1 \to S^2 \times S^1$ defined as follows. Let $r_t\colon S^2 \to S^2$, $t \in [0,1]$, be a single full rotation of S^2 around a fixed axis. For $(x, e^{2\pi i t}) \in S^2 \times S^1$, define

$$G(x, e^{2\pi i t}) = (r_t(x), e^{2\pi i t}).$$

This is the Gluck diffeomorphism of $S^2 \times S^1$. It acts as the identity on homology but is not homotopic to the identity.

PROBLEM 4.1 (4-dimensional Poincaré conjecture). *Is there a unique smooth structure on the 4-sphere?*

REMARKS.

(1) This appears as Problems 4.45 and 4.89 in [**Kir97**].

(2) This is one of the remaining open cases of the smooth *Poincaré Conjecture*. In dimensions at most three, uniqueness of smooth structures follows from work of Munkres [**Mun60**] and Moise [**Moi77a**]. In higher dimensions, the only odd-dimensional spheres with unique smooth structures are S^5 and S^{61} [**WX17**]; the situation for even-dimensional spheres is still open in general, but it is conjectured that the only even-dimensional spheres without exotic smooth structures in dimensions greater than 4 are S^6, S^{12}, and S^{56} [**WX17**].

(3) For context, we mention the history of the *topological* Poincaré conjecture: any manifold homotopy equivalent to S^n is homeomorphic to S^n. This is classical for $n \leqslant 2$, follows from Perelman's Geometrization Theorem for $n = 3$ [**Per02, Per03b, Per03a**], from Freedman's work for $n = 4$ [**Fre82**], and from Newman's work on topological engulfing for $n \geqslant 5$ [**New66**]. (When the manifold is smooth, the $n \geqslant 5$ case was first established as a consequence of Smale's h-cobordism theorem [**Sma62a**].)

(4) Some sources of potentially exotic homotopy 4-spheres are as follows.
   - Gluck twists on 2-knots in S^4. See Problem 4.9 for a discussion.
   - The Andrews–Curtis problem. See [**Kir78**, Problem 5.2] and Problem 5.10 for a discussion.

- Cappell–Shaneson homotopy spheres [**CS76**]. Many of their examples were proved to be standard in [**Akb10**] and [**Gom10**], but not all. More generally, one can consider the mapping torus of a 3-manifold diffeomorphism, and do surgery on a section; this sometimes gives a homotopy 4-sphere. One such example was proposed in [**Maz62**], and was shown standard in [**Zee65**]. One can construct many such examples where the diffeomorphism is obtained from the identity by a point push; e.g., when the 3-manifold is made by gluing two knot complements.
- Gabai's big dot carving. See Problem 4.24.
- Pairs of knots with the same 0-surgeries, such that exactly one knot is slice. See for example [**Akb93**], or [**MP23**].
- Cyclic branched covers of certain 2-knots. In general, the double branched cover of a 2-knot in S^4 is a rational homology sphere that need not be simply connected, but in certain cases the cover can be arranged to be a homotopy 4-sphere. For example, the double branched cover of the roll-spun $(-2, 3, 7)$ pretzel knot is a homotopy 4-sphere that is not known to be the standard 4-sphere (see [**Miy23**]).

(5) We mention some known potential approaches to the problem.
- If one wants to prove that the smooth structure on S^4 is unique, one can try to use geometric flows, such as the Ricci flow, or the mean curvature flow in $\mathbb{R}^5$ [**CMP15**].
- If one wants to put a nonstandard smooth structure on S^4 using pairs of knots with the same 0-surgeries, Rasmussen's s-invariant [**Ras10**] could be used to prove one of the knots is not slice. See the remarks on Problem 1.60.
- Let K be a topologically slice knot that is not smoothly slice, such as the Conway knot. If we could put a smooth structure on the complement of the interior of a tubular neighborhood of the topologically slice disk, then we could imitate the construction of an exotic $\mathbb{R}^4$ using S^4 and get a nontrivial homotopy sphere.
- Symplectic geometry gives a potential avenue to either proving or disproving the smooth 4-dimensional Poincaré conjecture. Given a homotopy 4-sphere X, one wants to either construct or obstruct the existence of an asymptotically standard symplectic structure on $X\backslash\{\text{point}\}$. See [**Ger21**] for a discussion.
- Finally, it is possible to rephrase Problem 4.1 purely in terms of group theory, using trisections [**AGK18**].

Scribed by: C. Manolescu, M. Miller, L. Piccirillo

PROBLEM 4.2. *Does every smooth, closed 4-manifold admit an exotic smooth structure? Infinitely many?*

REMARKS.

(1) The question is open for many families of 4-manifolds. See also Problems 4.3 and 4.4.

(2) A version of this problem in [**Kir97**, Problem 4.86] raises the same question for complex algebraic surfaces. Evident families in this class, still not known to admit exotic smooth structures, are the irrational ruled surfaces $(\Sigma_h \times S^2)\#^m\,\overline{\mathbb{CP}}^2$ and $(\Sigma_h \tilde{\times} S^2)\#^m\,\overline{\mathbb{CP}}^2$ for any $h, m \geqslant 1$.

(3) To date, on virtually any smooth, closed 4-manifold known to admit an exotic smooth structure, infinitely many distinct exotic smooth structures have been constructed. (See [**BSS24**] for the related discussion.) Many of the known methods for constructing infinitely many smooth structures require the presence of certain embedded tori of self-intersection zero [**FS98, FS11**].

Scribed by: I. Baykur, C. Manolescu

PROBLEM 4.3. *Are there exotic smooth structures on the following closed, simply-connected 4–manifolds?*

(a) $\#^k\,\mathbb{CP}^2$ *for any* $k \geqslant 1$.
(b) $\#^m(\mathbb{CP}^2\#\overline{\mathbb{CP}}^2)$ *and* $\#^n\,(S^2 \times S^2)$ *for* $m \leqslant 8$, $n \leqslant 10$.
(c) $\#^p\,\mathrm{K3}\,\#^q(S^2 \times S^2)$ *for any* $p \geqslant q + 5$.

If so, what are the smallest k, m *and* n*? The largest* p *and* q*?*

REMARKS.

(1) These are the main families of closed, simply connected 4–manifolds, besides S^4, not known to admit exotic smooth structures. Namely, those that are (a) definite, (b) $\sigma = 0$ with $e < 20$, and (c) spin with $c_1^2 = 2e + 3\sigma < -12$, respectively, where e is the Euler characteristic and σ is the signature.

In the non-spin case, advances in the geography of minimal symplectic 4–manifolds [**ABB$^+$10, AP10**], combined with blow-ups, connected sums with nontrivial Bauer–Furuta invariants, and orientation-reversals, yield exotic smooth structures on most 4-manifolds of the form $\#^a\mathbb{CP}^2\#^b\overline{\mathbb{CP}}^2$ with $ab(a-b) \neq 0$ and a, b not both even. Here, the gaps essentially stem from our lack of understanding related to Problem 4.17.

Less is known in the spin case, but the results in the geography of spin symplectic 4–manifolds [**PS00a, Par02**] similarly settle the existence of exotic smooth structures on most spin 4–manifolds with $c_1^2 \geqslant 0$ and $\sigma < 0$; i.e. on $\#^p\,\mathrm{K3}\,\#^q(S^2 \times S^2)$ for $0 < p \leqslant q + 1$. There are still some gaps, such as $\mathrm{K3}\,\#(S^2 \times S^2)$, which are again due to our lack of understanding related to Problem 4.17. Some exotic examples also exist in the range $q+2 \leqslant p \leqslant q+4$ (e.g. see Remark 5 below for $q = 0$), but none are known for $p \geqslant q + 5$ (i.e. $2e + 3\sigma < -12$).

(2) A potential source of exotic definite 4–manifolds arise from pairs of knots with the same 0–surgeries, where only one knot bounds a disk with a given self-intersection number in $\#^k\,\mathbb{CP}^2\backslash D^4$. See [**MP23**], [**Qin25**].

(3) There are infinitely many exotic smooth structures on $\#^m(\mathbb{CP}^2\#\overline{\mathbb{CP}}^2)$ and $\#^n\,S^2 \times S^2$ for any odd $m \geqslant 9$, $n \geqslant 11$ [**BH23**]. A possible strategy to extend these results to cover smaller m, n is through Lefschetz pencils [**Bay22**] and Fintushel–Stern reverse engineering [**FPS07**].

Here is another potential construction for exotic $S^2 \times S^2$ or $\mathbb{CP}^2 \# \overline{\mathbb{CP}}^2$, which could perhaps be detected using knot Floer homology: given a pair of knots in S^3 with the same n-surgery, by gluing their traces one obtains a manifold homeomorphic to either $S^2 \times S^2$ or $\mathbb{CP}^2 \# \overline{\mathbb{CP}}^2$ (depending on the parity of n). See [**LLP23**, Remark 6.13].

(4) Exotic definite 4-manifolds with finite fundamental groups can be derived from indefinite exotic 4-manifolds admitting free finite group actions. Examples with fundamental group $\mathbb{Z}/2$ have been announced in [**LLP23, SS24b**] as quotients of exotic $\mathbb{CP}^2 \#^m \overline{\mathbb{CP}}^2$ under free involutions, and with more general fundamental groups by Baykur, Stipsicz and Szabó in [**BSS24**]. The smallest definite example here has $b_2 = 1$. It is plausible that exotic signature zero 4-manifolds with finite fundamental groups and smaller b_2 can be derived similarly.

(5) The Bauer–Furuta invariants [**BF04**] are sometimes able to detect exotic smooth structures on connected sums. Bauer [**Bau04**] used these invariants to find exotic smooth structures on $\#^p \mathrm{K3}$ for $p \leqslant 4$, but the method stops working at $p = 5$ and is not applicable if any summand has $b_2^+ \equiv 1 \mod 4$.

Scribed by: I. Baykur, R. Lipshitz, C. Manolescu, L. Piccirillo

---

PROBLEM 4.4. *Is there an exotic smooth structure on some product* 4*-manifold* $S^1 \times Y^3$ *or* $\Sigma_g \times \Sigma_h$*? Do they all admit exotic smooth structures?*

REMARKS.

(1) A homeomorphism classification is unavailable for many of these 4–manifolds, making it challenging to determine whether a potential exotic copy falls within the desired homeomorphism class.

(2) Significant special cases, besides $S^2 \times S^2$, include $S^1 \times S^3$, $T^2 \times S^2$, and $S^1 \times Y^3$, for any Y^3 that is a T^2–bundle over S^1, for example $Y^3 = T^3$. The first three have good fundamental groups $\{1\}$, $\mathbb{Z}$, and $\mathbb{Z}^2$ respectively, and there are homeomorphism classifications of closed 4-manifolds with these fundamental groups [**FQ90, HKT09**]. The remaining manifolds $S^1 \times Y^3$ are aspherical infrasolvmanifolds with good fundamental groups, for which the Borel conjecture holds [**Hil02**]. Thus, for Y^3 as above, any homotopy equivalence of 4-manifolds $X^4 \xrightarrow{\simeq} S^1 \times Y^3$ is homotopic to a homeomorphism.

(3) For $S^1 \times S^3$, see Problems 4.22 and 4.65 about detecting potential exotic structures.

(4) For $S^2 \times T^2$, see [**Bay22**, Theorem 6] for a possible strategy to generate exotic smooth structures via Lefschetz pencils and some candidates.

Scribed by: I. Baykur

PROBLEM 4.5. *Does every connected, open 4-manifold admit uncountably many smooth structures?*

REMARKS.

(1) Recall that a manifold is said to be open if it has empty boundary and every component is noncompact. Any such 4-manifold is smoothable by work of Quinn [**Qui82**, Corollary 2.2.3] (see also [**FQ90**, Theorem 8.2]). However it is unknown whether every open 4-manifold has more than one smooth structure.
(2) Gompf showed that $M\backslash\{x\}$ for $x \in M$, with M an arbitrary topological 4-manifold (not necessarily compact or orientable), has uncountably many smooth structures [**Gom93**, Theorem 2.1].
(3) Euclidean 4-space has uncountably many smooth structures [**Tau87**], which form a monoid acting on the smooth structures on any noncompact 4-manifold. Part of the question is therefore whether there is a 4-manifold that somehow absorbs or unwinds the Euclidean structures.
(4) Recall that smooth structures on a manifold may be considered either up to diffeomorphism or up to isotopy. Isotopy implies diffeomorphism. This question is asking about smooth structures up to diffeomorphism. One could also ask the related question whether every open 4-manifold admits a smooth structure with uncountably many isotopy classes. Non-isotopic smooth structures can be produced by pulling back a smooth structure along a non-smoothable self-homeomorphism. See Problem 4.83.

Proposed for K3 and scribed by: A. Ray

PROBLEM 4.6. *Does every closed, orientable 3-manifold bound an absolutely exotic pair of smooth, orientable 4-manifolds?*

REMARKS.

(1) Yasui [**Yas12**], Etnyre–Min–Mukherjee [**EMM22**], and Iida-Mukherjee-Taniguchi [**IMT25**] gave sufficient conditions on a closed, connected, orientable 3-manifold Y to bound a compact 4-manifold X with $\partial X = Y$ such that X admits infinitely many pairwise non-diffeomorphic smooth structures. Suppose that at least one of the following conditions are satisfied:
    - Y carries a contact structure ξ such that its Heegaard Floer contact invariant does not vanish e.g. if Y is Seifert fibered;
    - (Y, ξ) has a weak symplectic filling, e.g. if Y has a taut foliation; in particular, if Y is irreducible with $b_1(Y) > 0$;
    - Y is a rational homology sphere that embeds as a separating hypersurface in a closed definite 4-manifold.

    Then there exists a manifold X as above. In some of the cases, one gets smooth structures that are absolutely exotic, i.e. cannot be related by any diffeomorphism. In other cases, it is only known that they cannot be related by a diffeomorphism restricting to the identity on the boundary.

(2) By arguments similar to those in [**Kre84a**], one can show that every closed, orientable 3-manifold bounds an exotic pair of nonorientable 4-manifolds, so the second "orientable" in the question is necessary for an open question. A related open question is as follows.

QUESTION. Does every closed, nonorientable 3-manifold bound an absolutely exotic pair of smooth 4-manifolds?

Scribed by: O. Plamenevskaya

PROBLEM 4.7.

(a) *If M_1, M_2 are two homeomorphic closed, oriented 4-manifolds, is $M_1 \# S^2 \times S^2$ diffeomorphic to $M_2 \# S^2 \times S^2$?*

(b) *Is there a fixed $n > 0$ such that $M_1 \#^n S^2 \times S^2$ is diffeomorphic to $M_2 \#^n S^2 \times S^2$ for every pair M_1, M_2 as above?*

REMARKS.

(1) By Gompf's generalization of Wall's theorem [**Wal64b, Gom84**], for any given pair M_1, M_2 as above, there exists $k \geqslant 0$ such that $M_1 \#^k S^2 \times S^2$ is diffeomorphic to $M_2 \#^k S^2 \times S^2$. A priori k depends on M_1 and M_2, but in most examples, $k = 1$ is known to suffice; see [**BS13**] and the references therein.

(2) In the case of nonempty boundary, it has been announced by S. Kang [**Kan22b**] that one stabilization is not enough. Kang proposes an example of two homeomorphic smooth contractible 4-manifolds M_1, M_2 with diffeomorphic boundaries, such that $M_1 \# S^2 \times S^2$ and $M_2 \# S^2 \times S^2$ are not diffeomorphic. The obstruction to existence of a diffeomorphism between $M_1 \# S^2 \times S^2$ and $M_2 \# S^2 \times S^2$ comes from involutive Heegaard Floer homology.

(3) In the closed case, detecting an exotic pair after stabilization is difficult because most known invariants vanish for $M \# S^2 \times S^2$. The only known exceptions are the Pin(2)-equivariant Bauer–Furuta invariant, and Fintushel–Stern's 2-torsion instanton invariants, which can survive up to two stabilizations. To date, neither have been shown to differ for double stabilizations of homeomorphic 4-manifolds.

(4) One idea for the closed case is to look at exotic surfaces in 4-manifolds that stay exotic after a weak internal stabilization, and take their double branched covers. See Problem 4.33.

(5) A related but easier question is to find an exotic pair of closed simply connected 4-manifolds X_1, X_2 such that both manifolds contain a homologically essential square 0 sphere, and there is a homeomorphism between X_1 and X_2 (or an isomorphism between their intersection forms) that takes the homology class of one square 0 sphere to the other. The complements of the square 0 sphere are not required to be simply connected. In this setting, X_1 and X_2 would be hard to distinguish for the same reason as above: due to the presence of a square 0 sphere, most known invariants vanish.

(6) A positive answer to (b) would follow from existence of a universal cork; see Problem 4.14.
(7) One can ask the same questions about stabilization by the twisted S^2-bundle over S^2. Kang's examples become diffeomorphic after one twisted stabilization [**HKM23**]. Twisted stabilizations are not spin, so it is harder to find invariants to distinguish them: the maps on involutive Heegaard Floer homology and the Pin(2)-equivariant Bauer–Furuta invariants do not work.

Scribed by: O. Plamenevskaya

PROBLEM 4.8. *Let X be a closed, simply connected, smooth 4-manifold, and T a smoothly embedded torus in X with $\pi_1(X - T) = 1$ and $[T]^2 = 0$. Let X_K be the result of Fintushel–Stern knot surgery on X along a knot K.*

If K_1 and K_2 are two prime knots such that X_{K_1} and X_{K_2} are diffeomorphic, does it follow that K_1 is either K_2 or its mirror?

REMARKS.

(1) Fintushel and Stern showed that X_K is homeomorphic to X, and $SW(X_K) = SW(X)\cdot\Delta_K$, where SW denotes the Seiberg-Witten series and Δ_K the Alexander polynomial.
(2) One can ask the question for the particular case when X is the K3 surface. In that case $SW(X) = 1$, so the two knots need to have the same Alexander polynomial. No other constraint is known. For $X =$ K3, the question in the problem was raised by Fintushel and Stern.
(3) Akbulut [**Akb02**] showed that, if $m(K)$ denotes the mirror of K, then X_K is diffeomorphic to $X_{m(K)}$. Using work of Finashin [**Fin02**], Akaho [**Aka06**] showed that $X_{K\#K}$ is diffeomorphic to $X_{K\#m(K)}$.
(4) If one allows $\pi_1(X) \neq 1$, then there are many examples of X_{K_1} and X_{K_2} that have the same Seiberg-Witten invariants (since $\Delta_{K_1} = \Delta_{K_2}$) but are not diffeomorphic, being distinguished by the Seiberg-Witten invariants of their finite covers [**FS99a, PY15**] or by invariants from Heegaard Floer theory [**LLP23**].

Scribed by: I. Baykur, C. Manolescu

PROBLEM 4.9. *Is every Gluck twist in S^4 standard?*

REMARKS.

(1) This is [**Kir78**, Problem 4.24]. Given an embedded 2-sphere in S^4, Gluck defined a surgery operation that yields a homotopy 4-sphere [**Glu62**], which is in fact homeomorphic to S^4 due to Freedman [**Fre82**]. However, it is unclear whether the resulting manifold is diffeomorphic to S^4.

(2) The following 2-knots are known to have trivial Gluck twists i.e. the result of Gluck surgery on the 2-knot yields S^4.
   - Any spun knot or more generally any ribbon 2-knots [**Glu62**].
   - Any 2-knot 0-concordant to a 2-knot with trivial Gluck twist. More generally, if two 2-knots are 0-concordant, then the Gluck twist of those two 2-knots are diffeomorphic [**Mel77, HMY00**].
   - Any twist spun knot [**Gor76**] or branched twist spun knot [**Pao78**].
   - Any 2-knot that can be unknotted by a single finger move followed by a single Whitney move, which includes any roll spin of a classical knot of unknotting number one [**NS22**].
   - Any even degree satellite 2-knot whose pattern has trivial Gluck twist, or any satellite 2-knot (of any degree) in which both the pattern and the companion 2-knots have trivial Gluck twist [**Kim20**].
   - Any 2-knot that can be decomposed as a union of two ribbon disks, one of which has undisking number one [**GNS25**].
   - Some other examples related to the Cappell-Shaneson homotopy 4-sphere (see [**AK79a, Gom91a**]) and any 2-knot obtained from gluing particular ribbon disks satisfying certain conditions (see [**NS12**]).

(3) A unit sphere in $\mathbb{C}P^2$ is a smoothly embedded sphere that intersects the standard $\mathbb{C}P^1$ transversely once. If a unit sphere is obtained from a 2-knot in S^4 by blowing up at a point, then we say a unit sphere is obtained from the 2-knot.

   QUESTION. Is every unit sphere in $\mathbb{C}P^2$ smoothly isotopic to the standard $\mathbb{C}P^1$?

   Melvin [**Mel77**] showed that the Gluck twist of a 2-knot K in S^4 is trivial if and only if the pair $(\mathbb{C}P^2, \Sigma)$, where Σ is the unit sphere obtained from K, is pairwise diffeomorphic to $(\mathbb{C}P^2, \mathbb{C}P^1)$. It is unknown whether such a pairwise diffeomorphism would imply a smooth isotopy from Σ to $\mathbb{C}P^1$. Some unit spheres are known to be smoothly isotopic to $\mathbb{C}P^1$ [**HKM20**].

Scribed by: S. Kim

PROBLEM 4.10.

(a) *Is every homotopy B^4 with boundary S^3 obtained by performing a Gluck twist on some knotted 2-sphere in B^4?*

(b) *Suppose a homotopy 4-ball X is obtained by performing a Gluck twist on a knotted 2-sphere in B^4. Is $X \times I$ necessarily diffeomorphic to B^5?*

REMARKS.

(1) Habiro, Marumoto and Yamada [**HMY00**] showed that if a homotopy 4-ball is obtained by performing Gluck twists on multiple 2-spheres in B^4, then it can be obtained by a Gluck construction on a single knotted 2-sphere in B^4; see [**Mel77**] for related ideas.

(2) Part (a) of the problem could also be asked for homotopy spheres; the reason for specifying a homotopy 4-ball is that part (b) is a potential route

to finding a counterexample to the Schoenflies Conjecture (Problem 4.23) as well as to the 4-dimensional Poincaré Conjecture (Problem 4.1). Part (b) is a weaker question than asking whether X is diffeomorphic to B^4 (Problem 4.9).

Proposed for K3 by: D. Gabai
Scribed by: R. Kirby, D. Ruberman

PROBLEM 4.11. *Let M be a smooth 4-manifold and let $f\colon S^2 \to M$ be a smooth embedding with trivial normal bundle. Then let M_f denote the result of Gluck twisting on M along f.*

(a) *Does there exist an orientable M, and an embedding f, such that M and M_f are homeomorphic but not diffeomorphic?*
(b) *Does there exist an orientable M, and smooth, homotopic embeddings $f, g\colon S^2 \to M$ with trivial normal bundle such that M_f and M_g are homeomorphic but not diffeomorphic?*

REMARKS.

(1) The version of this question with $M = S^4$ has received considerable interest (see Problems 4.9 and 4.10). It is possible that the general case might be attacked more readily and/or provide insight to the S^4 case.
(2) Akbulut showed that Gluck twisting along a (homotopically essential) sphere in a *nonorientable* 4-manifold can produce an exotic smoothing [**Akb88**] (see also [**Tor17**]).
(3) Akbulut and Yasui give a simple condition on a 4-manifold and an embedded sphere f to ensure that Gluck twisting along f does not change the diffeomorphism type in [**AY13**]. Note that Gluck twisting may change the homeomorphism type, e.g. Gluck twisting $S^2 \times \mathrm{pt}$ inside $S^2 \times S^2$ yields $S^2 \tilde{\times} S^2$. See [**GS99**, Exercise 5.2.7(a)].
(4) In [**KPR23**, Theorem 1.2], it was shown that M_f and M_g are simple homotopy equivalent when f and g are homotopic. Under the stronger hypothesis that f and g are concordant, M_f and M_g are s-cobordant, and therefore homeomorphic when $\pi_1(M)$ is a good group (see Problem 4.46). Thus in the oriented case, finding examples where M_f and M_g are an exotic pair might be easier than finding examples where M and M_f are. See [**KPR23**] for further discussion.

Scribed by: M. Powell, A. Ray

PROBLEM 4.12. *For X a closed simply connected smooth 4-manifold, let*

$$g_X\colon H_2(X) \to \mathbb{N}$$

denote the genus function, which assigns to every homology class the minimal genus of a smooth embedded surface representing that class.

(a) *Suppose* $f\colon X_1 \to X_2$ *is a homeomorphism between simply connected, smooth, closed 4-manifolds. Let* $f_*\colon H_2(X_1) \to H_2(X_2)$ *be the induced map on homology. If for all* x *in* $H_2(X_1)$ *we have* $g_{X_1}(x) = g_{X_2}(f_*(x))$, *is* X_1 *diffeomorphic to* X_2*?*

(b) *Suppose* $f\colon X_1 \to X_2$ *is a homeomorphism between simply connected smooth compact 4-manifolds with the same boundary* $\partial X_1 = \partial X_2$, *such that* f *is the identity on the boundary. Let* $f_*\colon H_2(X_1, \partial X_1) \to H_2(X_2, \partial X_2)$ *be the induced map on relative homology. If for all knots* K *and for all* x *in* $H_2(X_1, \partial X_1)$ *we have* $g_{X_1,K}(x) = g_{X_2,K}(f_*(x))$, *is* X_1 *diffeomorphic to* X_2 *rel. boundary?*

REMARKS.

(1) A version of this problem appeared in Stipsicz–Szabó [**SS24c**].

(2) There are several ways one might define a genus function for a 4-manifold with boundary. For concreteness, we present one. For X a simply connected, smooth 4-manifold with boundary and K a knot in ∂X, write $g_{X,K}\colon H_2(X, \partial X) \to \mathbb{N}$ for the relative genus function, that assigns to every relative homology class the minimal genus of a smooth embedded surface with boundary K representing that class.

(3) In the closed case, every isometry of the intersection form on H_2 is realized by a homeomorphism [**FQ90**, Theorem 10.1], so one could have stated the problem by starting with the isometry f_*.

(4) On the other hand, it is not the case that every isometry of the intersection form that respects the genus function is realized by a diffeomorphism. For example, the isomorphism $\phi\colon H_2(K3) \to H_2(K3)$ given by $\phi(a) = -a$ is not induced by a diffeomorphism; see [**DK90**, Corollary 9.1.4].

(5) A special case of part (a) is Problem 4.1, the smooth 4-dimensional Poincaré conjecture.

Scribed by: L. Piccirillo

PROBLEM 4.13.

(a) *Does every large* $\mathbb{R}^4$*-homeomorph lie in* $\mathcal{R}_K$ *for some* K *that is not smoothly slice?*

(b) *Does there exist an infinite sequence of knots* $\{K_i\}$ *such that* $\mathcal{R}_{K_i} \neq \mathcal{R}_{K_j}$ *whenever* $i \neq j$*?*

REMARKS.

(1) Given a knot K, let X_K denote its 0-trace, which is obtained, by definition, by attaching a 0-framed 2-handle to B^4 along K and smoothing corners. Reserve the symbol $\mathbb{R}^4$ to refer to 4-dimensional Euclidean space with its standard smooth structure. Following Gompf, we call a smooth manifold $\mathcal{R}$ an $\mathbb{R}^4$*-homeomorph* if it is homeomorphic to $\mathbb{R}^4$, but not necessarily diffeomorphic. This is to avoid the terminology 'exotic $\mathbb{R}^4$' in cases where we want to allow the standard smooth structure.

(2) An $\mathbb{R}^4$-homeomorph is said to be *large* if it contains a compact subset that does not admit a smooth embedding into $\mathbb{R}^4$, and *small* if it is not large.

(3) Given a knot K let $\mathcal{R}_K$ denote the set of $\mathbb{R}^4$-homeomorphs admitting a smooth embedding of X_K. Gompf [**Gom85**, Lemma 1.1] described a construction of elements in $\mathcal{R}_K$ using Quinn's smoothing theorem. By the trace embedding lemma, $\mathcal{R}_K$ is nonempty precisely if K is topologically slice (not necessarily smoothly slice), and K is smoothly slice if and only if $\mathbb{R}^4 \in \mathcal{R}_K$ if and only if $\mathcal{R}_K$ is the set of all $\mathbb{R}^4$-homeomorphs, including all small $\mathbb{R}^4$-homeomorphs.

(4) Let K_0 be the unknot, K_1 be the second iterated positive Whitehead double of the right-handed trefoil, and K_2 any knot with $\tau(K_2) < 0$, e.g. $-K_1$. Then $\mathcal{R}_{K_i} \neq \mathcal{R}_{K_j}$ for all $i \neq j$.

(5) An affirmative answer to the analogue of (b) for links follows from [**Gom85**, Lemma 1.2 and the proof of Theorem 1.3].

Proposed for K3 by: L. Piccirillo, A. Ray, D. Ruberman
Scribed by: A. Ray

PROBLEM 4.14. *Is there a universal cork? More precisely, does there exist some cork (C, f) such that given any pair W and W' of closed, simply connected 4-manifolds that are homeomorphic but not diffeomorphic, there exists a smooth embedding $C \hookrightarrow W$ such that $(W \setminus \operatorname{Int} C) \cup_f C$ is diffeomorphic to W'?*

REMARKS.

(1) A cork (C, f) is a pair of a compact, contractible, smooth 4-manifold C and a diffeomorphism $f\colon \partial C \to \partial C$, such that f does not extend to a diffeomorphism of C. Some authors require f to be an involution, or C to be Stein.

(2) Not all corks are universal: Ladu shows that the Akbulut cork is not universal [**Lad25**]. Let (C, f) denote the Akbulut cork. Briefly, the paper shows that since $-C$ has trivial Floer homology in the relevant grading, $(-C, f)$ cannot be used to go between two smooth 4-manifolds with differing Seiberg-Witten invariants, such as $K3\#\mathbb{CP}^2$ and $\#^4\mathbb{CP}^2\#^{19}\overline{\mathbb{CP}}^2$.But then twisting by (C, f) cannot be used to go between the exotic pair $-(K3\#\mathbb{CP}^2)$ and $-(\#^4\mathbb{CP}^2\#^{19}\overline{\mathbb{CP}}^2)$. This suggest an alternative definition of a universal cork, by allowing either a cork twist by C or $-C$. Under this definition it is still open whether the Akbulut cork is universal.

(3) One could ask the question only requiring W and W' to be compact with possibly nonempty boundary. Say a cork (C, f) is *∂-universal* if any pair of compact, simply connected 4-manifolds, which are homeomorphic but not diffeomorphic, is related by a cork twist along some embedding of C. The same preprint of Ladu mentioned above shows that there is no ∂-universal cork [**Lad25**].

(4) By Wall's theorem on h-cobordisms [**Wal64b**], if (C, f) is a cork then there exists n such that f extends to a diffeomorphism of $C\#^n S^2 \times S^2$. So, if there is a universal cork, then there is a fixed n so that any exotic pair W, W' as above are diffeomorphic after stabilization with n copies of $S^2 \times S^2$. See Problem 4.7. Note that if there is no bound on the number of $S^2 \times S^2$ summands needed to make a pair of exotic 4-manifolds diffeomorphic, there cannot be a closed universal cork.

This suggests another refinement of the terminology: say a cork (C, f) is *n-universal* if it relates every pair of homeomorphic simply connected 4-manifolds that become diffeomorphic after connected sum with at most n copies of $S^2 \times S^2$.

Scribed by: A. Ray

4.2. Existence of smooth 4-manifolds

The problems in this section ask for the existence of smooth 4-manifolds with certain properties, for example relating to the geography question for certain classes of 4-manifolds, which asks which values of the second Betti number $b_2(X)$, the signature $\sigma(X)$, and the Euler characteristic $\chi(X)$ can be attained for X within the given class. Another set of problems concern smoothability, that is, the existence of smooth structures on particularly interesting examples of topological 4–manifolds.

PROBLEM 4.15 (11/8 Conjecture). *Does every smooth, spin, closed 4-manifold X satisfy*

$$b_2(X) \geqslant \frac{11}{8}|\sigma(X)|,$$

where $\sigma(X)$ is the signature of the intersection form?

REMARKS.

(1) This is [**Kir97**, Problem 4.92], originally raised by Matsumoto.

(2) A positive resolution would answer the *Geography Question* 4.54 for the smooth simply connected case. By Rokhlin's theorem, the intersection form of a spin, smooth, closed 4-manifold is of the form

$$2p(\pm E_8) \oplus q \begin{pmatrix} 0 & 1 \\ 1 & 0 \end{pmatrix},$$

for $p, q \geqslant 0$ and a choice of sign $\pm$. The 11/8 conjecture can be rephrased as $q \geqslant 3p$. If an intersection form satisfies this bound, it is realized by the manifold $p(\pm K3)\#(q-3p)(S^2 \times S^2)$.

(3) Donaldson [**Don87b**] used Yang-Mills theory to prove that $p \geqslant 1$ implies $q \geqslant 3$, assuming $H_1(X; \mathbb{Z})$ has no 2-torsion.

Furuta [**Fur01**] used finite dimensional approximation of the Seiberg-Witten map to prove the inequality

$$b_2(X) \geqslant \frac{10}{8}|\sigma(X)| + 2$$

(i.e., $q \geqslant 2p + 1$) assuming $q \geqslant 1$.

Hopkins, Lin, Shi and Xu [**HLSX22**] refined Furuta's method to get, for $p \geqslant 2$,

$$q \geqslant \begin{cases} 2p+2 & \text{if } p \equiv 1, 2, 5, 6 \pmod 8 \\ 2p+3 & \text{if } p \equiv 3, 4, 7 \pmod 8 \\ 2p+4 & \text{if } p \equiv 0 \pmod 8 \end{cases}$$

They also showed this is the strongest possible inequality that can be proved using that method.

(4) There is a suggested strategy to prove the conjecture [**Bau12, Man14**]. The $11/8$ conjecture would be true if one had an invariant $f(Y)$ of oriented integral homology 3-spheres Y that satisfies the following property:

Let W be a compact oriented spin smooth cobordism from an integral homology 3-sphere Y_0 to an integral homology 3-sphere Y_1. The intersection form of W is of the form $rE_8 \oplus q\left(\begin{smallmatrix} 0 & 1 \\ 1 & 0 \end{smallmatrix}\right)$, where $q \geqslant 0$ and r can be any integer. The required property for f is that

$$f(Y_0) \leqslant f(Y_1) + q + r - 1.$$

(5) There is also a proposal to approach the conjecture using the Ricci flow, by reducing it to the Hitchin-Thorpe inequality for Einstein 4-manifolds. See [**Bam21b**, Section 5.6].

Scribed by: H. Konno, C. Manolescu

PROBLEM 4.16.

(a) *Do there exist closed, oriented, smooth, irreducible 4-manifolds with* $b_2^+ > 1$ *and* $c_1^2 := 2\chi + 3\sigma < 0$*?*
(b) *Is there an irreducible exotic smooth structure on* $\mathbb{CP}^2 \#^n \overline{\mathbb{CP}}^2$ *for* $n > 9$*?*

REMARKS.

(1) S^2–bundles over Σ_h are irreducible 4–manifolds with $b_2^+ = 1$, and for $h \geqslant 2$, they have $c_1^2 < 0$.
(2) An exotic $\mathbb{CP}^2 \#^n \overline{\mathbb{CP}}^2$ for $n > 9$ would have $c_1^2 < 0$. All known exotica in that range are blow-ups of exotic $\mathbb{CP}^2 \#^n \overline{\mathbb{CP}}^2$ for $n \leqslant 9$ (such as the Barlow or Dolgachev surface).

Proposed for K3 by: R. Fintushel, R. Stern
Scribed by: I. Baykur, C. Manolescu

PROBLEM 4.17. *Is there an irreducible, closed, simply connected, oriented 4–manifold with* b_2^+ *and* b_2^- *both even?*

REMARKS.

(1) The condition $b_2^+(X)$ and $b_2^-(X)$ both being even is equivalent to the holomorphic Euler characteristic $\chi_h(X) := \frac{1}{4}(e(X) + \sigma(X))$ not being an integer for either orientation on X. Note that $\chi_h(X)$ always lies in $\mathbb{Z}\left[\frac{1}{2}\right]$ and is in $\mathbb{Z}$ if and only if X admits an almost complex structure.

All current smooth invariants, such as the Seiberg–Witten invariants used for detecting irreducibility effectively, are defined for 4–manifolds with $b_1 = 0$ when $\chi_h \in \mathbb{Z}$. As a result, all known irreducible, closed, orientable 4–manifolds are almost complex under at least one orientation.

(2) This problem is discussed in [**Kir97**, Problem 4.97], where it is conjectured that no such X exists. A notable test case, suggested by Gompf [**Kir97**], involves taking two copies of $K3$, removing the tubular neighborhood of an embedded sphere with self-intersection -2 from each, and then gluing the complements along their boundaries with an orientation-reversing diffeomorphism of $\mathbb{RP}^3$. The resulting 4–manifold X, which has $\chi_h(X) \notin \mathbb{Z}$ for either orientation, is not known to be reducible.

Scribed by: I. Baykur

PROBLEM 4.18.

(a) *Does there exist a pair of smooth, closed 4-manifolds that are homotopy equivalent but not simple homotopy equivalent?*

(b) *Does there exist a pair of smoothly h-cobordant, smooth, closed 4-manifolds that are not smoothly s-cobordant?*

REMARKS.

(1) There exist examples of topologically h-cobordant 4-manifolds that are not topologically s-cobordant [**KPR22**], and examples of topological manifolds that are homotopy equivalent but not simple homotopy equivalent [**NNP23**].

(2) A potential way to answer the questions is as follows. Let M be a smooth, closed 4-manifold with $\pi := \pi_1(M)$. Let $x \in \mathrm{Wh}(\pi)$ be an element of the Whitehead group of π.

QUESTION. Does there exist a smooth h-cobordism $(W; M, N)$ between 4-manifolds with Whitehead torsion $\tau(W, M) = x$?

Let $w \colon \pi \to C_2$ be the orientation character of M. The Whitehead group admits an involution, sending a representative matrix to its conjugate-transpose. Elements are conjugated by the standard involution $x \to \bar{x}$ induced by extending $g \to w(g)g^{-1}$ linearly. The question is of particular interest when $x - \bar{x} \neq 0$, because then the induced homotopy equivalence $M \to N$ is not simple. This could lead to examples, if one can also compute the Whitehead torsions of all the homotopy self-equivalences.

(3) This problem is a little different from other similar smooth realization problems. In the process of building an h-cobordism, one adds 2-handles and then tries to add 3-handles. One can in fact represent the desired homotopy classes in the middle level by smoothly embedded, framed spheres. Unfortunately this does not suffice. The challenge is to find smooth embeddings in such a way that the inclusion-induced map $\pi_1(N) \to \pi_1(W)$ is an isomorphism.

For every $x \in \mathrm{Wh}(\pi)$, there exist $k \in \mathbb{N}$ such that x is smoothly realizable as the torsion $\tau(W, M\#^k S^2 \times S^2)$ of some smooth h-cobordism W based on $M\#^k S^2 \times S^2$ [**CS71**]; a preliminary construction appeared in [**Sta65**].

If one can find a topological h-cobordism $(W; M, N)$ with torsion x, then it might be possible to apply smoothing theory for 5-manifolds to improve W to a smooth h-cobordism.

(4) Another potential method for solving (a) was given by Kasprowski–Nicholson–Veselá in [**KNV24**]. Their proposed method, if it could be implemented, would give examples that are homotopy equivalent but not stably simple homotopy equivalent.

Proposed for K3 and scribed by: M. Powell, D. Ruberman

PROBLEM 4.19. *What are the possible Euler characteristics of closed, aspherical 4-manifolds? More specifically, we ask the following.*

(a) *Is it always the case that $\chi \geqslant |\sigma|$?*
(b) *What is the smallest Euler characteristic of a closed hyperbolic 4-manifold?*

REMARKS.

(1) The questions are motivated by the following conjecture.

CONJECTURE. *The Euler characteristic of X is non-negative.*

This is [**Kir97**, Problem 4.10]. One can ask the question for all aspherical, closed 4-manifolds, including non-smoothable ones.

Edmonds [**Edm15**] has proven the conjecture for Haken 4-manifolds, which were defined by Foozwell and Rubinstein in [**FR11**].

(2) Given that hyperbolic manifolds are aspherical, it is natural to ask the specific subquestion (b). By Chern–Gauss–Bonnet, for a closed hyperbolic 4-manifold X we have

$$\mathrm{vol}(X) = \frac{4}{3}\pi^2 \cdot \chi(X),$$

so we are equivalently asking for the smallest possible hyperbolic volume. The smallest known orientable hyperbolic 4-manifold was found by Conder and Maclachlan [**CM05**] and has $\chi = 16$; it covers a nonorientable hyperbolic manifold with $\chi = 8$. Notice that in the cusped case there are many known orientable examples with the minimal possible value $\chi = 1$ [**RT00**].

Scribed by: M. Powell

PROBLEM 4.20. *Is* $*\mathbb{RP}^4 \# *\mathbb{RP}^4$ *smoothable? Is* $*\mathrm{En} \# *\mathrm{En}$ *smoothable?*

REMARKS.

(1) The manifold $*\mathbb{RP}^4$ was first constructed by Ruberman [**Rub84**] (see also [**HKT94**]). It is homotopy equivalent to $\mathbb{RP}^4$ but not homeomorphic to $\mathbb{RP}^4$. In particular, it has nontrivial Kirby–Siebenmann invariant. Taking the connected sum of two copies yields a 4-manifold $*\mathbb{RP}^4 \# *\mathbb{RP}^4$ with trivial Kirby–Siebenmann invariant. Recall that a 4-manifold with trivial Kirby–Siebenmann invariant is *stably smoothable*, i.e. smoothable after connected sum with $k(S^2 \times S^2)$ for some $k \geqslant 0$.

(2) Recall that $*\mathbb{CP}^2$ denotes the *Chern manifold* constructed by Freedman in [**Fre82**]. It is homotopy equivalent to $\mathbb{CP}^2$ but not homeomorphic to $\mathbb{CP}^2$. In this case, we know that $*\mathbb{CP}^2 \# *\mathbb{CP}^2 \cong \mathbb{CP}^2 \# \mathbb{CP}^2$, and is therefore smoothable. Similarly, $*\mathbb{RP}^4 \# *\mathbb{CP}^2$ is smoothable [**RS97**], answering [**Kir97**, Problem 4.82]. Whether $*\mathbb{RP}^4 \# *\mathbb{RP}^4$ is smoothable is the next natural case of a 4-manifold with trivial Kirby–Siebenmann invariant that is not obviously smoothable.

(3) The smoothability of topological manifolds that are homotopy equivalent to $\mathbb{RP}^4 \# \mathbb{RP}^4$, arising from another construction using the action of L_5 on the structure set, is considered in Problem 4.22. By contrast, $*\mathbb{RP}^4 \# *\mathbb{RP}^4$ is detected in the normal invariants. See [**BDK07**] for the computation of the topological structure set of $\mathbb{RP}^4 \# \mathbb{RP}^4$.

(4) We use En to denote the Enriques surface, which by definition is a quotient of the K3 surface by an involution. The manifold $*\mathrm{En}$ is a *star partner* of En, as defined by Freedman–Quinn [**FQ90**, Section 10.4]. Once again, $*\mathrm{En} \# *\mathrm{En}$ has trivial Kirby–Siebenmann invariant, and it would be interesting to know whether or not it is smoothable, and similarly for $*\mathbb{RP}^4 \# *\mathrm{En}$.

Proposed for K3 by: M. Powell, A. Ray
Scribed by: A. Ray

PROBLEM 4.21. *Is every topological closed 4–manifold M the union of submanifolds $Y \cup Z$, where Y is smoothable, Z is acyclic, and $Y \cap Z$ is their common boundary, a homology sphere Σ?*

REMARKS.

(1) This is [**Kir97**, Problem 4.74].

(2) In the case that M is simply connected the result holds. Freedman [**Fre82**, Theorem 1.5] proved that the homeomorphism type of a simply connected closed 4–manifold is determined by its unimodular intersection form and its Kirby-Siebenmann invariant. The proof consists of an explicit construction of a manifold for each such form and possible Kirby-Siebenmann invariant. The construction begins with a smooth compact 4-manifold built from D^4 by adding 2-handles and with the desired intersection form. It then modifies the handle attachments as needed to achieve the desired Rokhlin invariant for Σ. Lastly, it caps the manifold off with a contractible topological manifold.

(3) There are compact topological manifolds that do not support such decompositions. We explain how to construct an example.

Let $K \subseteq S^3$ be a topologically slice knot that does not bound a smooth slice disk in any smooth compact acyclic 4–manifold with boundary S^3. Examples of such were first identified by Akbulut in unpublished work. For instance, the untwisted Whitehead double of the trefoil suffices. Let M denote the complement an open tubular neighborhood of a locally flat slice disk. Notice that $\partial M \cong S^3_0(K)$, 0–surgery on K. The claim is that M cannot have a decomposition of the desired type; an outline of a proof follows.

Suppose that $\Sigma \subseteq M$ is a locally flat homology 3–sphere that splits M as the union of topological manifold Y and an acyclic manifold Z. We have that $\partial Y \cong \Sigma \cup S^3_0(K)$.

Assume that there exists a smoothing A of Y; that is, A is a smooth manifold supporting a homeomorphism to Y. Attaching a smooth 2–handle to A along the meridian of K in $S^3_0(K)$, referred to as the trace of K in $S^3_0(K)$, yields a smooth homology $S^3 \times I$ with boundary $S^3 \cup \Sigma$. Attaching a smooth 4–ball results in a smooth acyclic 4–manifold A' with boundary Σ. We now see that the union $B = A' \cup_\Sigma A$ is a smooth homology $S^1 \times B^3$ with boundary $S^3_0(K)$. Appropriately attaching a 2–handle to B along the trace of $K \in S^3_0(K)$ results in a smooth acyclic 4–manifold bounded by S^3. The cocore of the 2–handle is a smooth slice disk for its boundary, which represents $K \subseteq S^3$.

Proposed for K3 by: D. Ruberman
Scribed by: D. Ruberman, A. Ray

PROBLEM 4.22. *Let π be a good group, and let X be a smooth 4-manifold with $\pi_1(X) = \pi$. Does $L^s_5(\mathbb{Z}[\pi])$ act on the smooth structure set of X, in such a way that the action reduces under the forgetful map to the Wall realization action on the topological structure set? In particular, can this be done when $\pi = \mathbb{Z}$, $\mathbb{Z} \times \mathbb{Z}_n$, or $\mathbb{Z}_2 * \mathbb{Z}_2$?*

REMARKS.

(1) The surgery exact sequence, valid for higher dimensional manifolds, ends with

$$\mathcal{N}(X \times I, \partial(X \times I)) \to L^s_{n+1}(\mathbb{Z}\pi) \xrightarrow{\Theta} \mathcal{S}(X) \to \mathcal{N}(X) \to L^s_n(\mathbb{Z}\pi).$$

Here $\mathcal{S}(X)$ is the smooth structure set and $\mathcal{N}(X)$ is normal bordism classes of degree one normal maps. The structure set consists of orientation-preserving homotopy equivalences $f\colon M \to X$ up to diffeomorphisms $F\colon M' \to M$ commuting up to homotopy with the homotopy equivalences. What is written as a map Θ is shorthand for an action of $L^s_{n+1}(\mathbb{Z}\pi)$ on $\mathcal{S}(X)$. The action of an element $A \in L^s_{n+1}(\mathbb{Z}\pi)$ on $(X, \mathrm{Id}\colon X \to X)$ produces, for $n+1 \geqslant 6$, a (normal) cobordism W from (X, Id) to another homotopy equivalence $f\colon X' \to X$. The problem asks if there is an action of $L^s_5(\mathbb{Z}[\pi])$ when X is a closed smooth 4-manifold, that reduces to the Wall realization action on the topological structure set $\mathcal{S}_{\mathrm{TOP}}(X)$.

(2) Cappell and Shaneson [**CS71**] showed that one can realize the action of L_5^s stably, i.e. after replacing X by its connected sum with some number of copies of $S^2 \times S^2$. For example, Scharlemann's construction [**Sch76**] realizes the action of the generator of $L_5^s(\mathbb{Z}[\mathbb{Z}])$ on $(S^1 \times S^3)\#(S^2 \times S^2)$ (this is also in [**CS71**] using an unpublished calculation of R. Lee).

In general, even when if one can realize the action of an element of L_5^s to produce a different element of the structure set, the resultant manifold X' might still be diffeomorphic to X. The difference in the structure set would then be due to the resultant self homotopy equivalence of X not being homotopic to a diffeomorphism. For instance, Akbulut [**Akb99**] showed that Scharlemann's manifold is diffeomorphic to $(S^1 \times S^3)\#(S^2 \times S^2)$.

In many cases of interest, in particular for $\pi = \mathbb{Z}$ and $\pi = \mathbb{Z}/2 * \mathbb{Z}/2$, the Whitehead group of π is trivial, in which case $L_5^s(\mathbb{Z}\pi) = L_5(\mathbb{Z}\pi)$ and one can ignore the s-decoration.

(3) The heart of the question is then whether one can realize the action without stabilization. There are three cases of particular geometric interest, corresponding to fundamental groups $\mathbb{Z}$, $\mathbb{Z}\times\mathbb{Z}/n$ (for $n > 1$), and $\mathbb{Z}/2*\mathbb{Z}/2$. All of these are good groups, so that realization works in the topological category; the issue is to find smooth realizations.

(i) Realizing the action of $L_5(\mathbb{Z}[\mathbb{Z}])$ on $X = S^1 \times S^3$. Every self-homotopy equivalence from X to itself is homotopic to a diffeomorphism, so realizing the action of the generator of $L_5(\mathbb{Z}[\mathbb{Z}])$ would produce an exotic $S^1 \times S^3$. It would be detected by the Rokhlin invariant of any spin manifold carrying the generator of H_3. A gauge-theoretic conjecture of Furuta-Ohta [**FO93**] would imply that no such manifold exists; see Problem 4.65 and [**RS05**] and [**MRS11**] for approaches to this conjecture.

QUESTION. Is there an exotic $S^1 \times S^3$ realizing the action of the generator of $L_5(\mathbb{Z}[\mathbb{Z}])$ on $X = S^1 \times S^3$?

(ii) Realizing the action of $\pi = \mathbb{Z} \times \mathbb{Z}/n$ where $L_5^s(\mathbb{Z}\pi)$ is large, and one would want to produce an action on the structure set of $S^1 \times L(n,q)$. In this setting, the manifold would again be fake, i.e. homotopy equivalent but not homeomorphic to $S^1 \times L(n,q)$, detected by codimension-one multisignature (or Atiyah-Singer invariants). In this case the topological manifold set, i.e. the structure set modulo homotopy self-equivalences of $S^1 \times L(n,q)$ is infinite. For a given n, only finitely many of them (corresponding to $S^1 \times L(n,q')$ where $L(n,q) \simeq L(n,q')$) are known to be smoothable, but as far as we know they could all be.

QUESTION. For any $n \geqslant 2$, is there a smooth structure on a fake $S^1 \times L(n,q)$ realizing the action of $L_5^s(\mathbb{Z}[\mathbb{Z} \times \mathbb{Z}/n])$ on $X = S^1 \times L(n,q)$, that is not of the form $S^1 \times L(n,q')$?

(iii) The group $\pi = \mathbb{Z}/2 * \mathbb{Z}/2$ is relevant to the problem of classifying manifolds homotopy equivalent to $\mathbb{RP}^4\#\mathbb{RP}^4$. The group $L_5(\mathbb{Z}\pi)$ contains [**Cap74**] an infinitely generated subgroup $\mathrm{UNil}_1(\mathbb{Z},\mathbb{Z}^-,\mathbb{Z}^-)$. The smooth action of an element in this subgroup would produce a homotopy equivalence from a manifold X to $\mathbb{RP}^4 \# \mathbb{RP}^4$ that is not homotopic to $f_1 \# f_2$ where the f_i are homotopy equivalences $X_i \to \mathbb{RP}^4$, $i = 1, 2$. A paper of Jahren-Kwasik [**JK06**] uses the topological realization of this Unil subgroup to construct topological 4-manifolds that are homotopy equivalent to $\mathbb{RP}^4 \# \mathbb{RP}^4$ that are not connected sums of homotopy $\mathbb{RP}^4$s. (See [**BDK07**] for the full topological classification.)

QUESTION. Are the non-splittable manifolds from [**JK06**] smoothable?

This is equivalent to asking about the smooth realizability of the corresponding Unil group elements.

Proposed for K3 and scribed by: M. Powell, D. Ruberman

4.3. Embedding 3-manifolds

The Schoenflies problem is perhaps second only to the smooth Poincaré conjecture in its historical significance. The remaining problems seek to encourage further study in the realm of embedding 3-manifolds in S^4.

PROBLEM 4.23 (Schoenflies problem). *If Σ is a smoothly embedded S^3 in S^4, then its closed complements are smooth 4-balls.*

REMARKS.

(1) This is the famous Schoenflies problem, which appears in [**Kir97**, Problem 4.32].

(2) Progress on this old and important conjecture has, sadly, been more aspirational than concrete. Scharlemann proved this conjecture in the special case there is a smooth function $f\colon S^4 \to \mathbb{R}$ whose restriction to the S^3 is Morse with k 0-handles and $\leqslant k+1$ 1-handles (so that the middle level has genus $\leqslant 2$) [**Sch84**]. Following Gabai's proof of *Property R* [**Gab87b**], Scharlemann [**Sch08**] used a reembedding process to extend this result in the previous update to the case of $(k+2)$ 1-handles, so the middle level has genus $\leqslant 3$. The argument explores further connections between the *Schoenflies Conjecture* and the *generalized Property R Conjecture* (see Problem 1.10). Other reembedding approaches have been proposed by Akbulut [**Akb14**] (twisting corks), Agol–Freedman [**AF15**], and Lambert-Cole [**LC21**]. Lambert-Cole employs Stein trisections and proposes this generalization of the Schoenflies Conjecture: any homotopy 4-ball in a compact Stein domain of complex dimension 2 is diffeomorphic to the standard 4-ball. Gabai [**Gab22**] has suggested an approach via pseudo-isotopy theory. In the other direction, Gabai, Naylor and Schwartz [**GNS25**] exhibit potential counterexamples – Schoenflies balls that are not known to be standard.

(3) This problem is equivalent to the corresponding one in the PL category. By Mazur [**Maz59**] both closed complements are topological 4-balls. They are also quasi-conformal [**Geh67**] and the Lipschitz [**LV77**] 4-balls. It is well known following Mazur [**Maz59**] and Cerf [**Cer68**] that the Schoenflies problem is equivalent to showing that if $f : S^1 \times S^3 \to S^1 \times S^3$ is a diffeomorphism, then after passing to a sufficiently large finite cover the lifted diffeomorphism can be isotoped to be supported in a 4-ball. By [**BG19**] it is also equivalent to showing that if $B \subset S^4$ is a 3-ball with ∂B the standard 2-sphere S, then after passing to a finite branched cover of S^4 branched over S, B is isotopic rel ∂ to the standard 3-ball. By [**Gab22**] it also would follow from a *carving/surgery* problem. It is not known whether or not a Schoenflies ball $\times I$ is always B^5.

Scribed by: D. Gabai, M. Miller, M. Scharlemann

PROBLEM 4.24. *Let K be a framed knot in $S^3 = \partial B^4$. Let U be a meridian of K. Does there exist a smoothly embedded disk D in $B^4 \cup_{\nu K} h^2$ bounded by U that intersects the cocore of h^2 algebraically zero times, such that the 4-manifold obtained by removing an open tubular neighborhood of the disk is not diffeomorphic to the 4-ball?*

REMARKS.

(1) We represent the situation of this problem in Figure 1. Note the "big dot" placed on U. An "ordinary dot" on an unknot is an instruction to carve the standard disk (in B^4) bounded by the unknot. An important variation is to carve a specific ribbon disk with boundary a ribbon knot (see [**Akb16**, Section 1.4] for examples and further discussion).

If the carved disk in the setting of this problem is a disk in the 0-handle (i.e. intersecting the cocore of h^2 geometrically zero times) then the 4-manifold $\mathcal{B}$ is B^4. The "big dot" indicates that we may carve along *any* disk bounded by U that intersects the cocore of h^2 algebraically zero times. In other words, we carve along any disk that represents the trivial class in H_2(0-handle $\cup$ 2-handles,∂; $\mathbb{Z}$) but do not restrict the carving disk

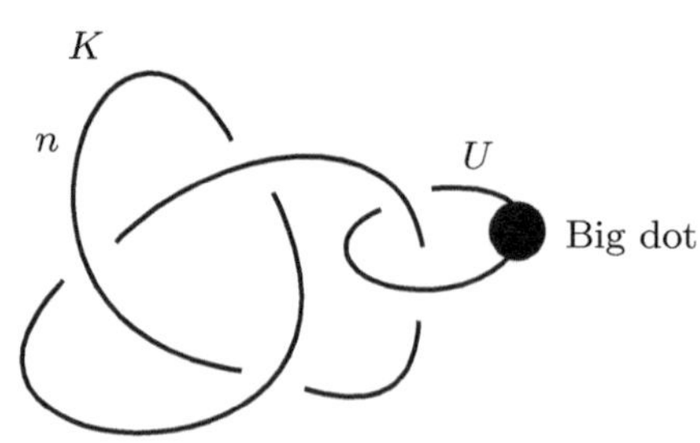

FIGURE 1. A diagram of a family of homotopy 4-balls. The resulting 4-manifold $\mathcal{B}$ depends on a choice of carving disk with boundary U. This disk is required to intersect the cocore of the pictured 2-handle algebraically zero times, causing the resulting 4-manifold $\mathcal{B}$ to be a homotopy 4-ball.

to be contained in the 0-handle. In the setting of this problem, the result of carving is a homotopy 4-ball $\mathcal{B}$ with boundary S^3. Thus, this construction is a natural extension of the idea of carving along nontrivial disks and yields potentially nonstandard homotopy 4-balls (see relevant discussion in Problem 4.1, i.e. the smooth 4-dimensional Poincaré Conjecture).

(2) The case that K is slice and $n = 0$ is particularly interesting, for then $\mathcal{B}$ smoothly embeds in B^4. Thus, the question of whether $\mathcal{B}$ is necessarily a 4-ball is related to the smooth 4-dimensional Schoenflies conjecture (Problem 4.23).

Proposed for K3 by: D. Gabai
Scribed by: R. Kirby, M. Miller, D. Ruberman

PROBLEM 4.25. *Under what conditions does a closed, orientable 3-manifold M smoothly embed in S^4? Is this question algorithmically decidable?*

REMARKS.

(1) This is in [**Kir78**, Problem 3.20]. As noted there, every homology 3-sphere embeds topologically in S^4 [**Fre82**]. Smoothly, even for homology spheres, there is an obstruction coming from the Rokhlin invariant, and more obstructions coming from gauge theory and its cousins, using the fact that if a homology sphere M embeds in S^4 then M bounds a homology ball, so is trivial in the homology cobordism group. (The relationship between embedding problems and homology cobordism was first noted in [**GL83**]; see, e.g., [**DHST23**] for a relatively recent summary of nontriviality results for the 3-dimensional homology cobordism group.) There are also homology spheres that bound homology balls but do not embed in S^4 [**McD22**]. For 3-manifolds with nontrivial homology, even topologically there are obstructions, starting with an obstruction coming from the homology [**Han38**]; see [**BB22**] for a relatively recent survey of such obstructions. In particular, it would also be interesting to understand the question of which non-homology spheres embed locally flatly in S^4; see [**Hil24**] for an extensive discussion. Another interesting special case is obstructing punctured homology 3-spheres from embedding in S^4; see Problem 4.26.

(2) Related problems include Problem 4.28 (about embeddings in connected sums of $S^2 \times S^2$) and Problem 3.27 (the computational complexity of the homeomorphism and recognition problems for 3-manifolds).

(3) Some other related questions include which closed, orientable 3-manifolds embed (smoothly or topologically flatly) in $K3$; and which homology 3-spheres embed in homology 4-spheres (in either the smooth or locally flat categories).

Scribed by: R. Lipshitz

PROBLEM 4.26. *If Y is a homology three-sphere, does the punctured manifold $Y_0 = Y \setminus \mathrm{Int}(B^3)$ smoothly embed in S^4?*

REMARKS.

(1) Every homology 3-sphere embeds topologically in S^4 (see [**FQ90**, Corollary 9.3C]). The Poincaré homology sphere does not embed smoothly in S^4, since it has nonzero Rokhlin invariant. Every homology 3-sphere Y embeds in a homology 4-sphere, given by spinning Y. It is constructed by surgery on $S^1 \times \{p\} \subseteq S^1 \times Y$ for some $p \in Y$.

(2) This problem is related to the group-theoretic question 3.39 of whether every finitely generated perfect group has weight one, i.e. is the normal closure of a single element [**KM14a**, Problem 5.52]. A twisted version of the spinning construction mentioned above shows that if Y is a homology sphere whose fundamental group has weight one, then Y_0 embeds in a homotopy 4-sphere. In some cases, one can show that the homotopy sphere is actually S^4. For instance, Larson [**Lar15**] proved that $1/n$-surgery on a knot in S^3 embeds, when punctured, into S^4.

(3) Using the twist-spinning construction for knots and a result of Zeeman [**Zee65**, Corollary 2], one can prove that every cyclic cover of S^3 branched over a knot embeds in S^4 after puncturing. This gives a lot of examples of $\mathbb{Z}$-homology 3-spheres that embed, after puncturing, in S^4. For example, the 2-fold cyclic branched cover on a knot of determinant one, the p-fold cyclic cover of the torus knot $T(q,r)$, with (p,q,r) coprime, and any finite cyclic branched cover on a knot of Alexander polynomial one.

Proposed for K3 and scribed by: C. Livingston

PROBLEM 4.27. *Find exotic 3-balls in S^4, considered up to isotopy rel. boundary. That is, find a pair of 3-balls B_1, B_2 smoothly embedded in S^4 with the same boundary such that B_1, B_2 are topologically but not smoothly isotopic rel. boundary.*

REMARKS.

(1) In [**BG19**, Section 9], Budney–Gabai observe that if no such 3-balls exist, then the smooth 4-dimensional Schoenflies Conjecture (Problem 4.23) is true.

(2) Budney–Gabai [**BG19**] find infinitely many 3-balls smoothly embedded in S^4 with the same boundary that are not topologically isotopic rel. boundary. These 3-balls are distinguished by considering corresponding automorphisms of $S^1 \times B^3$ coming from barbell diffeomorphisms. These maps are distinguished even as homeomorphisms.

QUESTION. Is there a smooth rel. boundary automorphism of $S^1 \times B^3$ that is topologically but not smoothly isotopic rel. boundary to the identity map?

Proposed for K3 by: S. Kim
Scribed by: S. Kim, M. Miller

PROBLEM 4.28. *Every closed, orientable 3-manifold embeds smoothly in some connected sum of copies of $S^2 \times S^2$. Given a closed 3-manifold M, let $s(M) \geqslant 0$ denote the smallest integer such that M embeds smoothly (respectively locally flatly) in a connected sum of $s(M)$ copies of $S^2 \times S^2$. Let $\widetilde{s}(M) \geqslant 0$ denote the smallest integer such that M embeds smoothly (respectively locally flatly) in a connected sum of $\widetilde{s}(M)$ copies of $S^2 \widetilde{\times} S^2$. Compute the functions $s(M)$ and $\widetilde{s}(M)$, in either the smooth or locally flat case, for interesting classes of 3-manifolds.*

Here are three more specific questions.

(a) *What are the functions s and $\widetilde{s}$ for lens spaces, and for Brieskorn spheres?*
(b) *Freedman [**Fre82**] proved that every $\mathbb{Z}$-homology sphere embeds topologically in S^4. Is there an integer homology 3-sphere such that (smoothly) $s(M) = 1$?*
(c) *Are the functions s or $\widetilde{s}$ algorithmically computable?*

REMARKS. Manifolds with $s(M) = 0$ are the topic of Problem 4.25. The function s was introduced and studied by [**AGL17**]. That paper includes a comprehensive summary of the literature. Note in particular that $s(M) = 1$ implies M bounds an integer homology ball, by an elementary Mayer–Vietoris argument. Edmonds proved that in the topological category, $\widetilde{s}(L_{p,q}) \leqslant 2$ for every p, q [**Edm05**].

Scribed by: M. Powell

4.4. Surfaces in 4-manifolds

A pair of surfaces embedded in a 4-manifold is *exotic* if the surfaces are topologically isotopic but not smoothly isotopic. Here we assume that topological isotopies are locally flat, or equivalently that they are ambient isotopies.

We present a range of problems on surfaces in 4-manifolds, starting with those concerning exotic surfaces, potentially non-smoothable surfaces, and the effect of stabilizations. We consider the algebraic topology of surface complements, the compelling Kinoshita conjecture, and generalizations of Hirose's theorem on extendable diffeomorphisms. After that come problems on concordance of surfaces, families of surfaces, and using surfaces to detect exotic structures.

PROBLEM 4.29. *Let Σ be a locally flat surface in S^4 with $\pi_1(S^4 \backslash \Sigma)$ cyclic.*

(a) *Prove that Σ is topologically unknotted.*
(b) *Assuming that Σ is smoothly embedded and orientable, prove that Σ is smoothly unknotted (or find an example that is not).*

REMARKS.

(1) The surface Σ is known to be topologically unknotted in many cases.
   - If $\Sigma = S^2$, the claim holds by work of Freedman–Quinn (see [**FQ90**]).
   - If Σ is orientable of genus more than 2, the claim holds by Conway–Powell [**CP23**].
   - If Σ is nonorientable of genus h (i.e. $\Sigma \cong \#^h \mathbb{RP}^2$), then the claim holds by Lawson [**Law84**] if $h = 1$; Conway–Orson–Powell [**COP23**] if $h \leqslant 3$ and if $h > 3$ and $|e(\Sigma)| < 2h$; Pencovitch [**Pen24**] if $h = 4, 5$ and $|e(\Sigma)| = 2h$.

Thus, the remaining open cases of (a) (in the topological category) are the following.

- Σ is an orientable surface of genus one or two,
- Σ is a nonorientable surface of genus $h \geqslant 6$ and with normal Euler number equal to $\pm 2h$.

(2) There are many instances of *non-orientable* surfaces in S^4 that are smoothly knotted while their complements have cyclic fundamental group. Finashin–Kreck–Viro [**FKV87**] constructed infinitely many genus 10 non-orientable surfaces that have complement with cyclic fundamental group (and are topologically unknotted [**Kre90**]) but are pairwise not smoothly isotopic. Finashin [**Fin09**] later produced an infinite family of genus 6 nonorientable surfaces that are pairwise not smoothly isotopic but are all topologically unknotted (via [**COP23**]). Matić-Öztürk-Stipsicz-Reyes-Urzúa [**MOR+24**] have announced the existence of a single exotically knotted $\#^5\mathbb{RP}^2$ and Miyazawa [**Miy23**] has announced the existence of an infinite family of exotically knotted projective planes (both papers focus on distinguishing surface smoothly, with the topological unknotting following respectively from [**COP23**] and [**Law84**]).

(3) A well-known question is the following.

QUESTION (i). Suppose that Σ is smooth and the radial function $h\colon S^4 \to \mathbb{R}$ restricts to Σ as a Morse function with exactly one local minimum. Is Σ smoothly unknotted? Is Σ topologically unknotted?

This question includes the case that Σ is a torus and $h|_\Sigma$ has exactly four critical points, which is Problem 4.30 in [**Kir78**]. (Note that if $\Sigma \cong S^2$ and $h|_\Sigma$ has four critical points then Σ is smoothly unknotted by Scharlemann [**Sch85a**]; if $\Sigma \cong \mathbb{RP}^2$ and $h|_\Sigma$ has three critical points then Σ is smoothly unknotted by Bleiler–Scharlemann [**BS88a**].)

This question also includes the case that Σ is a union of two (pushed-in) Seifert surfaces for a knot cross-section, or that Σ is a union of a ribbon surface and a Seifert surface for a knot cross-section.

(4) Question (i) itself has a well-known interesting subquestion.

QUESTION (ii). Suppose Σ is a union of two ribbon surfaces for an unknotted cross-section. Is Σ smoothly unknotted? Is Σ topologically unknotted?

The case that Σ is a 2-sphere appears in Suzuki's survey of problems in 2-knot theory [**Suz76**].

Scribed by: M. Miller

PROBLEM 4.30. *Does there exist a pair of closed, oriented surfaces in S^4 that are topologically but not smoothly isotopic? If such an exotic pair exists, does there exist an infinite family? Can the surfaces be taken to be 2-spheres? Surfaces of genus-g for any given $g \geqslant 0$?*

REMARKS.

(1) There are many examples of exotic oriented surfaces in general 4-manifolds. These include infinite families of pairwise exotic 2-spheres (existence from [**Wal64a**], with smooth obstruction from work of Donaldson [**Don87a, Don87b**]) in some simply-connected manifolds, but the usual smooth obstructions require the ambient 4-manifold to have $b_2^+ > 0$. These examples also feature surfaces that represent nontrivial homology classes. Exotic families of nullhomologous tori [**HS20**] and 2-spheres [**Tor25**] are known to exist in some simply connected 4-manifolds with large b_2, but there are currently no known examples of exotic pairs of higher-genus surfaces in a closed 4-manifold.

(2) In Problem 4.29, we discuss known constructions of pairwise exotic infinite families of nonorientable surfaces in S^4. The first such examples were given by Finashin–Kreck–Viro [**FKV87**]. These nonorientable surfaces are distinguished smoothly by the diffeomorphism classes of their double branched covers.

The double branched cover of an orientable surface in S^4 is a signature-zero manifold, so a first step toward constructing oriented examples might be finding involutions on known constructions of exotic signature-zero manifolds; see e.g. Baykur–Hamada [**BH23**].

(3) Compare this problem to Problem 4.29 (b), which asks whether a smooth, oriented surface in S^4 whose complement has cyclic fundamental group is smoothly unknotted. Such a surface is known to be topologically unknotted if it is a 2-sphere [**FQ90**] or has genus greater than two [**CP23**], so in these cases a smoothly non-standard example would solve both this problem and Problem 4.29(b).

Scribed by: M. Miller

PROBLEM 4.31. *Does every knot in S^3 bound an exotic pair of orientable surfaces in B^4?*

REMARKS.

(1) By an exotic pair, we mean two surfaces with the same knot as boundary, that are topologically but not smoothly isotopic *rel. boundary*.

(2) It is known that certain knots bound exotic genus one surfaces [**JMZ21**] and examples bounding exotic disks are announced in [**Hay21a**]. Improving the ability to produce exotic surfaces for a fixed boundary knot can be thought of as approaching the problem of producing closed oriented exotic surfaces in S^4 (which would follow from producing exotic surfaces with boundary the unknot); see Problem 4.30.

One can also ask the same question about exotic nonorientable surfaces.

Proposed for K3 and scribed by: M. Miller, L. Piccirillo

PROBLEM 4.32. *Does there exist a locally flat embedding $f\colon \Sigma \to S^4$ for some closed surface Σ such that f is not topologically ambiently isotopic to a smooth embedding? Particularly in the case that $\Sigma = S^2$?*

REMARKS.

(1) There are many examples of this behavior in other 4-manifolds. In particular, suppose there is a locally flat embedding of a genus-g surface into a smooth 4-manifold X such that there is no smooth genus-g surface representing the same homology class. Then the embedding is not smoothable. As an explicit example, Rudolph [**Rud84**] showed that for all $d > 6$, there is a locally flat surface in $\mathbb{CP}^2$ representing the homology class $d \in \mathbb{Z} = H_2(\mathbb{CP}^2; \mathbb{Z})$ of genus strictly less than $(d-1)(d-2)/2$. Lee–Wilczyński [**LW97**, Corollary 1.3] extended this to $d > 4$ (and decreased the realized genus of these locally flat surfaces). We deduce from the *Thom conjecture* [**KM94**] that there is a nonsmoothable embedding of a surface into $\mathbb{CP}^2$ representing any homology class $d \in \mathbb{Z}$ for $d > 4$. For this reason, we restrict the question to the ambient manifold S^4.

(2) There are examples of locally flat, non-smoothable embeddings of codimension two spheres in S^n for many $n > 4$. Lashof [**Las71**] gives examples for $n = 5$. In the introduction of [**Bri66**], Brieskorn shows that for $n \equiv 3 \pmod 8$ $(n > 3)$, there is a smooth embedding of an exotic n-sphere into S^{n+2}. Viewing this as a locally flat embedding of the standard n-sphere, the embedding is not smoothable.

Proposed for K3 by: A. Ray
Scribed by: M. Miller

PROBLEM 4.33. *Let S_1, S_2 be two topologically isotopic, smoothly embedded closed surfaces in a closed, oriented, smooth 4-manifold X. When do S_1 and S_2 in X become smoothly isotopic after one*

(a) *external stabilization?*
(b) *standard internal stabilization?*
(c) *internal stabilizations?*

Is there an upper bound depending on X on the number of such stabilizations required to make the surfaces smoothly isotopic?

REMARKS.

(1) We say that $(X, S_i)\#(S^2 \times S^2, \varnothing)$ is an *external stabilization* of (X, S_i), that is, we enlarge the topology of the surface complement by taking a connected sum with $S^2 \times S^2$. If S_1 and S_2 become smoothly isotopic after a finite number of external stabilizations then we say that S_1 and S_2 are *external-stably smoothly isotopic.*

(2) Let $T^2 \subseteq S^4$ be an unknotted torus. We say that $(X, S_i)\#(S^4, T^2)$ is a *standard internal stabilization* of S_i. If S_1 and S_2 become smoothly isotopic after a finite number of standard internal stabilizations then we say that S_1 and S_2 are *strongly internal-stably smoothly isotopic.*

(3) If S_i' is obtained from S_i by a surgery corresponding to attaching a 1-handle to S_i inside X, then we say that (X, S_i') is an *internal stabilization* of S_i. Note that the standard internal stabilization is the case where this 1–handle is attached in a local 4–ball neighborhood. If S_1 and S_2 become smoothly isotopic after a finite number of internal stabilizations then we say that S_1 and S_2 are *internal-stably smoothly isotopic.*
(4) A positive answer to part (b) implies a positive answer to (c), and similarly for the corresponding questions on upper bounds.
(5) Regarding part (a): Galvin [**Gal24**] showed that any two smooth, topologically isotopic surfaces in a simply connected 4-manifold X become smoothly isotopic after a sufficiently high number of external stabilizations of X. Galvin's proof does not control the number of external stabilizations required. A priori, the number of external stabilizations needed depends on X and on the surfaces, but one stabilization suffices in all examples successfully analyzed so far. When $\pi_1(X)$ is nontrivial, it is open whether or not all topologically isotopic surfaces are external-stably smoothly isotopic.
(6) It was shown in [**AKM$^+$19**] that in a smooth simply connected manifold X, two homologous connected surfaces of the same genus, and with simply connected complements, become smoothly isotopic after one external stabilization, as long as their homology class is not dual to $w_2(X)$, i.e. if the class is not characteristic.
(7) There are examples of exotic pairs of surfaces with boundary that require more than one external stabilization when X is not closed. Lin and Mukherjee [**LM25**] give an example of an exotic pair of disks in a punctured $K3$ that remain smoothly non-isotopic after one external stabilization; the disks are distinguished via a Pin(2)-equivariant family Bauer–Furuta invariant. Hayden–Kang–Mukherjee [**HKM23**] also announced examples involving closed surfaces in a manifold with boundary.
(8) Regarding parts (b) and (c): Baykur-Sunukjian [**BS16**] showed that any two homologous surfaces of the same genus become smoothly isotopic after a sufficiently high number of internal stabilizations. Baykur and Sunukjian also prove in [**BS16**] that the same result can be achieved with strong internal stabilizations when S_1, S_2 are exotically knotted surfaces with cyclic fundamental group complement, or more generally if the map $\pi_1(\overline{\partial \nu S_i}) \to \pi_1(X \backslash \nu S_i)$ is surjective. However, even the following question, which asks for a much weaker version of (b), is open.

    QUESTION. For every pair of topologically isotopic surfaces of the same genus, does there exist a sufficiently high finite number of strong internal stabilizations such that the surfaces become smoothly isotopic?

(9) In [**BS16**] it was shown that exotically knotted surfaces that are produced using many well-known (local) operations become smoothly isotopic after a single standard internal stabilization.
(10) In [**Auc23**], Auckly announced that for any given n, there are two homologically essential closed surfaces as above that remain non-smoothly isotopic after n strong internal stabilizations. Auckly notes that one can find examples where such surfaces are related by a diffeomorphism that is pseudo-isotopic to the identity, and one external stabilization suffices for

his examples. He also argues that this behavior exists under only mild constraints on the ambient manifold.

There are no such examples of *null-homologous surfaces* that are known to require more than one internal stabilization.

(11) This relates to versions of (b) and (c) in the relative setting, i.e. when the boundary is nonempty. The first examples of two topologically isotopic, non-smoothly isotopic disks D_1^n, D_2^n in B^4 that do not become smoothly isotopic after n internal stabilizations (for an arbitrarily given n) were presented by Guth [**Gut22**]; the disks are distinguished by Heegaard Floer techniques. Hayden–Kang–Mukherjee [**HKM23**] proposed examples of exotically knotted spheres in a 4-manifold with boundary that do not become smoothly isotopic after one internal or one external stabilization; their construction is based on a claim by Kang [**Kan22b**] about contractible 4-manifolds surviving a stabilization; see Problem 4.7. In [**Hay23**], Hayden announces an example of two topologically isotopic positive genus surfaces with boundary in B^4 where the absence of smooth isotopy, even after one weak internal stabilization, is detected by Khovanov homology invariants.

(12) See Problem 4.7 and Problem 4.79 for other "is one stabilization enough?" questions.

Scribed by: O. Plamenevskaya

Problem 4.34. *Let Σ be a surface embedded in S^4. Can Σ be unknotted by a sequence of torus surgeries in its complement, such that the ambient manifold is always S^4?*

Remarks.

(1) This problem may be addressed in the smooth or topological locally flat categories. This would be an analogue of the 3-dimensional fact that any knot in S^3 can be transformed into the unknot via ± 1 surgery on disjoint unknots.

(2) Without restricting the torus surgeries to preserve S^4, the answer to this question is "yes" by [**Iwa90**]. In fact, any surface Σ can be unknotted by at most two torus surgeries that do not preserve S^4 using Gabai's 4-dimensional light bulb theorem [**Gab20**], since a standard surgery along a torus centered about a meridian of Σ yields a surface $S^2 \times S^2 \# S^1 \times S^3$ that can be standardized using the light bulb theorem.

(3) Larson observed that spun 2-knots can be unknotted by such torus surgeries [**Lar18**, Theorem 3.9]. This argument also applies to twist-spun knots. Larson and Meier [**LM15**, Theorem 1.5] produced a different set of such torus surgeries unknotting 2-knots arising from spinning fibered classical knots.

(4) Note that Σ and the standard unknotted surface $\Sigma_0 \cong \Sigma$ become isotopic after sufficiently many ambient 1-handle surgeries [**BS13**], which roughly

correspond to excessive generators of complementary fundamental group. One way to approach this problem would be to realize these stabilizations and destabilizations as a sequence of certain torus surgeries.

Proposed for K3 and scribed by: M. Miller

PROBLEM 4.35.

(a) *Give an algebraic classification of groups that arise as the fundamental group of the complement of a smooth or locally flat 2-sphere in* S^4.
(b) *Let* $\Lambda = \mathbb{Z}[t, t^{-1}]$. *Classify the* Λ*-modules that arise as the Alexander module of a smooth or locally flat 2-sphere in* S^4.

REMARKS.

(1) This is [**Kir97**, Problem 1.48]. Hillman discusses this problem in the 2022 update of [**Hil02**, Part III, §14.6–14.10]. The analogous problem is also open for classical knots in S^3.
(2) Kervaire [**Ker65b**] gave necessary conditions for a group to be the fundamental group of the complement of a knotted $(n-2)$-sphere in S^n and showed these conditions are also sufficient for $n \geqslant 5$; see [**MW17**, Section 3].
(3) The answer to Problem (b) is known in dimension three when "$\mathbb{Z}$" is replaced by "$\mathbb{Q}$;" see [**Gor78**, Section 6]. One might try to extend this to dimension four, with the extra difficulty that Alexander ideals are generally not principal in this dimension (see [**Fox61**, Example 12]).
(4) If G is the fundamental group of $S^4 \backslash \Sigma$ for Σ a locally flat 2-sphere, then G has weight $w(G) = 1$, $H_1(G) = \mathbb{Z}$, and deficiency $d(G) \leqslant 1$.

    As a partial answer to Problem (a), when $w(G) = 1, H_1(G) = \mathbb{Z}$ and $d(G) = 1$ (rather than $d(G) \leqslant 1$), then by Kervaire [**Ker65a**], G is the fundamental group of the complement of a smooth 2-sphere in a homotopy 4-sphere and hence the complement of a locally flat 2-sphere in S^4 [**FQ90**].
(5) Gonzalez-Acuña [**GAn94**] has given a characterization of groups arising as the fundamental group of the complement of a smooth 2-sphere in S^4 in terms of group presentations. (Specifically, a group G satisfying the Kervaire conditions is such a group if and only if it admits a Wirtinger presentation.) This problem is specifically asking for an *algebraic characterization* along the lines of Kervaire's in dimension at least 5.

Scribed by: M. Miller

PROBLEM 4.36. *Are homotopy types of 2-knot complements determined by their homotopy 2-types?*

REMARKS.

(1) The homotopy 2-type of a topological space X is a triple $(\pi_1(X), \pi_2(X), k(X))$, where $k(X) \in H^3(\pi_1(X); \pi_2(X))$ is the first k-invariant of X [**MW50**]. Martins [**Mar09**] gave an algorithm to compute the homotopy 2-type of a 2-knot complement. Lomonaco [**Lom81**] proved that if

the third homology group of the universal cover of a knot complement is trivial (quasi-aspherical),then the homotopy 2-type determines the homotopy type of a 2-knot complement. However, not every 2-knot complement is quasi-aspherical [**Rat81**].

(2) Note that 2-knots are not even determined by the homeomorphism type of their exterior, since Gordon [**Gor76**] showed that there exist twist-spun knots K that are not equivalent to the knots K^* arising from Gluck twisting.

(3) Compare this question to Problem 4.55, which asks whether the homotopy type of a 4-manifold with finite fundamental group is determined by its quadratic 2-type, which includes the additional data of the equivariant intersection form. The fundamental groups of 2-knot complements are of course infinite, but not yet completely classified (see Problem 4.35).

Scribed by: S. Kim, M. Miller

PROBLEM 4.37 (Kinoshita conjecture). *Does every projective plane in S^4 decompose as a connected sum of a knotted 2-sphere and an unknotted projective plane?*

REMARKS.

(1) This question can be asked in either the smooth or the locally flat categories.

(2) The conjecture that the answer to this question is "yes," is known as the *Kinoshita Conjecture*. This is named after the work of Kinoshita [**Kin61**], who first constructed an infinite family of projective planes in S^4 that are pairwise not ambiently isotopic. Each of Kinoshita's projective planes decomposes as the connected sum of a knotted 2-sphere and an unknotted projective plane. We may refer to such projective planes as being of *Kinoshita type*. Then the question is: *is every projective plane in S^4 of Kinoshita type?*

The first instance of this problem appearing as "Kinoshita's Conjecture" in the literature is likely work of Katanaga–Saeki [**KS98**], who said in 1997 that this problem was already well-known to topologists in Japan.

(3) Kamada showed that the Kinoshita Conjecture is true for projective planes obtained by deform-spinning knots with finite summetry groups [**Kam92**].

(4) By Yoshikawa [**Yos98**], not every Klein bottle in S^4 admits an unknotted projective plane summand. The examples studied by Yoshikawa all have normal Euler number zero.

QUESTION (i). Does there exist a Klein bottle Σ in S^4 that does not admit an unknotted projective plane summand with $|e(\Sigma)| = 4$?

(5) Yoshikawa [**Yos94**] constructed 2-component links of projective planes in S^4 such that neither component admits an unknotted projective plane summand in the complement of the other component. In some sense, this can be interpreted as a failure of the Kinoshita Conjecture for multiple-component links. Yoshikawa's obstruction to these links decomposing was to show that the peripheral subgroup of each component was order 4, but

would be order 2 if the link admitted a trivial projective plane summand at that component. This motivates the following question.

QUESTION (ii). Given a projective plane P in S^4, must the kernel of the inclusion-induced homomorphism $\iota_* : \pi_1(\partial(S^4 \backslash \nu(P)) \to \pi_1(S^4 \backslash P)$ be order 4? (If not, then P is an example of a projective plane that does not decompose as a summand of a knotted 2-sphere and unknotted projective plane.)

(6) When Σ is a projective plane in a 4-manifold X with normal Euler number $e(\Sigma) = \pm 2$, we can perform *Price surgery* on X along Σ [**Pri77**]. When $X = S^4$, there is a preferred surgery whose result is a homotopy 4-sphere.
If Σ is obtained by tubing an unknotted projective plane to a 2-sphere S with $e(S) = 0$, then some Price surgery on Σ is diffeomorphic to the result of Gluck surgery on S [**KSTY99**]. In particular, when $X = S^4$ the homotopy 4-sphere obtained from Price surgery on Σ is diffeomorphic to the result of Gluck surgery on S. Thus, positively answering the proposed question (showing that all projective planes in S^4 are of Kinoshita type) would imply that Price surgeries in S^4 yield no more homotopy 4-spheres than Gluck surgeries.

Scribed by: M. Miller

PROBLEM 4.38. *Let $\Delta \subseteq B^4$ be a ribbon disk. Is $B^4 \backslash \Delta$ aspherical, i.e. is $\pi_i(B^4 \backslash \Delta) = 0$ for all $i > 1$?*

REMARKS.

(1) This is [**Kir97**, Problem 1.103]. Note that the *Whitehead asphericity conjecture* [**Whi41**, p. 428] (see also [**Ros07a, How85**] and Problem 5.11) predicts an affirmative answer. A generalized version, asking for the asphericity of ribbon concordances, was asked by Gordon in [**Gor81**, Conjecture 6.5].
(2) Howie showed in [**How85**] that the answer is yes when $\pi_1(B^4 \backslash \Delta)$ is locally indicable.
(3) One could equally well pose this question in the locally flat category, asking whether the complements of homotopy-ribbon disks are aspherical. (Recall that a homotopy-ribbon disk Δ for some knot $K \subseteq S^3$ is one such that the map $\pi_1(S^3 \backslash K) \to \pi_1(B^4 \backslash \Delta)$ is surjective.) In this case, it is open whether the complement is homotopy equivalent to a 2-complex.
(4) An affirmative answer to the question can be used to compute the algebraic 2-type of a 2-knot complement, as shown in [**Lom81**]. Note that the argument there is based on an incorrect proof asserting that ribbon disk complements are aspherical. See [**Mar09**] for an alternate approach to computing the algebraic 2-type of a 2-knot complement.

Proposed for K3 and scribed by: A. Conway, A. Ray

PROBLEM 4.39. *Let K be a closed surface smoothly embedded in a 4-manifold M^4. Describe the subgroup of the mapping class group* $\mathrm{Mod}(K)$ *of diffeomorphisms of K that extend to diffeomorphisms of M.*

Conversely, is it possible to characterize when a given subgroup of the mapping class group arises as the group of extendable diffeomorphisms for some K inside some M^4?

REMARKS.

(1) The problem is due to Hirose, who has studied instances of this problem in e.g. [**Hir93**, **Hir02**, **Hir05**, **Hir12**]. Some highlights of his results on this problem are summarized in the remarks to follow.
(2) In [**Hir02**], Hirose studies the case of an unknotted genus-g surface in S^4, showing that a mapping class is extendable in this setting if and only if it preserves the *Rokhlin quadratic form* (descending to a spin structure on the surface). In particular, this subgroup is of finite index in the mapping class group of K.

    In [**Hir02**], he extended this result to nonorientable unknotted surfaces in S^4, showing that in this setting a mapping class extends exactly when it preserves the *Guillou–Marin quadratic form* (descending to a Pin^- structure on the surface; see [**KS23**] for more discussion). Again, this subgroup is of finite index in the mapping class group of K.
(3) In [**Hir05**], Hirose classified which surface automorphisms extend from unknotted surfaces and certain complex curves to all of $\mathbb{CP}^2$. The question is open for all high-degree complex curves.
(4) In [**Hir93**], Hirose studied the case of knotted tori in S^4 arising by spinning (or turned spinning) knots in S^3. Here, the set of extendable mapping classes is always of infinite index. It would be interesting to characterize when the group of extendable mapping classes is of finite index in general.
(5) In [**HY08**], Hirose and Yasuhara showed that in many simply connected 4-manifolds (including $\mathbb{CP}^2, \overline{\mathbb{CP}}^2, S^2 \times S^2, E(n)$) there exist smoothly embedded surfaces of every topological type with the property that any surface automorphism extends to an ambient automorphism.

Proposed for K3 by: M. Miller, N. Salter
Scribed by: M. Miller

PROBLEM 4.40. *Is every link of 2-spheres in S^4 slice?*

REMARKS.

(1) A 2-link is *slice* if it bounds a disjointly embedded collection of 3-balls in the 5-ball B^5.
(2) Kervaire [**Ker65a**] proved that every 2-knot in S^4 is slice.
(3) This question can be asked in either category. If a smooth link of 2-spheres is topologically slice in B^5, then it is also smoothly slice by Daher–Powell [**DP23a**].

(4) Cochran [**Coc84**] gave a sufficient condition for a 2-link to be slice, generalizing the well-known fact that boundary 2-links (see Problem 1.62) are slice (following from Kervaire's proof for 2-knots). Not every 2-link is a boundary link [**Coc84**, §4].

Scribed by: M. Miller

PROBLEM 4.41. *Let Σ be a compact surface and let X be a connected 4-manifold. Let $f_0, f_1 \colon \Sigma \to X$ be π_1-negligible embeddings that agree on $\partial\Sigma$ and that are homotopic rel. boundary. Give computable invariants that decide whether there is a π_1-negligible concordance in $X \times I$ between f_0 and f_1.*

REMARKS.

(1) A surface is π_1-*negligible* if the complement of the surface included into X induces an isomorphism on fundamental groups [**FQ90**]. This assumption is equivalent to asking for an immersed dual sphere.
(2) The case that $\Sigma = S^2$ or that the dual sphere is framed is due to Freedman–Quinn [**FQ90**] and Stong [**Sto93**]. See also the papers of Klug–Miller [**KM21, KM22**].
(3) A strategy would be to extend the Freedman–Quinn and Stong invariants to arbitrary compact surfaces.

Proposed for K3 and scribed by: M. Powell, A. Ray

PROBLEM 4.42.

(a) *If X is a smooth, closed, simply connected* 4*-manifold with $b_2(X) \geqslant 2$, are all knots slice in X?*
(b) *In particular, are all knots slice in $\mathbb{C}\mathrm{P}^2\#\mathbb{C}\mathrm{P}^2$? In the K3 surface?*
(c) *Can the set of slice knots detect exotic smooth structures? That is, do there exist homeomorphic* 4*-manifolds X and X' and a knot $K \subset S^3$ that is slice in X but not in X'?*

REMARKS.

(1) A knot K is called *slice* in a closed 4-manifold X if it bounds a smoothly embedded disk in $X \setminus \mathrm{int}(B^4)$, where K lives on the boundary S^3. It is called *H-slice* if the disk is null-homologous (relative to the boundary).
(2) All knots are known to be slice in $S^2 \times S^2$ and $\mathbb{C}\mathrm{P}^2\#\overline{\mathbb{C}\mathrm{P}}^2$ [**Nor69, Suz69**], but not all knots are slice in $\mathbb{C}\mathrm{P}^2$ or in $\overline{\mathbb{C}\mathrm{P}}^2$ [**Yas91**].
(3) Part (a) is true in the topological category; see [**KPRT24**, Corollary 1.15].
(4) Concerning part (b), Marengon and Mihajlović [**MM25a**] proved that all knots with unknotting number at most 21 are slice in K3.
(5) The analogue of (c) for H-slice knots (instead of slice) was answered in [**MMP24**]: the right handed trefoil is H-slice in $\#3\mathbb{C}\mathrm{P}^2\#20\overline{\mathbb{C}\mathrm{P}}^2$ but not in $K3\#\overline{\mathbb{C}\mathrm{P}}^2$.
(6) A variant of part (c) is asked in Problem 4.12.

(7) Lidman and Piccirillo [**LP25**] announced the construction of a pair of 4-manifolds X and X' with the same integer cohomology ring, such that there exists a knot slice in X but not in X'. (However, X and X' are not homeomorphic.)

Scribed by: C. Manolescu

PROBLEM 4.43. *Let K be a knot on the boundary of $X\backslash\mathring{B}^4$, where X is a negative definite, smooth 4-manifold. Suppose K bounds a smoothly embedded, null-homotopic surface Σ in $X\backslash\mathring{B}^4$. Is the following inequality*

$$2g(\Sigma) \geqslant |s(K)| \tag{8}$$

always satisfied? Here $s(K)$ denotes Rasmussen's s-invariant [**Ras10**].

REMARKS.

(1) For $X = S^4$, the inequality (8) was proved by Rasmussen [**Ras10**]. When $X = \#^r\mathbb{C}P^2$, (8) was proved by Manolescu–Marengon–Sarkar-Willis [**MMSW23**].
(2) Freedman–Gompf–Morrison–Walker [**FGMW10**] proposed the following approach to disproving the smooth four-dimensional Poincaré conjecture: Find a homotopy 4-sphere X and a knot K that is slice in X (i.e. K bounds smoothly embedded disk in $X\backslash B^4$). Then prove that K is not slice in S^4 by showing that $s(K) \neq 0$. If this approach works, then (8) must fail for X.
(3) Using instanton Floer theory, Kronheimer-Mrowka [**KM13b**] defined a concordance invariant $s^\sharp(K)$ that satisfies a similar inequality
$$2g(\Sigma) \geqslant |s^\sharp(K)|.$$
However, it is known that $s(K) \neq s^\sharp(K)$ for many knots including the trefoil [**Gon21**]. One possible approach to proving (8) is to find a gauge theoretic interpretation of $s(K)$ by modifying Kronheimer-Mrowka's definition of $s^\sharp(K)$.
(4) One may ask whether (8) holds for other variants of Rasmussen's s-invariant, including $s^{\mathbb{F}_p}$ (defined using Khovanov homology over the field $\mathbb{F}_p$ [**MTV07, LS14**]) and s^{Sq^1} (the Lipshitz-Sarkar Sq^1 s-invariants [**LS14**]). It is known that their variants are different from $s(K)$ in general [**LS14**].

Proposed for K3 and scribed by: J. Lin

PROBLEM 4.44.

(a) *Let X be a closed, simply connected 4-manifold. Let $g \geqslant 0$ and $d \geqslant 0$ be integers. Fix $x \in H_2(X;\mathbb{Z})$. Does there exist a (smooth, locally flat) generic immersion $f\colon \Sigma_g \to X$ such that $f_*[\Sigma_g] = x$, with d transverse double points?*

(b) *An important special case is* $d = 0$. *Let* $g\colon H_2(X;\mathbb{Z}) \to \mathbb{N}_0$ *be the* genus function *assigning to* $x \in H_2(X;\mathbb{Z})$ *the minimal genus of any (smooth, locally flat) embedded surface in* X *whose fundamental class represents* x. *The problem is to compute the locally flat genus function for some 4-manifold* $X \neq S^4$, *and to compute the smooth genus function for new 4-manifolds.*

(c) *For a specific example, is the class* $(3, 2) \in H_2(\mathbb{C}\mathrm{P}^2\#\mathbb{C}\mathrm{P}^2;\mathbb{Z})$ *represented by a sphere?*

REMARKS.

(1) The smooth genus function is known for $\mathbb{C}\mathrm{P}^2$ by a celebrated result of Kronheimer–Mrowka [**KM94**]. The cases of $\mathbb{C}\mathrm{P}^2\#\overline{\mathbb{C}\mathrm{P}}^2$ and $S^2 \times S^2$ were resolved by Ruberman in [**Rub96**]. Partial results about the smooth genus problem in $\mathbb{C}\mathrm{P}^2\#\mathbb{C}\mathrm{P}^2$ appear in [**Bry98**], [**Nou14**], [**MMRS24**], [**ACM$^+$26**] By the symplectic Thom conjecture, proved in [**OS00**], any symplectic submanifold minimizes genus in its homology class; this gives a useful means of computing the value of the smooth genus function for certain homology classes.

(2) The locally flat genus function is not known except for S^4. In other 4-manifolds, much is known about *simple* locally flat embeddings by work of Lee-Wilczynski [**LW97, LW00**]. As a special case, every primitive class in a closed, simply-connected 4-manifold is represented by a locally flat torus [**LW97, KPRT24**]. Not much else is known.

(3) In the special case when the self-intersection number $Q(x, x) = 0$, we have the following interesting sub-problem. As usual we let $\mathbb{N}$ denote the positive integers.

QUESTION. For a given $x \in H_2(X;\mathbb{Z})$ such that $g(x) > 0$, consider the function $n_x\colon \mathbb{N} \to \mathbb{N}$ sending $m \mapsto g(mx)$. What is n_x in terms of $g(x)$? Is $n_x(m) = mg(x) - m + 1$?

Note that if $g(x) = 0$ then $g(mx) = 0$ for all x. The formula $mg(x) - m + 1$ comes from the following observation: if $[\Sigma] = x$ with $Q(x, x) = 0$, then there is a surface $\widetilde{\Sigma}$ that embeds in $\Sigma \times D^2$ representing mx such that the projection to Σ induces an m-sheeted covering.

The answer is negative for some 4-manifolds with boundary. Kawauchi [**Kaw09**] constructed 4-manifolds homotopy equivalent to S^2 (so $H_2(X;\mathbb{Z}) = \mathbb{Z}$), with the property that for x a generator, $g(mx) = 0$ for m even and $g(mx) \neq 0$ for m odd.

Scribed by: C. Manolescu, L. Piccirillo, M. Powell, A. Ray

PROBLEM 4.45. *Let* X *be a closed simply connected smooth 4-manifold and let* Σ *be a closed, orientable surface. Fix a smooth embedding of* $f\colon \Sigma \hookrightarrow X$ *to serve as a basepoint. For which* $k \geqslant 0$ *does the equality*

$$\begin{aligned} &\ker\left(i_*\colon \pi_k(\mathrm{Emb}^{\mathrm{DIFF}}(\Sigma, \mathring{X}); f) \to \pi_k(\mathrm{Emb}^{\mathrm{TOP}}(\Sigma, \mathring{X}); f)\right) \\ &= \ker\left(s_*\colon \pi_k(\mathrm{Emb}^{\mathrm{DIFF}}(\Sigma, \mathring{X}); f) \to \operatorname*{colim}_{N\to\infty} \pi_k(\mathrm{Emb}^{\mathrm{DIFF}}(\Sigma, \mathring{X}\#_N S^2 \times S^2); f)\right) \end{aligned} \tag{9}$$

hold? To make sense of this for $k = 0$, define $\ker i_*$ *and* $\ker s_*$ *to be the preimages under* i_* *and* s_* *respectively of* $[f]$.

REMARKS.

(1) Here $\mathring{X}$ denotes the punctured X, $\mathring{X} = X \setminus \mathrm{Int}(D^4)$, $\mathrm{Emb}^{\mathrm{DIFF}}(\Sigma, \mathring{X})$ is the space of smooth embeddings, $\mathrm{Emb}^{\mathrm{TOP}}(\Sigma, \mathring{X})$ is the space of locally flat embeddings, $i\colon \mathrm{Emb}^{\mathrm{DIFF}}(\Sigma, \mathring{X}) \to \mathrm{Emb}^{\mathrm{TOP}}(\Sigma, \mathring{X})$ is the inclusion, and $s\colon \mathrm{Emb}^{\mathrm{DIFF}}(\Sigma, \mathring{X}) \to \mathrm{Emb}^{\mathrm{DIFF}}(\Sigma, \mathring{X} \#_N S^2 \times S^2)$ is the stabilization map. Note that we consider X punctured so that the stabilization map s is defined on the space level.

(2) See Problem 4.77 for the analogous question on diffeomorphism and homeomorphism groups.

(3) For $k = 0$, the inclusion $\subseteq$ holds in the equality (9). That is, Galvin [**Gal24**] showed that any two smoothly embedded, topologically isotopic surfaces in a smooth, simply-connected 4-manifold X become smoothly isotopic after sufficiently many external stabilizations of X by copies of $S^2 \times S^2$. Is the converse true? What about when the complements are simply connected?

(4) There is considerable literature showing, under suitable conditions, that surfaces in 4-manifolds are topologically isotopic, especially those with simply connected complements, which might help. See e.g. [**LW90, LW93, LW01, Boy93, HK93b, LW97, Sun15, CP23**]. New techniques seem to be required for answering the question when $k > 0$.

Proposed for K3 and scribed by: H. Konno, M. Powell

4.5. Topological 4-manifolds

The first two questions below concern the topological surgery conjecture, relating to the central question of whether Freedman's disk embedding theorem applies to 4-manifolds with arbitrary fundamental group. The next few problems pertain to applications of surgery to the homeomorphism classification of 4-manifolds, before we end the section by considering homotopy classifications of 4-manifolds.

PROBLEM 4.46. *Are all groups good?*

REMARKS.

(1) The problem is due to Freedman and Quinn. The term 'good' group was first introduced in [**Fre84**] and [**FQ90**, p. 99] to refer to any group for which Freedman's disk embedding theorem holds. In those two sources it was shown that the latter theorem holds for any group belonging to the smallest class of groups that contains $\mathbb{Z}$ and finite groups and is closed under colimits, subgroups, extensions, and quotients. In [**Kir97**, Problem 4.6] a group is defined to be good if it belongs to this class of groups. The definition in terms of a capped grope was first formulated in [**FT95a**]. This has now become the accepted definition.

(2) A group Γ is said to be *good* if for every height 1.5 disc-like capped grope G, with some choice of basepoint, and for every group homomorphism $\phi\colon \pi_1(G) \to \Gamma$, there exists an immersed disk D in G whose framed boundary coincides with the attaching region of G, such that the double point loops of D, considered as fundamental group elements by making some choice of basing path, are mapped to the identity element of Γ by ϕ. For more details on the definition, including the definition of a capped grope, see [**FT95a**, **KOPR21a**], cf. [**Kir97**, Problem 5.9] (these definitions differ in the height of a capped grope that one starts with, but the two definitions are equivalent by a method called *grope height raising*). We limit ourselves to saying that a *capped grope* is a smooth 4-manifold, and its attaching region is a specified solid torus in its boundary. The core of the solid torus is null-homotopic in the capped grope (regardless of height). Therefore some immersed disk D always exists – the additional fact about the double point loops is the key feature of the definition. It is also worth observing that good groups are defined by a property that can be entirely stated in the smooth category.

(3) Almost every known result about topological 4-manifolds relies on the disk embedding theorem, and therefore the question of which groups are good is fundamental. Recall that any group is the fundamental group of some 4-manifold, but fundamental groups of compact 4-manifolds are necessarily finitely presented. Therefore, one could choose to ask the question only for finitely presented groups. However, as we will see in item 7 below, if all finitely presented groups were good, then all groups would be good, so there is no benefit to this restriction.

(4) By work of Freedman and Quinn [**FQ90**, Chapter 11] (see also [**OPR21b**]), the action of $L_5(\mathbb{Z}[\pi_1(X)])$ on the structure set $\mathcal{S}(X)$ is defined, and the surgery sequence is known to be exact at $\mathcal{S}(X)$ and $\mathcal{N}(X)$, for every 4-manifold X with $\pi_1(X)$ good. Therefore an affirmative answer would give the exactness of the topological surgery sequence for 4-manifolds in full generality (cf. [**Kir97**, Problem 4.6]). This is known in dimensions five and higher in the smooth, piecewise-linear, and topological categories [**Bro72**, **Nov64**, **Sul96**, **Wal99**, **KS77**]. See Problem 4.22 for more background on the surgery sequence.

(5) An affirmative answer would also imply the 5-dimensional s-cobordism theorem in full generality. The s-cobordism theorem for dimensions 6 and higher is true in both the smooth and topological settings [**Sma62a**, **Bar63**, **Maz63**, **Sta67**, **KS77**] (see also [**Mil65**, **RS72**]). The smooth 5-dimensional s-cobordism theorem is false by work of Donaldson [**Don87a**]. See also Problem 4.56.

(6) The exactness of the surgery sequence and the h-cobordism theorem is a key step in the classification of closed, simply connected 4-manifolds up to homeomorphism. Therefore the extension to general fundamental groups is likely to have several spectacular applications. We observe that even if all groups were good, classification results would not immediately follow. Nevertheless this would be a result of immense importance.

(7) As mentioned above, it was shown in [**FQ90**] that finite groups and the infinite cyclic group are good. Groups of subexponential growth were

shown to be good in [**KQ00, FT95a**]. The class of good groups is also known to be closed under subgroups, quotients, extensions, and direct limits [**FT95a, KOPR21a**]. It is an open question whether any nonabelian free group is good. Recall that every finitely generated group arises as a quotient of a subgroup of $\mathbb{Z} * \mathbb{Z}$ and every group is a colimit of its finitely generated subgroups. Therefore, all groups are good if and only if the free group on two generators is good. We state this as a question.

QUESTION. Is the group $\mathbb{Z} * \mathbb{Z}$ good?

(8) At present it is also open whether amenable groups are good. Addressing this would be a substantial step towards the general question.

QUESTION. Are amenable groups good?

(9) For further details on the relationships between the above problems and Problems 1.62 and 4.47, see [**KOPR21b**].

Scribed by: M. Powell, A. Ray

PROBLEM 4.47 (Round handle problem). *Is there a link $L \subseteq S^3$ with vanishing pairwise linking numbers that is not round handle slice?*

REMARKS.

(1) We define the terminology in the problem. A *round handle* is a copy of $D^1 \times D^2 \times S^1$, which is attached to the boundary of a 4-manifold along $S^0 \times D^2 \times S^1$. (So it is a 4–dimensional round 1–handle.) Given an m-component link $L \subseteq S^3$ we construct a 4-manifold $R(L)$ by attaching m round handles to D^4 as follows. For the i-th component L_i of L, let λ_i denote a 0-framed longitude and let μ_i denote a meridian, chosen so that the linking number $\operatorname{lk}(\lambda_i, \mu_i) = 0$. To attach the ith round handle, identify $\{-1\} \times S^1 \times D^2$ to λ_i and $\{+1\} \times S^1 \times D^2$ to μ_i, using the trivial framing in both cases. The resulting 4-manifold contains the link L in the boundary $\partial R(L)$. A link L is said to be *round handle slice* if $L \subseteq \partial R(L)$ is slice in $R(L)$, that is, if L is the boundary of a collection of locally flat pairwise disjoint embedded discs in $R(L)$.
(2) The round handle problem was proposed by Freedman and Krushkal in [**FK16**, Section 5.1], and is presented in detail in [**KPT21**].
(3) If the 4-dimensional topological surgery and s-cobordism conjectures hold for free groups, then every link with vanishing pairwise linking numbers is round handle slice. So this problem gives a way to potentially disprove the union of these conjectures, which are central open problems in 4-manifold topology. For more on those conjectures see Problems 4.46 and 1.62.

Scribed by: A. Ray

PROBLEM 4.48. *Let Δ be a contractible, compact 4-manifold. Is the space of homeomorphisms of Δ that fix the boundary pointwise, with the compact-open topology, a contractible space?*

REMARKS. This is true for $\Delta = D^4$, by the Alexander trick. By Perron-Quinn [**Per86, Qui86**], the space is connected. In dimensions at least 6, the space of homeomorphisms is contractible, as proven by Galatius–Randal-Williams [**GRW24**].

Scribed by: M. Powell

PROBLEM 4.49. *Give an effective necessary and sufficient condition for an open 4-manifold to be homeomorphic to the interior of a compact 4-manifold.*

REMARKS.

(1) In his thesis [**Sie65**], Siebenmann gave necessary and sufficient conditions for an n-dimensional open PL manifold M, for $n \geqslant 6$, to be compactifiable by adding a manifold boundary (see also [**BLL65**]). Recall that a manifold is said to be *open* if it has empty boundary and only non-compact components. The proof extends to the case where M is non-compact but has compact boundary. The case of possibly noncompact boundary is addressed in [**O'B83, GG20**]. The foundational results of Kirby-Siebenmann [**KS77**] on high-dimensional topological manifolds imply that Siebenmann's proof can also be applied in the purely topological setting.

(2) The cases $n = 2, 3$ are also well understood [**HP70**], [**Tuc74**], [**GG20**, Section 2].

(3) Note that if an n-manifold M is the interior of an n-manifold $\overline{M}$, then there is a neighborhood of the end of M of the form $Y \times [0, \infty)$ for some $(n-1)$-manifold Y. In this case we say that M has a *collared* end. Having a collared end imposes several homotopy-theoretical conditions on M. The prior results mentioned above roughly consist of showing that some collection of such homotopy-theoretic conditions is sufficient to conclude the existence of a collared end.

(4) An end of a manifold M is said to be *tame* if there is a closed neighborhood U of the end, and a proper map $U \times [0, \infty) \to M$ that is the inclusion on $U \times \{0\}$. Suppose M is a manifold with a tame connected end with finitely presented fundamental group π. Siebenmann defined an invariant $\sigma(\text{end}(M)) \in \widetilde{K}_0(\mathbb{Z}\pi)$, related to Wall's finiteness obstruction [**Wal65**]. Siebenmann's result from [**Sie65**] mentioned above states that if an n-manifold M, for $n \geqslant 6$, has a tame connected end with finitely presented fundamental group π, then $\sigma(\text{end}(M)) \in \widetilde{K}_0(\mathbb{Z}\pi)$ vanishes if and only if M has a collared end. This result was extended by Quinn to the case that $n = 5$ and π is *good* (see Problem 4.46). We emphasize that in the results above π is the fundamental group of the end of M, and not of M itself. See [**FQ90**, Section 11.9] for further details.

(5) The direct topological analogue of Siebenmann's theorem in dimension four fails by a result of Kwasik-Schultz [**KS88**, Theorem 2.1] and unpublished work of Weinberger. These latter authors constructed a class of 4-manifolds M such that the infinite cyclic cover $\widetilde{M}$ has an end with two components, each of which is tame, has good fundamental group,

and for which Siebenmann's obstruction vanishes, but such that $\widetilde{M}$ is not homeomorphic to $Y \times \mathbb{R}$, for any compact 3-manifold Y.

(6) By the result mentioned above, a characterization of when an open 4-manifold has collared end must be necessarily more complicated than it is in higher dimensions. An ideal characterization would be purely homotopy-theoretic. Partial results in dimension four in the topological setting are given in [**FQ90**, Section 11.9B], but there is not yet a candidate for a necessary and sufficient condition.

Scribed by: M. Powell, A. Ray

PROBLEM 4.50. *Classify closed topological 4-manifolds (orientable and not) with finite fundamental group, up to homeomorphism. The following infinite families of fundamental groups are of particular interest.*

(a) *Dihedral groups.*
(b) *Quaternionic groups.*
(c) *Abelian groups with two or three generators.*

A significant starting point would be to complete the classification for 4-manifolds with fundamental group of order at most eight.

REMARKS.

(1) The problem is deemed solved if the classification is reduced to algebraic invariants. An interesting subproblem is to complete the classification for smooth 4-manifolds only.
(2) The classification exists for trivial fundamental group by Freedman [**Fre82**], for cyclic fundamental groups in the orientable case by Hambleton-Kreck [**HK88, HK93a, HK93c, HK93b**], and for fundamental group $\mathbb{Z}/2\mathbb{Z}$ in the nonorientable case by Hambleton-Kreck-Teichner [**HKT94**]. See also [**HH23**].
(3) The homotopy classification is known in some cases: [**HK88, KPR24, KNR22, Hil23**]. In these cases one strategy is therefore to upgrade this to a homeomorphism classification using surgery theory.

Scribed by: M. Powell, A. Ray

PROBLEM 4.51. *Let X be a closed, smooth 4-manifold with fundamental group isomorphic to $\mathbb{Z}$. Is the $\mathbb{Z}[\mathbb{Z}]$-valued intersection form on $\pi_2(X)$ extended from $\mathbb{Z}$?*

REMARKS.

(1) We say that a Hermitian form on a free $\mathbb{Z}[\mathbb{Z}]$-module is *extended from* $\mathbb{Z}$ if it is represented by a matrix, with respect to some basis over $\mathbb{Z}[\mathbb{Z}]$, with integer entries.
(2) Hambleton–Teichner [**HT97**] exhibited a unimodular Hermitian form over $\mathbb{Z}[\mathbb{Z}]$ that is not extended over $\mathbb{Z}$.

(3) It was known from Freedman–Quinn [**FQ90**] that this form can be realized as the intersection form of closed topological manifold with fundamental group $\mathbb{Z}$. A manifold with such a form is not even homotopy equivalent to the connected sum of a homotopy $S^1 \times S^3$ with a simply connected 4-manifold.

(4) Conversely, if the intersection form of a 4-manifold with fundamental group $\mathbb{Z}$ is extended from $\mathbb{Z}$, then that 4-manifold is homeomorphic to a connected sum $M\#(S^1 \times S^3)$, where $\pi_1(M) = \{1\}$. Thus an affirmative answer to the question would imply that every closed, smooth, oriented 4-manifold with fundamental group $\mathbb{Z}$ splits topologically as such a connected sum.

(5) Friedl–Hambleton–Melvin–Teichner [**FHMT07**] showed, by an application of Donaldson's theorem [**Don83**] to the finite cyclic covers (of degree at least 3) of the manifold in [**HT97**] that no 4-manifold with this form can be smooth. Conjecture 1.3 in [**FHMT07**] states that the intersection form of any closed, smooth manifold with fundamental group $\mathbb{Z}$ is extended from the integers.

(6) Fintushel–Stern [**FS94**] showed that there are smooth 4-manifolds with fundamental group $\mathbb{Z}$ that do not have an $S^1 \times S^3$ connect summand smoothly. In fact, any symplectic 4-manifold will not smoothly decompose as a connected sum in this fashion.

Scribed by: D. Ruberman

PROBLEM 4.52. *Let M and N be closed, orientable topological 4-manifolds with $\pi_1(M) \cong \pi_1(N)$ a good group. Suppose that M and N are simple homotopy equivalent and stably homeomorphic. Is there a simple homotopy equivalence $N \to M$ that lies in the orbit of* $\mathrm{Id}\colon M \to M$ *under the Wall realization action of* $L_5^s(\mathbb{Z}[\pi_1(M)])$ *on the simple structure set of M?*

REMARKS.

(1) We can hope to classify 4-manifolds up to simple homotopy equivalence, and up to stable homeomorphism, for many groups. We can also hope to compute $L_5^s(\mathbb{Z}[\pi_1(M)])$. If the answer were yes, this would give a somewhat satisfying picture for the classification of 4-manifolds, for good groups. If the answer were no, this would uncover interesting new phenomena.

The answer is yes for simply connected 4-manifolds, and for those with fundamental group $\mathbb{Z}$ [**FQ90**] and $\mathbb{Z}/n\mathbb{Z}$ [**HK88, HK93a, HK93c, HK93b**].

(2) In the case of spin 4-manifolds, the program suggested by this question could potentially be made easier, only requiring the simple homotopy classification, if the answer to the following question is no.

QUESTION. Do there exist closed, spin 4-manifolds that are simple homotopy equivalent but not stably homeomorphic?

The latter question was part of [**Kir97**, Problem 4.84], due to Teichner. Note that closed, spin 4-manifolds that are homotopy equivalent have the same Kirby–Siebenmann invariant, so that invariant cannot be used here.

(3) Jim Davis [**Dav05**] proved that if the fundamental group is π and the map $\kappa_2\colon H_2(\pi;\mathbb{Z}/2\mathbb{Z}) \to L_4(\mathbb{Z}\pi)$ is injective, then homotopy equivalent 4-manifolds with that fundamental group and the same Kirby-Siebenmann invariant are stably homeomorphic.

(4) There are counterexamples known for non-spin 4-manifolds, even assuming they have the same Kirby-Siebenmann invariant [**Tei97**].

Proposed for K3 by: D. Kasprowski
Scribed by: M. Powell

PROBLEM 4.53. *Does there exist an algorithm that takes as input a closed, triangulated 4-manifold, and outputs in finite time whether or not that 4-manifold is homeomorphic to S^4?*

REMARKS.

(1) This is called the *recognition problem.* We say that we can *recognize* an n-manifold X if there exists an algorithm that takes as input a closed, triangulated n-manifold, and outputs in finite time whether or not that 4-manifold is homeomorphic to X.

(2) We assume that the input 4-manifolds are represented by the finite data of a triangulation, and hence they are smooth, since triangulated 4-manifolds are smooth. We are not assured that the input 4-manifold is simply-connected.

(3) The corresponding question has a positive answer in dimensions $\leqslant 3$ [**Tho94**], [**Rub95**], and a negative answer in dimensions $\geqslant 5$ [**VKF74**].

(4) Markov [**Mar58**] showed that, for some integer k, the corresponding question for the connected sum of k copies of $S^2 \times S^2$ has a negative answer. This raises the question of the minimal k for which this holds. It was shown that k could be taken to be 14 in [**Sht05**], 12 in [**Gor22**], and 9 in [**Tan23**]. See also the exposition in [**Kir20**].

(5) By Markov [**Mar58**], there exist infinite lists of group presentations $\{P_i\}$ such that there is no algorithm taking as input one of the P_i, and outputting in finite time whether or not the group $G(P_i)$ presented by P_i is trivial. One possible strategy to solve the problem is to construct a corresponding list $\{X_i\}$ of closed, triangulated 4-manifolds, whose 2-skeleta give rise to the P_i, so in particular $\pi_1(X_i) \cong G(P_i)$, with the property that X_i is homeomorphic to S^4 if and only if $G(P_i) = \{1\}$. The forward direction clearly holds, but it is not clear how to find $\{P_i\}$ and $\{X_i\}$ such that the backwards direction holds.

This strategy does work for $\#^k(S^2 \times S^2)$ in place of S^4, and was the basis for the proofs of [**Sht05, Gor22, Tan23**] for $k = 14, 12, 9$ respectively.

(6) One can also ask the analogous *smooth recognition problem* with 'diffeomorphic' in place of 'homeomorphic'. This is also open for S^4. As described in [**Kir20**], there exists a k such that $\#^k(S^2 \times S^2)$ is not smoothly recognizable, but there are no known upper bounds on the minimal k for which this holds.

Scribed by: M. Powell

PROBLEM 4.54. *The* quadratic 2 type *of a 4-manifold M is the data*

$$(\pi_1(M), \pi_2(M), \lambda_M, k_M)$$

of the fundamental group $\pi_1(M)$, the second homotopy group $\pi_2(M)$ considered as a $\mathbb{Z}[\pi_1(M)]$-module, the equivariant intersection form λ_M, and the k-invariant in $k_M \in H^3(B\pi_1(M); \pi_2(M))$.

Which quadratic 2-types are realized by closed, oriented topological 4-manifolds? Which are realized by closed, oriented, smooth 4-manifolds? Can this problem be solved for specific fundamental groups, for example for certain families of non-cyclic finite groups?

REMARKS.

(1) This is in [**Kir97**, Problem 4.1].
(2) We can also ask the question with the data of the Stiefel-Whitney classes w_1 and w_2. Which Stiefel-Whitney classes are realized within a given quadratic 2-type?
(3) The problem builds on the geography problem for simply connected topological 4-manifolds, where it restricts to the question of which intersection forms occur. In the smooth category, this is answered by *Donaldson's diagonalizability theorem* along with a positive resolution of the *11/8-conjecture*.
(4) For the topological case, Freedman [**Fre82**] showed that every nonsingular symmetric bilinear form is realized by a closed simply connected 4-manifold, so the problem is solved when $\pi_1 = 1$. A similar result is known for $\pi_1 = \mathbb{Z}$ [**FQ90**].

    For other good fundamental groups, a possible strategy in the topological category was introduced by Hambleton-Kreck in [**HK88**, Lemma 4.1], the paper where the quadratic 2-type initially arose. First, classify the quadratic 2-types that are stably realizable, meaning that they are realizable by topological 4-manifolds after taking the orthogonal sum with the quadratic 2-type of $S^2 \times S^2$. If a quadratic 2-type is stably realizable then it is realizable unstably by a topological 4-manifold, which can be shown using the sphere embedding theorem [**FQ90**, **BKK$^+$21**].

    Hambleton-Kreck [**HK88**, **HK93a**] used this strategy to solve the realization problem for finite cyclic groups, in the topological category.
(5) For non-simply-connected, smooth, oriented 4-manifolds, Donaldson's diagonalization theorem holds without any assumption on the fundamental group, so definite integral intersection forms must be diagonalizable. The

sphere embedding theorem cannot be applied in the smooth case, so the strategy described above of stably realizing and then destabilizing is not currently viable.

(6) The existence problem has a closely related uniqueness analogue. For finite cyclic groups [**HK88**], abelian groups with at most two generators [**KPR24**], dihedral groups [**KNR22**], and aspherical 3-manifold groups [**Hil23**], we know that the quadratic 2-type determines the 4-manifold up to homotopy equivalence. Hence if we could also understand the image of the invariants in one of these cases, i.e. the realization problem, we would have a fairly complete homotopy classification, at least modulo the algebraic problem of being able to reliably distinguish or identify given quadratic 2-types.

(7) Kirk and Livingston's paper [**KL09**] contains many related references and its own list of interesting related problems.

Scribed by: M. Powell, J. Van Horn-Morris

PROBLEM 4.55. *Let M and N be closed, orientable, connected 4-manifolds with isomorphic quadratic 2-types. If $\pi_1(M) \cong \pi_1(N)$ are finite, are M and N homotopy equivalent?*

REMARKS.

(1) This is known for π_1 trivial, finite cyclic groups [**HK88**], dihedral groups [**KPR24**], and abelian groups with at most two generators [**KNR22**]. It is known to be false in the nonorientable case, due to Kim-Kojima-Raymond [**KKR92**].

(2) A potentially interesting case is $\pi_1(M) = \mathbb{Z}/2 \times \mathbb{Z}/2 \times \mathbb{Z}/2$. As noted in [**KPR24**], in this case we can have torsion in $\mathbb{Z} \otimes_{\mathbb{Z}[\pi_1(M)]} \Gamma(\pi_2(M))$, which by [**HK88**] leads to *polarized* homotopically inequivalent Poincaré 4-complexes with the same quadratic 2-type. Are they homotopy equivalent? Do the homotopy types contain topological 4-manifolds?

(3) Let $L_{p,q}$ and $L_{p,q'}$ be lens spaces that are not homotopy equivalent. Then $S^1 \times L_{p,q}$ and $S^1 \times L_{p,q'}$ are also not homotopy equivalent but they do have isomorphic quadratic 2-type. So the problem is not true in general for infinite fundamental groups.

Proposed for K3 and scribed by: M. Powell

4.6. Cobordisms

The problems here fit under the theme of cobordisms, either 4-dimensional cobordisms between 3-manifolds, or 5-dimensional cobordisms between 4-manifolds.

PROBLEM 4.56 (4D s-cobordism conjecture). *Let $(W^4; M_0^3, M_1^3)$ be a smooth 4-dimensional s-cobordism between closed 3-manifolds. Is W diffeomorphic to $M_0 \times [0,1]$?*

REMARKS.

(1) Matumoto and Siebenmann found counterexamples to the topological analogue in [**MS78**], where both M_0 and M_1 are $\mathbb{RP}^2 \times S^1$ and W is not known to be smoothable. Specifically that paper showed that the s-cobordism theorem fails either in dimension four or five, providing specific s-cobordisms that would fail to be products. Later work of Freedman and Quinn [**FQ90**, Theorem 7.1A] showed that the potential 5-dimensional candidate is indeed a product. So the 4-dimensional candidate of Matumoto and Siebenmann must fail to be a product. Cappell and Shaneson later provided further topological counterexamples where M_0 and M_1 are orientable. In a subsequent paper [**CS87b**] they asserted these examples were smoothable, but this claim was later retracted [**CS87a**].
(2) A weaker version of the question asks whether s-cobordant, or possibly even simple homotopy equivalent, 3-manifolds are necessarily homeomorphic. Kwasik and Schultz showed, assuming geometrization, that every topological h-cobordism between closed, orientable 3-manifolds is an s-cobordism, and that simple homotopy equivalence implies homeomorphism for closed, orientable 3-manifolds [**KS92**, Theorem and Theorem 1.1]. The questions appear to be open in the nonorientable setting; in particular geometrization is not yet known for nonorientable 3-manifolds. Whether there exists an h-cobordism between 3-manifolds with nontrivial Whitehead torsion appeared as Problem 4.9 on [**Kir97**].
(3) The s-cobordism theorem for dimensions 6 and higher is true in the smooth, piecewise linear, and topological settings [**Sma62a, Bar63, Maz63, Sta67, KS77**] (see also [**Mil65, RS72**]). The s-cobordism theorem in dimension five is false in the smooth (and equivalently piecewise linear) settings, by work of Donaldson [**Don87a**], and known to be true in the topological setting for good fundamental groups [**FQ90**, Theorem 7.1A] (see also [**OPR21a**]). See Problem 4.46.

Scribed by: A. Ray

PROBLEM 4.57. *Let X and Y be closed, oriented, smooth 4-manifolds with the same Euler characteristic and signature. Is there a torus link L in X with trivial normal bundle such that some choice of torus surgery along L transforms X into a manifold diffeomorphic to Y? What if we only require that the result be homeomorphic to Y?*

REMARKS.

(1) These questions were discussed in [**Ste06, BS13, FS11**].
(2) Let T be a torus in a 4-manifold W with trivial normal bundle. We say that a 4-manifold W' is obtained from W by torus surgery along T if $W' = W \backslash \nu(T) \cup (T^2 \times D^2)$ for some choice of gluing map $\partial\nu(T) \to \partial(T^2 \times D^2)$.
(3) It is a theorem of Iwase [**Iwa90**] that there is a 4-manifold Z and links of tori $L_X \subset X, L_Y \subset Y$ so that every component of L_X, L_Y has trivial normal bundle and there exists some choice of torus surgeries on L_X, L_Y transforming X, Y into manifolds equivalent to Z. Here, if X, Y are

smooth, then L_X, L_Y may be taken to be smooth, and "equivalent" means "diffeomorphic." Otherwise, surfaces are locally flat, and "equivalent" means "homeomorphic."

(4) In Iwase's paper, he actually proves that X can be transformed into $\#^a\mathbb{C}\mathrm{P}^2\#^b\overline{\mathbb{C}\mathrm{P}}^2\#^c S^1 \times S^3$ by surgery on a torus link in X for sufficiently large a with $b = a - \sigma(X)$, $c = (a + b + 2 - \chi(X))/2$. That is, we may take Z in the above discussion to be a connected sum of copies of $\mathbb{C}\mathrm{P}^2$s, $\overline{\mathbb{C}\mathrm{P}}^2$s, and $S^1 \times S^3$s. Iwase's argument holds in both categories.

A generalization of this result, extended over to the nonorientable 4–manifolds, is announced in a recent preprint of Baykur and Morgan [**BM25**], which states that any closed smooth 4–manifold is obtained by a surgery along a link of tori in a Z that is a connected sum of copies of $S^2 \times \mathbb{R}\mathrm{P}^2$, $\mathbb{R}\mathrm{P}^4$, $\mathbb{C}\mathrm{P}^2$s, $\overline{\mathbb{C}\mathrm{P}}^2$s, and $S^1 \times S^3$s.

(5) The problem asks whether, instead of X and Y being related by a *sequence* of two surgeries on torus links, all necessary torus surgeries transforming X into Y can be performed simultaneously.

(6) The problem seeks the 4-dimensional analog of the fact that any two closed, oriented 3-manifolds M, N are related by Dehn surgery along a link, rather than a sequence of Dehn surgeries. In dimension three, these two facts are clearly equivalent by dimensionality, but since surfaces generically intersect in 4-manifolds the situation is different.

(7) Fintushel–Stern [**FS11**] asked the following version of this question in the case the 4–manifolds are simply connected.

QUESTION ([**FS11**, §9]). Can a simply-connected closed, smooth 4-manifold always be obtained from torus surgery on a link of tori in a connected sum of $\mathbb{C}\mathrm{P}^2, S^2 \times S^2$ and $K3$ summands, taken with either orientations?

See [**BS13**] for an approach to this question via 5-dimensional round handles. Round 2-handle attachments correspond to certain torus surgeries, and the work of [**BS13**] falls short at the same point as discussed above; the authors build cobordisms made out of only round 2-handles, but it is not clear if there is always a cobordism where these round 2-handles can be attached independently.

(8) Fintushel–Stern noted that many interesting simply connected 4-manifolds arise from a single null-homologous torus surgery, e.g. for $n = 2, \ldots, 7, 9$ there are infinitely many exotic $\mathbb{C}\mathrm{P}^2\#^n\overline{\mathbb{C}\mathrm{P}}^2$ that each arises from a single torus surgery on a null-homologous torus in $\mathbb{C}\mathrm{P}^2\#^n\overline{\mathbb{C}\mathrm{P}}^2$ [**FS11**, Theorem 6].

QUESTION (Fintushel and Stern). If X and Y are homeomorphic, simply-connected, smooth 4-manifolds, is it possible to obtain Y from surgery on a single torus in X? If so, can we arrange for the torus to be null-homologous?

Proposed for K3 and scribed by: S. Kim, M. Miller

PROBLEM 4.58.

(a) *Which Seifert fibered homology spheres $\Sigma(a_1, \ldots, a_n)$ bound acyclic manifolds? Are there any examples with four or more singular fibers that bound acyclic manifolds?*
(b) *Which Seifert fibered homology spheres bound contractible manifolds? Is there an example that bounds an acyclic manifold but not a contractible manifold?*
(c) *Are there any Seifert fibered homology spheres that arise as cork boundaries?*

REMARKS.

(1) The second part of Problem (a) (regarding the number of fibers) is in [**Kir97**, Problem 4.123] in a different guise, with a different motivation.
(2) Many Seifert fibered homology spheres bound acyclic (and, in fact, contractible) manifolds (see for example [**AK79b, CH81, Fic84, Şav20**]) but a complete characterization is unknown. All known such examples have three singular fibers. Amongst spheres $\Sigma(p, q, r)$ with three singular fibers, there is no closed characterization of which bound acyclic manifolds in terms of p, q, and r.
(3) It is a longstanding conjecture that no Seifert fibered homology sphere with four or more singular fibers bounds an acyclic manifold; see [**FS87b, Kol08**]. This is related to the Montgomery-Yang conjecture, which states that every pseudofree action of S^1 on S^5 has at most 3 non-free orbits. One can also ask the related question of which Seifert fibered homology spheres embed in $\mathbb{R}^4$. In the setting of symplectic topology, it is known that no Seifert fibered homology sphere (of any number of fibers) occurs as a hypersurface of contact type in $(\mathbb{R}^4, \omega_{std})$ [**MT22**]. A recent preprint [**AC24**] builds on this work and announces that no (standardly-oriented) Seifert fibered homology sphere bounds a Stein rational ball.
(4) One can also ask about the difference between bounding an acyclic manifold and bounding a contractible manifold. In general, these notions differ: Taubes' periodic ends theorem implies that $\Sigma(2, 3, 5) \# - \Sigma(2, 3, 5)$ bounds no contractible manifold [**Tau87**], whereas this trivially bounds an acyclic manifold. Many other examples can be obtained through instanton Floer theory. However, no such example consisting of an *individual* Seifert fibered homology sphere is known.
(5) A similar question is whether or not any Seifert fibered homology sphere Y forms a cork boundary. Here, recall that Y is a cork boundary if there exists a contractible manifold W with boundary Y, together with a self-diffeomorphism of Y that does not extend over W (as a diffeomorphism). It is known that the standard cyclic group actions on any Brieskorn sphere $\Sigma(p, q, r)$ do not extend (smoothly) as group actions to any contractible manifold with boundary $\Sigma(p, q, r)$ [**AH16, AH21**]. However, these do extend as diffeomorphisms. Current Floer-theoretictechniques devoted to

establishing corks [**AKS20, DHM23**] are known to fail for Seifert fibered homology spheres. Note that several authors take W to be Stein in the definition of a cork; the notion defined here is sometimes called a loose cork.

Scribed by: I. Dai, J. Van Horn-Morris

PROBLEM 4.59. *Are lens spaces topologically homology cobordant if and only if they are homeomorphic?*

REMARKS.

(1) Livingston conjectures that the statement is correct.
(2) If $L(a_1, b_1)$ and $L(a_2, b_2)$ are homology cobordant, then $a_1 = a_2$. The problem can be restated: if $L(n, b_1)$ and $L(n, b_2)$ are homology cobordant, then $L(n, b_1)$ and $L(n, b_2)$ are homeomorphic. Gilmer and Livingston [**GL83**] proved this in the case that n is a prime power, using Atiyah–Singer signature invariants $\rho_\alpha(L)$ associated to characters $\alpha : \pi_1(L) \to U(1)$ of prime-power order. The simplest unknown case is the pair they identified, $L(231, 53)$ and $L(231, 86)$. It remains unknown whether these lens spaces are homology cobordant.
(3) In the smooth category, the conjecture was proved for n even by Fintushel–Stern [**FS87a**], with later generalizations by Matić [**Mat88**] and Ruberman [**Rub88**]. The conjecture in the smooth setting can also be proved using Heegaard Floer theory; see [**DW15**].
(4) For higher dimensional lens spaces, Cappell and Ruberman [**CR88**] showed that the ρ_α-invariants for α of prime-power order give the homology cobordism classification. This uses homology surgery theory [**CS74, Vog82**] which is known [**Akb79**] to fail in dimension 4, even topologically.

    It is conceivable to try to construct a topological homology cobordism W between non-diffeomorphic $L(n, b_1)$ and $L(n, b_2)$ with n composite using ordinary surgery theory. A first step might be to find an appropriate homotopical model for such a cobordism. The Gilmer–Livingston argument implies that the n-fold cyclic $\widetilde{W}$ would have to have nontrivial b_1, so that $\pi_1(W)$ would have to be large in this sense; see [**AGL18**] for more information on this.

Proposed for K3 and scribed by: C. Livingston

PROBLEM 4.60.

(a) *Let X be an open, spin, smooth 4-manifold. Does X have a proper smooth embedding in $\mathbb{R}^6$?*
(b) *By choosing a proper exhaustion function on W,* (a) *would follow from an affirmative answer to the following. Let $(W; M, N)$ be a compact, smooth spin cobordism with a smooth embedding f of M in S^5. Is there a smooth*

embedding

$$F\colon (W; M, N) \to (S^5 \times I; S^5 \times \{0\}, S^5 \times \{1\})$$

whose restriction to M coincides with f?

(c) *Let $(W; M, N)$ be a compact, smooth spin cobordism with smooth embeddings f of M in S^5 and g of N in S^5 such that $\sigma(N, S^5) - \sigma(M, S^5) = \sigma(W)$. Is there a smooth embedding*

$$F\colon (W; M, N) \to (S^5 \times I; S^5 \times \{0\}, S^5 \times \{1\})$$

whose restriction to M coincides with f and whose restriction to N coincides with g?

REMARKS.

(1) The spin condition is necessary in both parts of the problem.
(2) One motivation for part (a) is to understand the *embedding dimension* for a Stein surface (of real dimension four). By definition, this is the minimal d for which X has a proper holomorphic embedding in $\mathbb{C}^d$. General results about embedding dimension due to Eliashberg and Gromov [**EG92**] imply that a Stein 4-manifold has a proper holomorphic embedding into $\mathbb{C}^4$. Since a 4-manifold properly embedded in $\mathbb{C}^3 = \mathbb{R}^6$ is spin, this is the best possible result for non-spin Stein 4-manifolds, but it is conceivable that one could get embeddings in $\mathbb{C}^3$ for spin Stein 4-manifolds. Part (a) is a topological version of that question.
(3) Part (b) asks for a relative version of the result, announced by Cappell-Shaneson in [**CS79**] and proved by Ruberman in [**Rub82**], that a closed spin 4-manifold embeds in $\mathbb{R}^6$ if and only if its signature is 0.
(4) For part (c), note that an embedding of a 3-manifold M in S^5 has a well-defined signature $\sigma(M, S^5)$, given by the signature of any 4-manifold that M bounds in S^5. The equality $\sigma(N, S^5) - \sigma(M, S^5) = \sigma(W)$ is necessary for W to be a cobordism as in (b). It follows, in the setting of part (b), that one cannot specify the embedding of both M and N in advance. It is not clear if there are further obstructions, so part (c) represents a sharpening of part (b).

Proposed for K3 and scribed by: D. Ruberman

4.7. Smooth 4-manifold invariants

Here we present a few problems on diffeomorphism invariants of 4-manifolds, coming from gauge theory and Khovanov homology.

PROBLEM 4.61. *What do different 4-manifold gauge theories see?*

REMARKS.

(1) It is conjectured that the Donaldson, Seiberg–Witten, and Heegaard Floer invariants of closed 4-manifolds coincide (after organizing the Donaldson invariant as a generating function [**KM95**]). Yet each of these theories

taken more broadly seem to see different geometric and topological properties. This problem asks how to understand some of these properties proved in one theory via one of the other theories.

(2) Is it possible to prove the existence of uncountable many exotic structures on $\mathbb{R}^4$ using Seiberg–Witten or Heegaard Floer theory? One might try to do this by adapting Taubes' periodic end gauge theory [**Tau87**] to the Seiberg–Witten setting. This can be done but to date requires rather strong hypotheses such as positive scalar curvature on the periodic end. Another possibility is to use limit invariants of ends such as the Heegaard Floer end invariant introduced by Gadgil [**Gad10**].

(3) Is there a Seiberg–Witten proof of the Donaldson–Sullivan [**DS89**] results that there are 4-manifolds without quasiconformal (and hence Lipschitz) structure, and that 4-manifolds can admit more than one such structures? A seemingly fundamental difficulty here is whether Lipschitz or quasiconformal manifolds have something like a Dirac operator; see the discussion in [**Sul87, Sul99**] and also Problem 4.131.

(4) Donaldson and Seiberg-Witten theory have parameterized versions that can be used to study invariants of diffeomorphisms and families of 4-manifolds as well as families of symplectic structures [**Rub98, BK22, Kon21, Kro97**]. Are there family versions of Heegaard Floer theory that would be useful for such applications?

(5) Seiberg-Witten invariants can be extended to give the Bauer–Furuta invariant [**BF04**] living in (equivariant) stable homotopy groups. Are there similar stable homotopy theoretic invariants coming from Donaldson or Heegaard Floer theory?

(6) Seiberg–Witten theory can be used to show that certain 4-manifolds admit no Riemannian metric of positive scalar curvature (PSC) [**Wit94**] and to distinguish path components in the space of PSC metrics [**Rub01**]. Find a way to do this using Donaldson or Heegaard Floer theory.

(7) All three theories [**Frø02, Frø96, OS03a**] give statements about the definite intersection forms of 4-manifolds with given boundary. Are these statements equivalent?

(8) Hambleton and Lee [**HL95**] used an equivariant version of Donaldson's original argument for his definite manifolds theorem to study smooth group actions on a simply connected (positive) definite 4-manifold. Among other results, they showed that a homologically trivial cyclic group action has the same fixed-point data and tangential isotropy representations as an equivariant connected sum of linear actions on $\mathbb{C}P^2$. Are there Seiberg–Witten or Heegaard Floer proofs of their results?

Proposed for K3 and scribed by: D. Ruberman

PROBLEM 4.62. *Let X be a smooth, closed, connected, oriented 4-manifold with $b_2^+(X) > 1$.*

(a) *Does X have Donaldson simple type?*
(b) *Does X have Seiberg–Witten simple type?*

REMARKS.

(1) The Simple Type Conjecture holds that the answer is affirmative. Part (a) is in [**Kir97**, Problem 4.131], and was raised by Kronheimer and Mrowka in [**KM95**] for simply connected 4–manifolds.
(2) A 4-manifold is said to have Donaldson simple type if the Donaldson polynomials q_k for principal $SU(2)$-bundles with $c_2 = k$ satisfy

$$q_{k+1}(\nu, \Sigma_1, \ldots, \Sigma_d) = 4q_k(\Sigma_1, \ldots, \Sigma_d)$$

where $\nu = \mu(1)$ ($\mu\colon H_0(X;\mathbb{Z}) \to H_4(M_{k+1})$), $\Sigma_i \in H_2(X;\mathbb{Z})$, and $2d = \dim M_k$.
(3) Manifolds which have Donaldson simple type [**KM95**] include:
   - complete intersections,
   - elliptic surfaces,
   - any manifold with a Gompf nucleus,
   - manifolds with a smoothly embedded surface F satisfying $2(\text{genus}(F)) - 2 = F \cdot F > 0$.
(4) The Kronheimer–Mrowka structure theorem from [**KM95**] says that, for manifolds of simple type, the Donaldson invariants are determined by finitely many basic classes $K_1, \ldots, K_s$ and rational numbers $\beta_1, \ldots, \beta_s$.
   When $b_2^+ = 1$, some manifolds, e.g. $\mathbb{C}\mathrm{P}^2$, $S^2 \times S^2$, $\mathbb{C}\mathrm{P}^2 \# \overline{\mathbb{C}\mathrm{P}}^2$, do not have Donaldson simple type.
(5) A Seiberg–Witten *basic class* $\mathfrak{s}$ is a spin^c structure $\mathfrak{s}$ with nonzero Seiberg-Witten invariant $\mathrm{SW}_X(\mathfrak{s})$. A 4-manifold has *Seiberg-Witten simple type* if the virtual dimension of the Seiberg-Witten moduli space

$$d_X(\mathfrak{s}) = \frac{1}{4}(c_1(\mathfrak{s})^2 - 2\chi(\mathfrak{s}) - 3\sigma(X))$$

is zero for every basic class $\mathfrak{s}$.
   Note that, when $d_X(\mathfrak{s}) = 0$, the Seiberg–Witten invariants are defined by a signed count of monopoles. When $d_X(\mathfrak{s}) > 0$, they are defined by evaluating a higher degree cohomology class on the moduli space of monopoles; the conjecture says that, in such cases, the evaluation is always zero.
(6) We list some important progress.
   - Taubes' equivalence between the Seiberg–Witten invariants and the Gromov-Taubes invariants implies that all symplectic 4-manifolds are Seiberg–Witten simple type [**Tau96**].
   - Kato-Nakamura-Yasui proved the conjecture for the mod 2 Seiberg-Witten invariants under a mild condition on the homology ring [**KNY23**].
   - Baraglia [**Bar23a**] proved that the mod 2 Seiberg–Witten simple type conjecture holds for spin structures without any extra assumptions.

Just as in the Donaldson case, there are manifolds with $b_2^+ = 1$ that are not of simple type, because of the wall-crossing formula.

(7) Witten's conjecture [**Wit94**] states that if X has Seiberg–Witten simple type, then it also has Donaldson simple type, and there is a precise relation between its Donaldson and Seiberg–Witten invariants. The conjecture was proved in many cases by Feehan and Leness; see [**FL15**], [**FL18**].

Scribed by: C. Manolescu

PROBLEM 4.63. *How many independent basic classes can a simply connected smooth* 4*-manifold* X *have, as measured by* $\mathrm{br}(X)$*, the rank of the span of the basic classes?*

(a) *Is there an upper bound for* $\mathrm{br}(X)$ *in terms of topological invariants of* X*?*
(b) *In particular, is* $\mathrm{br}(X) \leqslant b_2^+(X)$ *for all simply connected* X *with* $b_2^+(X)$ *odd?*
(c) *Is there a smooth indefinite, simply connected 4-manifold* X *with* $b_2(X) \geqslant 3$ *for which the image of* $\mathrm{ev}_*\colon \mathrm{Diff}(X) \to \mathrm{Aut}(Q_X)$ *is finite, or even trivial?*

REMARKS.

(1) Here we define

$$\mathrm{br}(X) = \mathrm{Rank}\,(\mathrm{Span}\{c_1(\mathfrak{s}) \mid \mathfrak{s} \text{ is a Seiberg-Witten basic class on } X\})$$

if X has any basic classes, and 0 otherwise.
(2) Knot surgery [**FS98**] on n disjoint and homologically independent tori in a manifold with nontrivial Seiberg-Witten invariant would create a manifold X for which $\mathrm{br}(X) = n$. For example, starting with an elliptic surface, the construction in [**GM93**] produces examples of manifolds X for which $\mathrm{br}(X) = b_2^+(X)$.
(3) Questions (a) and (b) are relevant to the study of the map $\mathrm{ev}_* : \mathrm{Diff}(X) \to \mathrm{Aut}(Q_X)$ giving the action of a diffeomorphism on the intersection form. Since any diffeomorphism must permute the basic classes up to sign, the presence of many basic classes can restrict the size of the image of ev_* .
(4) The restriction to indefinite manifolds and $b_2 > 2$ is to rule out intersection forms with finite automorphism groups. If $\mathrm{br}(X) = b_2(X)$, then $\mathrm{ev}_*(\mathrm{Diff})$ is contained in a finite permutation group and hence is finite.

Proposed for K3 by: D. Auckly
Scribed by: D. Ruberman

PROBLEM 4.64. *Find an irreducible, closed, smooth* 4*-manifold with nontrivial Bauer–Furuta invariant but with trivial Seiberg–Witten invariant.*

REMARKS. The Bauer–Furuta invariant Ψ [**BF04**] is a stable cohomotopy refinement of the Seiberg–Witten invariant [**Wit94**], building on Furuta's proof of the 10/8-theorem [**Fur01**]. There are examples for which the Bauer–Furuta invariant is strictly stronger than the Seiberg–Witten invariant; for instance, Ψ can be used to distinguish between certain connected sums of homotopy K3 surfaces (which have vanishing Seiberg–Witten invariant). Such examples rely on a gluing formula due to Bauer [**Bau04**]; no irreducible examples are known.

Proposed for K3 by: M. Stoffregen
Scribed by: I. Dai

PROBLEM 4.65. *Suppose X is a smooth 4-manifold with the homology of $S^1 \times S^3$ whose infinite cyclic cover $\widetilde{X}$ has $H^1(\widetilde{X}) = 0$. Furuta and Ohta* [**FO93**] *define an invariant $\lambda_{FO}(X)$ as 1/4 of the signed count of irreducible flat* SU(2) *connections on X; this requires an orientation of X and a specified generator of $H^1(X)$.*

(a) *Does the following hold? The invariant $\lambda_{FO}(X)$ is an integer, and $\lambda_{FO}(X) \equiv \rho(Y, \mathfrak{s})$, where $\rho(Y, \mathfrak{s})$ is the Rokhlin invariant of an oriented spin* 3 *manifold Y that is Poincaré dual to the generator of $H^1(X)$.*
(b) *Mrowka–Ruberman–Saveliev* [**MRS11**] *give an approach, defining an invariant $\lambda_{SW}(X)$ by counting solutions to the Seiberg-Witten equations and adding an index-theoretic correction term. Is*

$$\lambda_{FO}(X) = -\lambda_{SW}(X)?$$

REMARKS.

(1) Part (a) was conjectured by Furuta–Ohta.
(2) A solution to (a) would imply that the Wall group $L_5(\mathbb{Z}[\mathbb{Z}])$ does not act on the smooth structure set of $S^1 \times S^3$; compare the discussion in Problem 4.22. By construction, $\lambda_{SW}(X)$ is an integer, and it is shown in [**MRS11**] that it reduces mod 2 to $\rho(Y, \mathfrak{s})$. So a positive answer to (b) implies a positive answer to (a).
(3) The invariant $\lambda_{FO}(X)$ is defined in greater generality; one could require only that X is a homology $S^1 \times S^3$ whose twisted cohomology $H^1(X; \mathbb{C}_\alpha)$ vanishes for any homomorphism $\alpha : \pi_1(X) \to \mathrm{U}(1)$. In this setting, neither part of (a) holds, but (b) is still plausible. The paper [**LRS21**] shows that (b) holds for mapping tori of all orientation preserving diffeomorphisms of homology spheres generating a semifree finite cyclic group action. For involutions, the result also follows from [**LRS23b**].

Scribed by: D. Ruberman, N. Saveliev

PROBLEM 4.66. *Can the skein lasagna module detect exotic smooth structures on closed 4-manifolds?*

REMARKS.

(1) The skein lasagna module is an extension of Khovanov homology. It is an invariant of 4-manifolds with (possibly empty) boundary and a framed link in their boundary. It was defined by Morrison, Walker and Wedrich in [**MWW22**]. Ren and Willis [**RW24**] gave examples of exotic compact 4-manifolds with boundary that are detected by the skein lasagna module. For closed 4-manifolds, the computations so far are limited; see [**MN22**],

[**MWW23**], [**RW24**]. The invariant is nonvanishing for S^4, $S^1 \times S^3$ and $\overline{\mathbb{CP}}^2$. On the other hand, it vanishes for manifolds that contain a smoothly embedded sphere with positive self-intersection (cf. Theorem 1.3 in [**RW24**]); e.g. for $\mathbb{CP}^2$ or $-K3$.

(2) The invariant is multiplicative under connected sums, and it takes the value 0 on $S^2 \times S^2$. Thus, it has a chance of detecting exotic smooth structures on simply connected, closed 4-manifolds, even though such structures become standard after sufficiently many stabilizations.

Proposed for K3 and scribed by: C. Manolescu

4.8. Symmetries of 4-manifolds

Symmetries of 4-manifolds refers primarily to the topological groups of self-homeomorphisms Homeo(X) or self-diffeomorphisms Diff(X) of a fixed 4-manifold X; it can also refer to group actions on the manifold. The smooth (resp. topological) *mapping class group of* X is the group of isotopy classes of these diffeomorphisms (resp. homeomorphisms).

We begin with problems asking for exotic diffeomorphisms, namely diffeomorphisms that are topologically but not smoothly isotopic to the identity. After that, the next few problems consider higher families of symmetries, for instance the homotopy groups of Diff(X) or Homeo(X). Then come a series of problems on stabilizations. After a few more problems, on non-smoothability, the Nielsen problem, uniform perfectness, and 5-dimensional mapping tori, we end the section with a collection of problems on group actions on 4-manifolds.

We collect a few conventions here. For connected X, we write $\mathring{X} := X \backslash \mathring{D}^4$ for a punctured X obtained by removing an open 4-ball from the interior of X. In the following definitions, we use morphism to refer to either homeomorphisms or diffeomorphisms, depending on the category. The groups $\mathrm{Diff}^+(X)$ and $\mathrm{Homeo}^+(X)$ denote the respective subgroups of orientation-preserving morphisms. The notation $\mathrm{Diff}_\partial(X)$ and $\mathrm{Homeo}_\partial(X)$ denotes morphisms of X that fix the boundary ∂X pointwise. A pseudo-isotopy is a morphism $F\colon X \times I \to X \times I$ that restricts to the identity on $(X \times 0) \cup (\partial X \times I)$. We say that the morphism $F|_{X \times 1}$ is pseudo-isotopic to the identity.

PROBLEM 4.67.

(a) *Compute* $\pi_0(\mathrm{Diff}^+(S^4))$. *Do we have* $\pi_0(\mathrm{Diff}^+(S^4)) = \{1\}$?

(b) *In particular, does some implantation of the barbell map provide a nontrivial element in* $\pi_0(\mathrm{Diff}^+(S^4))$?

REMARKS.

(1) The *4-dimensional generalized Smale conjecture* asked whether the inclusion $SO(5) \hookrightarrow \mathrm{Diff}^+(S^4)$ is a homotopy equivalence (see Problems 4.34 and 4.126 in [**Kir97**]). Watanabe [**Wat19**] disproved this conjecture, by proving that $\pi_k(\mathrm{Diff}_\partial(D^4))$ is nontrivial for many k, including $k = 1, 4, 8$.

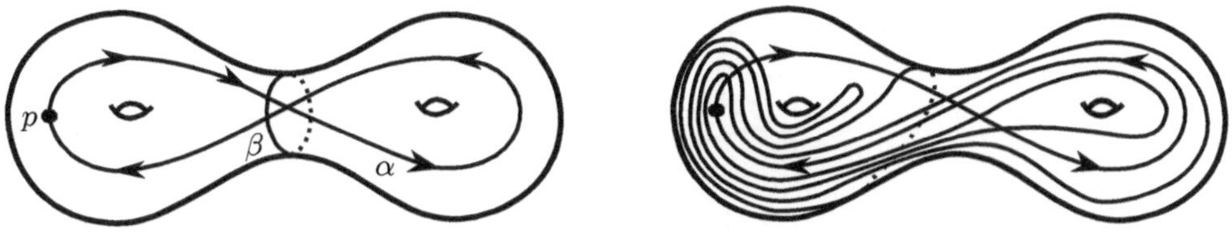

(a) Before pushing (b) After pushing

FIGURE 2. Point push map on a surface

A fundamental issue remaining to understand is $\pi_0(\mathrm{Diff}^+(S^4))$, the orientation-preserving mapping class group of S^4, and it is presently unknown whether or not this group is trivial.

Some sources of possibly nontrivial diffeomorphisms of S^4 are suggested in [**Gay25**], [**BG19**] and [**GGH$^+$23**]. The Pin(2)-equivariant family Bauer-Furuta invariant could potentially detect such nontrivial diffeomorphisms; see [**LM25**].

(2) The *barbell map* is defined by Budney and Gabai in [**BG19**] (see also [**BG25**] for the 'X-resolution') using isotopy extension, via a generalization of Birman's 'push map' [**Bir69**]. We give an exposition here.

Birman's map can be described by pushing a point p (see Figure 2(a)) along a path α in a closed surface S and back to p. This isotopy of the surface can be assumed to fix a disk D centered at p at the end of the isotopy. Removing the interior of D, we get a diffeomorphism of the punctured surface that is the identity on ∂D and also the identity outside a nice neighborhood of the arc α. Figure 2(b) shows what happens to the loop β under this diffeomorphism.

In general, when pushing a 0-dimensional point p along a 1-dimensional loop α in a 2-dimensional surface, the point crosses a loop (say β) and bulldozes it over D.

Consider a pair of 2-spheres in $\mathbb{R}^3$ centered at $(\pm 2, 0, 0)$ of radius one, which are joined by the arc in the x-axis $[-1, 1]$. We can use the z-axis to define the equator $(x^2 + y^2 = 1)$, latitudes, and the two poles. Thicken this slightly in $\mathbb{R}^3$, to what might be called a (hollow) barbell. Now cross with $[-3, 3]$ to get a four-dimensional analogue called B. We will find an interesting diffeomorphism β of B that is the identity on ∂B.

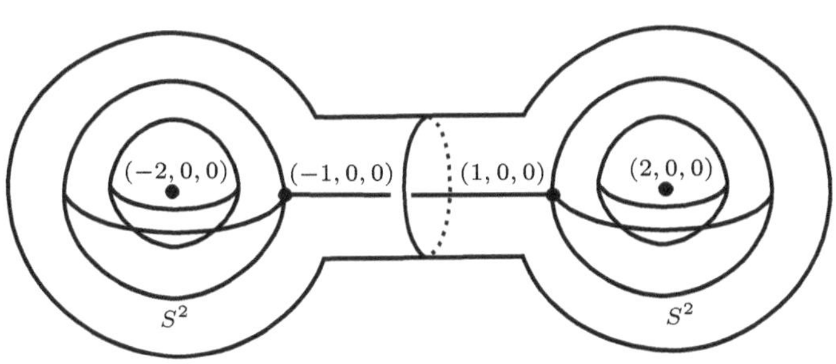

FIGURE 3. Barbell

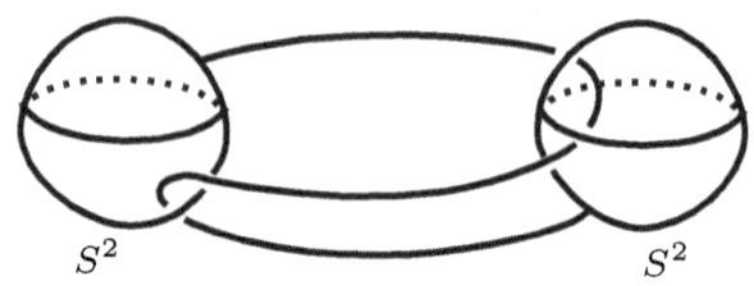

FIGURE 4. Embedding of barbell

Notice the two line segments $(\pm 2, 0, 0) \times [-3, 3]$. These can be thickened so that they fill in the "hollows", the pair of $S^2 \times [-1, 1]$s, so that we have a solid 4-dimensional barbell, which is obviously B^4. Figure 3 may help.

Now focus just on the left line segment and move a portion of it, namely $(-2, 0, 0) \times [-1, 1]$, around the 2-sphere on the right side, sort of like lassoing a horse's head. We will do this with a 1-parameter family of Birman push maps, parameterized by $t \in [-3, 3]$. The push maps occur on a 2-dimensional surface given by fixing t and z.

The push map for $t = 0$ will push the point $p = (-2, 0, 0) \times 0$ over to and then around the equator ($z = 0$) of the 2-sphere at $t = 0$ and back to p; for $t \in (-1, 1)$, p goes over to and around the latitude at $z = t$ and back to p; for $t = \pm 1$, p goes over to a pole and then back to p; for $t \in (-2, -1) \cup (1, 2)$, p goes partway to a pole and then back to p; finally, for $t \in (-3, -2) \cup (2, 3)$, p does not move.

This isotopy of the left line segment extends to an isotopy of B^4 that is the identity on ∂B^4 and on the two line segments. But if this isotopy fixes the two line segments, then it can also be made to fix the above thickenings of the segments. Therefore, the end of this isotopy is a diffeomorphism β taking the 4-dimensional hollow barbell B back to itself, and fixing its boundary.

Note that the lasso can go over the horse's head with two possible orientations. Also there is a rotation ρ switching the two balls in the barbell. It can be checked that $\rho\beta\rho^{-1} = \beta^{-1}$.

The 4-dimensional barbell, B, can be embedded, by f, in a 4-manifold X in many ways, and these are called implantations and the induced barbell map can be called $f_*\beta$.

(3) A specific example of a barbell diffeomorphism to consider is as follows.

QUESTION. Does the following embedding f induce a nontrivial diffeomorphism of S^4, where the embedding is given by mapping the two 2-spheres to separate 2-spheres in S^4, and the bar goes from the left sphere over to link the right sphere and then back again to link the first and finally attaching to the right sphere? See Figure 4.

A positive answer has been announced by Gabai–Gay–Hartman [**GGH25**].

(4) We can also ask the following structural question.

QUESTION. Is $\mathrm{Diff}^+(S^4)$ (finitely) generated up to isotopy by a composition of $f_*\beta$s?

Scribed by: H. Konno, R. Kirby

PROBLEM 4.68. *Does every closed smooth 4-manifold admit an exotic diffeomorphism? How about the following special cases?*

(a) *Is there a definite smooth closed 4-manifold that admits an exotic diffeomorphism?*
(b) *Does $S^2 \times S^2$ or $K3$ admit an exotic diffeomorphism?*

REMARKS.

(1) A self-diffeomorphism $f\colon X \to X$ of a smooth manifold X is called *exotic* if it is topologically but not smoothly isotopic to the identity. The first examples of exotic diffeomorphisms of 4-manifolds were given by Ruberman [**Rub98**]. Most known examples of closed 4-manifolds confirmed to admit exotic diffeomorphisms are of the form $\#^m\mathbb{CP}^2\#^n\overline{\mathbb{CP}}^2$ for $m, n > 0$ [**Rub98**], $\#^m K3\#^n S^2 \times S^2$ for $m, n > 0$ [**BK20**] or $m = 0$ [**AR25**], and $K3\#K3$ [**KM20**]. There exist irreducible 4-manifolds that admit exotic diffeomorphisms [**BK24a**], but the proof in [**BK24a**] does not apply to S^4, $S^2 \times S^2$, $\mathbb{CP}^2$, or $K3$.
(2) In the literature, the smallest (in term of second Betti number) closed 4-manifold that is known to admit an exotic diffeomorphism is $\#^2\mathbb{CP}^2\#^{10}\overline{\mathbb{CP}}^2$, announced in [**Qiu24**]. It is natural to ask how small a closed 4-manifold with an exotic diffeomorphism can be. In particular, whether such a diffeomorphism exists on S^4 is the π_0 case of the Smale conjecture; see Problem 4.67.

   For 4-manifolds with boundary, there exists an example of a contractible (hence, definite) 4-manifold that has an exotic diffeomorphism (relative to the boundary) [**KMT23a, KMPW24, KPT26, KLMME24**].
(3) For a simply connected closed smooth 4-manifold X, every exotic diffeomorphism of X is smoothly isotopic to the identity after sufficiently many stabilizations by $S^2 \times S^2$. This fact follows by combining work of Kreck [**Kre79**] and either Quinn [**Qui86**] (cf. [**GGH$^+$23**]) or Gabai [**Gab22**]. It is an interesting question to determine how many stabilizations are needed to kill the exotic property of a given diffeomorphism. Many known examples of exotic diffeomorphisms, such as those in [**Rub98, BK20**], are smoothly isotopic to the identity after only one stabilization [**AKMR15**]. On the other hand, Lin [**Lin23**] proved that the exotic diffeomorphism of $K3\#K3$ from [**KM20**] stays exotic after one stabilization. There is no known upper bound for how many stabilizations are needed to trivialize this diffeomorphism. See also Problem 4.79.

Proposed for K3 and scribed by: H. Konno

PROBLEM 4.69. *Does there exist a diffeomorphism of a closed* 3*-manifold $f\colon M \to M$ such that f is topologically but not smoothly pseudo-isotopic to the identity?*

REMARKS.

(1) Friedman–Witt [**FW86**] constructed diffeomorphisms f that are homotopic but not isotopic to the identity. These are also topologically pseudo-isotopic to the identity [**KS96**].

QUESTION. Are the Friedman–Witt diffeomorphisms smoothly pseudo-isotopic to the identity?

(2) The following observation yields a potentially useful reformulation. Consider the embedding

$$i_{1/2}\colon M \to M \times \{1/2\} \hookrightarrow M \times [0,1]$$

where the first map is the canonical identification. Then f is smoothly (topologically) pseudo-isotopic to the identity if and only if $i_{1/2} \circ f$ and $i_{1/2}$ are smoothly (topologically) isotopic as embeddings.

The proof of this observation, which we give next, was provided separately by Hatcher and Igusa. It applies in all dimensions.

If f were pseudo-isotopic to the identity via $F\colon M \times I \to M \times I$, then $i_{1/2} \circ f$ and $i_{1/2}$ would isotopic as embeddings. To see this, note that by translation it suffices to show that i_0 and $i_1 \circ f$ are isotopic. But $g_t\colon M \to M$ defined by $g_t(x) = F(x,t)$ gives such an isotopy.

Conversely, if $i_{1/2}$ and $i_{1/2} \circ f$ are isotopic as embeddings, then apply isotopy extension rel. $M \times \{0,1\}$ to the isotopy, to obtain a diffeomorphism of $M \times [0,1/2]$ that restricts to the identity on $M \times \{0\}$ and to f on $M \times \{1/2\}$ (and similarly in $M \times [1/2,1]$). Thus after rescaling we obtain a pseudo-isotopy from f to the identity.

Proposed for K3 and scribed by: I. Agol

PROBLEM 4.70. *Do there exist $k \geqslant 0$ and a smooth closed 4-manifold X such that the map $\pi_k(\mathrm{Diff}(X)) \to \pi_k(\mathrm{Homeo}(X))$ induced by the inclusion $\mathrm{Diff}(X) \hookrightarrow \mathrm{Homeo}(X)$ is an isomorphism?*

REMARKS.

(1) Lin–Xie [**LX23**] proved that, for every orientable compact smooth 4-manifold X, at least one of the following holds:
- $\pi_1(\mathrm{Diff}(X)) \to \pi_1(\mathrm{Homeo}(X))$ is not injective.
- $\pi_2(\mathrm{Diff}(X)) \to \pi_2(\mathrm{Homeo}(X))$ is not surjective.

They prove also that, if $\partial X \neq \varnothing$ or the signature of X is non-zero, then there are many degrees k for which $\pi_k(\mathrm{Diff}(X)) \to \pi_k(\mathrm{Homeo}(X))$ is not an isomorphism. However, it is still possible that for some 4-manifold X and some degree k, the map $\pi_k(\mathrm{Diff}(X)) \to \pi_k(\mathrm{Homeo}(X))$ is an isomorphism.

The above result of Lin–Xie is based on Watanabe's work [**Wat19**] for $X = S^4$. Watanabe proved that nontrivial elements in the kernel of $\pi_k(\mathrm{Diff}(S^4)) \to \pi_k(\mathrm{Homeo}(S^4))$ exist whenever $\mathcal{A}_{k+1} \neq 0$. Here $\mathcal{A}_{k+1}$ is the degree $(k+1)$-part of a specific graph cohomology.

(2) Many other negative results are obtained by (mainly family) gauge theory, such as [**AR25, FM88, Don90, MS97, Rub98, BK20, KKN21b, BK22, Bar21, BK23, KN23, KM20, Lin23, KT22b, IKMT25, KMT23a, GL25**].

(3) As a positive result, a classical theorem by Wall [**Wal64a**] shows that there are many 4-manifolds X for which the map $\pi_0(\mathrm{Diff}(X)) \to \pi_0(\mathrm{Homeo}(X))$ is surjective. However, there also exist examples where this map is not surjective; see [**FM88, Don87b**].

Proposed for K3 and scribed by: H. Konno

PROBLEM 4.71.

(a) *Do there exist $k \geqslant 0$ and a smooth closed orientable 4-manifold X such that $\pi_k(\mathrm{Diff}(X))$ is finitely generated?*

(b) *Do there exist $k > 0$ and a smooth closed orientable 4-manifold X such that $H_k(B\,\mathrm{Diff}(X); \mathbb{Z})$ is finitely generated?*

REMARKS.

(1) For each $k > 0$, there exist simply-connected closed smooth 4-manifolds X where $H_k(B\,\mathrm{Diff}(X); \mathbb{Z})$ are not finitely generated [**Kon24b**]. Also, Auckly–Ruberman [**AR25**] proved that, for each $k > 0$, there exist simply-connected closed smooth 4-manifolds X where $\pi_k(\mathrm{Diff}(X))$ is not finitely generated. In dimension $\neq 4$, there are several finiteness results. See [**Kup19b, BKK24**]. See also Problem 4.72 for the analogous question in the topological category.

(2) If X is oriented, we can ask an analogous question for $\mathrm{Diff}^+(X)$, the orientation-preserving diffeomorphism group, in place of $\mathrm{Diff}(X)$. (Note that $\mathrm{Diff}(X) = \mathrm{Diff}^+(X)$ if X has non-zero signature.) For homotopy groups $\pi_k(\mathrm{Diff}(X))$ and $\pi_k(\mathrm{Diff}^+(X))$, finite generation of $\pi_k(\mathrm{Diff}(X))$ and that of $\pi_k(\mathrm{Diff}^+(X))$ are equivalent. However, for $H_k(B\,\mathrm{Diff}(X); \mathbb{Z})$ and $H_k(B\,\mathrm{Diff}^+(X); \mathbb{Z})$, the questions may not be equivalent. For example, when $\mathrm{Diff}(X) \neq \mathrm{Diff}^+(X)$, and if we take rational coefficients, the covering map $\mathbb{Z}/2\mathbb{Z} \to B\,\mathrm{Diff}^+(X) \to B\,\mathrm{Diff}(X)$ induces an isomorphism
$$H_k(B\,\mathrm{Diff}^+(X); \mathbb{Q})^{\mathbb{Z}/2} \cong H_k(B\,\mathrm{Diff}(X); \mathbb{Q}),$$
where the superscript $\mathbb{Z}/2$ indicates the monodromy invariant part. Thus finite generation of $H_k(B\,\mathrm{Diff}^+(X); \mathbb{Q})$ implies that of $H_k(B\,\mathrm{Diff}(X); \mathbb{Q})$, but the converse may not be true in general.

(3) As sets, the homotopy and homology groups of diffeomorphism groups of compact manifolds are always countable.

(4) The diffeomorphism groups of non-compact 4-manifolds can have finitely generated homotopy and homology groups, e.g. $\mathrm{Diff}(\mathbb{R}^4)$, in the weak C^∞-topology, is homotopy equivalent to $O(4)$. Nevertheless, there also exist exotic $\mathbb{R}^4$'s with infinitely generated mapping class groups [**Gom18**].

(5) The problem is open even for $X = S^4$ or D^4.

Proposed for K3 and scribed by: H. Konno, M. Powell

PROBLEM 4.72. *Let X be a closed orientable topological 4-manifold with finite $\pi_1(X)$.*

(a) *Is $\pi_k(\mathrm{Homeo}(X))$ finitely generated for every $k \geqslant 0$?*
(b) *Is $H_k(B\,\mathrm{Homeo}(X);\mathbb{Z})$ finitely generated for every $k \geqslant 0$?*

REMARKS.

(1) For $\pi_1(X) = 1$, it follows from a result by Perron [**Per86**] and Quinn [**Qui86**] (cf. [**GGH$^+$23**]) that $\pi_0(\mathrm{Homeo}(X))$ is finitely generated, which implies that $H_1(B\,\mathrm{Homeo}(X);\mathbb{Z})$ is also finitely generated. There is no known finiteness result on $\pi_k(\mathrm{Homeo}(X))$ for $k \geqslant 1$ and $H_k(B\,\mathrm{Homeo}(X);\mathbb{Z})$ for $k \geqslant 2$. See Problem 4.71 for the analogous question in the smooth category. The following question is closely related.

QUESTION. Does Top(4) have finitely generated homotopy groups in each degree?

Indeed, Top(4) has finitely-generated homotopy groups if and only if $\mathrm{Homeo}(S^4)$ has finitely-generated homotopy groups, because we have a fibration

$$\mathrm{Top}(4) \to \mathrm{Homeo}(S^4) \to S^4,$$

and the homotopy groups of S^4 are finitely generated. The question on the homotopy groups of Top(4) is itself closely related to Problem 4.73 on the Morlet correspondence in dimension 4.

(2) Here is another closely related question, on topological embedding spaces.

Let Σ be a compact surface, and consider a compact 4-manifold X. Fix a locally flat embedding $\iota\colon \partial\Sigma \hookrightarrow \partial X$. Let $\mathrm{Emb}^t_\partial(\Sigma, X)$ be the space of locally flat embeddings of Σ extending ι. This is defined as the geometric realization of a semi-simplicial set, where the p-simplices are locally flat embeddings $\Sigma \times \Delta^p \to X \times \Delta^p$ over the projection to Δ^p, and extending $\iota \times \mathrm{Id}\colon \partial\Sigma \times \Delta^p \to X \times \Delta^p$. Let $f_0\colon \Sigma \hookrightarrow X$ be a 0-simplex.

QUESTION. Suppose that $\pi_1(X \setminus f_0(\Sigma))$ is finite, and fix $k > 0$. Is $\pi_k(\mathrm{Emb}^t_\partial(\Sigma, X), f_0)$ finitely generated?

Randal-Williams [**RW**] has announced that the answer to the analogous question is no, in a case where the fundamental group of the surface complement is infinite. Using [**BG19, BG25**], Randal-Williams deduced that $\pi_4(\mathrm{Emb}^t(S^2, S^4), U)$ is infinitely generated, where U is the trivial 2-knot.

The homotopy groups spaces of embeddings are closely related to the homotopy groups of the homeomorphism groups of both the ambient space and of the exterior of the basepoint embedding. The space of thickenings of a fixed embedding to a closed tubular neighborhood plays an important rôle as well. As demonstrated by Randal-Williams' note, information about any of these characters often leads to information about the others.

Proposed for K3 and scribed by: H. Konno, M. Powell

PROBLEM 4.73. *Does the Morlet correspondence*

$$\mathrm{BDiff}_\partial(D^n) \simeq \Omega^n_0(\mathrm{Top}(n)/O(n)) \tag{10}$$

hold for $n = 4$?

REMARKS.

(1) Here $\mathrm{BDiff}_\partial(D^n)$ denotes the classifying space of the diffeomorphism group of D^n relative to its boundary. We write $\mathrm{Top}(n)$ for the group of homeomorphisms on $\mathbb{R}^n$ that fix the origin, and $\mathrm{Top}(n)/O(n)$ for the homotopy fiber of $\mathrm{BO}(n) \to \mathrm{BTop}(n)$. Finally let $\Omega^n_0(\mathrm{Top}(n)/O(n))$ denote the unit component of the loop space $\Omega^n(\mathrm{Top}(n)/O(n))$.
(2) In dimension $n \neq 4$, the weak equivalence (10) follows from smoothing theory [**BL74, KS77**]. It is known that smoothing theory fails in dimension 4. For example, the manifold $E_8 \# E_8$ has a formal smooth structure (i.e. a vector bundle structure on its tangent microbundle) but has no smooth structure. However, (10) may still hold for $n = 4$.
(3) Watanabe [**Wat19**] disproved the 4-dimensional Smale conjecture by showing that $\pi_k(\mathrm{BDiff}_\partial(D^4)) \otimes \mathbb{Q} \neq 0$ for many values of k including $2, 5, 9$. It is known that the group $\pi_{k+4}(\mathrm{Top}(n)/O(n)) \otimes \mathbb{Q}$ is also nonvanishing for these values of k [**LX23**].
(4) Gauge theory can distinguish non-diffeomorphic smooth structures on a closed 4-manifold that are isomorphic as formal smooth structures. So gauge theory could potentially be used to disprove (10).
(5) The question is closely related to Problem 4.67. If there is a diffeomorphism of D^4 not isotopic to the identity, and this is detected using gauge theory, then this could show that the Morlet correspondence does not hold.

Proposed for K3 and scribed by: J. Lin

PROBLEM 4.74. *Does there exist a closed, smooth 4-manifold X and a diffeomorphism $f\colon X \xrightarrow{\cong} X$ such that f is smoothly pseudo-isotopic to the identity, but f is not stably smoothly isotopic to the identity?*

REMARKS.

(1) We can stabilize a diffeomorphism by making a choice of isotopy to one that is the identity on a 4-ball, connect summing with $\#^k S^2 \times S^2$ using that 4-ball, for some k, and then extending by the identity on the new $\#^k S^2 \times S^2$. If some stabilization of f (for some choice of isotopy and for some k) is smoothly isotopic to the identity, then we say that f is smoothly stably isotopic to Id.
(2) Gabai [**Gab22**] proved that there is a smooth pseudo-isotopy F with vanishing Hatcher-Wagoner pseudo-isotopy obstruction $\Sigma(F) \in \mathrm{Wh}_2(\pi_1(X))$ if and only if f is smoothly stably isotopic to Id. The question is whether there exists an f such that $\Sigma(F)$ is nontrivial for *all* pseudo-isotopies F from f to Id. Or perhaps, for any pair (f, F), there is always a choice of F with $\Sigma(F) = 0$ restricting to the same f. Singh [**Sin25**] proved that the

Hatcher–Wagoner obstruction Σ can be stably realized, which could be useful if one could control the diffeomorphism produced by his realization procedure.

Proposed for K3 and scribed by: M. Powell

PROBLEM 4.75. *Let X be a connected smooth 4-manifold with nonempty boundary, with finite $\pi_1(X)$, and let $k \geqslant 0$. Let $\mathrm{Diff}_\partial(X)$ denote the group of diffeomorphisms of X that are the identity near ∂X. Is the image of the natural map*

$$s_*\colon \pi_k(\mathrm{Diff}_\partial(X)) \to \operatorname*{colim}_{N\to\infty} \pi_k(\mathrm{Diff}_\partial(X \#_N S^2 \times S^2)) \tag{11}$$

finitely generated? What about the analogous problem in the topological category?

REMARKS.

(1) Fixing a model of the interior connected sum $X\#S^2 \times S^2$ defined by

$$X\#S^2 \times S^2 = X \cup ((\partial X \times [0,1])\#S^2 \times S^2),$$

we have a well-defined stabilization map

$$s\colon \mathrm{Diff}_\partial(X) \to \mathrm{Diff}_\partial(X\#S^2 \times S^2),$$

extending diffeomorphisms by the identity map on $(\partial X \times [0,1])\#S^2 \times S^2$.

(2) In the topological category, Problem 4.72 asks the analogous question without stabilization.

(3) We take a closer look at the problem for $k = 0$, $\pi_1(X) = 1$, and connected boundary. A diffeomorphism $f\colon X \to X$ determines a Poincaré variation [**Sae06**]. The group of Poincaré variations is denoted $\mathcal{V}(H_2(X), \lambda_X)$, where λ_X is the intersection form of X. Saeki [**Sae06**, Theorem 3.7] proved that the stable mapping class group of X is isomorphic to the group of stable Poincaré variations $S\mathcal{V}(H_2(X), \lambda_X)$. Let

$$V := \mathrm{Im}(\theta\colon \pi_0 \,\mathrm{Diff}_\partial(X) \to \mathcal{V}(H_2(X), \lambda_X)).$$

The image of s_* is thus isomorphic to the image of V under the algebraic stabilization map $\mathcal{V}(H_2(X), \lambda_X) \to S\mathcal{V}(H_2(X), \lambda_X)$. This latter map is injective, so in fact $\mathrm{Im}(s_*) \cong V$. We need to decide whether V is finitely generated.

By [**Sae06**, Section 4], the group $\mathcal{V}(H_2(X), \lambda_X)$ sits in an exact sequence

$$0 \to \wedge^2 H^1(\partial X; \mathbb{Z}) \to \mathcal{V}(H_2(X), \lambda_X) \to \mathrm{Aut}_\partial(H_2(X;\mathbb{Z}), \lambda_X).$$

The first group is finitely generated, and the image of the second map is a finite index subgroup of $\mathrm{Aut}_\partial(H_2(X;\mathbb{Z}), \lambda_X)$, the automorphisms of the intersection form of X whose algebraic boundary is trivial. The latter group is arithmetic, so is finitely generated. It follows that $\mathcal{V}(H_2(X), \lambda_X)$ is finitely generated, since finite index subgroups and extensions of finitely generated groups are again finitely generated. Is V finitely generated?

In the special case that $\partial X = S^3$

$$\mathcal{V}(H_2(X), \lambda_X) \cong \operatorname{Aut}(H_2(X;\mathbb{Z}), \lambda_X).$$

By capping off with a 4-ball we can apply Wall's theorem [**Wal64a**] to see that θ is surjective when X is of the form $X = M \# S^2 \times S^2$, where M is indefinite or $b_2(M) < 9$. Thus in these cases the image of s_*, for $k = 0$, is known to be finitely generated. In other cases, such as for $X = K3 \setminus \mathring{D}^4$, V is finite index in $\mathcal{V}(H_2(X), \lambda_X) \cong \operatorname{Aut}(H_2(X;\mathbb{Z}), \lambda_X)$, so is finitely generated. However there are also examples, such as for X the punctured Dolgachev surface, where V is infinite index [**FM88**] in $\operatorname{Aut}(H_2(X;\mathbb{Z}), \lambda_X)$. In such cases, is V finitely generated?

(4) Without stabilization, it is known that $\pi_k(\operatorname{Diff}(X))$ need not be finitely generated, even for $\pi_1(X) = 1$. This was proven by Ruberman [**Rub99b**] for $k = 0$, with later proofs by Baraglia [**Bar23c**] and Konno [**Kon24b**]. For $k = 1$ it was proven by Baraglia[**Bar23b**] and Lin [**Lin22**]. For $k \geqslant 1$ this is announced in recent work of Auckly–Ruberman [**AR25**]. Different 4-manifolds are used in each work. All of these results are proven using gauge theory for families.

(5) For 4-manifolds with infinite fundamental group, Budney–Gabai [**BG19**] and Watanabe [**Wat23**] proved (without using gauge theory) that some 4-manifolds with infinite fundamental group (e.g. $S^1 \times S^3$ in [**BG19**], and hyperbolic manifolds of dimension at least 4 in [**BG25**]) have infinitely generated mapping class groups. However the diffeomorphisms they construct are pseudo-isotopic to the identity, and hence are stably isotopic to the identity by Gabai's theorem [**Gab22**]. It would be interesting to know whether there are counterexamples when $\pi_1(X)$ is infinite.

(6) Since $\pi_k(\operatorname{Diff}_\partial(X)) \cong \pi_{k+1}(B\operatorname{Diff}_\partial(X))$, the problem can be described in terms of $B\operatorname{Diff}_\partial(X)$. If one considers the analogous question for the homology groups of $B\operatorname{Diff}_\partial(X)$, the answer is often positive. This is because it follows from work by Galatius and Randal-Williams [**GRW17**] that

$$\operatorname*{colim}_{N\to\infty} H_k(B\operatorname{Diff}_\partial(X \#^N S^2 \times S^2)) \tag{12}$$

is finitely generated for all k and many X, such as for simply-connected X. In particular, any subgroup of (12) such as the image of the stabilization map is also finitely generated. However, this is not necessarily strong evidence to hope for a positive solution to the problem for homotopy groups, since there are many spaces with finitely generated homology groups but infinitely generated homotopy groups (for example, $H_2(S^1 \vee S^2) = \mathbb{Z}$ but $\pi_2(S^1 \vee S^2) = \mathbb{Z}^\infty$).

(7) In even, higher dimensions, the answer to the analogous problem is positive, and indeed this holds without taking the image in a colimit. For a compact smooth manifold M of dimension $2n \geqslant 6$ with finite fundamental group, $\pi_k(\operatorname{Diff}_\partial(M))$ is finitely generated, due to Bustamante–Krannich–Kupers [**BKK24**, Theorem 6.1].

Proposed for K3 and scribed by: H. Konno, M. Powell

Problem 4.76. *Let X be a closed, oriented, simply connected, smooth 4-manifold and fix $k > 0$. Is there $N \geqslant 0$ such that, for every $n \geqslant N$, the natural map*

$$\pi_k(\mathrm{Diff}^+(X\#^n S^2 \times S^2)) \to \pi_k(\mathrm{Homeo}^+(X\#^n S^2 \times S^2))$$

is surjective?

Remarks.

(1) If $k = 0$, the answer to the analogous question is affirmative. First, for $N = 2$, it follows from Wall's theorem [**Wal64a**] that the natural map

$$\pi_0(\mathrm{Diff}^+(X\#^n S^2 \times S^2)) \to \mathrm{Aut}(H_2(X\#^n S^2 \times S^2;\mathbb{Z}), \lambda_{X\#^n S^2\times S^2})$$

is surjective for every $n \geqslant N$. Here $\mathrm{Aut}(H_2(X;\mathbb{Z}), \lambda_X)$ denotes the automorphism group of the intersection form. On the other hand, work of Freedman [**Fre82**], Kreck, [**Kre79**], Perron [**Per86**], and Quinn [**Qui86**] (plus [**GGH$^+$23**]), implies that the natural map

$$\pi_0(\mathrm{Homeo}^+(X\#^n S^2 \times S^2)) \to \mathrm{Aut}(H_2(X\#^n S^2 \times S^2;\mathbb{Z}), \lambda_{X\#^n S^2\times S^2})$$

is an isomorphism. Thus, considering the obvious commuting triangle, we have

$$\pi_0(\mathrm{Diff}^+(X\#^n S^2 \times S^2)) \to \pi_0(\mathrm{Homeo}^+(X\#^n S^2 \times S^2))$$

is surjective.

(2) For the analogous problem obtained by replacing "surjective" with "injective", the answer is negative for $k = 0$. Indeed, for any simply-connected closed smooth 4-manifold X, one can find a strictly increasing divergent sequence $0 < N_1 < N_2 < \cdots \to \infty$ such that

$$\pi_0(\mathrm{Diff}(X\#^{N_i} S^2 \times S^2)) \to \pi_0(\mathrm{Homeo}(X\#^{N_i} S^2 \times S^2))$$

is not injective for every $i \geqslant 0$ (*cf.* [**KL23c**, Theorem 1.5]).

Proposed for K3 by: H. Konno, M. Powell
Scribed by: H. Konno

Problem 4.77. *For which $k \geqslant 0$ and closed smooth 4-manifold X does the equality*

$$\begin{aligned} &\ker(i_*\colon \pi_k(\mathrm{Diff}_\partial(\mathring{X})) \to \pi_k(\mathrm{Homeo}_\partial(\mathring{X}))) \\ &= \ker(s_*\colon \pi_k(\mathrm{Diff}_\partial(\mathring{X})) \to \operatorname*{colim}_{n\to\infty} \pi_k(\mathrm{Diff}_\partial(\mathring{X}\#^n S^2 \times S^2))) \end{aligned} \tag{13}$$

hold?

Remarks.

(1) Here $i\colon \mathrm{Diff}_\partial(\mathring{X}) \to \mathrm{Homeo}_\partial(\mathring{X})$ is the inclusion, and $s\colon \mathrm{Diff}_\partial(\mathring{X}) \to \mathrm{Diff}_\partial(\mathring{X}\#^n S^2 \times S^2)$ is the stabilization map from Problem 4.75.

(2) For $k = 0$ and $\pi_1(X) = 1$, the equality (13) is known to hold: by a result by Quinn [**Qui86**] and Perron [**Per86**], the kernels in (13) coincide with the group of mapping classes of diffeomorphisms that act trivially on homology.

(3) See Problem 4.45 for the analogous question on smooth and topological embedding spaces of surfaces in 4-manifolds.
(4) Gabai [**Gab22**] proved that diffeomorphisms that are pseudo-isotopic, via a pseudo-isotopy with trivial Wh_2 obstruction, are smoothly stably isotopic. If we knew a topological analogue of this result, that could help to find counterexamples for $k = 0$, and be an interesting development in its own right.

QUESTION. Suppose that a self-homeomorphism $f\colon X \to X$ of a compact 4-manifold X is topologically pseudo-isotopic to the identity via a pseudo-isotopy F with vanishing primary topological Hatcher–Wagoner invariant, $\Sigma(F) = 0 \in \mathrm{Wh}_2(\pi_1(X))$. Is f topologically stably isotopic to Id_X?

See [**GN25**] for a definition of the primary topological Hatcher–Wagoner invariant.

Proposed for K3 and scribed by: H. Konno

PROBLEM 4.78. *Let X be a simply connected closed smooth 4-manifold. Does $B\,\mathrm{Diff}(X)$ satisfy homological stability over $\mathbb{Q}$?*

REMARKS.
(1) As before, let

$$s\colon\ \mathrm{Diff}_\partial(\mathring{X}) \to \mathrm{Diff}_\partial(\mathring{X}\#S^2 \times S^2)$$

be the stabilization map from Problem 4.75. The question asks whether, for each $k \geqslant 0$, the induced maps

$$s_*\colon H_k(B\,\mathrm{Diff}_\partial(\mathring{X}\#^N S^2 \times S^2);\mathbb{Q}) \to H_k(B\,\mathrm{Diff}_\partial(\mathring{X}\#^{N+1} S^2 \times S^2);\mathbb{Q})$$

are isomorphic for all $n \gg 0$ large enough relative to k.
(2) For a simply connected compact manifold of dim $= 2n \neq 4$, an analogous stability holds over any (untwisted) coefficients due to work by Harer [**Har85**] for dim $= 2$ Galatius and Randal-Williams [**GRW18**] for dim $\geqslant 6$. The colimit

$$\underset{n\to\infty}{\mathrm{colim}} H_k(B\,\mathrm{Diff}_\partial(\mathring{X}\#^N S^n \times S^n))$$

is called the stable homology, which has been extensively studied, especially with $\mathbb{Q}$ coefficients: indeed, the stable homology over $\mathbb{Q}$ coefficient has been determined [**MW07, GRW18**].
(3) With $\mathbb{Z}$ coefficients, an analogous stability in dimension 4 was shown to fail by Konno and Lin [**KL23c**]. However, the unstable homology classes detected in [**KL23c**] are 2-torsion, so the result in [**KL23c**] does not imply rational instability.

Proposed for K3 and scribed by: H. Konno

PROBLEM 4.79. *Given $k > 0$, is there a closed, simply connected, smooth 4-manifold X and a nonzero homotopy class $\alpha \in \pi_k(\mathrm{Diff}_\partial(\mathring{X}))$ such that*

$$\alpha \in \ker(i_* : \pi_k(\mathrm{Diff}_\partial(\mathring{X})) \to \pi_k(\mathrm{Homeo}_\partial(\mathring{X}))) \tag{14}$$

and

$$\alpha \notin \ker(s_* : \pi_k(\mathrm{Diff}_\partial(\mathring{X})) \to \pi_k(\mathrm{Diff}_\partial(\mathring{X}\# S^2 \times S^2)))? \tag{15}$$

REMARKS.

(1) Here we use the notation i, and s of Problem 4.75 and the previous two problems.
(2) The answer to an analogous statement for $k = 0$ is known to be affirmative: Kronheimer–Mrowka [**KM20**] proved that the Dehn twist on $X = K3\#K3$ along S^3 gives non-zero class α that lies in (14), and Lin [**Lin23**] proved that this α satisfies (15).

Proposed for K3 and scribed by: H. Konno

PROBLEM 4.80. *Is there $n > 2$ and a smooth closed simply connected 4-manifold X for which there is an element of order n in the subgroup* $\ker(\pi_0 \mathrm{Diff}(X) \to \pi_0 \mathrm{Homeo}(X))$*? More generally, is there a subgroup of order n in* $\ker(\pi_0 \mathrm{Diff}(X) \to \pi_0 \mathrm{Homeo}(X))$*?*

REMARKS.

(1) An element of $\ker(\pi_0 \mathrm{Diff}(X) \to \pi_0 \mathrm{Homeo}(X))$ is called an *exotic diffeomorphism*. Examples of exotic diffeomorphisms by Ruberman [**Rub98**] and Baraglia–Konno [**BK20**] are of infinite order in $\ker(\pi_0 \mathrm{Diff}(X) \to \pi_0 \mathrm{Homeo}(X))$ A Dehn twist on $K3\#K3$ along the connected sum S^3, detected by Kronheimer–Mrowka [**KM20**], is of order 2. These exotic diffeomorphisms are detected by $\mathbb{Z}$ or $\mathbb{Z}/2$-valued invariants defined using parameterized ASD Yang–Mills or Seiberg-Witten theory. It would seem that to detect an exotic diffeomorphism of order n one would need $\mathbb{Z}/n$-valued invariants of this type.

   Note that, for a simply connected, closed, oriented 4-manifold, the orientation-preserving topological mapping class group $\pi_0(\mathrm{Homeo}^+(X)))$ is isomorphic to $\mathrm{Aut}(H_2(X;\mathbb{Z}), \lambda_X)$ by work of Freedman [**Fre82**], Kreck, [**Kre79**], Perron [**Per86**], and Quinn [**Qui86**].
(2) One could consider the following special cases.

   QUESTION.
   (i) Is there a smooth, closed, simply connected 4-manifold X with a subgroup isomorphic to $\mathbb{Z}/2 \times \mathbb{Z}/2$ in $\ker(\pi_0 \mathrm{Diff}(X) \to \pi_0 \mathrm{Homeo}(X))$?
   (ii) In particular, for $X = K3\#K3\#K3$, do the Dehn twists on the two connected sum copies of S^3 in X generate such a subgroup?

(iii) More generally, is there a copy of $(\mathbb{Z}/2)^{n-1}$ in $\ker(\pi_0 \operatorname{Diff}(X) \to \pi_0 \operatorname{Homeo}(X))$ when X is a connected sum of n copies of the $K3$ surface?

Proposed for K3 and scribed by: H. Konno, D. Ruberman

PROBLEM 4.81. *Is there a smooth, closed, simply connected 4-manifold X for which the group* $\ker(\pi_0(\operatorname{Diff}(X)) \to \pi_0(\operatorname{Homeo}(X)))$ *is finitely generated?*

REMARKS.

(1) Let $\operatorname{TDiff}(X)$ be the group of diffeomorphisms that act trivially on $H_*(X;\mathbb{Z})$. Quinn [**Qui86**] and Perron [**Per86**] proved that, for $\pi_1(X) = \{1\}$, the group $\ker(\pi_0(\operatorname{Diff}(X)) \to \pi_0(\operatorname{Homeo}(X)))$ is isomorphic to $\pi_0(\operatorname{TDiff}(X))$, called the Torelli group.

(2) Ruberman [**Rub99b**] proved that there exist simply connected closed 4-manifolds X for which $\pi_0(\operatorname{TDiff}(X))$ is not finitely generated.

(3) In dimension 2, for the closed oriented surface Σ_g of genus $g \geqslant 2$, Johnson [**Joh83**] proved that the Torelli group is finitely generated for $g > 2$, and McCullough–Miller [**MM86**] proved that the Torelli group is not finitely generated for $g = 2$.

Scribed by: H. Konno

PROBLEM 4.82. *Let φ be a self-diffeomorphism of a closed, simply-connected, smooth 4-manifold X. Suppose that for every smooth surface Σ in X, the surfaces Σ and $\varphi(\Sigma)$ are smoothly isotopic.*

(a) *Is φ necessarily smoothly isotopic to the identity map?*

(b) *Is φ smoothly isotopic to a diffeomorphism that is supported on a $B^4 \subseteq X$?*

REMARKS.

(1) In dimension three, a self-homeomorphism f of an irreducible 3-manifold M^3 that preserves free homotopy classes of loops is isotopic to the identity map [**ABD$^+$20**]. The analogous result holds for simply connected 4-manifolds in the topological category by Perron [**Per86**] and independently by work of Quinn [**Qui86**] combined with that of Gabai–Gay–Hartman–Krushkal–Powell [**GGH$^+$23**]. (These authors show that in the topological category, the weaker hypothesis that φ induces the identity map on $H_2(X;\mathbb{Z})$ implies that φ is topologically isotopic to the identity map.)

This question essentially asks whether these theorems have an analogue in dimension four in the smooth category. The stronger hypothesis is necessary, as there are many examples of exotic self-diffeomorphisms of simply connected 4-manifolds, e.g. the Dehn twist $\phi\colon K3\#K3 \to K3\#K3$ is topologically but not smoothly isotopic to the identity map [**KM20**]. In this case, the induced map from ϕ clearly preserves $H_2(K3\#K3;\mathbb{Z})$, but it is not clear whether $\phi(\Sigma)$ is smoothly isotopic to Σ for every surface Σ inside $K3\#K3$.

(2) The first question, in the special case $X = S^4$, would imply that every orientation-preserving diffeomorphism of S^4 is isotopic to the identity, answering Problem 4.67. To see that every diffeomorphism $\varphi\colon S^4 \to S^4$ satisfies the hypothesis of the problem, isotope φ to fix some B^4, and then isotope Σ into that B^4. The second question decouples this problem from Problem 4.67.

(3) Another variation on the problem allows the stronger hypothesis that for every $g \geqslant 0$, every smooth embedding $h\colon \Sigma_g \hookrightarrow X$ is smoothly isotopic to the embedding $\varphi \circ h$.

(4) Assuming instead that φ becomes isotopic to the identity after connected sum with $S^2 \times S^2$ (which in particular implies that φ induces the identity on $H_2(X;\mathbb{Z})$), then Krushkal-Mukherjee-Powell-Warren [**KMPW24**] showed that φ can be isotoped so as to be supported on a contractible submanifold. Another point of view on the question asks whether the assumption that Σ and $\varphi(\Sigma)$ are smoothly isotopic for every Σ, enables us to show that the contractible supporting manifold can be assumed to be a 4-ball.

Proposed for K3 and scribed by: J. Chaidez, M. Miller

PROBLEM 4.83. *For which 4-manifolds does there exist a smooth structure such that there exists a non-smoothable homeomorphism with respect to that smooth structure?*

REMARKS.

(1) We say that a homeomorphism is *smoothable* if it is rel. boundary isotopic to a diffeomorphism. If this does not hold, we say that the homeomorphism is *non-smoothable.*

(2) This question is interesting for closed 4-manifolds, compact 4-manifolds with boundary, and for open 4-manifolds.

(3) Recent progress for compact simply connected 4-manifolds with boundary was made by Galvin–Ladu [**GL25**] and Konno–Taniguchi [**KT22b**]. Here is an open question for such 4-manifolds.

QUESTION (i). Let X be a simply connected, smooth, spin 4-manifold with $\partial X = Y_1 \sqcup Y_2$ having two connected components that do not admit generalized Dehn twists. Consider a boundary-fixing homeomorphism that has trivial Poincaré variation but that acts nontrivially on the relative spin structures of X (there are two such relative spin structures). Is it isotopic to a diffeomorphism?

See [**Sae06, OP25**] for the definition of a Poincaré variation. If the Y_i are hyperbolic, then they do not admit generalized Dehn twists. The answer is also yes in the case that there is a separating embedding of a 3-manifold Z, with Y_1 and Y_2 in different connected components of $X \backslash Z$, where Z admits a generalized Dehn twist.

(4) One can wonder whether there is a relationship with the existence of exotic smooth structures.

QUESTION. Does there exist a 4-manifold that admits a non-smoothable homeomorphism, but no exotic smooth structure? Or vice versa, a 4-manifold admitting exotic smooth structures for which every self-homeomorphism is smoothable?

See e.g. the work of Donaldson [**Don90**], Friedman–Morgan [**FM88**], and Baraglia [**Bar21**].

Proposed for K3 and scribed by: M. Powell

PROBLEM 4.84. *Is there a closed oriented smooth 4-manifold X for which every finite subgroup G of the mapping class group $\pi_0(\mathrm{Diff}^+(X))$ can be realized by a finite group action on X?*

REMARKS.

(1) This question relates to the study of the algebraic properties of diffeomorphism groups as opposed to their homotopy type.

If there is a group-theoretic section $G \to \mathrm{Diff}^+(X)$ of the quotient map $\mathrm{Diff}^+(X) \to \pi_0(\mathrm{Diff}^+(X))$ over G, we say that G is *realized* by a finite group action. The question whether a given G is realized is known as the Nielsen realization problem.

(2) For orientable surfaces X, Kerckhoff [**Ker83**] proved that every finite subgroup of $\pi_0(\mathrm{Diff}^+(X))$ is realized. On the other hand, in dimension 4, there are several known examples of 4-manifolds X for which there are non-realizable finite subgroups of $\pi_0(\mathrm{Diff}^+(X))$ [**RS77, BK23, FL24a, Kon24a, KMT23b, AB25, Bar23c**].

(3) One may also ask the analogous questions for the extended mapping classes group of X, which also contains the orientation-reversing diffeomorphisms, and for nonorientable X.

(4) The existence of *asymmetric manifolds*, i.e. manifolds that do not admit any effective action of a finite group, has been studied in higher dimensions. See e.g. [**Pup07, CR72**]. Related to the Nielsen realization problem, one may ask whether there is a smooth asymmetric 4-manifold where the mapping class group has a nontrivial finite subgroup.

(5) Here is a variant of the Nielsen realization problem, and for some specific X, several non-realizability results are known. Let $\mathrm{Aut}(H_2(X;\mathbb{Z}))$ denote the automorphism group of the intersection form and set

$$I(X) := \mathrm{Im}(\mathrm{Diff}^+(X) \to \mathrm{Aut}(H_2(X;\mathbb{Z}))).$$

QUESTION. Which X admits a (finite) group G and a homomorphism $\varphi\colon G \to I(X)$ that cannot be realized by a smooth action on X?

Here we say that φ is *realized* by a smooth action if there a homomorphism $\tilde{\varphi}\colon G \to \mathrm{Diff}^+(X)$ that descends to φ.

Scribed by: H. Konno

PROBLEM 4.85. *Is there a closed orientable smooth 4-manifold X for which the identity component* $\mathrm{Diff}_0(X)$ *of the diffeomorphism group is not uniformly perfect?*

REMARKS.

(1) By Mather and Thurston [**Mat71, Mat74, Thu74a**], the group $\mathrm{Diff}_0(X)$ is perfect for every closed orientable manifold X.
(2) For a given group G, the *commutator length* of an element $g \in [G,G]$ is defined to be the minimal number of factors in expressions of g as products of commutators. The commutator length has been studied for Lie groups, groups of automorphisms of (topological, smooth, or symplectic) manifolds, and mapping class groups in contexts of dynamics and bounded cohomology. It is known to be difficult to compute the commutator length of a given element in general, which leads us to consider the following qualitative property for a perfect group (a group that is equal to its commutator subgroup). Given a perfect group G, we say that G is *uniformly perfect* if the commutator lengths of elements of G are uniformly bounded.
(3) For $\dim X \neq 2, 4$, $\mathrm{Diff}_0(X)$ is known to be uniformly perfect due to work by Burago–Ivanov–Polterovich [**BIP08**] and Tsuboi [**Tsu08, Tsu12**]. Their works also show that $\mathrm{Diff}_0(X)$ is uniformly perfect for $X = S^2$ and S^4. On the other hand, Bowden–Hensel–Webb [**BHW22**] proved that, for $\dim X = 2$, $\mathrm{Diff}_0(X)$ is not uniformly perfect if the genus of X is positive. Nothing is known in dimension 4 except for $X = S^4$.

Proposed for K3 and scribed by: H. Konno

PROBLEM 4.86. *Is it the case that for every closed, smoothable topological* 4-*manifold X, there exists a locally linear finite group action on X, such that for every smooth structure on X, the action is non-smoothable?*

REMARKS.

(1) Given a topological manifold X and a smooth structure $\mathcal{O}$ on X, we say that a locally linear topological action φ of a group on X is *non-smoothable* with respect to $\mathcal{O}$ if φ is not conjugate to a smooth action on X.
(2) A variant of the problem is the following.

    QUESTION. If we fix a smooth structure $\mathcal{O}$ on X, is there a locally linear finite group action on X that is non-smoothable with respect to $\mathcal{O}$?

(3) There are many examples of locally linear finite group actions on closed 4-manifolds that are non-smoothable with respect to any smooth structure, such as [**KL88, KL93, HL95, Bry98, HT04, Nak09, Bar19, Bar21, Kat22**].

Scribed by: H. Konno

PROBLEM 4.87. *Is there an exotic action of $\mathbb{Z}/n$ on S^4 with 0-dimensional fixed point set? 1-dimensional? 2-dimensional?*

REMARKS.

(1) In the case that the fixed set is empty, any nontrivial symmetry is orientation-reversing, so $n = 2$. In this case Cappell and Shaneson [**CS76**] constructed an exotic smooth $\mathbb{RP}^4$, whose double cover is S^4 [**AK79a, Gom91b**]. Another construction was given by Fintushel and Stern [**FS81**]. It is not known whether these examples are diffeomorphic, or more generally whether there is more than one exotic $\mathbb{RP}^4$. (In the topological setting there are exactly two homeomorphism classes of 4-manifolds that are homotopy equivalent to $\mathbb{RP}^4$ according to a calculation from the surgery exact sequence [**Wal99**, Chapter 14]). These two manifolds are distinguished by their Kirby-Siebenmann invariant. As a point of interest, we refer the reader to Ruberman's explicit construction of the non-smoothable homotopy $\mathbb{RP}^4$ [**Rub84**, Section 2].)

(2) In the case that the fixed set is 0-dimensional or 1-dimensional, very little is known in the smooth setting. In the case that the fixed set is 2-dimensional, there exist actions with knotted fixed point set [**Gif66, Gor74, Sum75**], but the tools for establishing this are topological. The case that the fixed point set has dimension 3 has been extensively studied (see e.g. [**Maz61**]); examples arise from gluing two copies of a cork along the boundary.

Scribed by: K. Hendricks

PROBLEM 4.88. *Let $\tau\colon S^4 \to S^4$ be a free (hence orientation-reversing) involution. Is there an embedded $S^2 \subseteq S^4$ that is invariant under τ? This is equivalent to the existence of an embedded $\mathbb{RP}^2$ in $S^4/\tau \simeq \mathbb{RP}^4$, carrying the nontrivial class in $H_2(\mathbb{RP}^4; \mathbb{Z}/2)$. This condition on the homology class will be assumed for the rest of the problem.*

REMARKS.

(1) This is an instance of the codimension two splitting problem [**CS74**] and may be asked in the smooth or topological category (where one would be looking for an $\mathbb{RP}^2$ with a normal bundle). Another classic phrasing is that the problem is asking if the involution desuspends twice.

(2) In the topological category, there is only one 4-manifold homotopy equivalent but not homeomorphic to $\mathbb{RP}^4$, which is sometimes denoted $*\mathbb{RP}^4$.

QUESTION (i). Does $*\mathbb{RP}^4$ have an embedded $\mathbb{RP}^2$ with a normal bundle?

(3) There are two known constructions of exotic $\mathbb{RP}^4$s in the smooth category, due to Cappell–Shaneson [**CS76**] and Fintushel–Stern [**FS81**]. By construction, the Cappell–Shaneson $\mathbb{RP}^4$s all contain an embedded $\mathbb{RP}^2$.

QUESTION (ii). Do the Fintushel–Stern $\mathbb{RP}^4$s contain smoothly embedded $\mathbb{RP}^2$s?

A negative answer would show that the Fintushel–Stern $\mathbb{R}P^4$s are not diffeomorphic to those obtained by the Cappell–Shaneson construction; see also Problem 4.87.

Proposed for K3 and scribed by: D. Ruberman

PROBLEM 4.89. *Classify smooth, effective circle actions on simply connected 4-manifolds with boundary.*

(a) *Classify simply connected 4-manifolds with boundary that admit circle actions.*
(b) *For each such 4-manifold, classify the circle actions up to conjugation by diffeomorphisms.*
(c) *Given a 4-manifold X with a circle action on its boundary, when does the action extend over X? When does it extend uniquely?*

REMARKS.

(1) Information about the smooth classification of the closed 4-manifolds admitting a circle action was obtained by Fintushel and Pao [**FP77**] via Pao's replacement trick [**Pao77**]. (These papers assumed the truth of the 3-dimensional Poincaré conjecture, an assumption that is now known to be satisfied.) Part (a) asks for similar results in the bounded case.
(2) The classification of circle actions on *closed* simply connected 4-manifolds was carried out in the 1970s by Fintushel [**Fin77, Fin78**] in terms of orbit data. Part (b) is asking for an analogous result in the bounded case. Note that in the bounded case, one necessarily has a Seifert-fibered space on the boundary, and so the orbit data is somewhat more complicated than in the closed case.
(3) There are obstructions to the extension problem in part (c) coming from gauge theory. Konno–Mallick–Taniguchi [**KMT23a**] studied, for M equal to one of the Milnor fibers $M(2,3,7)$ or $M(2,3,11)$, the loop in $\pi_1(\mathrm{Diff}(\partial M))$ corresponding to the circle action coming from the Seifert-fibred structure of the boundary. This gives rise to a *boundary generalized Dehn twist*, a diffeomorphism of M supported in a collar neighborhood of ∂M. This diffeomorphism is isotopic to the identity rel. boundary if and only if the element of $\pi_1(\mathrm{Diff}(\partial M))$ extends to a loop in $\pi_1(\mathrm{Diff}(M))$. Konno–Mallick–Taniguchi showed that there is no such extension, and hence there is no corresponding circle action. Montague [**Mon23**] drew the same conclusion for the Gompf nuclei $N(2n)$ and simple plumbing $P(2n)$ with boundary $-\Sigma(2,3,12n-5)$, by proving non-extension results for $\mathbb{Z}/p$ actions embedded in the circle action on the boundary. All of these results continue to hold when M is replaced by $M\#S^2 \times S^2$. Further results along these lines have been announced in [**KLMME24, KPT26**].

QUESTION. Does there exist a compact 4-manifold X, and a circle action on ∂X, such that the latter extends to a loop of diffeomorphisms in $\pi_1(\mathrm{Diff}(X))$ but not to a circle action on X?

(4) If ∂X admits a unique circle action, then an answer to (c) would follow from an answer to (a). But there could exist distinct circle actions on ∂X such that one extends over X and one does not.

(5) A complementary obstruction to extending would be via a classical theorem of Atiyah–Hirzebruch [**AH70**] that restricts the characteristic classes of closed *spin* manifolds of dimension $4k$ admitting a circle action; they show in this case that the $\widehat{A}$ genus of the manifold must vanish. In dimension 4, this means that the signature vanishes.

QUESTION. Find an analogue of the Atiyah–Hirzebruch result for 4-manifolds (or more generally $4k$-manifolds) with nonempty boundary.

It is easy to see that the signature does not necessarily vanish for spin manifolds with circle actions if the boundary is nonempty. For instance, one could take the disk bundle over S^2 with even, nonzero, Euler class. It has signature ± 1 but supports an S^1 action. From such examples, it seems plausible that there could be some sort of boundary correction to the Atiyah–Hirzebruch argument, presumably involving an S^1-equivariant η-invariant [**Don78**].

Proposed for K3 and scribed by: D. Ruberman

4.9. Symplectic and complex structures

This section focuses on problems concerning symplectic and complex structures on 4–manifolds, including geography and botany questions, the realization and diversity of symplectic representatives in fixed homology classes, and the existence of distinct Stein fillings and Weinstein structures. A *symplectic structure* on a smooth, oriented 4–manifold X is a closed, nondegenerate 2-form $\omega \in \Omega^2(X)$; that is, $d\omega = 0$ and $\omega^2 > 0$ everywhere, compatible with the orientation on X. An important homotopy invariant of a symplectic 4–manifold (X, ω) is its *canonical class* $K = -c_1(X, \omega)$, the first Chern class of any almost complex structure J compatible with the symplectic form ω, which is unique up to homotopy. When equipped with the associated Kähler form, a *compact complex algebraic surface* — that is, a complex smooth projective surface—becomes a symplectic 4-manifold where ω is compatible with an integrable complex structure J.

A *Weinstein domain* is a quadruple (W, ω, ϕ, J) where:

(i) W is a compact, smooth 4-manifold with boundary;
(ii) λ is a 1-form on W such that the symplectic form $\omega = d\lambda$;
(iii) the associated Liouville vector field V defined by $\iota_V d\lambda = \lambda$ is gradient-like for a Morse function $\phi : W \to \mathbb{R}$ with ∂W as a regular level set;
(iv) J is an almost complex structure compatible with $d\lambda$.

A *(compact) Stein surface* is a complex 2-dimensional manifold S with boundary, admitting a proper, strictly plurisubharmonic function ϕ: $W \to \mathbb{R}$ such that S is exhausted by compact sublevel sets. (We often drop the adjective 'compact' not to avoid irritating algebraic geometers — for whom Stein manifolds are by definition non-compact — but because it will be clear from the context that we are restricting attention to Stein subdomains with contact-type boundary.) A Stein surface S

admits the structure of a Weinstein domain, with $\lambda = -d^c\phi$ and symplectic form $d\lambda = -dd^c\phi$, and the complex structure on it is Kähler.

PROBLEM 4.90. *Do the Chern numbers c_1^2 and c_2 of every closed, symplectic 4–manifold X that is not a ruled surface satisfy the following?*

(a) $c_1^2 \leqslant 3\,c_2$.
(b) $c_1^2 = 3\,c_2 > 0$ *if and only if* $X = \mathbb{C}\mathrm{P}^2$ *or* X *is a complex ball quotient.*
(c) $c_2 \geqslant 0$.

REMARKS.

(1) These invariants depend only on the underlying homotopy type of the 4–manifold X, satisfying the identities $c_1^2 = 2\chi + 3\sigma$ and $c_2 = \chi$ where χ and σ are the Euler characteristic and the signature of X.
(2) All hold true for Kähler surfaces [**BPVdV84**]; part (a) is the well-known *Bogomolov–Miyaoka–Yau inequality* (BMY) and part (b) follows from Yau's celebrated solution of the Calabi conjecture [**Yau78**].
(3) The inequality (c) was conjectured to hold by Gompf. If the non-ruled symplectic 4–manifold X is *minimal*, then $c_1^2 \geqslant 0$ by [**Tau94**, **Tau95**, **Tau96**, **LL95**]; so part (c) is implied by part (a) in this case.
(4) These geographic constraints would have strong consequences. For instance, if true, part (b) would imply that there is no symplectic 4–manifold X homeomorphic but not diffeomorphic to $\mathbb{C}\mathrm{P}^2$; part (a) or (c) would imply the same for any ruled surface over Σ_h, with $h \geqslant 2$.
(5) If there exist symplectic 4–manifolds with $c_1^2 = 3c_2$ that are not diffeomorphic to any Kähler surface, a further line of inquiry for part (b) would be whether there is a symplectic analog of Yau's theorem for them; e.g., are they always $K(G,1)$s? See also Problem 4.91.
(6) One approach to an affirmative solution for part (a) is via branched coverings. Auroux proved that every symplectic 4-manifold admits a simple branched covering to $\mathbb{C}\mathrm{P}^2$, where the branch locus in $\mathbb{C}\mathrm{P}^2$ is an immersed symplectic surface with nodes and simple cusps [**Aur00**]. One can compute the Chern numbers in terms of the degree of the branch locus, the number of cusps, the genus of the branch curve, and the number of sheets of the cover (see [**Aur06b**, p.266]). Thus, one can translate the existence of an example on or above the BMY line into the existence of a braided singular surface in $\mathbb{C}\mathrm{P}^2$ with a coloring with certain constraints.
(7) One can ask part (a) more generally as follows (see also Problem 4.16):

   QUESTION. Does any closed 4–manifold X with $b_2^+ > 1$ and with nontrivial Seiberg–Witten invariants satisfy the BMY inequality?

   Recall that by the work of Taubes [**Tau94**, **Tau95**], if such a 4–manifold X admits a symplectic structure, then it has a Seiberg-Witten basic class.

   Feehan and Leness have announced a program to answer this broader question affirmatively [**Fee22**, **FL23**, **FL24b**]. Their approach is to adapt Hitchin's Morse theory analysis of the moduli space of Higgs monopoles on a rank-2 Hermitian vector bundle over a Riemann surface [**Hit87**] to the moduli space of non-Abelian monopoles (A, Φ) on a rank-2 Hermitian vector bundle E.

Scribed by: I. Baykur, L. Starkston

PROBLEM 4.91. *Present a topological construction of symplectic fake projective planes. Does there exist a symplectic fake projective plane that is not a complex ball quotient?*

REMARKS.

(1) A *fake projective plane* (FPP) is a closed 4–manifold X with the same rational cohomology ring as the complex projective plane $\mathbb{C}P^2$ but not diffeomorphic to it.

By Yau's solution of the Calabi Conjecture [**Yau78**], every *complex* FPP is a torsion-free quotient of the complex unit ball by a discrete co-compact subgroup of $PU(1,2)$. The first example of a complex FPP was given by Mumford [**Mum79**] (hence the alternate name *Mumford surface* for a complex FPP) using p-adic uniformization, with further examples later given by Ishida and Kato [**IK98**] and Keum [**Keu06**]. Notably, Prasad and Yeung [**PY07**], with the help of computer-assisted calculations by Cartwright and Steeger, established that there are exactly 50 diffeomorphism types for complex FPPs [**PY07, PY10, CS10**].

In fact, these 50 Kähler surfaces are the only known examples of symplectic FPPs to date and their constructions involve arithmetic geometry. (Keum's work [**Keu06, Keu11**] may be largely reinterpreted using topological arguments but also relies on arithmetic geometric results [**Ish88**].) The problem asks if there are "softer" constructions via symplectic topology.

(2) A reconstruction of even the known complex FPPs using topological methods may lead to the further discovery of non-Kähler, symplectic examples on the Bogomolov–Miyaoka–Yau line. See Problem 4.90.

(3) Every complex FPP is smoothly irreducible [**BSS24**, Proposition 6.1]; this fact extends to any other potential symplectic FPP under mild assumptions on its fundamental group.

Indeed, Fintushel and Stern raised the more general question of whether there exist smoothly irreducible FPPs besides the complex ones and what could be said about their π_1. (For reducible examples, one can simply take a connected sum of any rational homology 4-sphere with $\mathbb{C}P^2$.) Irreducible FPPs with various fundamental groups were recently constructed in [**BSS24**]; for instance, any finite abelian group with a $\mathbb{Z}_2$ factor is claimed to be realized as the π_1 of an irreducible FPP. None of these FPPs admit symplectic structures, and in fact, it can be shown that the approach in [**BSS24**] falls short of producing *symplectic* examples.

(4) The existence of an (irreducible) FPP with trivial π_1, which amounts to an exotic smooth structure on $\mathbb{C}P^2$, is a particularly important open question. See Problem 4.2.

Let G be a finitely presented group and Q be an integral intersection form. Baldridge and Kirk conjectured that the diffeomorphism type of a closed symplectic 4–manifold that minimizes the Euler characteristic among all with $\pi_1 = G$ and intersection form Q is unique [**BK07**, Conjecture 23]. A *symplectic* FPP with trivial π_1 would be a counter-example to a special case of this conjecture, known as the *Symplectic Poincaré Conjecture.*

Proposed for K3 by: R. Fintushel, R. Stern
Scribed by: I. Baykur

PROBLEM 4.92. *Is every symplectic Calabi-Yau surface diffeomorphic to either the K3 surface, the Enriques surface or a T^2-bundle over T^2?*

REMARKS.

(1) Here a *symplectic Calabi-Yau surface* (SCY) is defined as a closed symplectic 4-manifold with torsion first Chern class.

(2) To date, the only smooth 4–manifolds known to support a symplectic structure with torsion c_1 are the ones listed in the problem. The problem asks whether this list indeed provides a complete classification of the diffeomorphism types of SCYs.

(3) T-J Li [**Li06a, Li06b**] and Bauer [**Bau08**] established that any SCY has the rational homology type of one of these standard 4-manifolds. (This extends the earlier results of Morgan and Szabó [**MS97**] when $b_1 = 0$ and Ruberman and Strle [**RS00**] when $b_1 = 4$.)

 Any SCY with $b_2^+ > 1$ has only one Seiberg-Witten basic class, namely the (trivial) canonical class, and its SW invariant is one. When $b_2^+ = 1$, the SW invariant of an SCY is determined by the wall crossing formula, and only depends on the cohomology ring structure.

(4) The list has been verified to be complete in some specific cases. Any SCY that smoothly fibers over a circle, a surface, or a 3–manifold is standard [**BF15, Bay14, FV13, LN14, Ni17b**]. The same holds when the SCY admits certain finite symplectic group actions [**Che20a, Che20b**].

 Moreover, many well-known construction techniques—such as Luttinger surgery, generalized fiber sums, knot surgery, and the simplest rational blow-downs—have been shown not to yield new SCYs from standard symplectic 4-manifolds; see [**Li19**] for a detailed survey and references.

(5) One approach towards an affirmative answer is to first detect the existence of a (possibly singular) T^2–fibration. In the case of a homology K3, if further assuming that it has a winding family of symplectic forms (a symplectic generalization of a hyperkähler family), then a parameterized version of Taubes' SW $\Rightarrow$ GW theorem, including a parametrized wall-crossing formula, produces a 2–dimensional family of embedded symplectic tori in the winding family [**Li10**, Section 7.4].

(6) One may probe the existence of new SCYs via symplectic Lefschetz pencils and multisections by analyzing certain positive factorizations in mapping class groups. New constructions of SCYs realizing all possible rational

homology types are given in [**BH16b, Bay22, BHM23**] adapting this approach. Only a handful of these SCYs are confirmed to be standard so far [**HH18a, Ful23**].

Due to the shortcomings of gauge-theoretical invariants in distinguishing SCYs (up to diffeomorphism) within the same homotopy type, the possibility of detecting a new SCY hinges on identifying a new SCY group. An intriguing, not-yet-ruled-out possibility is $\pi_1 = \mathbb{Z}^2$ [**FV13, Bay22**].

Proposed for K3 and scribed by: T.J. Li, I. Baykur

PROBLEM 4.93. *Is every symplectic form on the standard K3 surface symplectomorphic to a Kähler form?*

REMARKS.

(1) This question has several notable generalizations. First, since every symplectic form on the K3 surface is cohomologous to a Kähler form [**Li08**, 4.10], this is a special case of the following broader question [**Don06**]:

QUESTION (Donaldson). Let (X, ω_0) be a closed Kähler surface. Is any other symplectic form ω on X with $[\omega] = [\omega_0]$ and $c_1(X, \omega) = c_1(X, \omega_0)$ symplectomorphic to ω_0?

As every closed hyperkähler surface is diffeomorphic to either the K3 surface or T^4, the problem also constitutes a special case of the following conjecture.

CONJECTURE. *Let X be a closed hyperkähler surface and let $a \in H^2(X;\mathbb{R})$ be such that $a^2 > 0$. Then the space of symplectic forms on X representing the class a is connected.*

Finally, the problem is a significant instance of the more general Problem 4.96.

(2) Donaldson proposed using the almost Kähler Calabi-Yau equation given in [**Don06**] to get a family of cohomologous symplectic forms connecting the given symplectic form ω to the Kähler form ω_0. A partial a priori estimate for the almost Kähler Calabi-Yau equation was obtained in [**TWY08**].

(3) There are also approaches to this problem via Donaldson's geometric flow [**Don00, KS19**] and hypersymplectic flow [**FY19**].

Scribed by: T.J. Li

PROBLEM 4.94. *Are homotopy equivalent Horikawa surfaces in different deformation classes diffeomorphic as 4–manifolds? Are they symplectomorphic?*

REMARKS.

(1) A *Horikawa surface* is a minimal (non-singular) complex projective surface of general type whose Chern numbers satisfy $5c_1^2 = c_2 - 36$ (i.e. it lies on the *Noether line*), or equivalently, $c_1^2 = 2p_g - 4$, where $p_g \geqslant 3$ is the geometric genus. They are all simply connected.

These surfaces were studied extensively by Horikawa in [**Hor76b**, **Hor76a**, **Hor78**, **Hor79**], who classified them up to deformation. For each $r \geqslant 2$, there are two deformation classes of surfaces with $c_1^2 = 8r - 8$ and $c_2 = 40r - 4$. The smooth 4–manifolds $H(r)$ and $H'(r)$ in the two respective deformation classes are distinguished by their intersection form when r is even; whereas, when r is odd, they are homotopy equivalent. They have the same Donaldson polynomials and the same Seiberg-Witten invariants. Some 50 years after Horikawa, it remains unknown whether any pair of the latter surfaces are diffeomorphic.

(2) The question on the number of diffeomorphism classes was originally raised by Horikawa in [**Hor76b**] and it is in [**Kir97**, Problem 4.101(A)].

If two compact complex manifolds X and X' are deformation equivalent, then there exists a diffeomorphism $f: X \to X'$ such that

$$f^*(c_1(X, \omega)) = c_1(X', \omega').$$

The bold speculation of the time, namely that the converse holds in complex dimension two (the refined *"DEF = DIFF Conjecture"*), was disproved by Manetti in [**Man01**], and there are simply connected counterexamples as well [**CW07**].

A minimal complex surface of general type has a canonical symplectic structure, unique up to symplectomorphism, which is invariant under smooth deformation [**Cat09**]. Catanese showed that Manetti surfaces are indeed symplectomorphic [**Cat09**], so they also constitute counterexamples to the elusive *"DEF = SYMP Conjecture"*.

Nonetheless, it is still open whether DEF equivalence is more strict than DIFF or SYMP equivalence for the Horikawa surfaces.

(3) Let $\mathbb{F}_{2r}$ denote the Hirzebruch surface with fiber f and sections Δ_0 and Δ_∞ with self-intersections $\Delta_0^2 = 2r$ and $\Delta_\infty^2 = -2r$. The Horikawa surface $H(r)$ is the double cover of $\mathbb{F}_0$ branched over a smoothing of $6\Delta_0 + 4rf$ and $H'(r)$ is the double cover of $\mathbb{F}_{2r}$ branched over a disconnected branch locus that is a smoothing of $5\Delta_0 + \Delta_\infty$. Based on this, one can describe $H(r)$ and $H'(r)$ as smooth 4–manifolds using Kirby diagrams for branched covers, and as symplectic 4–manifolds via monodromy factorizations for compatible Lefschetz fibrations/pencils [**Ful98**, **Aur06a**]. In [**Aur06a**], Auroux compared the canonical symplectic Lefschetz pencils of the same genus on $H(3)$ and $H'(3)$ and observed that they are related through fibered Luttinger surgeries. As suggested by the status of the problem, there has been otherwise no success in relating such diagrams via Kirby calculus or the pencils via monodromy manipulations to prove DIFF or SYMP equivalence.

(4) Another possible strategy to obtain a diffeomorphism between $H(r)$ and $H'(r)$ is based on the observation of Manetti, that a common degeneration with Wahl singularities would prove that they are diffeomorphic [**Man01**][Theorem 1.5]. Recently in [**MNU24**], through an extensive study of complex surfaces with Wahl singularities, the authors invalidated

this approach. On the other hand, a different common degeneration was recently found in [**RR24**], without any interpretation of moving from one deformation component to the other as diffeomorphism.

Scribed by: I. Baykur, G. Urzúa

PROBLEM 4.95.

(a) *Is there a closed hyperbolic oriented 4-manifold that admits a symplectic structure?*
(b) *Do the Seiberg–Witten invariants vanish on every closed hyperbolic 4-manifold?*

REMARKS.

(1) In a preprint [**Sto25**], Stover announces the construction of a hyperbolic orbifold structure on $\mathbb{C}P^2$. This resolves an orbifold version of this problem.
(2) By work of Taubes [**Tau94**], a positive answer to (a) would imply a negative answer to (b); see [**Rei06**, Proposition 4.5]. By Donaldson [**Don99**] and Gompf–Stipsicz [**GS99**, Theorem 10.2.18], (a) is equivalent to asking whether there is a closed hyperbolic oriented 4-manifold that admits a smooth Lefschetz pencil. Compare this to Problem 2.23, which asks whether there exists a surface bundle over a surface whose total space is a hyperbolic 4-manifold.
(3) In [**LeB02**, Conjecture 1.1], LeBrun conjectures that the answer to (b) is, 'yes,' and hence the answer to (a) is 'no.' LeBrun provides evidence towards the moduli spaces being empty (for suitable perturbations) and proposes that one should consider the generalization of Seiberg-Witten invariants involving the evaluation of cohomology classes in $\Lambda^* H^1(X;\mathbb{Z})\otimes\mathbb{Z}[U]$ on higher dimensional moduli spaces. The first examples of manifolds for which all these invariants vanish were provided in [**AL20**], followed by more concrete examples in [**BFS24**]. It is worth noting that there are no known examples of closed hyperbolic 4-manifolds with $b_2^+ \leqslant 1$, cf. Problem 4.19.

Proposed for K3 and scribed by: F. Lin, B. Martelli

PROBLEM 4.96. *Does there exist a pair of symplectic 4–manifolds (X_1, ω_1) and (X_2, ω_2), where there is a diffeomorphism $f\colon X_1 \to X_2$ such that $f^*(c_1(X_2, \omega_2)) = c_1(X_1, \omega_1)$ and $f^*([\omega_2]) = [\omega_1] \in H^2(X_1, \mathbb{R})$, but (X_1, ω_1) is not symplectomorphic to (X_2, ω_2)?*

REMARKS.

(1) It may be difficult to detect such a difference with current invariants due to Taubes' result relating Gromov–Witten invariants and Seiberg–Witten invariants [**Tau96**].
(2) There is also a relative version of this problem for fillings.

QUESTION. Does there exist a pair of Stein manifolds filling the same contact 3-manifold that are diffeomorphic where the diffeomorphism takes the first Chern class of one to that of the other, but such that the Stein manifolds are not symplectomorphic?

One could also ask whether there is such a pair of Stein manifolds that are not Weinstein homotopic. See also Problem 5.17.

Scribed by: L. Starkston, J. Van Horn-Morris

PROBLEM 4.97. *Let $\lambda := c_1^2/c_2$ be the Chern slope of a closed, almost complex 4–manifold X. What is the supremum of λ as X ranges over the following families?*

(a) *Symplectic Σ_g–bundles over Σ_h, with $g, h \geqslant 2$.*
(b) *Holomorphic Σ_g–bundles over Σ_h, with $g, h \geqslant 2$.*
(c) *Symplectic Lefschetz fibrations over S^2.*
(d) *Holomorphic Lefschetz fibrations over S^2.*

(Here, the Lefschetz fibrations are assumed to have critical points.)

REMARKS.

(1) Recall that the Chern numbers of X are determined by its homotopy type. Any Σ_g–bundle over Σ_h with $g \geqslant 2$ or a Lefschetz fibration (with nonempty critical set) over S^2 can be made symplectic, but not always holomorphic. See e.g. [**Joh86, Bay12b**]. In fact, the sole obstruction to making a surface bundle over a surface holomorphic is whether the smooth 4–manifold X admits a complex structure [**Hil00, Kot99**].
(2) The first line of research in this direction, pioneered by Atiyah, Kodaira, and Hirzebruch in the late 1960s, and Endo in the late 1990s, is to understand for which pairs of $g \geqslant 3$ and $h \geqslant 2$ one can have a Σ_g–bundle over Σ_h with $\sigma > 0$. (The signature vanishes when $g \leqslant 2$ or $h \leqslant 1$ [**Mey73**].) This geography problem was nearly resolved in the recent work of Baykur and Korkmaz [**BK24b**], who showed that for all but 19 possible pairs (g, h), there are symplectic Σ_g–bundles over Σ_h with $\sigma > 0$. However, the remaining few cases are quite significant for symplectic geography; see below.
(3) The Bogomolov–Miyaoka–Yau inequality for complex surfaces is encoded as $\lambda \leqslant 3$, providing an upper bound on the slopes of X in (b) and (d). Since no complex ball quotient can smoothly fiber over a surface, it is moreover known that any X in (b) satisfies $\lambda < 3$.

    Two of the unsettled cases in [**BK24b**] have direct implications on symplectic geography: any example with $(g, h) = (3, 2)$ and $\sigma > 0$ is a symplectic X of general type violating the BMY inequality, whereas one with $(g, h) = (4, 2)$ and the smallest possible positive signature would give a symplectic X on the BMY line and cannot possibly be complex. See [**BK24b**, Question 2] and Problem 4.90. Do such surface bundles over surfaces exist?

An unpublished preprint by Hamenstädt [**Ham20**] announces that $\lambda \leqslant 3$ more generally for any X in (a).

(4) Presumably, these questions will all have different answers. Nonetheless, the largest currently known slope for (a) and (b) coincide: these are the examples by Catanese and Rollenske with $\lambda = 8/3$ [**CR09**]. There exist holomorphic Σ_g–bundles over Σ_h with positive signatures, for $h = 2$ (and large g), as shown by Bryan and Donagi in [**BD02**], and with $g = 3$ (and unspecified, presumably very large h), as shown by Kazuhiro Konno (in works only available in Japanese). The situation for (c) and (d) is more mysterious; for instance, (symplectic) Lefschetz fibrations over S^2 with $\sigma > 0$ were only recently discovered in [**BH24c**].

There are fewer constraints on the slopes of Lefschetz fibrations over higher genera surfaces; e.g. the Cartwright-Steger surface on the BMY line admits a holomorphic Lefschetz fibration over T^2 [**KY21**], realizing the maximal possible slope $\lambda = 3$ in this case.

Proposed for K3 and scribed by: I. Baykur

PROBLEM 4.98. *Does every closed symplectic 4-manifold admit inequivalent Lefschetz pencils with the same fiber genus g, for sufficiently large g? How about infinitely many?*

REMARKS.

(1) Here Lefschetz pencils are assumed to have base points, whereas Lefschetz fibrations do not. Equivalence is defined by a diffeomorphism of the symplectic 4–manifold commuting with any pair of pencil/fibration maps.

(2) Any symplectic 4–manifold can be equipped with Lefschetz pencils with arbitrarily high fiber genera by increasing the degree in Donaldson's construction. However, the number of base points also increases, meaning the Lefschetz fibrations derived by blowing up the base points would be on birational yet different 4–manifolds.

Except for rational and ruled surfaces, the number of base points in a pencil is bounded above by $2g - 2$, where g is the genus of the fiber. Therefore, a positive answer to the second question would imply the existence of infinitely many genus–g Lefschetz pencils with the same number of base points. By blowing up the base points of such examples, one can obtain inequivalent Lefschetz fibrations over S^2 with the same fiber genus, but the converse is not always feasible.

(3) By [**Bay16**], any symplectic 4–manifold that is not a rational or ruled surface, possibly after blow-ups, admits arbitrarily many non-isomorphic Lefschetz pencils with the same genus, the same number of base points, and matching topological types of singular fibers. A slightly weaker result also holds for rational and ruled surfaces [**Bay19**]. These constructions require blow ups, addressing the problem only up to birational equivalence.

Other examples of inequivalent Lefschetz pencils and fibrations on a few specific 4–manifolds are given in [**PY09, PY17, BH16b, BHM23, Ham17**]. The monodromy invariants and arguments used in these works do not distinguish more than finitely many classes; for a positive resolution of the problem, one needs finer monodromy invariants.

Proposed for K3 by: I. Baykur
Scribed by: I. Baykur, N. Salter, L. Starkston

PROBLEM 4.99. *Let X be a closed symplectic 4-manifold. Let $T \subset X$ be a symplectic submanifold that is diffeomorphic to a 2-dimensional torus such that $[T]^2 = 0$. Let X_K be a manifold obtained by Fintushel–Stern knot surgery on T using a knot $K \subset S^3$. If X_K has a symplectic structure, must K be a fibered knot?*

REMARKS.

(1) Fintushel and Stern [**FS98**] showed that X_K has a symplectic structure if K is a fibered knot, and raised the question in the problem.
(2) One can ask the question for the particular case when X admits a Lefschetz fibration whose regular fibers are tori and T is a regular fiber. This particular case was explicitly stated in [**Ni17a**]. An interesting example is the case when X is the K3 surface, then $SW(X)$ is essentially the Alexander polynomial Δ_K of K [**FS98**], so we know Δ_K should be monic if X_K has a symplectic structure [**Tau94, Tau95**]. No other constraint is known.
(3) The answer to this problem is "Yes" when $X = T^2 \times S^2$ and $T = T^2 \times \{\text{point}\}$ by Friedl–Vidussi [**FV11**], and when X is a torus bundle over a closed surface with homologically essential fibers and T is a fiber by Ni [**Ni17a**].

Scribed by: Y. Ni

PROBLEM 4.100. *Given a closed, connected, symplectic 4-manifold (X, ω) and $c \in H_2(X, \mathbb{Z})$ represented by an embedded, connected, oriented, smooth surface S such that*

(i) $\langle [\omega], c\rangle > 0$, *and*
(ii) $\langle c_1(X, \omega), c\rangle = \chi(S) + S^2$,

is c represented by an embedded, connected, symplectic surface?

REMARKS.

(1) This is a special case of the question of which homology classes in a symplectic manifold are represented by symplectic surfaces. By the adjunction formula, the homology class of every embedded symplectic surface satisfies both (i) and (ii). The *Symplectic Thom Conjecture* [**OS00**] states that symplectic surfaces minimize genus in their homology classes. In their proof of this conjecture, Ozsváth and Szabó showed that every class satisfies the adjunction inequality, $-\chi(S) \geqslant S^2 + |\langle c_1(X, \omega), c\rangle|$, provided $b_2^+(X) > 1$. (The result extends to $b_2^+(X) = 1$ unless X is a rational or ruled surface.) The question is whether this is a sufficient condition to conclude the existence of a symplectic representative. See also [**DLW18**].

(2) By Donaldson [**Don99**], if $[\omega] \in H_2(X, \mathbb{Z})$, then all large enough multiples of $[\omega]$ are represented by symplectic submanifolds. (When $b_2^+(X) = 1$, one can say a bit more; see e.g. [**Li08**, Proposition 3.18].) When $b_2^+(X) > 1$, Taubes [**Tau96**] showed this is also true for the Poincaré dual of the canonical class $c_1(X, \omega)$; also see [**DS03**].

Scribed by: I. Baykur, T.J. Li, J. Van Horn-Morris

PROBLEM 4.101. *Is every smooth symplectic surface in $(\mathbb{C}\mathrm{P}^2, \omega_{\mathrm{FS}})$ symplectically isotopic to a complex curve? Equivalently, is there a unique symplectic isotopy class of smoothly embedded symplectic surfaces of degree d in $(\mathbb{C}\mathrm{P}^2, \omega_{\mathrm{FS}})$?*

REMARKS.

(1) Here ω_{FS} denotes the standard Fubini–Study symplectic form on $\mathbb{C}\mathrm{P}^2$. A symplectic surface in $\mathbb{C}\mathrm{P}^2$ is said to have degree d if it represents the homology class $dh \in H_2(\mathbb{C}\mathrm{P}^2; \mathbb{Z})$ where h is the generator of $H_2(\mathbb{C}\mathrm{P}^2; \mathbb{Z}) \cong \mathbb{Z}$ represented by a complex projective line. Note that symplectic surfaces always represent a *positive* multiple of h because they have positive symplectic area.

(2) A complex projective algebraic plane curve of degree d is the zero set of a homogeneous polynomial of degree d in $\mathbb{C}\mathrm{P}^2$. Such a curve represents $dh \in H_2(\mathbb{C}\mathrm{P}^2; \mathbb{Z})$. If the zero set is regularly cut out, the complex algebraic curve is smooth. Any two smooth complex algebraic curves of the same degree are isotopic, i.e. connected through a family of smooth complex algebraic curves. This can be seen by observing that the set of singular curves of degree d (corresponding to homogeneous polynomials that do not regularly cut out 0) is a complex subvariety of positive complex codimension (cut out by the discriminant) in the space of all degree-d curves (a high dimensional complex projective space parametrized by the coefficients of the monomials in a degree-d homogeneous polynomial). Thus the smooth curves form a path connected space, as it is the complement of a subset of real codimension at least 2. Thus, the two questions in the problem statement are equivalent and are known as the "symplectic isotopy problem."

(3) Key progress on the symplectic isotopy problem relies on the fact that every symplectic surface can be realized as a J-holomorphic curve for some almost complex J compatible with the symplectic form. The development of pseudoholomorphic curves originates from Gromov's seminal paper [**Gro85**]. It is proven in Gromov's work that the answer to the symplectic isotopy question is yes in degrees 1 and 2. Extensive further work using Gromov's strategy of pseudoholomorphic curves shows that the answer is yes for degrees less than or equal to 17 [**She00, Sik03, ST05**]. After degree 17, we have been unable to rule out the possibility that a family of J_t-holomorphic curves may degenerate to a singular curve with unreduced components, and it is unknown whether such degenerations can produce an unavoidable codimension-1 "wall" in the moduli space of pseudoholomorphic curves.

(4) Another approach to this problem is via quasipositive factorizations in the braid group. Given a smooth symplectic surface in $\mathbb{C}P^2$ realized as a J-holomorphic curve, one can generate a J-holomorphic linear pencil on $\mathbb{C}P^2$, $\pi : \mathbb{C}P^2 \backslash \{p\} \to \mathbb{C}P^1$. Restricting the pencil to the symplectic surface gives a simple branched covering from the surface to $\mathbb{C}P^1$, assuming the pencil point p is chosen sufficiently generically. The branch points correspond to places where the fibers of the pencil are tangent to the symplectic surface. Looking at the preimages under π of loops in $\mathbb{C}P^1$ gives a braid monodromy presentation that fully encodes the symplectic surface [**MT88, MT91**]. Through this, the symplectic isotopy problem can be related to the following problem in the braid group.

QUESTION. Let Δ^2 denote the full twist on d strands in the braid group B_d, and let σ_i denote the standard generators of the braid group (a half twist of two adjacent strands). If $\rho_1 \dots \rho_k = \Delta^2$ and each ρ_j is a conjugate of some σ_i, then is there a sequence of Hurwitz moves and global conjugations one can perform on $\rho_1 \dots \rho_k$ to turn it into $(\sigma_1 \dots \sigma_{d-1})^d$?

For more background on this perspective and the definition of Hurwitz moves, see this survey by Auroux [**Aur06b**].

(5) Another strategy to attack the symplectic isotopy problem is proposed in [**Sta20**], which uses deformations of the smooth symplectic surface to a singular surface (a symplectic line arrangement) to show that the symplectic isotopy problem is equivalent to the existence of certain Lagrangian disks with boundary on the symplectic surface. Finding or obstructing the existence of such Lagrangians could potentially be approached using Floer theoretic/Fukaya categorical techniques.

(6) Note that there are smooth surfaces in $\mathbb{C}P^2$ that are not smoothly isotopic to any complex curve [**Fin02, Kim06**]. However, one can mix the symplectic and smooth categories for the following weakening of the symplectic isotopy problem that is currently equally open.

QUESTION. Are any two symplectic surfaces in $\mathbb{C}P^2$ of the same degree smoothly isotopic?

A strategy for answering this weaker version of the problem using transverse bridge trisections was proposed by Lambert-Cole [**LC23**].

(7) There are examples of symplectic surfaces which are homologous but not symplectically isotopic in other closed symplectic 4-manifolds [**FS99b**]. However, the question appears to also be open more generally for rational and ruled surfaces, and their blow-ups (symplectic manifolds of Kodaira dimension $-\infty$).

In ruled surfaces $S^2 \times S^2$, $\mathbb{C}P^2 \# ^n \overline{\mathbb{C}P}^2$, or an S^2 bundle over a higher-genus surface, in a fixed homology class, is there a unique symplectic isotopy class of symplectic surfaces?

Scribed by: L. Starkston

PROBLEM 4.102. *Is every symplectic rational cuspidal curve in* $(\mathbb{CP}^2, \omega_{\text{FS}})$ *equisingularly symplectically isotopic to a complex curve? More generally, which types of singular symplectic surfaces in* $(\mathbb{CP}^2, \omega_{\text{FS}})$ *(where type is specified by the genus of each irreducible component and the topological types of the singularities), admit symplectic representatives that are not equisingularly symplectically isotopic to complex curves?*

REMARKS.

(1) This is a singular version of the symplectic isotopy problem (Problem 4.101). Note that "curve" here refers to a real 2-dimensional surface in analogy with the terminology for complex/pseudoholomorphic curves. A rational cuspidal curve is a singular curve homeomorphic to S^2.

(2) We say that two singular symplectic surfaces are *equisingularly symplectically isotopic* if there is a family of symplectic surfaces connecting them such that the topological type of each singularity remains constant through the isotopy. (The topological type is the isotopy class of the link of the singularity.)

(3) Although one could define symplectic surfaces whose singularities are a cone on any transverse knot, for there to be any chance that the symplectic curve is equisingularly isotopic to a complex curve, its singularities must be modeled on the singularities appearing in complex plane curves. By a result of McDuff [**McD92**] (reproved by Micallef–White [**MW95**] through different methods), these are precisely the singularity types that can be realized in J-holomorphic curves for some almost complex structure J compatible with the standard symplectic form on $\mathbb{CP}^2$. Thus these types of singular symplectic surfaces can be studied with pseudoholomorphic techniques.

(4) There are a number of works which approach this problem for certain classes of singular surfaces and are able to prove in some cases that the symplectic surfaces with certain specified singularity types are symplectically isotopic to complex curves. For example Baurraud established symplectic isotopy results for nodal symplectic spheres [**Bar99**] and sufficiently generic line arrangements [**Bar00**]. Shevchishin proved results for nodal symplectic surfaces of sufficiently low genus [**She04**], conjecturing this holds more generally with a positivity assumption on the Chern number. Ohta–Ono proved uniqueness up to symplectic isotopy of a cuspidal cubic [**OO05**]. More generally, unicuspidal single Puiseux pair families and low degree examples of symplectic rational cuspidal curves were shown to always be equisingularly symplectically isotopic to complex curves in [**GS22b, GK23**], motivating the plausibility of a positive answer to the first question in the problem statement.

(5) There are a number of ad hoc examples of singular symplectic surfaces in $\mathbb{CP}^2$ which are *not* equisingularly isotopic to any complex curve. Possibly the first such examples appeared in Moishezon's work [**Moi94**]. These examples were high degree and contained a very large number of nodes (positive transverse double points) and simple cusps (modeled on $z_1^3 = z_2^2$). An important tool in constructing and detecting such examples is braid monodromy [**MT88, KK03**]. An easier and lower-degree example is the "fake Pappus" line arrangement, which cannot be realized by complex

projective lines, but which can be realized symplectically (see [**RS19**]). For an irreducible example of degree 8 with locally reducible singularities see [**GS22b**, Section 8]. However, we do not currently have any examples where the surface is irreducible and the singularities are locally irreducible, a.k.a., "cuspidal".

(6) One could also ask whether every pair of equisingular singular symplectic surfaces in $\mathbb{CP}^2$ are equisingularly symplectically isotopic to each other. In contrast to the smooth symplectic isotopy problem, this is not equivalent to asking whether every singular symplectic surface is equisingularly symplectically isotopic to a complex curve. In fact, it is well known that there are examples of equisingular complex curves in $\mathbb{CP}^2$ that are not equisingularly symplectically isotopic. Additionally there are examples of singular symplectic surfaces such that there does not exist any complex curve with the same singularities as mentioned above.

Equisingular complex algebraic curves that are not equisingularly isotopic are often known as "Zariski pairs" (due to the first example being a pair of sextics discovered by Zariski [**Zar29**]) and are of great interest in the study of complex algebraic plane curves. (Note that Zariski pair sometimes is used to refer to examples that are not related by a weaker equivalence than equisingular isotopy, such as the fundamental groups of the complements being non-isomorphic; this implies they are not symplectically isotopic.) See [**ABCT08**] for an in-depth survey on Zariski pairs. For singular symplectic curves, it is interesting to ask when there are further "Zariski pairs" that can be realized symplectically than can be realized complexly.

Proposed for K3 by: M. Golla, L. Starkston
Scribed by: L. Starkston

PROBLEM 4.103.

(a) *What polynomials can occur as the Alexander polynomials of complex plane algebraic curves?*

(b) *More generally, what are the conditions that must be satisfied by a finitely presented group G, so that there exist a plane algebraic curve having G as the fundamental group of its complement?*

REMARKS.

(1) Let C be an algebraic curve in $\mathbb{C}^2$. The Alexander polynomial of C relative to a surjection $\pi_1(\mathbb{C}^2\backslash C) \to \mathbb{Z}$ is the characteristic polynomial the automorphism of the homology of the associated infinite cyclic cover $H_1(\widetilde{\mathbb{C}^2\backslash C}, \mathbb{C})$ induced by a generator of the group of deck transformations of the cover.

(2) Since a constraint on the Alexander polynomial is also a constraint on the fundamental group, part (a) is a very special case of part (b). One of many other questions underlying the second part is: which *finite* groups can occur as the fundamental groups of the complements to irreducible algebraic curves in $\mathbb{CP}^2$?

(3) A root of the Alexander polynomial of C must be a root of the Alexander polynomial of the link of at least one of the singularities of C and also a root of the Alexander polynomial of the link at infinity, i.e., the intersection of C with a 3-sphere in $\mathbb{C}^2$ of a sufficiently large radius [**Lib83**, **Lib21**]. In particular the Alexander polynomial of an algebraic curve is cyclotomic but degrees of the factors that can occur are unknown. For example, for an irreducible plane curve having only ordinary cusps and nodes as singularities (i.e., locally homeomorphic to $x^2 + y^3 = 0$ and $x^2 + y^2 = 0$ respectively) the Alexander polynomial has the form $(t^2 - t + 1)^r$ and the largest known value for r is 4; see [**CAL14**]. Is the set of possible degrees of the Alexander polynomials of curves C of arbitrary degrees but with fixed local types of singularities bounded?

(4) For a recent survey of this circle of questions and mentioned properties of the Alexander polynomials, see [**Lib82**]. Further developments discussed there include characteristic varieties providing a multivariable generalization of Alexander polynomials, the Alexander polynomials of the complements to ample singular divisors on projective simply connected surfaces, and the role of the Alexander polynomials in the study of Zariski pairs (see Problem 4.102).

(5) It would be interesting to understand a symplectic analog of the above problems. The Alexander polynomials of the fundamental groups of the complements to pseudoholomorphic curves with isolated singularities can be similarly defined, but their characterization, and the question "are the classes of realizable polynomials different in the symplectic or algebraic context?" are open. In particular, it is unknown if the classes of the fundamental groups of the complements to plane algebraic and pseudoholomorphic curves are different. Recently, a symplectic analog of the divisibility theorem mentioned in (3) was announced in the symplectic context; see [**AG24**].

Proposed for K3 and scribed by: A. Libgober

PROBLEM 4.104. *Does there exist a transverse link $L \subset (S^3, \xi_{std})$ bounding a pair of complex curves in $B^4 \subset \mathbb{C}^2$ that are isotopic through embedded smooth surfaces but not through complex curves? Smoothly isotopic but not symplectically? Can there be infinitely many of them?*

REMARKS.

(1) By Rudolph [**Rud83**] and Boileau-Orevkov [**BO01**], L should be a quasipositive link and the complex curves can be realized as quasipositive surfaces in B^4. The boundary braids of the latter do not necessarily need to be the same quasipositive braid representatives of K, but after Markov stabilization one can assume they are.

In fact, there are transverse knots in (S^3, ξ_{std}) with quasipositive braid representatives that bound infinitely many quasipositive surfaces that are not smoothly isotopic [**BVHM18**], but these are distinguished by the topology of their complements. There are also pairs that are topologically isotopic but not smoothly isotopic, announced in [**Hay21a**].

(2) This can be regarded as a relative version of the symplectic isotopy problem; see Problem 4.101. The case of $K = T(d,d)$, the (d,d) torus link, is equivalent to the symplectic isotopy problem. This is because the symplectic isotopy classes of a non-singular surface of degree d are in bijection with the equisingular symplectic isotopy classes of the union of the non-singular degree d curve with a generically intersecting line [**GS22b**, Proposition 5.1]. The complement of a neighborhood of the line is symplectomorphic to B^4 and the degree d symplectic surface will intersect the boundary S^3 in a transverse (d,d) torus link. There may be some generalizations of this to other algebraic links besides the (d,d) torus link, by allowing the degree d surface to intersect the line tangentially or by studying degree d surfaces with singularities.

(3) There are infinitely many examples of a Legendrian link $L \subset (S^3, \xi_{std})$ bounding infinitely many (exact) Lagrangians in $B^4 \subset \mathbb{C}^2$ that are smoothly isotopic but not Hamiltonian isotopic [**CG22**].

Proposed for K3 by: I. Baykur, J. Van Horn-Morris
Scribed by: I. Baykur, L. Starkston, J. Van Horn-Morris

PROBLEM 4.105. *Does there exist a planar contact 3-manifold that has infinitely many distinct Stein fillings?*

REMARKS.

(1) A contact 3-manifold is *planar* if it is supported by a planar open book decomposition.

(2) There are many interesting classes of planar contact 3-manifolds, for example lens spaces and links of rational surface singularities with reduced fundamental cycle. By Wendl [**Wen10**], and Wendl and Niederkrüger [**NW11**], any minimal weak symplectic filling of a planar contact manifold is in fact Stein, fills any given planar open book, and hence is determined by a positive Dehn twist factorization of the monodromy of that open book. Additionally, by Plamenevskaya [**Pla12**] and separately Wand [**Wan12**], there are only finitely many "homological types" of positive factorizations of the monodromy of a planar open book. (See [**BVHM18**] for an explicit statement.) Lisca [**Lis08**] classified the diffeomorphism types of Stein fillings of the standard contact structure on lens spaces, showing that there are only finitely many.

(3) The adjective "distinct" could be interpreted in many ways and most are interesting. One strong "no" result would be to show there exist only finitely many Stein fillings up to diffeomorphism. Stronger would be to prove this up to symplectomorphism and deformation. Lisi and Wendl conjecture that there are only finitely many deformation classes of minimal symplectic fillings [**LW21**].

(4) Such examples are known when the page genus is 1 and higher [**OS04a**] [**BVHM18**]. Moreover, for genus 2 and higher the Stein fillings can have arbitrarily large b_2. This is thus related to the *support genus* of the contact manifold; see Problem 3.45.

Scribed by: J. Van Horn-Morris

PROBLEM 4.106. *Is the exact symplectomorphism type of T^*X^4 sensitive to the smooth structure on a 4-manifold X, or does it depend only on the simple-homotopy or homeomorphism type of X?*

REMARKS.

(1) Recall that the cotangent bundle T^*X has a canonical 1-form λ_{can}; the canonical symplectic form on T^*X is $d\lambda_{can}$. A diffeomorphism $\phi\colon T^*X \to T^*Y$ is an *exact symplectomorphism* if $\phi^*\lambda_{can} = \lambda_{can} + df$ for some function $f\colon X \to \mathbb{R}$.
(2) The problem is a version of Arnol'd's *Nearby Lagrangian Conjecture*, which states that any exact Lagrangian submanifold of a cotangent bundle is Hamiltonian isotopic to the 0-section. In particular, a positive solution to the Nearby Lagrangian Conjecture would imply that the exact symplectomorphism type of T^*X determines the diffeomorphism type of X.
(3) Evidently, the symplectomorphism type of T^*X determines the homotopy type of T^*X and hence of X. In fact, deep results in Floer theory show that it determines the simple homotopy type of X [**AK18**].
(4) In higher dimensions, the symplectic structure on the cotangent bundle can detect (some) exotic smooth structures [**Abo12, EKS16**]. In dimension 3, a relative version of this construction—considering the unit conormal bundle to a knot—gives a strong invariant of knots [**Ng05, ENS18**], though of course the homotopy type of a knot complement is itself a strong knot invariant. It would be interesting to know whether this conormal construction detects exotic embeddings of surfaces in 4-manifolds.

Scribed by: R. Lipshitz

PROBLEM 4.107. *Problems about contact hypersurfaces:*

(a) *Let (Y,ξ) be a contact manifold and $(\mathbb{R}\times Y,\omega)$ its symplectization. Let $f: Y \to \mathbb{R}\times Y$ be a smooth embedding such that f induces an isomorphism on homology. When can we isotope f so that its image is a contact type hypersurface?*
(b) *Special case: Is every embedded 3-sphere $Y \subset (\mathbb{R}^4,\omega_{std})$ smoothly isotopic to a hypersurface of contact type?*
(c) *Is there any contact rational homology sphere other than (S^3,ξ_{std}) which embeds as a contact type hypersurface in $(\mathbb{R}^4,\omega_{std}))$?*

REMARKS.

(1) There are some geometric obstructions to a hypersurface Y being isotopic to a contact type hypersurface in the ambient symplectic 4–manifold (M,ω); e.g. [**Sul76**] provides a necessary condition formulated in terms of a characteristic foliation of ω on Y and [**Cie98**] shows that the existence of certain concentric annuli is an obstruction.
(2) Question (b) has a bearing on the Schoenflies conjecture. If it is possible to isotope a smooth embedding of S^3 to make it a contact hypersurface in $(\mathbb{R}^4,\xi_{std})$ then it bounds a symplectic filling which is necessarily the 4-ball by Gromov [**Gro85**]. A strategy towards trying to isotope a smoothly embedded S^3 to a contact type hypersurface has been proposed by Lambert-Cole [**LC21**].
(3) It was conjectured by Gompf [**Gom13**] that no Brieskorn sphere with either orientation admits a pseudoconvex embedding in $\mathbb{C}^2$. This has a bearing on question (c). Mark and Tosun proved half of this conjecture in [**MT22**] that no positively oriented Brieskorn sphere (with any number of singular fibers) embeds as a contact type hypersurface of $(\mathbb{R}^4,\omega_{std})$.

Proposed for K3 by: J. Chaidez, B. Tosun
Scribed by: J. Chaidez, L. Starkston, J. Van Horn-Morris

PROBLEM 4.108. *Let W_+ and W_- be two 4-dimensional Liouville domains with a contactomorphism $\Phi : \partial W_- \simeq \partial W_+$. This determines an $\mathbb{R}$-invariant contact structure ξ on $\mathbb{R}\times X$ where X is the gluing*

$$X = W_+ \cup_\Phi W_-$$

Moreover, $0\times X \subset \mathbb{R}\times X$ is a convex hypersurface in the sense of Giroux.

(a) *Is there a 4–dimensional version of Giroux's criterion in the case where W_+ and W_- are Weinstein? That is, a necessary and sufficient topological criterion on (Φ, W_+, W_-) for ξ to be overtwisted.*
(b) *Does every 4-manifold X admit a decomposition $W_+ \cup_\Gamma W_-$ so that the corresponding contact structure ξ is tight (i.e. not overtwisted)?*
(c) *Is there a 4-dimensional Liouville domain W and a contactomorphism $\Phi : \partial W \to \partial W$ that extends to a diffeomorphism Ψ of W, but so that the contact structure corresponding to (Φ, W, W) is overtwisted?*

REMARKS.

(1) A hypersurface $\Sigma \subset Y$ of a contact manifold (Y,ξ) is *convex* if there is a contact vector-field V that is transverse to Σ. A contact manifold is *overtwisted* if there is an embedded codimension one disk $D_{\rm ot}$ [**BEM15**] with characteristic foliation determining a standard overtwisted contact structure in a neighborhood.
(2) Overtwistedness is essentially characterized by an h-principle, and the overtwisted-tight dichotomy is emblematic of the general "flexible-rigid" dichotomy in symplectic topology. The theory of convex surfaces in contact 3-manifolds was pioneered by Giroux [**Gir02, GM03**].

(3) The foundational work on convex surfaces was extended to higher dimensions by several works of Honda-Huang [**HH18b, HH19**], Breen–Honda–Huang [**BHH23**] and Eliashberg–Pancholi [**EP23**]. When W_+ and W_- are Weinstein, the data (Φ, W_+, W_-) can be described using a variant of the Weinstein-Kirby diagrams due to Gompf (c.f. Breen-Christian [**BC24c**]). In particular, many of the outstanding open problems in higher dimensional convex surface theory may be particularly interesting to study in dimension four.

(4) On part (a): given a splitting $\Sigma = W_+ \cup_\Phi W_-$ of a closed, oriented, connected surface, *Giroux's criterion* states that the resulting contact structure on $\mathbb{R} \times \Sigma$ is overtwisted if and only if either $\Sigma \simeq S^2$ and Γ has more than one component or $\Sigma \not\simeq S^2$ and Γ contains a contractible curve. A higher dimensional, algebraic analogue of this criterion has been obtained by Avdek [**Avd23**], but a true topological analogue is still unknown.

(5) On Part (b): a celebrated result of Etnyre–Honda [**EH01b**] gives an example of a closed contact 3-manifold with no tight contact structures. Part (b) of this problem may be viewed as a 4-manifold analogue of that result.

(6) On Part (c): this problem is motivated by the goal of finding diffeomorphisms fixed at the boundary that act nontrivially on the space of symplectic structures. Indeed, the extension Ψ of Φ in part (c) would constitute such a map, since the double $W \cup_{Id} W$ is always tight by Avdek [**Avd23**]. Moreover, such a map may be usable as a sort of symplectic cork twist - by embedding W into a closed symplectic manifold X, removing W and regluing it by Ψ to acquire a new symplectic manifold X', one may be able to find diffeomorphic, non-symplectomorphic closed 4-manifolds (see Problem 4.96).

Scribed by: J. Chaidez

PROBLEM 4.109. *If $\Sigma \subseteq (X, \omega)$ is a symplectic surface in a closed symplectic 4-manifold with $[\Sigma] = PD(k[\omega])$, $k \in \mathbb{Z}$, does $(X \backslash \Sigma, \omega)$ support a Weinstein structure?*

REMARKS.

(1) In particular, one could ask this for surfaces in $(\mathbb{CP}^2, \omega_{\text{std}})$. A negative answer to the question in this case would yield a counterexample to the symplectic isotopy problem (Problem 4.101).

(2) Donaldson [**Don96**] and Giroux [**Gir02, Gir17**] prove that for suitably large k there is some symplectic surface for which this is true. However, there are not currently known bounds on k, or known examples where $k = 1$ is not enough. In principle, large upper bounds could be obtained by carefully tracking the proof in Donaldson's construction. Such bounds are likely much larger than needed, but any explicit bound in terms of the

symplectic manifold would represent progress on this question. Obtaining more effective bounds through new techniques would be particularly interesting.

Scribed by: L. Starkston, J. Van Horn-Morris

PROBLEM 4.110. *Does there exist a 2-handlebody W that admits an exact symplectic structure with convex contact boundary that does not admit a Weinstein structure filling the same contact boundary?*

REMARKS.

(1) So far, all known examples of exact symplectic manifolds that we can obstruct from having a Weinstein structure are known to have handle structures requiring 3-handles [**McD91**, **Bow12**].
(2) We have limited tools for obstructing an exact filling from being Weinstein when it has the topology of a 2-handlebody. Bowden's argument in [**Bow12**] utilized a theorem of Eliashberg that says that a Weinstein filling of a connected sum is necessarily a boundary sum of Weinstein fillings of the summands [**Eli90**]. A potential strategy to generalize this obstruction to other examples may be to use Menke's "mixed tori" [**CM19**], which gives a different decomposition criterion for Weinstein fillings.

Scribed by: L. Starkston, J. Van Horn-Morris

4.10. Trisections of 4-manifolds

A *trisection* of a 4–manifold X, introduced by Gay and Kirby in [**GK16**], is a decomposition $X = Z_1 \cup Z_2 \cup Z_3$ such that, for each $i \in \mathbb{Z}_3$,

(1) Z_i is a 4–dimensional 1–handlebody;
(2) $H_i = Z_i \cap Z_{i-1}$ is a 3–dimensional handlebody; and
(3) $\Sigma = Z_1 \cap Z_2 \cap Z_3$ is a closed, genus g surface.

We say that such a trisection has *genus* g. If $Z_i = \natural^{k_i}(S^1 \times B^3)$, we say the trisection has *complexity* $(g; k_1, k_2, k_3)$, or (g, k) when $k_i = k$ for all k.

PROBLEM 4.111. *Is trisection genus additive? In other words, must it be the case that $g(X\#X') = g(X) + g(X')$.*

REMARKS.

(1) This is Conjecture 1.6 of [**LCM22**].
(2) The *trisection genus* of a 4–manifold X is

$$g(X) = \min\{g \mid X \text{ admits a genus } g \text{ trisection}\}.$$

(3) If $\mathfrak{T}$ and $\mathfrak{T}'$ are trisections for X and X' of genus g and g', respectively, then $\mathfrak{T}\#\mathfrak{T}'$ is a trisection of genus $g + g'$ for $X\#X'$. It follows that trisection genus satisfies:

$$g(X\#X') \leqslant g(X) + g(X').$$

Equality holds when the lower-bound on $g(X)$ and $g(X')$ coming from standard algebraic topology is sharp; see Problem 4.117.

(4) A positive resolution to this problem would have sweeping consequences. If trisection genus is additive, then no manifold from the following set has an exotic copy:

$$\mathcal{M} = \{S^4, \mathbb{CP}^2, S^1 \times S^3, 2\mathbb{CP}^2, \mathbb{CP}^2\#\overline{\mathbb{CP}}^2, S^2 \times S^2\}.$$

This follows from the classification of trisections up to genus two [**GK16**, **MZ17b**] and the observation that if trisection genus is additive, homeomorphic smooth four-manifolds have the same trisection genus; see Proposition 1.7 of [**LCM22**], where modest evidence for this phenomenon is given; see also Remark 5 of Problem 4.113.

In particular, if trisection genus is additive, then any exotic copy of $\mathbb{CP}^2\#2\overline{\mathbb{CP}}^2$ admits a genus three trisection. Such exotic copies are known to exist [**AP10**, **FS11**]. This motivates the important problem of classifying genus three trisections; see Problem 4.113.

(5) There is an analogous problem for orientable surface-links in S^4 that is related to the *Meridional Rank Conjecture* for orientable surface-links [**JP25**]; see Problem 1.18. For more details and precise definitions, see [**MZ17a**] and [**AAD$^+$23**, Question 6.1]. Given an orientable surface-link $\mathcal{K} \subset S^4$, let $p(\mathcal{K})$ denote its *patch number*, the minimum value p such that $\mathcal{K}$ admits a $(b; p, p', p'')$–bridge trisection.

QUESTION. Is patch number (-1)–additive for orientable surface-links? In other words, must it be the case that

$$p(\mathcal{K}_1\#\mathcal{K}_2) = p(\mathcal{K}_1) + p(\mathcal{K}_2) - 1$$

for orientable surface-links $\mathcal{K}_1$ and $\mathcal{K}_2$?

Proposed for K3 and scribed by: J. Meier

PROBLEM 4.112. *Is every trisection of the 4-sphere with positive genus a stabilization of the genus zero trisection?*

REMARKS.

(1) This is Conjecture 3.11 of [**MSZ16**]. The analogous result regarding Heegaard splittings of the 3–sphere is *Waldhausen's Theorem* [**Wal68a**]. Connections between this problem, the *Generalized Property R Conjecture* (Problem 1.10), and the *Andrews–Curtis Conjecture* (Problem 5.10) are detailed in [**MZ18**] and elaborated in [**MZ22**]. In particular, if every positive-genus trisection of S^4 is stabilized, then well-known potential counter-examples (dating back to [**AK85**]) to the two conjectures mentioned above are not, in fact, counter-examples.

(2) The following related question may be more tractable and of independent interest [**MZ17a**]. It is a four-dimensional analog of the classical result of Otal that says the unknot has a unique bridge splitting for each bridge number [**Ota82**].

QUESTION. *Is every b–bridge trisection of the unknotted 2–sphere in S^4 with $b > 1$ a perturbation of the 1–bridge trisection?*

Proposed for K3 and scribed by: J. Meier

PROBLEM 4.113. *Which closed, oriented, smooth 4–manifolds admit genus–3 trisections? Which ones admit genus–3 simplified trisections? How about genus–4?*

REMARKS.

(1) There is a complete classification of trisections of genus $g \leqslant 2$ and only a handful of standard 4–manifolds admit them: S^4 for $g = 0$; $\mathbb{CP}^2$, $\overline{\mathbb{CP}}^2$, or $S^1 \times S^3$ for $g = 1$; and $S^2 \times S^2$, or connected sums of $\mathbb{CP}^2$, $\mathbb{CP}^2$ and $S^1 \times S^3$ with two summands, for $g = 2$ by [**GK16, MZ17b**].

(2) While the *simplified trisections* (see [**BS23b**] for a precise definition) constitute a subclass of Gay-Kirby trisections, the following question [**BS18c**, Question 2] is still open:

QUESTION. Is there a closed 4–manifold that admits a trisection, but not a simplified trisection of the same genus?

The classification of genus $g \leqslant 2$ simplified trisections coincide with that of standard trisections [**BS18c, Hay20**]. So, the genus $g = 3$ is the lowest genus where a discrepancy may appear.

(3) Some partial results are known: A trisection with complexity $(3; k_1, k_2, k_3)$ and $k_i \geqslant 2$ for some $i \in \mathbb{Z}_3$ is reducible [**MSZ16**]. Recall that a trisection that is the connected sum of smaller trisections is called *reducible*; otherwise, it is *irreducible*.

Infinitely many, pairwise homotopy inequivalent 4–manifolds admit genus–3 trisections; e.g. every spun lens space admits a (simplified) genus–3 trisection [**Mei18, BS18c**]. Similarly, there are genus–4 (simplified) trisections of 3–manifold bundles over S^1 with lens space or $S^1 \times S^2$ fibers [**Mei18, BS18c**].

(4) The classification of low genera trisections is intimately related to that of *simplified broken Lefschetz fibrations* [**BS23b, BS18c**]. The classification for the latter is complete for $g \leqslant 1$ and spans a larger class of 4–manifolds [**BK15a, Bay12a, Hay11, Hay14**].

(5) In the case of (broken) Lefschetz fibrations, the following exotic phenomenon already appears when we hit $g = 2$: there are pairs of homeomorphic but not diffeomorphic 4–manifolds that both admit genus–2 Lefschetz fibrations; for instance, let $X_i := E(1)_{K_i}$, $i = 1, 2$, be the knot surgered rational elliptic surface for K_1 a trefoil knot and K_2 the figure eight knot [**FS04, Bay09**]. There are analogous results for (simplified) trisections when $g = 20$ [**BS23b**] and many other, larger genera, starting at $g = 23$, where one of the exotic pairs is an algebraic surface [**ST18, LCM22**]. It is reasonable to expect this exotic phenomenon to manifest for much smaller g. (This is related to the question on the additivity of triseciton genus under connected sum; see Problem 4.111.)

QUESTION. What is the smallest g for which there is an exotic pair of closed, oriented 4–manifolds admitting genus–g trisections?

(6) While the problem is formulated only for orientable 4–manifolds, the classification of small-genera (simplified) trisections and (simplified) broken Lefschetz fibrations on *nonorientable* 4–manifolds is also within reach, and suggests analogous classification schemes; see [**MN24, ST22, BM25**].

Proposed for K3 and scribed by: I. Baykur, J. Meier

PROBLEM 4.114. *For a given Heegaard splitting of a closed orientable* 3*–manifold, classify self-indexing Morse functions that give the given Heegaard splitting, up to* C^∞ *right-left (or right) equivalence. Likewise, for a given (simplified) trisection of a closed orientable* 4*–manifold, classify generic maps that give the given trisection, up to right-left (or right) equivalence.*

REMARKS.

(1) Let $f_i\colon M_i \to N_i$ be C^∞ maps between smooth manifolds, $i = 0, 1$. We say that they are C^∞ *right-left equivalent* if there exist diffeomorphisms $\varphi\colon M_0 \to M_1$ and $\psi\colon N_0 \to N_1$ such that $f_1 = \psi \circ f_0 \circ \varphi^{-1}$. If $N_0 = N_1$ and ψ can be taken to be the identity, we say that f_0 and f_1 are C^∞ *right equivalent.*

$$\begin{array}{ccc} M_0 & \xrightarrow{f_0} & N_0 \\ \cong\downarrow & & \downarrow\cong \\ M_1 & \xrightarrow{f_1} & N_1 \end{array} \qquad\qquad \begin{array}{ccc} M_0 & \xrightarrow{f_0} & N_0 \\ \cong\downarrow & \nearrow_{f_1} & \\ M_1 & & \end{array}$$

FIGURE 5. Two commutative diagrams. Left: the maps f_0, f_1 are right-left equivalent. Right: the maps f_0, f_1 are right-equivalent.

It is known that to a self-indexing Morse function on a closed orientable 3–manifold is canonically associated a Heegaard splitting. Likewise, to a certain generic map, called *Morse 2–function*, on a closed orientable 4–manifold is canonically associated a trisection [**GK16**] or a simplified trisection [**BS23b, BS18c**].

(2) The 4–dimensional problem can be rephrased as follows. Let M be a closed orientable 4–manifold and let $f\colon M \to \mathbb{R}^2$ a C^∞ be a stable map that corresponds to a (simplified) trisection. If two such maps are C^∞ right-left equivalent, then the resulting trisections are naturally equivalent. Does the converse hold? Namely, if the resulting (simplified) trisections are equivalent, then are the original C^∞ stable maps C^∞ right-left equivalent? If not, how different are they?

Some related results are obtained in [**Hay20**] and [**Asa23**].

(3) It is known that Morse functions on closed surfaces can be classified by means of Kronrod-Reeb graphs up to C^∞ right-left (or right) equivalence (for example, see [**Mak05**]).

Proposed for K3 and scribed by: O. Saeki

PROBLEM 4.115.

(a) *Find two diffeomorphic but non-isotopic trisections of the same 4–manifold.*
(b) *Find two non-diffeomorphic balanced trisections of the same genus on a closed simply connected 4–manifold.*

REMARKS.

(1) Part of the difficulty of Problem (a) is that it is connected to the problem of understanding the smooth mapping class groups of 4–manifolds. In particular, if a 4–manifold X has trivial smooth mapping class group, so that all (orientation preserving) diffeomorphisms are isotopic, then all diffeomorphic trisections are obviously isotopic. On the other hand, 4–manifolds with nontrivial homology can have diffeomorphisms that act nontrivially on homology, and one might imagine that one could apply such a diffeomorphism to a given trisection to get a diffeomorphic but non-isotopic trisection. However this diffeomorphism might be isotopic to a diffeomorphism that is not the identity but that fixes the initial trisection (i.e. fixes the sectors setwise), in which case the new trisection would in fact be isotopic to the initial one.
(2) Problem (b) seems to be easier than Problem (a), with the first vague approximation being to come up with examples of trisections that look at first glance like they might be diffeomorphic but that in fact are not. Examples of non-diffeomorphic same-genus trisections of a 4-manifold were first constructed by Islambouli [**Isl21**]. However, Islambouli's techniques rely on the 4-manifold having nontrivial fundamental group G, as the trisections are distinguished by the Nielsen classes of associated presentations of G. This mirrors an argument in the 3–dimensional Heegaard splitting setting [**Eng70**]
(3) In principle, invariants of trisections up to diffeomorphism should be easier to find than invariants up to isotopy. Meier and Lambert-Cole construct various genus–22 trisections of K3 that may or may not be diffeomorphic and/or isotopic by viewing K3 as a branched cover in different ways [**LCM22**]. See Problem 4.112 (and examples of Meier and Zupan [**MZ18**]) for a related problem regarding the uniqueness of trisections of S^4.

Proposed for K3 by: D. Gay
Scribed by: D. Gay, J. Meier

PROBLEM 4.116. *Is there an algorithm to compute 'distance' in the cut complex of a trisection? Is the L-invariant computable?*

REMARKS.

(1) A Heegaard splitting for a 3-manifold M is defined by two "cut systems" of curves on a Heegaard surface Σ. Various notions of 'distance' (e.g. the *Hempel distance*) for Heegaard splittings have been extensively studied and yielded valuable geometric information about M. A trisected

smooth closed 4-manifold X is defined by three "cut systems" of curves on a surface Σ. One can define various analogous notions of 3-manifold Heegaard distance for a trisected X by measuring the length of a shortest loop in the cut complex (or pants complex) that includes a representative of each of the three specified cut systems. Given a specific trisection, the first question asks whether any such distance is computable. The *L-invariant* [**KT22a**] uses this idea in a limit to define a 4-manifold invariant.

(2) When L is zero and X is a homology sphere, X is diffeomorphic to the 4-sphere. Hence a positive answer to the second question is related to recognizability of the 4-sphere.
(3) In a preprint, Asano–Naoe–Ogawa [**ANO24**] gave a lower bound on the L-invariant of a 4-manifold X in terms of the first Betti number of X.

Proposed for K3 and scribed by: A. Thompson

PROBLEM 4.117. *Let X be a closed, orientable, smooth 4-manifold, with $g(X)$ the trisection genus of X. Does*

$$g(X) = \chi(X) - 2 + 3\mathrm{rk}(\pi_1(X))?$$

REMARKS.

(1) Chu and Tillmann proved that $\chi(X) - 2 + 3\,\mathrm{rk}(\pi_1(X))$ is a lower bound for $g(X)$ [**CT19**]. It is reasonable to expect there to exist manifolds for which this inequality is strict, but at present, Chu and Tillmann's result remains the only known lower bound on trisection genus. Following the classification of trisections up to genus two [**MZ17b**], if the problem has an affirmative answer, then there do not exist exotic copies of any X such that $g(X) \leqslant 2$, including S^4, $S^1 \times S^3$, $\mathbb{C}P^2$, and $S^2 \times S^2$. (See also Problem 4.111.)
(2) Like many problems in trisections, there is an analogous problem related to Heegaard splittings of 3-manifolds, the rank-genus conjecture. The now-disproved rank-genus conjecture asks whether $g(Y) = \mathrm{rk}(\pi_1(Y))$ for a closed, orientable 3-manifold Y. Originally posed by Waldhausen [**Wal78**], this conjecture was also called the *Generalized Poincaré Conjecture*, since it implies the 3-dimensional Poincaré conjecture [**Hak70**]. The first examples of a 3-manifold Y for which $g(Y) > \mathrm{rk}(\pi_1(Y))$ were exhibited by Boileau and Zieschang [**BZ84**], while the first hyperbolic counterexamples to the conjecture were produced by Li [**Li13**].
(3) A related problem for bridge trisections of a knotted surface $\mathcal{K} \subset S^4$ involves the *meridional rank* of the group $\pi_1(S^4 \backslash \mathcal{K})$, denoted $\mathrm{mrk}(\pi_1(S^4 \backslash \mathcal{K}))$, the smallest number of meridians needed to generate the group, and the bridge number $b(\mathcal{K})$, the minimal b such that $\mathcal{K}$ admits a b-bridge trisection.

QUESTION. Let $\mathcal{K}$ be an orientable knotted surface in S^4. Does

$$b(\mathcal{K}) = -\chi(\mathcal{K}) + 3\mathrm{mrk}(\pi_1(S^4 - \mathcal{K}))?$$

As in the case of trisections, the question is suggested by a straightforward (and the only currently known) lower bound. To derive the bound, note that if $\mathcal{K}$ admits a $(b; c_1, c_2, c_3)$-bridge trisection, then $\chi(\mathcal{K}) = c_1 + c_2 + c_3 - b$ and $\mathrm{mrk}(\pi_1(S^4 \backslash \mathcal{K})) \leqslant c_i$ for all i [**MZ17a**, Corollary 5.3]). Thus,

$$b = -\chi(\mathcal{K}) + c_1 + c_2 + c_3 \geqslant -\chi(\mathcal{K}) + 3\,\mathrm{mrk}(\pi_1(S^4 - \mathcal{K})).$$

This problem can be compared to the meridional rank conjecture ([**Kir97**, Problem 1.11] and Problem 1.18), an unsolved problem that asks whether the bridge number of a knot K in S^3 is equal to the minimal number of meridional generators required to generate its knot group. See also Problem 4.111.

(4) Meier and Zupan showed that if $\mathcal{K}$ is the spin of a classical knot K that satisfies the meridional rank conjecture in dimension three, then $\mathcal{K}$ satisfies the equality in the problem above [**MZ17a**]. Note that a positive answer to the problem would imply that any 2-sphere or torus in S^4 with infinite cyclic fundamental group has bridge number one or three, respectively, and as such is smoothly unknotted by the classification of b-bridge trisections with $b \leqslant 3$ [**MZ17a**]. Miyazawa's recent construction of an exotic $\mathbb{RP}^2$, announced in [**Miy23**], yields a negative answer to the problem in the case of nonorientable surfaces.

Proposed for K3 and scribed by: A. Zupan

4.11. Other structures on 4-manifolds

The problems in this final subsection ask about various structures on 4-manifolds not hitherto considered, including low-complexity handle decompositions, CW-structures, achiral pencils, open books, branched covers, foliations, hyperbolic structures, PSC metrics and Lipschitz structures.

PROBLEM 4.118. *Does every simply connected, closed, smooth 4-manifold admit a handle decomposition without any 1-handles? Without 1-handles and 3-handles?*

REMARKS.

(1) A manifold is called *geometrically simply connected* if it admits a handle decomposition without 1-handles. Any simply connected closed manifold of any dimension other than four is geometrically simply connected; this follows from the celebrated works of Smale in higher dimensions [**Sma62b**] and of Perelman in dimension three [**Per02, Per03b, Per03a**].

(2) This is [**Kir97**, Problem 4.18]. Some candidates for counter-examples, such as the Dolgachev surface $E(1)_{2,3}$, have since been shown to admit handle decompositions without any 1- and 3-handles [**Yas08, Akb12**].

(3) If one allows the 4-manifold to have boundary, the answer is negative. Indeed, there are many contractible 4-manifolds that require 1-handles, by the following argument due to Casson: If a compact, contractible 4–manifold X can be built without 1-handles, then, turning the handlebody

upside down, we can also build X from ∂X by adding the same number of 1- and 2–handles and a 4-handle. So $\pi_1(\partial X)$ can be killed by adding the same number of generators and relators. However, a finitely presented group with a nontrivial representation to a compact connected Lie group cannot be trivialized by adding the same number of generators and relators by [**GR62**] and there are contractible X where $\pi_1(\partial X)$ is such a group. In fact, geometrization now implies all nontrivial 3-manifold groups admit nontrivial finite quotients by [**Hem87**], so any contractible 4-manifold with boundary other than S^3 requires 1-handles, such as the Mazur manifold with boundary $\Sigma(2,3,13)$ [**AK79b**].

(4) If a geometrically simply connected closed 4-manifold X has $b_2^+(X) > 1$ and $b_2^-(X) = 0$, then all the stable cohomotopy Seiberg-Witten invariants of X vanish [**Yas19**]; so, e.g. it cannot admit a symplectic structure—see also [**HL19**].

    In the same paper, Yasui shows that if X is geometrically simply connected, then every $\alpha \in H_2(X;\mathbb{Z})$ has a neighborhood W diffeomorphic to a 2-handle attached to a 4-ball, where α is the image of the generator of $H_2(W;\mathbb{Z}) \cong \mathbb{Z}$ under the inclusion induced homomorphism. Any simply connected X with some $\alpha \in H_2(X;\mathbb{Z})$ that does not admit such a neighborhood would be a counter-example.

(5) Admitting a handle decomposition without 1- and 3-handles has strong implications. For instance, if this is true for every homotopy S^4 or homotopy $\mathbb{CP}^2$, then there are no exotic copies of these 4-manifolds. The conclusion for homotopy $\mathbb{CP}^2$s follows from [**GL89**].

Scribed by: I. Baykur, M. Powell

PROBLEM 4.119. *Is every topological 4-manifold homeomorphic to a CW complex?*

REMARKS.

(1) It follows from the work of Kirby and Siebenmann [**KS77**] that every topological manifold M has the homotopy type of a CW complex, and moreover there is a canonical simple homotopy type of CW complexes homotopy equivalent to M.
(2) Every smooth manifold is triangulable and hence homeomorphic to a CW complex. See [**Cai35**], [**Whi40**]. In particular, non-compact 4-manifolds are smoothable [**Qui82**]; therefore, the question has a positive answer for those.
(3) Every topological manifold of dimension $n \neq 4$ has a handlebody structure, and hence is homeomorphic to a CW complex. See [**Moi77a**] for $n = 3$, [**KS77**, p.104] for $n \geqslant 6$ and [**Qui86**] for $n = 5$.
(4) Any 4-manifold with a handlebody structure is smooth. Indeed, when a handle is attached to a smooth 4-manifold, the attaching map is in dimension three, where everything is smoothable.

(5) Topological, non-smoothable 4-manifolds (such as the E_8 manifold) are known to not be homeomorphic to simplicial complexes. See [**AM90**] and [**Man13**, Remark 4.2].

Scribed by: C. Manolescu

PROBLEM 4.120. *Which closed, smooth 4–manifolds admit achiral Lefschetz pencils? Does every simply connected 4–manifold have one?*

REMARKS.

(1) Achiral Lefschetz pencils are generalizations of Lefschetz pencils, where the local models for nodal singularities and base points are allowed to reverse orientations. So, they are also defined on nonorientable 4–manifolds. Here we allow achiral pencils to possibly have no critical points and/or no base points.
(2) There are only a couple of known obstructions to the existence of an achiral Lefschetz pencil on a given closed, oriented 4–manifold X; see [**GS99**, Theorem 8.4.13] and [**Sco03**, Theorem 4.15]. These obstructions rule out definite 4–manifolds with $b_2 + 1 < b_1$, such as $\#^m (S^1 \times S^3)$, for $m \geqslant 2$. A curious question is: *Are there homotopy equivalent smooth 4–manifolds X and X', where X admits an achiral pencil but X' does not?*
(3) Any closed, orientable X admits an achiral Lefschetz fibration (without base points) after surgery along a curve, and specifically, $X\#(S^2 \times S^2)$ always admits one when X is simply connected [**EF06**].
(4) Just like Lefschetz pencils can be equipped with certain symplectic forms making all the fibers symplectic, achiral Lefschetz pencils can be equipped with certain *folded-symplectic* forms. Furthermore, a variation of achiral Lefschetz pencils, defined in the complement of a 1–manifold, support *folded-Kähler* forms on all closed, oriented, smooth 4–manifolds. See [**Bay06, Hit16**].

Proposed for K3 and scribed by: I. Baykur

PROBLEM 4.121. *Which closed, smooth 4–manifolds admit open book decompositions? In particular, does every closed, simply connected 4–manifold with signature zero admit one?*

REMARKS.

(1) An *open book decomposition* of an n-dimensional manifold X is given by a smooth fibration $f\colon X\backslash L \to S^1$, where the *binding* $L \subset X$ is an $(n-2)$–dimensional embedded submanifold with a trivial normal bundle, and there is a neighborhood $N(L) \cong L \times D^2$ such that f conforms to the local model $f(x,(r,\theta)) = \theta$ for $r \neq 0$, where $x \in L$ and $(r,\theta) \in D^2$ are the polar coordinates.

(2) If a closed, oriented n–dimensional manifold X admits an open book, its signature vanishes. An extension of this necessary condition is the vanishing of the *asymmetric signature*; see [**Ran98**] for a definition and related discussion. Vanishing signature is also a sufficient condition for
   - any odd $n \geqslant 3$, by the works of Alexander, Lawson and Quinn [**Law78, Qui79**];
   - any even $n \geqslant 6$, when X is simply connected, by Winkelnkemper [**Win73**] for $n \geqslant 8$, and for $n = 6$ by Quinn [**Qui79**].

(3) Kastenholz [**Kas25**] claims that the simplicial volume vanishes for 4-manifolds that admit an open book, and gives examples of non-simply-connected 4-manifolds with vanishing asymmetric signature that do not admit open book decompositions.

(4) Open books on 4–manifolds are related to Engel structures. If a closed, oriented 4–manifold X admits an open book with a binding that is a link of tori and a monodromy that preserves a framing on the fiber, then X admits an Engel structure [**CPV18**].

(5) It would be interesting to see if there are *smooth* obstructions to admitting open books in dimension four. If X_1 and X_2 admit open book decompositions, then so does $X_1 \# X_2$. There are natural open books on any Σ_g–bundle over S^2, where the binding is a pair of fibers. Combining these general constructions, one gets an open book on every $\#^m\,(S^2 \times S^2)$ and $\#^n(\mathbb{C}\mathrm{P}^2 \# \overline{\mathbb{C}\mathrm{P}}^2)$. Do their exotic copies always admit open books?

Scribed by: I. Baykur

PROBLEM 4.122. *Is there a universal branching surface $S \subseteq S^4$ such that every closed, orientable 4-manifold W admits a branched covering $W \to S^4$ with branching locus S?*

REMARKS.

(1) This question originated in the introduction of [**PZ05**] by Piergallini–Zuddas.

(2) Using the signature, it can be shown that a universal branching surface in S^4 is necessarily disconnected [**Vir84, IP02**].

(3) There exists an orientable ribbon surface $F \subseteq B^4$, consisting of the disjoint union of one annulus and two discs, such that every compact orientable 4-manifold M constructed by adding only 1-handles and 2-handles to B^4 admits a branched covering $M \to B^4$ with branching locus F [**PZ05**].

   QUESTION ([**PZ05**, Question 2]). Does there exist a *connected* universal branching surface in B^4?

(4) A link $L \subseteq S^3$ is said to be *a universal branching link* if every closed, orientable 3-manifold Y admits a branched covering $Y \to S^3$ with branching locus L. The figure eight knot, the 9_{46} knot, the Whitehead link, the Borromean rings, and various other knots and links are known to be universal branching links [**HLM83a, HLM83b, HLM85**]. The first universal branching link was found by Thurston in unpublished work.

QUESTION ([**PZ05**, Question 4]). Which universal links in S^3 are boundaries of universal surfaces in B^4?

Scribed by: A. Ray

PROBLEM 4.123. *Is every closed leaf of a two dimensional co-orientable smooth taut foliation of an oriented 4-manifold smoothly genus-minimizing in its homology class?*

REMARKS.

(1) This question is due to Kronheimer [**Kro98**, Question 7.12], and is the natural generalization of the corresponding three-dimensional result of Thurston [**Thu86**]. Here we say that the foliation $\mathcal{F}$ is *taut* if there exists a 2-form ω such that:
  - for every leaf L, $\omega|_L$ is an area form;
  - for every $v_1, v_2 \in T_{\mathcal{F}}$ and $z \in T_M$, $d\omega(v_1, v_2, z) = 0$.

  This directly generalizes the definition in dimension three (in which case the second condition simply says that ω is closed). As Kronheimer points out, the question is also interesting when one allows foliations with singularities with a suitable local model, e.g. the foliations defined by the vanishing of a holomorphic 1-form.
(2) If the foliation is calibrated to a symplectic form, then every closed leaf is a symplectic subsurface and hence genus-minimizing by the symplectic Thom conjecture [**OS00**].
(3) If "taut," is eliminated from the hypotheses, then the answer is "no." Constructions of non-genus-minimizing compact leaves of coorientable foliations of 4-manifolds were constructed by Mitsumatsu–Vogt [**MV08**] and Bowden [**Bow11**].
(4) The following specific subquestion would be an interesting first step: *Must a compact leaf of a coorientable smooth taut foliation of $S^2 \times S^2$ be a 2-sphere?*

Proposed for K3 by: P. Kronheimer
Scribed by: F. Lin

PROBLEM 4.124. *Does there exist a hyperbolic integer homology four-sphere? What about an arithmetic one? Homology four-spheres have Euler characteristic 2, so it makes sense to ask more generally if there exist any closed hyperbolic four-manifold with Euler characteristic 2.*

REMARKS.

(1) If any closed, hyperbolic manifold with Euler characteristic 2 exists, there are only finitely many. This is because the Gauss-Bonnet Theorem says that $Vol(M) = \frac{4}{3}\pi^2\chi(M)$ for a hyperbolic four-manifold M, and there are only finitely many manifolds with volume below any fixed value. Note that in general, the Euler characteristic of a closed, orientable hyperbolic

four-manifold is always even, since such manifolds have zero signature (see [**LR00**]), so these would be the ones with smallest volume. The best known example seems to be the Conder-Maclachlan manifold with Euler characteristic 16 [**CM05**]. For comparison, there are one-cusped orientable hyperbolic four-manifolds with Euler characteristic one [**RT00**].

(2) It is worth pointing out that there are various constructions of aspherical 4-manifolds with Euler characteristic 2. For example, Luo constructed an aspherical rational homology 4-sphere [**Luo88**] and Tschantz constructed aspherical integer homology 4-spheres [**RT05**] (answering [**Kir97**, Problem 4.17]). The latter examples even admit metrics of non-positive curvature.

(3) If there exists an arithmetic hyperbolic integer homology sphere (or, more generally, an arithmetic, closed, hyperbolic manifold with Euler characteristic 2), then by Belolipetsky [**Bel07**, Theorem 5.5'] (see also [**Bel04**, Theorem 5.5]) it has to be an index-28800 cover of the 4-dimensional hyperbolic reflection group with the below Coxeter diagram.

Proposed for K3 by: A. Reid
Scribed by: T. Lidman

PROBLEM 4.125. *Is there a noncompact, finite volume, orientable hyperbolic four-manifold without a spin structure?*

REMARKS. All compact orientable manifolds with dimension at most three admit spin structures. Sullivan observed that every finite volume hyperbolic n-manifold has a finite cover that admits a spin structure [**Sul79b**, p.533]. Reid and Long showed in [**LR20**] that there are orientable, finite volume, non-compact hyperbolic n-manifolds with $n \geqslant 5$ that do not admit spin structures. Martelli-Riolo-Slavich showed that there are closed orientable hyperbolic four-manifolds that do not admit spin structures [**MRS20**].

Proposed for K3 by: A. Reid
Scribed by: T. Lidman

PROBLEM 4.126.

(a) *If M is a closed, orientable hyperbolic 4-manifold then it always has signature 0, because its Pontryagin class vanishes* [**Che55**]. *This implies that M bounds a compact, orientable 5-manifold. Is M always a geometric boundary?*

(b) *For general n, suppose that M is a closed hyperbolic n-manifold. Is M a geometric boundary?*

REMARKS.

(1) Following [**LR00**], we say that M^n is a *geometric boundary* if it is the totally geodesic boundary of a compact hyperbolic manifold W^{n+1}.

(2) Perhaps a good example to start with for (a) is the Davis manifold [**Dav85**].

(3) Every closed, orientable surface of genus at least 2 has a hyperbolic metric that is the totally geodesic boundary of a compact, orientable, hyperbolic 3-manifold by [**Fuj90**]. However, some closed, hyperbolic, orientable 3-manifolds are not the geodesic boundary of any compact, orientable, hyperbolic 4-manifold [**LR00**]. It is still open if there is a hyperbolic rational homology 3-sphere that is the totally geodesic boundary of a compact, orientable, hyperbolic 4-manifold (see Problem 3.77).

(4) We restrict to the setting of orientable 4-manifolds as a closed, nonorientable, hyperbolic 4-manifold may have odd Euler characteristic. In this case, the 4-manifold cannot bound any compact 5-manifold, without mention of geometry. Even in the setting of nonorientable 4-manifolds, it may be interesting to ask this question with the additional hypothesis that the nonorientable 4-manifold does bound some compact (nonorientable) 5-manifold.

(5) For (b), the general problem can be posed either in the orientable or nonorientable setting; one might assume that M is null-bordant to start with. In the orientable case, Long and Reid [**LR00**] observe that when $n = 4k - 1$, the η-invariant [**APS75a**] of M would have to be integral. They give examples of orientable hyperbolic 3-manifolds with non-integral η-invariant, which are therefore not geometric boundaries in the oriented category. The oriented version in higher dimensions could similarly be answered by finding hyperbolic $(4k - 1)$-manifolds with non-integral η-invariant for $k > 1$. Some constructions of higher-dimensional orientable hyperbolic manifolds that are geometric boundaries are given in [**LR01**].

(6) In the nonorientable case, J. Chen [**Che25**] claims to construct, in all dimensions $n \geqslant 4$ not of the form $n = 4k - 1$, examples of nonorientable closed hyperbolic n-manifolds that are not the boundary of any compact $(n+1)$-manifold (not assuming any geometric condition). The question of whether there are nonorientable hyperbolic manifolds that are boundaries but not geometric boundaries remains open.

Proposed for K3 by: A. Reid
Scribed by: T. Lidman

PROBLEM 4.127. *Given an aspherical closed (or compact and bounded by flat 3-manifolds) 4-manifold M and a self-diffeomorphism f of M, find necessary and sufficient conditions on f so that the resulting 5-dimensional mapping torus M_f admits a hyperbolic structure.*

REMARKS.

(1) By hyperbolization, in dimension two the sufficient and necessary condition is that f is pseudo-Anosov (see Thurston [**Thu82, Thu22**] and Otal [**Ota01**] for a complete proof). There are some analogues in strictly higher dimensions, e.g. [**Far72**, Theorem 6.4] gives necessary and sufficient conditions for a manifold of dimension at least six to fiber over S^1. If such a theorem worked in dimension 5, then one could potentially

check such a condition against hyperbolic 5-manifolds. However, similar higher-dimensional techniques have not yet been successfully applied in the context of 5-dimensional hyperbolic manifolds.

(2) Italiano–Martelli–Migliorini recently found examples of (M, f) with M 4-dimensional that produce a hyperbolic 5-manifold M_f [**IMM23**].

Proposed for K3 and scribed by: B. Martelli

PROBLEM 4.128. *What is the structure of* 4*-manifolds that admit a Riemannian metric of positive scalar curvature? There are variations of this problem for different classes of manifolds.*

(a) *Is every closed simply connected PSC* 4*-manifold diffeomorphic to a connected sum of copies of* $\mathbb{C}\mathrm{P}^2$, $\overline{\mathbb{C}\mathrm{P}}^2$, *and* $S^2 \times S^2$*?*
(b) *What is the structure of non-simply connected closed PSC* 4*-manifolds?*
(c) *Which* 4*-manifolds with boundary have a PSC metric?*
(d) *Which non-compact* 4*-manifolds have a complete PSC metric with uniformly positive scalar curvature?*

REMARKS.

(1) Let us say that a smooth manifold is a *PSC manifold* if it admits a Riemannian metric of positive scalar curvature.
(2) The corresponding problem in dimension 2 is easy by the Gauss-Bonnet theorem, and is solved in dimension 3 as a consequence of Perelman's work. In particular, every orientable closed PSC 3-manifold is a connected sum of spherical space forms and copies of $S^1 \times S^2$. In dimensions at least 5 the problem is completely solved for simply connected manifolds via index theory and surgery theory, and there is a well-developed obstruction theory in the non-simply connected case; see the survey articles [**Ros07b**, **RS01**]. The classical Lichnerowicz obstruction [**Lic63**] states that a spin PSC $4k$-manifold has vanishing $\widehat{A}$-genus; in dimension 4 this is equivalent to having vanishing signature.

   Witten [**Wit94**] showed that PSC 4-manifolds with $b_2^+ > 1$ have vanishing Seiberg-Witten invariants. This implies, for instance, that the existence of a PSC metric depends on the underlying smooth structure. In the non-simply connected case, there are further obstructions based on Rokhlin's theorem [**RS07**] and gauge theory [**Lin19**, **Kon19**, **KT20**, **KT23**].
(3) It is conjectured that the answer to (a) is positive. This seems wildly optimistic, but there are no known counterexamples. A weaker version would allow an exotic S^4 with a PSC metric as a summand in the connected sum decomposition. The conjecture was stated as a question in [**Kir97**, Problem 4.143].
(4) Some standard examples of non-simply connected PSC 4-manifolds are $S^1 \times Y$ for Y a spherical space form, as well as $S^2 \times \Sigma$ and $\mathbb{R}\mathrm{P}^2 \times \Sigma$ for any closed surface Σ. Some other constructions are described in the survey [**MT21**]. The decomposition theorem for PSC 4-manifolds in [**BLM23**] reproduces some portion of the picture in dimension 3. Problem 4.129 has a discussion of a decomposition question for non-simply

connected PSC 4-manifolds, which would reduce the general problem to the classification of PSC 4-dimensional orbifolds with finite orbifold fundamental group.

(5) One standard setting for part (c) requires that the metric be a product near the boundary, in which case the boundary would have positive scalar curvature. There are obstructions to the existence of PSC metrics on bounding 4-manifolds coming from index theory [**BG95**] and Seiberg-Witten theory. Formulating a good conjecture here would be welcome.

Rosenberg-Weinberger [**RW23**] discuss a different boundary condition, and conjecture that a manifold has a PSC metric whose boundary has positive mean curvature (with respect to the outward normal) if and only if its double has a PSC metric.

(6) By [**Ros07b**, Theorem 0.1] every non-compact 4-manifold admits a PSC metric, typically not complete, so one needs to have additional constraints on the geometry at infinity. Using Gromov's notion of a *μ-bubble*, Chodosh-Maximo-Mukherjee [**CMM24**] show that there are exotic $\mathbb{R}^4$s that do not admit a complete PSC metric with uniformly positive scalar curvature. A particular case of interest, asked by A. Mukherjee, is whether the punctured $K3$ surface admits such a metric.

Scribed by: D. Ruberman

PROBLEM 4.129. *Given a closed, 4-dimensional PSC manifold M, is there a (possibly disconnected) 4-dimensional orbifold M' with isolated singularities such that the following hold.*

(I) *The orbifold M' also admits a metric of positive scalar curvature.*
(II) *All components of M' have finite orbifold-fundamental group.*
(III) *M can be obtained from M' by a series of 0 and 1 surgeries. Here we also allow 0-surgeries between two orbifold points of the same type, which amounts to a removal of two subsets of the form D^4/Γ and an addition of a copy of $S^3/\Gamma \times [0,1]$.*

REMARKS.

(1) Because the orbifold M' has isolated singularities, its orbifold fundamental group is just the fundamental group of its regular part.
(2) The following converse statement is true due to the work of Gromov–Lawson [**GL80**]. If M' satisfies Property (I) and M can be obtained from M' as in Property (III), then M admits a PSC metric. So if the answer to the problem is 'yes', then this would reduce the study of PSC 4-manifolds to the study of PSC 4-orbifolds with finite fundamental group. See problem 4.128 for a conjectural picture in the simply connected case.
(3) The examples of non-simply connected closed 4-manifolds admitting a PSC metric described in Question (b) of Problem 4.128 satisfy properties (I)–(III). For example, a bundle over S^1 with fiber a PSC 3-manifold can be obtained from the unreduced suspension of Y by 0-surgery at the two orbifold points. Likewise, $S^2 \times \Sigma_g$ can be obtained from a connected sum of $2g$ copies of $S^1 \times S^3$ by surgery on a circle.

(4) It seems likely that the problem can be approached using 4-dimensional Ricci flow once there is a reasonable construction of Ricci flow with surgery in this dimension. Here the 0 and 1 surgeries would correspond to geometric surgeries that excise cylindrical singularities. Other singularities, for example those modeled on non-cylindrical shrinking solitons, would contribute components to M' with finite fundamental group [**Bam21a**]. Partial progress to the problem was made via minimal surfaces in [**BLM23**], where Property (II) was proved with the weaker conclusion that every component of M' has vanishing first Betti number.

(5) Properties (I)–(III) impose nontrivial topological restrictions on M. For example, they imply that M cannot be aspherical; note, however, that non-asphericity was already shown in [**CL24**]. More generally, Properties (I)–(III) imply that any cover of M must have finite 2-dimensional Urysohn width [**Gro88**] (for the lift of one and thus any Riemannian metric on M). Here we say that a metric space (X, d) has k-dimensional Urysohn width of at most W if there is a continuous map $p : X \to A$ to a k-dimensional simplicial complex such that the diameter of any fiber $p^{-1}(a)$ is at most W. The property that any cover of a has finite 2-dimensional Urysohn width imposes a restriction on homotopy type of the manifold. See also [**CLL23, LM23a, Bol09**] for related results.

Proposed for K3 by: R. Bamler
Scribed by: D. Ruberman

PROBLEM 4.130. *Does longitudinal knot surgery using a knot K, along a fiber in a K3 surface always yield a reducible 4–manifold? A completely decomposable 4–manifold?*

REMARKS.

(1) *Longitudinal knot surgery* is a variation of Fintushel-Stern knot surgery [**FS98**] with a different gluing map. The original Fintushel-Stern version is defined using a knot K and a square-0 torus T in a 4-manifold X. Then X_K is defined as

$$(X - T \times D^2) \cup_\varphi (S^1 \times E(K)).$$

Here $E(K)$ is the exterior of the knot, and the gluing map φ sends the longitude of K to the boundary of a meridional disk of T. Under appropriate hypotheses, Fintushel and Stern show that this operation multiplies the Seiberg-Witten invariant of X by the Alexander polynomial of K. In particular, if the Seiberg-Witten invariant of X is nontrivial, the same holds for X_K.

(2) In longitudinal knot surgery, the S^1 factor in $S^1 \times E(K)$ is identified with the boundary of the meridional disk of T. Denote the result by X_K^λ. Taubes shows [**Tau16**] that in contrast to the standard surgery, the Seiberg-Witten invariant of X_K^λ vanishes, even if K is nontrivial. This raises the question of whether X_K^λ splits as a connected sum, or is completely decomposable, i.e. is diffeomorphic to a connected sum

$\#^m\mathbb{C}\mathrm{P}^2 \# \#^n\overline{\mathbb{C}\mathrm{P}}^2$. Taubes reports, based on communications with Akbulut, Baykur, and Fintushel, that when K is an unknot, then $(K3)^\lambda_K$ completely decomposes.

(3) One motivation comes from the search (starting with [**Poo86**], albeit with opposite orientation conventions) for 4-manifolds that admit a Riemannian metric whose self-dual Weyl curvature W_+ vanishes. Such metrics are called conformally anti-self-dual. Taubes [**Tau16**] shows that when K is hyperbolic, X is a $K3$ surface, and T is a fiber in an elliptic fibration coming from the Kummer construction, the manifold X^λ_K admits a Riemannian metric with W_+ arbitrarily small. After repeated blowing up by connected sum with $\overline{\mathbb{C}\mathrm{P}}^2$, it will have [**Tau92**] a conformally anti-self-dual metric; for 2–bridge knots, three blowups suffice. Hence it would be of interest to identify the manifold $(K3)^\lambda_K$. If K is hyperbolic and $(K3)^\lambda_K$ is completely decomposable, then $\#^3\mathbb{C}\mathrm{P}^2 \#^{19} \overline{\mathbb{C}\mathrm{P}}^2$ would admit a conformally anti-self-dual metric. Such metrics are known on $\#^3\mathbb{C}\mathrm{P}^2\#^N\overline{\mathbb{C}\mathrm{P}}^2$ when $N \geqslant 30$ by work of LeBrun [**LeB95**] and Rollin-Singer [**RS09**].

Proposed for K3 by: C. Taubes
Scribed by: D. Ruberman

PROBLEM 4.131. *Does every Lipschitz 4-manifold admit a smooth structure? Is this smooth structure unique if so?*

Some more specific, related questions are as follows.

(a) *Is there a topological, spin, closed, indefinite 4-manifold X that admits a Lipschitz structure and violates the 10/8-inequality*

$$b_2(X) \geqslant \frac{5}{4}|\sigma(X)| + 2$$

by Furuta [**Fur01**], *or the "10/8 + 4" inequality*

$$b_2(X) \geqslant \frac{5}{4}|\sigma(X)| + 4$$

by Hopkins–Lin–Shi–Xu [**HLSX22**] *for $X \neq S^4, S^2 \times S^2, K3$.*

(b) *Let $X \subset \mathbb{R}^4$ be a Lipschitz embedded 4-manifold with boundary $\partial X \simeq S^3$. Does X admit a unique smooth structure?*

REMARKS.

(1) In dimension $n \neq 4$, every topological n-manifold X admits a Lipschitz structure by a theorem of Sullivan [**Sul79b**]. On the other hand, Donaldson and Sullivan [**DS89**, Theorem 2] showed that there exist closed topological 4-manifolds X that admit more than one inequivalent Lipschitz structure. Donaldson and Sullivan established that the simplest numerical invariants of smooth 4-manifolds (due to Kotschick [**Kot89**]) are quasiconformal invariants of smooth 4-manifolds.

(2) The inequalities in Problem (a) are proved for smooth 4-manifolds using the variation of Seiberg–Witten theory introduced in [**Fur01**]. Hence a Lipschitz manifold violating either of those inequalities would suggest that there is no extension of this version (or perhaps other versions) of

Seiberg–Witten theory to the setting of Lipschitz 4-manifolds. One reason to suspect that there is no such extension is that the Seiberg–Witten equations on X are defined in terms of the Dirac operator associated to a Spin^c structure on X. Sullivan has conjectured [**Sul99, Sul95**] that the existence of a Dirac operator on X (as part of a full 'Dirac package') implies that X is in fact smoothable.

(3) The 4-dimensional Schoenflies conjecture (Problem 4.23) is known [**Sul79b**] to hold for Lipschitz embeddings $\phi\colon S^3 \to \mathbb{R}^4$. Thus a positive solution to Problem (b) would imply the Schoenflies conjecture.

Proposed for K3 by: J. Chaidez, D. Gabai, H. Konno
Scribed by: J. Chaidez

CHAPTER 5

Miscellany

Chapter 5 Editors: Robion Kirby and Daniel Ruberman

This short last chapter contains problems about manifolds of any dimension that interact with low-dimensional topology and geometry. We have not sought out problems on complex surfaces, although some interesting problems in this area, described in the corresponding chapter of [**Kir97**], remain unsolved.

Section 5.1 begins with two problems from the days of "analysis situs", one on the fixed point property of cellular sets and one on convergence of homeomorphisms. This latter problem is connected to the Hilbert–Smith problem and the question of whether homeomorphism groups are ANRs. We then continue in this topological vein with questions about higher dimensional pairwise connected sums, branched covers, aspherical 5-manifolds, and the classic Montgomery-Yang problem about exotic circle actions on the 5-sphere.

These are followed in Section 5.2 by three hallowed conjectures of Andrews–Curtis, Whitehead and Zeeman from [**Kir97**] (see also [**HAMS93, HHAMR20**]) relating to 2- and 3-complexes and their elementary expansions and collapses. It concludes with a problem about triangulability of aspherical 5-manifolds, and a wide-ranging problem investigating differential topology through the lens of singularity theory.

Section 5.3 contains problems of a more geometric character. This includes two questions asking about the extension to higher dimensions of known results about hyperbolic 3-manifolds: the existence of one-cusped manifolds, and whether Thurston's 'virtual' conjectures hold in higher dimensions. This section concludes with the 4 and 5-dimensional versions of the Milnor conjecture about the fundamental group of a complete Riemannian manifolds of non-negative Ricci curvature.

The final section, 5.4 has three questions related to symplectic and contact geometry: the first about the differential graded algebras that appear in contact geometry, followed by a problem about homotopies of Weinstein structures. The final problem asks about higher-dimensional analogues of lattice cohomology [**Ném05, Ném08**], currently defined for certain 3-manifolds.

5.1. Topology problems

In this section, we consider a number of problems relating to the basic topology of manifolds and their continuous automorphisms. The Hilbert–Smith Conjecture, with roots going back to Hilbert's Fifth Problem, is perhaps the most famous of these. It seeks to characterize the (locally compact) topological groups acting effectively on a topological manifold, and is related to other fundamental problems about homeomorphism groups.

Other basic problems have to do with the well-definedness of pairwise connected sum in the topological category, and the question (inspired by old work of Alexander) of whether every n-manifold is an n-fold branched cover of the n-sphere. We also discuss the Montgomery–Yang problem about circle actions on the 5-sphere, which was the inspiration for much early work on the homology cobordism classification of Brieskorn homology spheres, and address the last case for finding aspherical nontriangulable manifolds. Finally, we pose several questions about smooth maps between smooth manifolds where the maps have singularities with various restrictions on their types.

PROBLEM 5.1. *Does every cellular set in the plane have the fixed point property?*

REMARKS.

(1) A space has the fixed point property (FPP) if every self-map has a fixed point. A set in an n-manifold is cellular if it is the intersection of a countable sequence of embedded closed n-cells, each contained in the interior of the preceding one. A subset of the plane is cellular if and only if it is non-empty, compact, connected and non-separating (meaning that its complement is connected). Since a (planar) cellular set is the intersection of cells, and since the FPP holds for cells, it is natural to ask if it also has the FPP.
(2) The Mandelbrot set [**BM81**, **Man80**] in the plane is cellular, but it is unknown whether it has the FPP.
(3) The answer in dimensions$\geqslant 3$ is no; a counterexample is due to Kinoshita [**Kin53**]. See the expository article by Bing [**Bin69**].

Proposed for K3 by: R. Edwards
Scribed by: R. Kirby, D. Ruberman

PROBLEM 5.2 (Doubly-Small Morphisms of Manifolds).

(a) *Suppose that $h\colon \mathbb{R}^n \to \mathbb{R}^n$ is a homeomorphism (or diffeomorphism) that satisfies two smallness hypotheses:*
   (i) *every orbit of h (of any $x \in \mathbb{R}^n$, under all powers of h) is uniformly bounded in diameter (by 1 say), and*
   (ii) *some subsequence of powers of h converges to $\mathrm{Id}_{\mathbb{R}^n}$ in (say) the compact-open topology.*

   Then must h be the identity map?
(b) *A special case of this question is: let h be a homeomorphism of B^n that is the identity on ∂B^n. If there is a subsequence of powers of h that converge to the identity, must h itself be the identity?*

REMARKS.

(1) The case $n = 1$ is trivial, $n = 2$ seems likely to be true, as a consequence of results of Brouwer and Cartwright-Littlewood (nicely and succinctly reproved in [**Bro84**] and [**Bro77**]), and $n \geqslant 3$ is open.

(2) The question is meant to be local in nature, i.e. the question can be adapted to any open subset of $\mathbb{R}^n$. It also applies to any manifold, where in Condition (i) one would assume that every orbit of h has diameter less than some ϵ in the compact-open topology.
(3) The answer is 'yes' if h is periodic. This is Newman's Theorem, with an excellent exposition in [**Dre69**].
(4) For h a diffeomorphism, and using C^∞ convergence, the answer seems likely to be yes.
(5) Since a homeomorphism $h\colon \mathbb{R}^n \to \mathbb{R}^n$ generates a homomorphism $\phi\colon \mathbb{Z} \to \text{Homeo}(\mathbb{R}^n)$ (and vice-versa) by $m \mapsto h^m$, the Question can be rephrased in terms of such a ϕ. Condition (i) becomes: Assume that image(ϕ) lies in a suitably small neighborhood of $\text{Id}_{\mathbb{R}^n}$, and Condition (ii) becomes: Assume that ϕ accumulates at $\text{Id}_{\mathbb{R}^n}$. And the Question becomes: Must ϕ be the trivial homomorphism?
(6) The question is 'stronger' than the Hilbert–Smith Conjecture, discussed in Problem 5.3 below. That is, an affirmative answer to it would imply the Hilbert–Smith Conjecture. The Hilbert–Smith Conjecture (in its Question form) is equivalent to the Question above if in addition one assumes that the closure of the union of the powers of h in $\text{Homeo}(\mathbb{R}^n)$ is compact.

Proposed for K3 by: R. Edwards
Scribed by: R. Edwards, R. Kirby, D. Ruberman

PROBLEM 5.3 (Hilbert–Smith Conjecture).

(a) *The Hilbert–Smith Conjecture* [**Smi41**] *asserts that a locally compact subgroup of the homeomorphism group of a connected manifold must be a Lie group.*
(b) CONJECTURE: *The free-action subset of any Cantor group action on an ENR is a homology-Z-subset (of the ENR).*

REMARKS.

(1) We first define the terms in Conjecture (b).
 (i) A *Cantor group* is a topological group G whose underlying space is a Cantor set (space). That is, G is a profinite group that is non-finite and 2nd countable (hence metrizable). A universal example of such a group is the direct product of all (of the countably many) finite groups (or, if you wish, the finite symmetry groups). Other important examples of Cantor groups are the p-adic integers.
 (ii) An *ENR* is a Euclidean Neighborhood Retract, that is, a subset of some $\mathbb{R}^n$ that has a neighborhood that retracts onto it. Such subsets are characterized as stably having manifold mapping cylinder neighborhoods (like tubular neighborhoods for manifolds, or regular neighborhoods for polyhedra).
 (iii) A subset A of X is a *homology-Z-set* (in X) if for any $a \in A$ and any open neighborhood U of a in X, the relative homology $H_*(U, U - A)$ is 0. If X is a manifold, the only such A are subsets of ∂X.

(2) Pardon [**Par13a**] proved the Hilbert–Smith Conjecture for dimension 3. The conjecture has been reduced to whether the locally compact subgroup in question can be the p-adic integers. See Pardon's papers [**Par13a, Par19**] for more discussion and additional references.
(3) Conjecture (b) is known as the *Free-Set (is a) Z-Set Conjecture* (FSZSC) and is due to R. Edwards. It is a natural and stronger version of the Hilbert–Smith Conjecture. As a special case, the FSZSC asserts that a Cantor group cannot act freely on an ENR. Like the HSC, the FSZSC has been reduced to the case of proving it for the p-adic integers.
(4) The Hilbert–Smith Conjecture is closely related to the preceding problem 5.2, as discussed above.

Proposed for K3 by: R. Edwards
Scribed by: R. Edwards, R. Kirby, D. Ruberman

PROBLEM 5.4. *Is the homeomorphism group of a manifold an absolute neighborhood retract (ANR)?*

REMARKS.

(1) This is [**Kir97**, Problem 5.27] and is also listed in [**Wes90**]. In dimension 2 it was shown to hold by Mason [**Mas71**] and Luke-Mason [**LM72**]; it is unknown for manifolds of dimension greater than 2.
(2) It is not easy to distinguish between ANRs and more general locally contractible spaces, so it is worth recalling that the homeomorphism group of a manifold is locally contractible [**EK71, Čer69**].

Proposed for K3 by: R. Edwards
Scribed by: R. Kirby, D. Ruberman

PROBLEM 5.5. *Is the connected sum of (locally flat) pairs*

$$(M_1^{n+k}, N_1^n)\#(M_2^{n+k}, N_2^n)$$

well-defined in the topological category?

REMARKS. When $k = 1$, this is easily true from Brown's paper showing that locally flat codimension one submanifolds are flat [**Bro62**]. According to [**Liv24**], it is well-defined in any dimension when $k = 2$. This uses the uniqueness of normal bundles in codimension 2, which fails in higher codimensions; see [**Liv24**, Appendix C]. However this question appears to be open for codimension > 2 and $n + k > 4$. It would follow from a pairwise version of the 'torus trick' [**KS77**] if a pairwise version of Wall's non-simply connected surgery [**Wal99**] were known and in print.

Proposed for K3 and scribed by: C. Livingston

PROBLEM 5.6.

(a) *Does every closed, PL, orientable n-manifold admit an n-fold branched covering map over S^n?*

Assuming that the answer to this problem is "yes", the following follow-up questions would be natural to ask.

(b) *Does every n-manifold admit an n-fold branched covering over S^n with branch locus a codimension-2 embedded submanifold?*
(c) *If the answer to* (b) *is "yes", then one could naturally ask if the cover can additionally be taken to be* simple, *meaning the covering monodromy sends every meridian of the embedded submanifold to a transposition. For $n = 2, 3, 4$, the above papers show the answer is, "yes".*
(d) *If the answer to* (b) *is "yes", then in a separate direction one could ask when the branch locus can be taken to be orientable.*

REMARKS.

(1) A classical theorem of Alexander [**Ale20**] says that every n-manifold admits a branched covering map over S^n with no restriction on degree of the covering. An easier version of this problem would be to show that for every n, there exists some natural number m_n such that every n-manifold admits an m_n-fold branched covering over S^n. Certainly m_n cannot be smaller than n: for example, because the n-torus T^n has reduced cohomology ring of length n, any branched covering $f : T^n \to S^n$ has degree at least n [**BE78**].
(2) It is well-known that every orientable surface is a 2-fold branched cover over S^2. Hilden [**Hil74**], Hirsch [**Hir74**], and Montesinos [**Mon74**] showed that every 3-manifold is a 3-fold branched cover over S^3. Piergallini [**Pie95**] showed that every PL 4-manifold is a 4-fold branched cover over S^4. The answer to this question is unknown in all higher dimensions.
(3) Note by Berstein–Edmonds [**BE78**] the answer to Question (b) is "no" if we additionally require that this submanifold be locally flat. For $n = 2, 3$ the answer is known to be "yes," but for $n = 4$, the best known result to date, by Piergallini in [**Pie95**], produces branched loci that are immersed surfaces. Iori–Piergallini [**IP02**] later showed that every 4-manifold admits a 5-fold simple branched covering over S^4 with branch locus an embedded surface. Whether every 4-manifold admits a 4-fold cover over S^4 with embedded branch locus remains open.
(4) The answer to Question (d) is generally negative – for example, when $n = 4k$ the existence of such a covering implies the n-manifold has signature zero [**Vir84**]. For this question, it may be simpler to restrict to the case that the n-manifold is null-cobordant.

Scribed by: M. Miller

PROBLEM 5.7 (Montgomery–Yang problem). *Does there exist a pseudo-free, smooth, S^1 action on S^5 with more than three multiple orbits?*

REMARKS.

(1) This is Problem 4.123 in [**Kir97**]; see the surveys by Kollár [**Kol08**] and Şavk [**Şav24**]. Recall that an S^1 action is pseudo-free if there are no fixed points and the orbits of finite isotropy are isolated.
(2) There is a good understanding of the situation for other odd dimensional spheres: S^1 actions on S^3 are linear by Seifert [**Sei33**], whereas every homotopy $2k-1$ sphere (for $k \geqslant 4$) admits pseudo-free S^1 actions with arbitrarily many multiple orbits by Montgomery-Yang [**MY72**] (for $k=4$) and Petrie [**Pet75**] (for $k>4$).
(3) The answer is positive if a Seifert fibered homology 3–sphere Σ with more than 3 multiple fibers bounds an acyclic 4-manifold W, where the induced homomorphism by the inclusion of boundary is surjective on the π_1. Then $\Sigma \times D^2 \cup W \times S^1 = S^5$ and S^1 acts diagonally on $\Sigma \times B^2$ and trivially on W. See also Problem 4.58.

Proposed for K3 by: R. Fintushel, R. Stern
Scribed by: I. Baykur

PROBLEM 5.8. *Is there a closed aspherical* 5*-manifold that is not triangulable?*

REMARKS.

(1) In this context, 'triangulable' means homeomorphic to a simplicial complex. Davis and Januszkiewicz [**DJ91**] constructed aspherical non-triangulable 4-manifolds, by applying the process of 'hyperbolization' to Freedman's E_8 manifold. Subsequent to Manolescu's (negative) solution to the triangulation conjecture [**Man16b**] in dimensions $\geqslant 5$, Davis-Fowler-Lafont [**DFL14**] used a hyperbolization procedure to construct aspherical non-triangulable n-manifolds for all $n \geqslant 6$. They explain why their procedure breaks down in dimension 5, and explicitly asked [**DFL14**, §3] about the 5-dimensional case.
(2) It follows from [**Sie70**] (using the double suspension theorem [**Edw06**, **Can79**]; compare [**GS80**, **Mat78**]) that any orientable 5-manifold is triangulable. Hence any aspherical non-triangulable 5-manifold would have a triangulable cover. This suggests the following.

    QUESTION. Are there examples of non-triangulable aspherical 4-manifolds in dimension 4 that are virtually triangulable, i.e., which admit a finite cover that can be triangulated?

    The examples of [**DJ91**, **DFL14**] have residually finite fundamental groups, so they would be a good place to start.

Scribed by: D. Ruberman

PROBLEM 5.9. *Let M_1 and M_2 be smooth manifolds of dimension n. Suppose M_1 admits an $\mathcal{S}$-map into $\mathbb{R}^p$. If M_2 is homeomorphic to M_1, then does M_2 admit an $\mathcal{S}$-map as well?*

REMARKS.

(1) For this problem, we consider C^∞ maps $f\colon M \to \mathbb{R}^p$ of smooth n–dimensional manifolds into $\mathbb{R}^p$ with $n \geqslant p \geqslant 1$. Let $\mathcal{S}$ be a set of (equivalence classes of) singularities of smooth map germs $(\mathbb{R}^n, 0) \to (\mathbb{R}^p, 0)$. A smooth map $f\colon M \to \mathbb{R}^p$ is called an *$\mathcal{S}$-map* if all its singularities are equivalent to a singularity belonging to $\mathcal{S}$.

(2) If M_2 is diffeomorphic to M_1, then by composing a diffeomorphism $M_2 \to M_1$ and an $\mathcal{S}$-map $M_1 \to \mathbb{R}^p$, we get an $\mathcal{S}$-map on M_2. Therefore, if the answer to the above question is negative, then M_2 is not diffeomorphic to M_1; in other words, (M_1, M_2) is an exotic pair of manifolds.

   In this sense, for $n \leqslant 3$, the answer is always affirmative.

(3) Let $\mathcal{S}$ be the singleton consisting of the definite fold singularity represented by the map germ

$$(x_1, x_2, \ldots, x_{p-1}, x_p^2 + x_{p+1}^2 + \cdots + x_n^2).$$

   In this case, an $\mathcal{S}$-map is called a *special generic map*. Then there are many examples of manifold pairs (M_1, M_2) for which the answer is negative as follows.

   (i) For $n \geqslant 7$ and $p = n-1, n-2$ and $n-3$, the pair of the standard n–sphere and an exotic n–sphere is such an example [**Sae93b**].

   (ii) For $n = 4$ and $p = 1, 2$ and 3, the pair of the standard $\mathbb{R}^4$ and an exotic $\mathbb{R}^4$ is such an example, provided that we consider *proper* special generic maps [**Sae10**].

   (iii) For $n = 4$ and $p = 3$, there are quite a few such examples (M_1, M_2), where M_1 is a connected sum of $S^2 \times S^2$ and $\mathbb{C}\mathrm{P}^2 \sharp \overline{\mathbb{C}\mathrm{P}}^2$ [**Sae93a**, **SS99**].

(4) According to the solution to the Poincaré Conjecture in high dimensions due to Smale [**Sma61**], for special generic functions with $p = 1$, the answer is affirmative for $n \geqslant 5$. In other words, only by the existence of a special generic function with $n \geqslant 5$ and $p = 1$ one cannot detect exotic differentiable structures. Note also that for $n = 4$ and $p = 1$, the problem for special generic functions is equivalent to the 4-dimensional smooth Poincaré Conjecture.

   On the other hand, for $p = 1$, let $\mathcal{S}$ be the set of non-degenerate critical points of indices in the set $\{0, 2, 3, \ldots, n-2, n\}$. Then, for $n \geqslant 5$, by Smale's result [**Sma62a**], the answer to the problem is always affirmative, whereas for $n = 4$, the problem seems to be still open.

(5) Let us consider C^∞ stable maps of closed 4–manifolds into $\mathbb{R}^4$. They have, in general, fold, cusp, swallowtail, butterfly and umbilic singularities. It is known that when the 4–manifold is oriented, each umbilic singularity can be given a sign $+1$ or -1 and the total number of umbilic points counted with signs is equal to 3 times the signature. Furthermore, if the signature vanishes, then all the umbilic singularities can be eliminated by homotopy

[**And82, Sti95**]. Hence, for $\mathcal{S}$ consisting of fold, cusp, swallowtail and butterfly singularities, the answer to the above question is affirmative in this case.

(6) Let F be a set of (equivalence classes) of singular fibers of C^∞ maps in the following sense [**Sae04**]. Let $f_i : M_i \to N_i$ be smooth maps, $i = 0, 1$. For $y_i \in N_i$, we say that the fibers over y_0 and y_1 are equivalent if for some open neighborhoods U_i of y_i there exist diffeomorphisms $\tilde{\varphi}\colon f_0^{-1}(U_0) \to f_1^{-1}(U_1)$ and $\varphi\colon U_0 \to U_1$ with $\varphi(y_0) = y_1$, which make the following diagram commutative:

$$\begin{array}{ccc} (f_0^{-1}(U_0), f_0^{-1}(y_0)) & \xrightarrow{\tilde{\varphi}} & (f_1^{-1}(U_1), f_1^{-1}(y_1)) \\ {\scriptstyle f_0}\big\downarrow & & \big\downarrow{\scriptstyle f_1} \\ (U_0, y_0) & \xrightarrow[\varphi]{\qquad\qquad} & (U_1, y_1) \end{array}$$

When the fibers over y_0 and y_1 are equivalent, we also say that for $i = 0, 1$, the map germs $f_i : \big(M_i, f_i^{-1}(y_i)\big) \to (N_i, y_i)$ are right-left equivalent. A smooth map $f : M \to \mathbb{R}^p$ is called an F-map if all its singular fibers are equivalent to a singular fiber belonging to F. Then, we can ask the same question as Problem 5.9 for F-maps.

(7) It is known that for maps of smooth closed oriented 4–manifolds into $\mathbb{R}^3$, each singular fiber of type III^8 can be given a sign $+1$ or -1, and for a certain class of generic maps (so-called C^∞ stable maps), the number of III^8–fibers counted with signs coincides with the signature of the source 4–manifold [**SY06**]. Therefore, it is a natural question if singular fibers of type III^8 with opposite signs can be eliminated by homotopy. So far, we do not know if there exists a smooth closed oriented 4–manifold M with zero signature such that an arbitrary C^∞ stable map $M \to \mathbb{R}^3$ necessarily has a singular fiber of type III^8. Though for connected sums of copies of $S^2 \times S^2$ and $\mathbb{CP}^2 \sharp \overline{\mathbb{CP}}^2$, the answer is known to be affirmative.

(8) In the problem, one can also consider the problem by replacing "homeomorphic to" by "homotopy equivalent to". For example, for $n \geqslant p \geqslant 1$, let $\mathcal{S}$ be the set of fold singularities of all (absolute) indices. Then, the answer to the homotopy version of the problem is affirmative for closed orientable 4–manifolds and $1 \leqslant p \leqslant 4$ [**Sae03, Sad04**].

Proposed for K3 and scribed by: O. Saeki

5.2. Andrews–Curtis, Whitehead, and Zeeman Conjectures

The conjectures and questions discussed in this section touch on the interaction between group theory and the topology of low-dimensional cell complexes. The Andrews–Curtis Conjecture posits that a certain collection of algebraic operations suffices for passing between finite presentations of the trivial group; it has intriguing relations to the 4-dimensional Poincaré Conjecture. Whitehead's Asphericity Question is disarmingly simple, and asks if subcomplexes of an aspherical 2-complex are always aspherical. A positive answer would imply, among other things, that ribbon disk complements are aspherical. Zeeman's Conjecture posits that a (finite) contractible 2-complex '3-collapses', i.e., collapses to a point after taking the product with the interval. It is closely connected to the Andrews–Curtis conjecture, and to older approaches to the Poincaré Conjecture.

PROBLEM 5.10. *The Andrews–Curtis Conjecture* [**AC65**] *for the trivial group: a presentation of the trivial group can be changed to the trivial presentation by Andrews-Curtis moves.*

REMARKS.

(1) This problem appears in Problem 5.2 in [**Kir97**].
(2) Let $\mathcal{P}$ be a finite presentation of a given group π. The stabilized - Andrews–Curtis moves (abbreviated AC moves) change the presentation $\mathcal{P} = \{x_1, \ldots, x_n : R_1, \ldots, R_m\}$ as follows:
 (i) $R_i \to R_i^{-1}$,
 (ii) $R_i \to R_i R_j, i \neq j$
 (iii) $R_i \to w R_i w^{-1}, w$ any word,
 (iv) add generator x_{n+1} and relation $w x_{n+1}$.
 The AC Conjecture for the trivial group states that the presentation $\mathcal{P}$ can be changed to the trivial presentation $(x_1, \ldots, x_n : x_1, \ldots, x_n)$ by AC moves.
 Note that redundant relations cannot be added, so that $m - n$ is unchanged. Also note that the broader conjecture that any two presentations of an arbitrary finitely presented group are equivalent by AC moves is false for some nontrivial groups, e.g., the trefoil group [**HAMS93**]. (See also [**AC66**].)
(3) Given a presentation $\mathcal{P}$ of the trivial group, we can construct a 5-dimensional handlebody Y from a 0-handle, 1-handles for each generator, and 2-handles for each relation; Y is unique because the attaching maps are isotopic because they are homotopic.
 Then if the AC Conjecture is true, Y is diffeomorphic to the 5-ball, because the AC moves correspond to handle moves (in particular the move $R_i \to R_i R_j, i \neq j$ corresponds to sliding the ith handle over the jth handle).

Scribed by: R. Kirby

PROBLEM 5.11 (Whitehead's Asphericity Question). *Is every subcomplex of an aspherical 2-complex aspherical?*

REMARKS.

(1) This was problem 5.4 in [**Kir97**]; a thorough discussion of older progress on this problem, also known as the Whitehead Conjecture, may be found in [**HAMS93**, Chapter X]. It was stated as a question on page 428 in [**Whi41**], with the implicit assumption that the complex is finite, but the question still makes sense without that assumption.

(2) The fundamental group of a finite-dimensional aspherical complex is torsion free, so an a priori easier question is whether the fundamental group of a subcomplex of an aspherical two-complex is necessarily torsion-free. For partial results relating to fundamental group structure, see [**Coc54, Ada55, How79**].

(3) Bestvina and Brady [**BB97**] showed that the Whitehead conjecture and the Eilenberg–Ganea conjecture cannot both be true. The Eilenberg–Ganea conjecture [**EG57**] is that a group with cohomological dimension 2 has a 2-dimensional Eilenberg-Mac Lane space.

More concretely, Bestvina and Brady showed that the (Bestvina–Brady) group H_L associated with a flag triangulation L of a spine of the Poincaré homology sphere is either a counterexample to the Eilenberg–Ganea conjecture, or there exists a contractible 2-complex Y that contains a non-aspherical subcomplex [**BB97**, Theorem 8.7]. Note that the potential counterexample Y to the Whitehead conjecture that they construct is infinite (since the group H_L acts freely and cellularly on it).

(4) In [**How79**], J. Howie reduced the problem to finding counterexamples $K \subset L$ of two types: (i) L is finite and contractible and $K = L - e$ for some 2-cell e; or (ii) L is the union of an infinite chain of non-aspherical subcomplexes $K = K_0 \subset K_1 \subset \ldots$ such that each inclusion is null-homotopic.

According to Howie's results in [**How79**] and [**How83**], if the Andrews–Curtis conjecture (Problem 5.10) holds, then the standard 2-complexes associated with LOT presentations account for all test cases of type (i). Recall that a version of the Andrews–Curtis conjecture asserts that every finite contractible 2-complex can be 3-deformed to a point. A LOT presentation is a group presentation described by a (finite) labeled oriented tree, and the associated 2-complex has the homotopy type of a ribbon disc complement [**How83**].

It remains unknown whether all LOT complexes are aspherical, though significant progress has been made in the last decades, establishing the asphericity of various subfamilies. For instance, Harlander and Rosebrock [**HR17**] showed that alternating ribbon disk-complements, the ones that can be encoded by injective labeled oriented trees, are aspherical.

(5) Costa and Farber [**CF17**] give a model for random simplicial complexes in which aspherical 2-complexes satisfy the Whitehead conjecture with probability 1.

Scribed by: R. Kirby

PROBLEM 5.12 (Zeeman Conjecture). *If K is a finite contractible 2-complex, then $K \times I$ collapses to a point* [**Zee64**, Conjecture (1)].

REMARKS. This problem appears as part of Problem 5.2 in [**Kir97**].

The Zeeman Conjecture for a special polyhedron that is the spine of a compact 3-manifold is equivalent [**GR83**] to the Poincaré Conjecture, and thus true by Perelman. The Zeeman conjecture for special polyhedra that do not embed in compact 3-manifolds is equivalent to the Andrews–Curtis Conjecture [**Mat87**]. A nice discussion of this general area may be found in Chapters I, XI and XII of [**HAMS93**] and the exposition in [**Kup21**].

Scribed by: R. Kirby

5.3. Geometry

The first two problems in this section are concerned with hyperbolic manifolds of dimension greater than 3. Thurston's 'virtual' conjectures (see [**Ago08**] and Problem 3.29) asking about properties of finite covers of hyperbolic 3-manifolds, have been highly influential in the study of 3-manifolds in recent years, and it is natural to ask the corresponding questions for hyperbolic manifolds of any dimensions. Similarly, there are many explicit examples of one-cusped hyperbolic 3-manifolds, and we ask for such examples in higher dimensions.

Milnor's conjecture, discussed in the next problem, proposed that the fundamental group of a manifold admitting a complete Riemannian metric of nonnegative Ricci curvature is finitely generated. It holds for 2- and 3-manifolds, and there are counterexamples in every dimension starting with 6. So we ask about the status of this conjecture in the remaining dimensions 4 and 5.

PROBLEM 5.13. *Let M be a finite-volume hyperbolic n-manifold.*

(a) *Does it always have a finite cover with $b_1 > 0$?*
(b) *Does it always have a finite cover with fundamental group that surjects onto a non-abelian free group?*
(c) *Does it have a finite cover with cubulated fundamental group?*
(d) *When n is odd, does it always have a finite cover that fibers over the circle?*

REMARKS.

(1) When $n = 3$, these questions all have positive answers, by work of Agol [**Ago13**] and Wise [**Wis21**]. However, their methods break down when $n > 3$. The crucial steps in their argument are as follows. By work of Kahn-Markovic [**KM12**], $\pi_1(M^3)$ contains lots of surface subgroups. These are 'codimension 1' subgroups and hence $\pi_1(M^3)$ has a finite index subgroup that is the fundamental group of a compact non-positively curved cube complex, i.e., it is *cubulated*. Agol showed that such cubulated groups are virtually special, and Wise showed that virtually special groups have many excellent virtual properties. In particular, they have a finite index subgroup with $b_1 > 0$ and indeed a finite index subgroup that surjects onto a non-abelian free group. Using a 3-dimensional argument involving sutured manifolds, Agol [**Ago08**] was able to show that hyperbolic 3-manifolds virtually fiber. An alternative and more general argument using group rings was given by Kielak [**Kie20**].
(2) This argument fails at the first step when $n > 3$. The methods of Kahn-Markovic have been extended to all odd dimensions by Hamenstädt [**Ham15**], and hence $\pi_1(M)$ is known to contain many surface subgroups. However, these groups are not codimension one, and therefore do not

establish that the group is cubulated. However, it is known that some hyperbolic n-manifolds are cubulated when $n > 3$. Indeed, any arithmetic hyperbolic n-manifold containing a totally geodesic $(n-1)$-dimensional (possibly immersed) submanifold is cubulated. When n is even, this includes all arithmetic hyperbolic n-manifolds.

(3) There are further results for arithmetic hyperbolic n-manifolds. There are 3 types of arithmetic lattices in $\mathrm{SO}(n,1)$: those arising from quadratic forms (type I), those arising from quaternion algebras (type II), and those arising from octonion algebras, and type III - trialitairian lattices. The first type appears in all dimensions and are known to be cubulated after work of Bergeron–Wise [**BW12**]. The second type are not known to be cubulated, but only appear in even dimensions. The third type only occurs in dimension 7; for these it is known [**BC17a**] that the first Betti number of every congruence subgroup equals 0. Moreover, it is still open if the lattices have the congruence subgroup property.

(4) If a manifold fibers over the circles, its Euler characteristic is zero. In even dimensions, the volume of a hyperbolic manifold is proportional to its Euler characteristic, and hence its Euler characteristic is necessarily non-zero. This explains the restriction to n odd in the final question above. The first examples of hyperbolic 5-manifolds that fiber over the circle were given by Italiano-Martelli-Migliorini [**IMM23**].

(5) When a hyperbolic 3-manifold fibers over the circle, the universal cover $\widetilde{F}$ of the fiber has a circle at infinity, and the inclusion $\widetilde{F} \to \mathbb{H}^3$ is known to extend continuously to a map from the circle at infinity of $\widetilde{F}$ to the sphere at infinity of $\mathbb{H}^3$. This forms a space-filling curve. This was established by Cannon and Thurston [**CT07**]. It would be interesting to know whether there is any analogue of this in higher dimensions. The fiber will not in general have word hyperbolic fundamental group, and so part of the challenge would be to define its space at infinity appropriately.

(6) By work of Delzant and Gromov [**DG05**], complex hyperbolic manifolds do *not* have cubulated fundamental group.

Proposed for K3 and scribed by: M. Lackenby

PROBLEM 5.14. *Does there exist a 1-cusped finite-volume hyperbolic n-manifold for any $n \geqslant 5$?*

REMARKS.

(1) Hyperbolic n-manifolds with one cusp are abundant when $n = 2$ or 3, and Kolpakov–Martelli [**KM13a**] constructed 1-cusped hyperbolic 4-manifolds. However, there are far fewer constructions of hyperbolic n-manifolds when $n \geqslant 4$ and they tend to produce manifolds with a large number of cusps. There are no known examples in dimensions $\geqslant 5$ with a single cusp; the question of their existence was raised in [**LR02**].

(2) Stover [**Sto13**] proved that there are no 1-cusped arithmetic hyperbolic n-dimensional orbifolds when $n > 30$. In fact, he showed that for each $m \geqslant 1$, there is a $c_m \geqslant 1$ such that there are no m-cusped arithmetic hyperbolic n-dimensional orbifolds when $n > c_m$.

Proposed for K3 by: A. Reid
Scribed by: M. Lackenby

PROBLEM 5.15. *Suppose M is a manifold with a complete Riemannian metric with nonnegative Ricci curvature. Is the fundamental group of M finitely generated?*

REMARKS.

(1) This is a conjecture of Milnor, following on his paper [**Mil68a**]. The conjecture was known to hold in dimension 2 by work of Cohn-Vossen [**CV35**] and in dimension 3 by work of Liu [**Liu13**]. Bruè-Naber-Semola construct counterexamples in dimensions at least 7 [**BNS25**], and 6 [**BNS23**]. The issue is therefore to determine the status of the conjecture in dimensions 4 and 5. This question is explicitly posed in [**BNS23**, Question 1.1], along with some other interesting open problems in the area.

Previous work of Wilking [**Wil00**] shows that the fundamental group of any counterexample could be assumed to be abelian; the fundamental group of the example from [**BNS25**] in dimension 7 is $\mathbb{Q}/\mathbb{Z}$.

(2) A crucial point in the Bruè-Naber-Semola paper [**BNS25**, Lemma 9.1] is that the orbit of the mapping class group of $S^3 \times S^3$ acting on the standard product metric lies in a single path component of the space of positive Ricci curvature metrics on $S^3 \times S^3$. This raises several questions.

(i) What can be said about the action of the mapping class group of $S^2 \times S^2$ on the path components of the space of positive Ricci curvature metrics on $S^2 \times S^2$?
(ii) Is the space of positive Ricci curvature metrics on $S^2 \times S^2$ path connected?
(iii) One could ask the same questions for $S^2 \times S^3$; see [**BNS23**, §6] for a particular diffeomorphism of $S^2 \times S^3$ such that the pullback of the standard metric is connected to the standard metric through Ricci curvature metrics.

Proposed for K3 and scribed by: D. Ruberman

5.4. Symplectic and contact topology

These last three problems deal with aspects of symplectic and contact topology in dimensions greater than 4 (respectively, 3).

PROBLEM 5.16. *Let R be $\mathbb{Z}$ or a field. Let A and B be differential graded algebras so that either:*

- *As a graded algebra, A (respectively B) is isomorphic to the free, non-commutative R-algebra on finitely many homogeneous generators x_i, i.e., to the tensor algebra on the free R-module generated by $x_1, \ldots, x_n$.*
- *As a graded algebra, A (respectively B) is isomorphic to a free graded-commutative algebra on finitely many homogeneous generators x_i. (So, if the x_i have even gradings, A is a polynomial algebra.)*

(Different generators can have different gradings, A and B may have different numbers of generators, and gradings of generators may be negative.)

Are the following questions decidable?

(a) *Is A stable tame isomorphic to B?*
(b) *Is A quasi-isomorphic to B?*
(c) *Is A derived Morita equivalent to B?*

REMARKS.

(1) This question is inspired by contact topology, where many invariants take the form of a finitely-generated differential graded algebras up to a notion of equivalence.
(2) The kinds of dgas described in the problem are often called *semifree* in the literature. The information in a semifree dga A consists of the grading of the generators $x_1, \ldots, x_n$ and the elements $d(x_i)$, which are either polynomials in $x_1, \ldots, x_n$ (in the commutative case) or linear combinations of words in $x_1, \ldots, x_n$ (in the non-commutative case).
(3) A dga homomorphism is a *quasi-isomorphism* if it induces an isomorphism on homology; two dgas are *quasi-isomorphic* if there is a dga C and quasi-isomorphism $C \to A$ and $C \to B$. *Stable tame isomorphism* of semifree dgas was introduced in [**Che02**] (see also [**ENS02**, Section 3.3] and [**EN22**, Section 3.2]); it allows introducing a pair of canceling generators (stabilizing) and isomorphisms sending some x_i to x_i plus a word in the other variables. Two dgas are *derived Morita equivalent* if the derived categories of differential modules over them are equivalent (as triangulated categories). Stable tame isomorphism implies quasi-isomorphism implies derived Morita equivalence.
(4) This question arises from contact homology and related invariants. For example, to a Legendrian knot Λ in $\mathbb{R}^3$, one can associate a dga $\mathcal{A}_\Lambda$, the *Legendrian contact dga*, whose stable tame isomorphism class is an isotopy invariant of Λ (see the citations above). The Legendrian contact dga can also be defined for knots in other manifolds and higher-dimensional Legendrian knots. Contact homology can also be defined for (certain) closed contact manifolds; in this case, one is forced to work with a commutative dga. In practice, it seems to be hard to tell whether two dgas are stable tame isomorphic or quasi-isomorphic.
(5) One strategy that has been developed to distinguish dgas is to study the set of *augmentations* of A, i.e., dga maps $A \to R$ (where R lies in grading 0 and has trivial differential); this set is called the *augmentation*

variety of A. Given an augmentation of A, one can form the linearized homology with respect to that augmentation, which is a finitely generated R-module [**Che02**].

(6) For Legendrian knots in $\mathbb{R}^3$, the set of augmentations form the objects of a category, the *augmentation category*, which in the case of Legendrian contact homology is equivalent to a certain category of sheaves [**NRS+20**]. The set of augmentations and the linearized contact homology depend only on the abelianization of the dga, but the augmentation category needs the non-commutative version.

(7) Contact homology can also be applied to study smooth objects in low-dimensional topology, for instance by considering the unit cotangent bundle or unit conormal bundle. An instance is Ng's knot contact homology [**Ng05**] (a variant of which is a complete knot invariant [**ENS18**]); another is given in Problem 4.106. It is possible that a result along these lines could also be applied to decision problems in symplectic or smooth topology.

(8) In some of the applications, one actually works over $\mathbb{Z}\{t, t^{-1}, x_1, \dots, x_n\}$, say, but this seems unlikely to affect the question.

(9) In the special case that A and B are commutative with all of their generators in positive gradings, Sullivan's theory of minimal models gives an algorithm for answering the question; in particular, that case is well studied in rational homotopy theory. Note also that there is a unique augmentation in this case; in particular, the dgas arising from contact topology rarely have this property.

(10) Given that the question has some similarity to Hilbert's tenth problem, the answer might be different for R a finite field from $R = \mathbb{Q}$ or $\mathbb{Z}$, say.

Proposed for K3 and scribed by: R. Lipshitz

PROBLEM 5.17. *Let (W, ω, V_i, ϕ_i) be two Weinstein structures on a fixed symplectic manifold (W, ω) (or equivalently consider two Weinstein handle decompositions). Is there a Weinstein homotopy from (W, ω, V_1, ϕ_1) to (W, ω, V_2, ϕ_2)?*

REMARKS.

(1) The convex combination $V_t = (1-t)V_2 + tV_2$ is a Liouville vector field positively transverse to the boundary, but it may fail to be gradient-like for some t. The question is whether we can find a family of convex Liouville vector fields V_t which are all gradient-like.

(2) There are various additional hypotheses one can add that will ensure a Weinstein homotopy exists. For example, by [**CE12**, Proposition 11.22], if we have a fixed complex Stein structure J on (W, ω), any two J-convex (plurisubharmonic) functions ϕ_1 and ϕ_2 will yield Weinstein homotopic Weinstein structures. There are numerous other specific Weinstein homotopies constructed in [**CE12**, Chapter 12], that assume some specific properties or estimates about the Liouville form and/or gradient-like function.

For Weinstein manifolds of dimension strictly greater than 4, there is a notion of "flexible Weinstein manifolds." In this case any two flexible Weinstein structures are Weinstein homotopic [**CE12**, Theorem 14.5].

(3) A special case, which may be easier, is the following. Suppose (W,ω) admits a symplectomorphism that restricts to a contactomorphism on the boundary, but that is not isotopic to the identity in the class of symplectomorphisms that are contact on the boundary. Let λ be one Liouville form on (W,ω), corresponding to a Liouville vector field V_1 that is gradient-like for some Morse function ϕ_1. Consider the Liouville form $f^*\lambda$ and denote its corresponding Liouville vector field by V_2. Then V_2 is gradient-like for $\phi_2 = f^*\phi_1$. We can ask if (W,ω,V_1,ϕ_1) Weinstein homotopic to (W,ω,V_2,ϕ_2) in this case.

(4) It may be helpful to think about the Liouville form λ_i rather than the Liouville vector field V_i (which are related by $\iota_{V_i}\omega = \lambda_i$). While λ_i is not closed (since $d\lambda_i = \omega$ by the Liouville condition), $\lambda_1 - \lambda_2$ is closed and thus represents an element of $H^1(W)$. It may be easier to construct a counterexample for manifolds where $H^1(W)$ is nontrivial, in which case, one should refine the question to consider manifolds where $H^1(W) = 0$, or further that W is simply connected.

(5) This question is open in arbitrary dimensions > 2, but dimension 4 could be a good place to start.

Proposed for K3 by: Y. Eliashberg
Scribed by: L. Starkston

PROBLEM 5.18.

(a) *In higher dimensions, find the 'non-analytic' cohomology module* $\mathbb{H}^*_?$ *analogous to the analytic lattice cohomology* $\mathbb{H}^*_{an}$, *and a natural functor* $\mathbb{H}^*_{an} \to \mathbb{H}^*_?$ *connecting them.*

(b) *Define a version ECH* of Embedded Contact Homology for isolated complex singularity links (in any dimension) associated with the canonical contact structure of the link, together with a natural graded* $\mathbb{Z}[U]$*–module morphism* $\mathbb{H}^*_{an} \to \mathrm{ECH}^*$. *More precisely,*
   (i) *Show that this morphism is injective.*
   (ii) *Fix the diffeomorphism type of a link, and also a contact structure on it that can be realized as the canonical contact structure associated with some singularity analytic structure. Then characterize the family* $\{\mathbb{H}^*_{an}\}$, *indexed by all the possible analytic germs inducing this contact link, via those graded sub–*$\mathbb{Z}[U]$*–modules of* ECH^* *that satisfy certain specific properties.*

REMARKS.

(1) As background, consider a complex normal surface singularity with a rational homology sphere link. The (topological) lattice cohomology $\mathbb{H}^*_{top}$, associated with the link (which is a plumbed 3–manifold associated with a connected negative definite graph), was introduced by Némethi in [**Ném05, Ném08**]. It has an analytic analogue $\mathbb{H}^*_{an}$ constructed recently in [**ÁN21a, ÁN21b**]. Both theories are multigraded, $\mathbb{H}^*_{top} = \oplus_{q\geqslant 0}\mathbb{H}^q_{top}$,

and $\mathbb{H}^q_{top}$ is a $2\mathbb{Z}$–graded $\mathbb{Z}[U]$–module, with an additional grading indexed by the spin^c structures of the link. The same is valid for $\mathbb{H}^*_{an}$ as well. Even more, the existence of a graded $\mathbb{Z}[U]$–module morphism $\mathbb{H}^*_{an} \to \mathbb{H}^*_{top}$ was also established.

Following on calculations in [**OS03b**], it was conjectured in [**Ném08**] that the topological version can be identified with Heegaard Floer homology. Further results from [**Ném08, OSS14b**] support the conjecture, whose full proof was announced in [**Zem25**].

In particular, it is also isomorphic with any other (co)homology theory of 3-manifolds that agree with Heegaard Floer homology, e.g. with Embedded Contact Homology (introduced in [**Hut10, Hut14**], for its equivalence with the Heegaard Floer homology see [**CGH20**]).

One of the spin^c structures (determined from the analytic structure) is distinguished; it is called the 'canonical spin^c structure'. The analytic lattice cohomology associated with this canonical structure has an extension to any higher dimension, for complex isolated singularities (again, with certain restrictions).

(2) Part (b) addresses the important issue that in higher dimension the link of the isolated singularity contains essentially less information than might be needed to construct such an invariant (e.g. the link can be even the usual sphere). In particular, in the above context, the 'non-analytic' version shouldn't be 'topological' or 'smooth'. However, the smooth structure enhanced with its canonical contact structure (induced by the analytic structure of the germ) might produce the desired cohomology.

A good candidate for this is some version of Embedded Contact Homology associated with the canonical contact structure of the link. For 3-dimensional singularity links the canonical contact structure can be uniquely determined from the link itself [**CNPP06**]. However, in higher dimensions, it is an essential enhancement of the smooth structure. For results regarding ECH in higher dimensions see e.g. [**CHT24**].

Proposed for K3 and scribed by: A. Némethi

Bibliography

[AAD⁺23] Wolfgang Allred, Manuel Aragón, Zack Dooley, Alexander Goldman, Yucong Lei, Isaiah Martinez, Nicholas Meyer, Devon Peters, Scott Warrander, Ana Wright, and Alexander Zupan, *Tri-plane diagrams for simple surfaces in S^4*, J. Knot Theory Ramifications **32** (2023), no. 6, Paper No. 2350041, 28, DOI 10.1142/S0218216523500414. MR4611803 Cited in Problems 1.18 and 4.111.

[AALM00] V. Arnold, M. Atiyah, P. Lax, and B. Mazur (eds.), *Mathematics: frontiers and perspectives*, American Mathematical Society, Providence, RI, 2000. MR1754762 Cited in Problem 1.24.

[AB95] D. M. Austin and P. J. Braam, *Morse-Bott theory and equivariant cohomology*, The Floer memorial volume, Progr. Math., vol. 133, Birkhäuser, Basel, 1995, pp. 123–183. MR1362827 Cited in Problem 3.59.

[AB23] Mihail Arabadji and R. İnanç Baykur, *Spin Lefschetz fibrations are abundant*, Pacific J. Math. **326** (2023), no. 1, 1–16, DOI 10.2140/pjm.2023.326.1. MR4678147 Cited in Problem 2.6.

[AB25] Mihail Arabadji and R. İnanç Baykur, *Nielsen realization in dimension four and projective twists*, Adv. Math. **463** (2025), Paper No. 110112, 22, DOI 10.1016/j.aim.2025.110112. MR4854456 Cited in Problem 4.84.

[AB24a] Antonio Alfieri and Fraser Binns. Is the geography of Heegaard Floer homology restricted or the L-space conjecture false?, 2024. `arXiv:2404.00490`. Cited in Problem 3.55.

[AB24b] Antonio Alfieri and Keegan Boyle, *Strongly invertible knots, invariant surfaces, and the Atiyah-Singer signature theorem*, Michigan Math. J. **74** (2024), no. 4, 845–861, DOI 10.1307/mmj/20226183. MR4792563 Cited in Problem 3.76.

[ABB⁺10] Anar Akhmedov, Scott Baldridge, R. İnanç Baykur, Paul Kirk, and B. Doug Park, *Simply connected minimal symplectic 4-manifolds with signature less than -1*, J. Eur. Math. Soc. (JEMS) **12** (2010), no. 1, 133–161, DOI 10.4171/JEMS/192. MR2578606 Cited in Problem 4.3.

[ABCT08] Enrique Artal Bartolo, José Ignacio Cogolludo, and Hiro-o Tokunaga, *A survey on Zariski pairs*, Algebraic geometry in East Asia—Hanoi 2005, Adv. Stud. Pure Math., vol. 50, Math. Soc. Japan, Tokyo, 2008, pp. 1–100, DOI 10.2969/aspm/05010001. MR2409555 Cited in Problem 4.102.

[ABD⁺20] Paolo Aceto, Corey Bregman, Christopher W. Davis, JungHwan Park, and Arunima Ray. Isotopy and equivalence of knots in 3-manifolds, 2020. `arXiv:2007.05796`. Cited in Problems 1.19 and 4.82.

[ABDS22] Antonio Alfieri, John A. Baldwin, Irving Dai, and Steven Sivek, *Instanton Floer homology of almost-rational plumbings*, Geom. Topol. **26** (2022), no. 5, 2237–2294, DOI 10.2140/gt.2022.26.2237. MR4520305 Cited in Problem 3.59.

[Abe17] Michael Abel, *HOMFLY-PT homology for general link diagrams and braid-like isotopy*, Algebr. Geom. Topol. **17** (2017), no. 5, 3021–3056, DOI 10.2140/agt.2017.17.3021. MR3704251 Cited in Problem 1.34.

[Abo12] Mohammed Abouzaid, *Framed bordism and Lagrangian embeddings of exotic spheres*, Ann. of Math. (2) **175** (2012), no. 1, 71–185, DOI 10.4007/annals.2012.175.1.4. MR2874640 Cited in Problem 4.106.

[AC65] J. J. Andrews and M. L. Curtis, *Free groups and handlebodies*, Proc. Amer. Math. Soc. **16** (1965), 192–195, DOI 10.2307/2033843. MR173241 Cited in Problem 5.10.

[AC66] J. J. Andrews and M. L. Curtis, *Extended Nielsen operations in free groups*, Amer. Math. Monthly **73** (1966), 21–28, DOI 10.2307/2313917. MR195928 Cited in Problem 5.10.

[AC24] Antonio Alfieri and Alberto Cavallo. Holomorphic curves in Stein domains and the tau-invariant, 2024. `arXiv:2310.08657`. Cited in Problem 4.58.

[Ace14] Paolo Aceto, *Symmetric ribbon disks*, J. Knot Theory Ramifications **23** (2014), no. 9, 1450048, 9, DOI 10.1142/S0218216514500485. MR3268984 Cited in Problem 1.51.

[ACF+06] C. Adams, A. Colestock, J. Fowler, W. Gillam, and E. Katerman, *Cusp size bounds from singular surfaces in hyperbolic 3-manifolds*, Trans. Amer. Math. Soc. **358** (2006), no. 2, 727–741, DOI 10.1090/S0002-9947-05-03662-7. MR2177038 Cited in Problem 3.11.

[ACM+26] Paolo Aceto, Nickolas A Castro, Maggie Miller, JungHwan Park, and András Stipsicz, *Slice Obstructions From Genus Bounds in Definite 4-Manifolds*, Int. Math. Res. Not. IMRN **2** (2026), Paper No. rnaf377, DOI 10.1093/imrn/rnaf377. MR5016767 Cited in Problems 1.50 and 4.44.

[AD19] Akram Alishahi and Nathan Dowlin, *The Lee spectral sequence, unknotting number, and the knight move conjecture*, Topology Appl. **254** (2019), 29–38, DOI 10.1016/j.topol.2018.11.020. MR3894208 Cited in Problem 1.28.

[Ada55] J. F. Adams, *A new proof of a theorem of W. H. Cockcroft*, J. London Math. Soc. **30** (1955), 482–488, DOI 10.1112/jlms/s1-30.4.482. MR76335 Cited in Problem 5.11.

[Ada87] Colin C. Adams, *The noncompact hyperbolic 3-manifold of minimal volume*, Proc. Amer. Math. Soc. **100** (1987), no. 4, 601–606, DOI 10.2307/2046691. MR894423 Cited in Problem 3.1.

[Ada88] Colin C. Adams, *Volumes of N-cusped hyperbolic 3-manifolds*, J. London Math. Soc. (2) **38** (1988), no. 3, 555–565, DOI 10.1112/jlms/s2-38.3.555. MR972138 Cited in Problem 3.1.

[Ada96] Colin C. Adams, *Hyperbolic 3-manifolds with two generators*, Comm. Anal. Geom. **4** (1996), no. 1-2, 181–206, DOI 10.4310/CAG.1996.v4.n2.a1. MR1393562 Cited in Problem 3.11.

[ADMT23] Antonio Alfieri, Irving Dai, Abhishek Mallick, and Masaki Taniguchi. Involutions and the Chern-Simons filtration in instanton Floer homology, 2023. `arXiv:2309.02309`. Cited in Problem 3.74.

[AdSSQ24] Raimundo N. Araújo dos Santos and Eder L. Sanchez Quiceno. On real algebraic links in the 3-sphere associated with mixed polynomials. *Res. Math. Sci.*, 11(2):Paper No. 22, 22, 2024. `doi:10.1007/s40687-024-00424-3`. Cited in Problem 1.87.

[AET04] Greg Aloupis, Günter Ewald, and Godfried Toussaint, *More classes of stuck unknotted hexagons*, Beiträge Algebra Geom. **45** (2004), no. 2, 429–434. MR2093175 Cited in Problem 1.72.

[AF15] Ian Agol and Michael Freedman, *Simplifying 3-manifolds in* $\mathbb{R}^4$ (English, with English and French summaries), Ann. Fac. Sci. Toulouse Math. (6) **24** (2015), no. 5, 1079–1101, DOI 10.5802/afst.1476. MR3485327 Cited in Problem 4.23.

[AFMW23] Menny Aka, Peter Feller, Alison Beth Miller, and Andreas Wieser. Seifert surfaces in the four-ball and composition of binary quadratic forms, 2023. `arXiv:2311.17746`. Cited in Problem 1.6.

[AFP17] Javier Aramayona, Ariadna Fossas, and Hugo Parlier, *Arc and curve graphs for infinite-type surfaces*, Proc. Amer. Math. Soc. **145** (2017), no. 11, 4995–5006, DOI 10.1090/proc/13608. MR3692012 Cited in Problem 2.38.

[AFW15] Matthias Aschenbrenner, Stefan Friedl, and Henry Wilton, *3-manifold groups*, EMS Series of Lectures in Mathematics, European Mathematical Society (EMS), Zürich, 2015, DOI 10.4171/154. MR3444187 Cited in Problem 3.37.

[AG22] Tarik Aougab and Jonah Gaster. Combinatorially random curves on surfaces, 2022. `arXiv:2209.11309`. Cited in Problem 2.11.

[AG24] Hanine Awada and Marco Golla. Alexander polynomials of symplectic curves and divisibility relations, 2024. `arXiv:2412.15792`. Cited in Problem 4.103.

[AGK18] Aaron Abrams, David T. Gay, and Robion Kirby, *Group trisections and smooth 4-manifolds*, Geom. Topol. **22** (2018), no. 3, 1537–1545, DOI 10.2140/gt.2018.22.1537. MR3780440 Cited in Problem 4.1.

[AGL17] Paolo Aceto, Marco Golla, and Kyle Larson, *Embedding 3-manifolds in spin 4-manifolds*, J. Topol. **10** (2017), no. 2, 301–323, DOI 10.1112/topo.12010. MR3653313 Cited in Problem 4.28.

[AGL18] Paolo Aceto, Marco Golla, and Ana G. Lecuona, *Handle decompositions of rational homology balls and Casson-Gordon invariants*, Proc. Amer. Math. Soc. **146** (2018), no. 9, 4059–4072, DOI 10.1090/proc/14035. MR3825859 Cited in Problem 4.59.

[Ago00] Ian Agol, *Bounds on exceptional Dehn filling*, Geom. Topol. **4** (2000), 431–449, DOI 10.2140/gt.2000.4.431. MR1799796 Cited in Problem 3.11.

[Ago08] Ian Agol, *Criteria for virtual fibering*, J. Topol. **1** (2008), no. 2, 269–284, DOI 10.1112/jtopol/jtn003. MR2399130 Cited in Problems 3.7, 3.29, 5.12, and 5.13.

[Ago10] Ian Agol, *The minimal volume orientable hyperbolic 2-cusped 3-manifolds*, Proc. Amer. Math. Soc. **138** (2010), no. 10, 3723–3732, DOI 10.1090/S0002-9939-10-10364-5. MR2661571 Cited in Problems 3.1 and 3.2.

[Ago11] Ian Agol, *Ideal triangulations of pseudo-Anosov mapping tori*, Topology and geometry in dimension three, Contemp. Math., vol. 560, Amer. Math. Soc., Providence, RI, 2011, pp. 1–17, DOI 10.1090/conm/560/11087. MR2866919 Cited in Problem 3.21.

[Ago13] Ian Agol, *The virtual Haken conjecture*, Doc. Math. **18** (2013), 1045–1087, DOI 10.4171/DM/421. With an appendix by Agol, Daniel Groves, and Jason Manning. MR3104553 Cited in Problems 3.5, 3.7, 3.26, 3.29, 3.30, 3.37, and 5.13.

[Ago22] Ian Agol, *Ribbon concordance of knots is a partial ordering*, Commun. Am. Math. Soc. **2** (2022), 374–379, DOI 10.1090/cams/15. MR4520779 Cited in Problems 1.54 and 1.85.

[AGPS17] Tarik Aougab, Jonah Gaster, Priyam Patel, and Jenya Sapir, *Building hyperbolic metrics suited to closed curves and applications to lifting simply*, Math. Res. Lett. **24** (2017), no. 3, 593–617, DOI 10.4310/MRL.2017.v24.n3.a1. MR3696596 Cited in Problem 2.11.

[AH70] Michael Atiyah and Friedrich Hirzebruch, *Spin-manifolds and group actions*, Essays on Topology and Related Topics (Mémoires dédiés à Georges de Rham), Springer, New York-Berlin, 1970, pp. 18–28. MR278334 Cited in Problem 4.89.

[AH16] Nima Anvari and Ian Hambleton, *Cyclic group actions on contractible 4-manifolds*, Geom. Topol. **20** (2016), no. 2, 1127–1155, DOI 10.2140/gt.2016.20.1127. MR3493101 Cited in Problems 3.75 and 4.58.

[AH21] Nima Anvari and Ian Hambleton, *Cyclic branched coverings of Brieskorn spheres bounding acyclic 4-manifolds*, Glasg. Math. J. **63** (2021), no. 2, 400–413, DOI 10.1017/S0017089520000269. MR4244205 Cited in Problems 3.75 and 4.58.

[AH25] José Ayala and Joel Hass. Gordian unlinks, 2025. `arXiv:2502.08499`. Cited in Problem 1.71.

[AHK17] Karim Adiprasito, June Huh, and Eric Katz, *Hodge theory of matroids*, Notices Amer. Math. Soc. **64** (2017), no. 1, 26–30, DOI 10.1090/noti1463. MR3586249 Cited in Problem 1.7.

[Ahl06] Lars V. Ahlfors, *Lectures on quasiconformal mappings*, 2nd ed., University Lecture Series, vol. 38, American Mathematical Society, Providence, RI, 2006. With supplemental chapters by C. J. Earle, I. Kra, M. Shishikura and J. H. Hubbard, DOI 10.1090/ulect/038. MR2241787 Cited in Problem 2.40.

[AJK24] Soheil Azarpendar, András Juhász, and Tamás Kálmán. On Fox's trapezoidal conjecture, 2024. `arXiv:2406.08662`. Cited in Problem 1.7.

[AJOT13] Tetsuya Abe, In Dae Jong, Yuka Omae, and Masanori Takeuchi. Annulus twist and diffeomorphic 4-manifolds. *Math. Proc. Cambridge Philos. Soc.*, 155(2):219–235, 2013. `doi:10.1017/S0305004113000194`. Cited in Problem 1.57.

[AK79a] Selman Akbulut and Robion Kirby, *An exotic involution of S^4*, Topology **18** (1979), no. 1, 75–81, DOI 10.1016/0040-9383(79)90015-6. MR528237 Cited in Problems 4.9 and 4.87.

[AK79b] Selman Akbulut and Robion Kirby, *Mazur manifolds*, Michigan Math. J. **26** (1979), no. 3, 259–284. MR544597 Cited in Problems 4.58 and 4.118.

[AK81] S. Akbulut and H. King. All knots are algebraic. *Comment. Math. Helv.*, 56(3):339–351, 1981. `doi:10.1007/BF02566217`. Cited in Problem 1.87.

[AK85] Selman Akbulut and Robion Kirby, *A potential smooth counterexample in dimension 4 to the Poincaré conjecture, the Schoenflies conjecture, and the Andrews-Curtis conjecture*, Topology **24** (1985), no. 4, 375–390, DOI 10.1016/0040-9383(85)90010-2. MR816520 Cited in Problem 4.112.

[AK02] Tadashi Ashikaga and Kazuhiro Konno, *Global and local properties of pencils of algebraic curves*, Algebraic geometry 2000, Azumino (Hotaka), Adv. Stud. Pure Math., vol. 36, Math. Soc. Japan, Tokyo, 2002, pp. 1–49, DOI 10.2969/aspm/03610001. MR1971511 Cited in Problem 2.20.

[AK18] Mohammed Abouzaid and Thomas Kragh, *Simple homotopy equivalence of nearby Lagrangians*, Acta Math. **220** (2018), no. 2, 207–237, DOI 10.4310/ACTA.2018.v220.n2.a1. MR3849284 Cited in Problem 4.106.

[Aka06] Manabu Akaho, *A connected sum of knots and Fintushel-Stern knot surgery on 4-manifolds*, Turkish J. Math. **30** (2006), no. 1, 87–93. MR2215509 Cited in Problem 4.8.

[Akb77] Selman Akbulut. On 2-dimensional homology classes of 4-manifolds. *Math. Proc. Cambridge Philos. Soc.*, 82(1):99–106, 1977. doi:10.1017/S0305004100053718. Cited in Problem 1.57.

[Akb79] Selman Akbulut, *A note on homology surgery and the Casson-Gordon invariant*, Math. Proc. Cambridge Philos. Soc. **85** (1979), no. 2, 335–344, DOI 10.1017/S0305004100055754. MR516093 Cited in Problem 4.59.

[Akb88] Selman Akbulut, *Constructing a fake 4-manifold by Gluck construction to a standard 4-manifold*, Topology **27** (1988), no. 2, 239–243, DOI 10.1016/0040-9383(88)90041-9. MR948185 Cited in Problem 4.11.

[Akb91] Selman Akbulut, *A fake compact contractible 4-manifold*, J. Differential Geom. **33** (1991), no. 2, 335–356. MR1094459 Cited in Problem 3.74.

[Akb93] S. Akbulut, *Knots and exotic smooth structures on 4-manifolds*, J. Knot Theory Ramifications **2** (1993), no. 1, 1–10, DOI 10.1142/S0218216593000027. MR1209316 Cited in Problems 1.57 and 4.1.

[Akb99] Selman Akbulut, *Scharlemann's manifold is standard*, Ann. of Math. (2) **149** (1999), no. 2, 497–510, DOI 10.2307/120972. MR1689337 Cited in Problem 4.22.

[Akb02] Selman Akbulut, *Variations on Fintushel-Stern knot surgery on 4-manifolds*, Turkish J. Math. **26** (2002), no. 1, 81–92. MR1892802 Cited in Problem 4.8.

[Akb08] Erol Akbas, *A presentation for the automorphisms of the 3-sphere that preserve a genus two Heegaard splitting*, Pacific J. Math. **236** (2008), no. 2, 201–222, DOI 10.2140/pjm.2008.236.201. MR2407105 Cited in Problem 2.15.

[Akb10] Selman Akbulut, *Cappell-Shaneson homotopy spheres are standard*, Ann. of Math. (2) **171** (2010), no. 3, 2171–2175, DOI 10.4007/annals.2010.171.2171. MR2680408 Cited in Problem 4.1.

[Akb12] Selman Akbulut, *The Dolgachev surface. Disproving the Harer-Kas-Kirby conjecture*, Comment. Math. Helv. **87** (2012), no. 1, 187–241, DOI 10.4171/CMH/252. MR2874900 Cited in Problem 4.118.

[Akb14] Selman Akbulut, *Cork twisting Schoenflies problem*, J. Gökova Geom. Topol. GGT **8** (2014), 35–43, DOI 10.1017/s0020269x00003418. MR3310571 Cited in Problem 4.23.

[Akb16] Selman Akbulut, *4-manifolds*, Oxford Graduate Texts in Mathematics, vol. 25, Oxford University Press, Oxford, 2016, DOI 10.1093/acprof:oso/9780198784869.001.0001. MR3559604 Cited in Problem 4.24.

[AKM+19] Dave Auckly, Hee Jung Kim, Paul Melvin, Daniel Ruberman, and Hannah Schwartz, *Isotopy of surfaces in 4-manifolds after a single stabilization*, Adv. Math. **341** (2019), 609–615, DOI 10.1016/j.aim.2018.10.040. MR3873547 Cited in Problem 4.33.

[AKMR15] Dave Auckly, Hee Jung Kim, Paul Melvin, and Daniel Ruberman, *Stable isotopy in four dimensions*, J. Lond. Math. Soc. (2) **91** (2015), no. 2, 439–463, DOI 10.1112/jlms/jdu075. MR3355110 Cited in Problem 4.68.

[AKS20] Antonio Alfieri, Sungkyung Kang, and András I. Stipsicz, *Connected Floer homology of covering involutions*, Math. Ann. **377** (2020), no. 3-4, 1427–1452, DOI 10.1007/s00208-020-01992-9. MR4126897 Cited in Problems 3.57, 3.74, and 4.58.

[AL03] Ian Agol and Tao Li, *An algorithm to detect laminar 3-manifolds*, Geom. Topol. **7** (2003), 287–309, DOI 10.2140/gt.2003.7.287. MR1988287 Cited in Problem 3.31.

[AL18] Selman Akbulut and Kyle Larson, *Brieskorn spheres bounding rational balls*, Proc. Amer. Math. Soc. **146** (2018), no. 4, 1817–1824, DOI 10.1090/proc/13828. MR3754363 Cited in Problem 3.72.

[AL20] Ian Agol and Francesco Lin, *Hyperbolic four-manifolds with vanishing Seiberg-Witten invariants*, Characters in low-dimensional topology, Contemp. Math., vol. 760, Amer. Math. Soc., [Providence], RI, [2020] ©2020, pp. 1–8, DOI 10.1090/conm/760/15283. MR4193918 Cited in Problem 4.95.

[Ale20] James W. Alexander, *Note on Riemann spaces*, Bull. Amer. Math. Soc. **26** (1920), no. 8, 370–372, DOI 10.1090/S0002-9904-1920-03319-7. MR1560318 Cited in Problem 5.6.

[All23] Samantha Allen, *Nonorientable surfaces bounded by knots: a geography problem*, New York J. Math. **29** (2023), 1038–1059. MR4646147 Cited in Problem 1.47.

[Alt20] Tülin Altunöz, *The number of singular fibers in hyperelliptic Lefschetz fibrations*, J. Math. Soc. Japan **72** (2020), no. 4, 1309–1325, DOI 10.2969/jmsj/82988298. MR4165934 Cited in Problem 2.21.

[AM90] Selman Akbulut and John D. McCarthy, *Casson's invariant for oriented homology 3-spheres*, Mathematical Notes, vol. 36, Princeton University Press, Princeton, NJ, 1990. An exposition, DOI 10.1515/9781400860623. MR1030042 Cited in Problem 4.119.

[AM17] Miguel Abreu and Leonardo Macarini, *Dynamical convexity and elliptic periodic orbits for Reeb flows*, Math. Ann. **369** (2017), no. 1-2, 331–386, DOI 10.1007/s00208-017-1532-4. MR3694649 Cited in Problem 3.47.

[AM20] Mohammed Abouzaid and Ciprian Manolescu, *A sheaf-theoretic model for* $\mathrm{SL}(2,\mathbb{C})$ *Floer homology*, J. Eur. Math. Soc. (JEMS) **22** (2020), no. 11, 3641–3695, DOI 10.4171/jems/994. MR4167016 Cited in Problems 3.65 and 3.66.

[amd22] amd1234. Bing sling isotopy to unknot, 2022. by user `https://mathoverflow.net/users/170240/amd1234`. See `https://mathoverflow.net/q/437106` (version: 2022-12-28). Cited in Problem 1.81.

[AMU06] Jørgen Ellegaard Andersen, Gregor Masbaum, and Kenji Ueno, *Topological quantum field theory and the Nielsen-Thurston classification of* $M(0,4)$, Math. Proc. Cambridge Philos. Soc. **141** (2006), no. 3, 477–488, DOI 10.1017/S0305004106009698. MR2281410 Cited in Problem 2.49.

[AN09] Yinghua Ai and Yi Ni, *Two applications of twisted Floer homology*, Int. Math. Res. Not. IMRN **19** (2009), 3726–3746, DOI 10.1093/imrn/rnp070. MR2539188 Cited in Problem 1.13.

[AN12] Miklós Abért and Nikolay Nikolov, *Rank gradient, cost of groups and the rank versus Heegaard genus problem*, J. Eur. Math. Soc. (JEMS) **14** (2012), no. 5, 1657–1677, DOI 10.4171/JEMS/344. MR2966663 Cited in Problem 3.35.

[ÁN21a] Tamás Ágoston and András Némethi. Analytic lattice cohomology of surface singularities, 2021. `arXiv:2108.12294`. Cited in Problem 5.18.

[ÁN21b] Tamás Ágoston and András Némethi. Analytic lattice cohomology of surface singularities, II (the equivariant case), 2021. `arXiv:2108.12429`. Cited in Problem 5.18.

[ANC20] Macarena Arenas and Max Neumann-Coto, *Measuring complexity of curves on surfaces*, Geom. Dedicata **204** (2020), 25–41, DOI 10.1007/s10711-019-00443-3. MR4056691 Cited in Problem 2.11.

[And82] Yoshifumi Ando. Elimination of certain Thom-Boardman singularities of order two. *J. Math. Soc. Japan*, 34(2):241–267, 1982. `doi:10.2969/jmsj/03420241`. Cited in Problem 5.9.

[And06] Jørgen Ellegaard Andersen, *Asymptotic faithfulness of the quantum* $\mathrm{SU}(n)$ *representations of the mapping class groups*, Ann. of Math. (2) **163** (2006), no. 1, 347–368, DOI 10.4007/annals.2006.163.347. MR2195137 Cited in Problem 2.49.

[And07] Jorgen Ellegaard Andersen. Mapping class groups do not have Kazhdan's property (t), 2007. `arXiv:0706.2184`. Cited in Problem 2.1.

[ANO24] Nobutaka Asano, Hironobu Naoe, and Masaki Ogawa, *Some lower bounds for the Kirby-Thompson invariant*, Math. Res. Lett. **31** (2024), no. 6, 1611–1637, DOI 10.4310/mrl.250210225353. MR4862351 Cited in Problem 4.116.

[AP04] Marta M. Asaeda and Józef H. Przytycki, *Khovanov homology: torsion and thickness*, Advances in topological quantum field theory, NATO Sci. Ser. II Math. Phys. Chem., vol. 179, Kluwer Acad. Publ., Dordrecht, 2004, pp. 135–166, DOI 10.1007/978-1-4020-2772-7_6. MR2147419 Cited in Problem 1.30.

[AP10] Anar Akhmedov and B. Doug Park, *Exotic smooth structures on small 4-manifolds with odd signatures*, Invent. Math. **181** (2010), no. 3, 577–603, DOI 10.1007/s00222-010-0254-y. MR2660453 Cited in Problems 4.3 and 4.111.

[APS75a] M. F. Atiyah, V. K. Patodi, and I. M. Singer. Spectral asymmetry and Riemannian geometry. I. *Math. Proc. Cambridge Philos. Soc.*, 77:43–69, 1975. `doi:10.1017/S0305004100049410`. Cited in Problems 3.63 and 4.126.

[APS75b] M. F. Atiyah, V. K. Patodi, and I. M. Singer. Spectral asymmetry and Riemannian geometry. II. *Math. Proc. Cambridge Philos. Soc.*, 78(3):405–432, 1975. `doi:10.1017/S0305004100051872`. Cited in Problem 3.63.

[APS76] M. F. Atiyah, V. K. Patodi, and I. M. Singer. Spectral asymmetry and Riemannian geometry. III. *Math. Proc. Cambridge Philos. Soc.*, 79(1):71–99, 1976. `doi:10.1017/S0305004100052105`. Cited in Problem 3.63.

[APS04] Marta M. Asaeda, Józef H. Przytycki, and Adam S. Sikora, *Categorification of the Kauffman bracket skein module of I-bundles over surfaces*, Algebr. Geom. Topol. **4** (2004), 1177–1210, DOI 10.2140/agt.2004.4.1177. MR2113902 Cited in Problems 1.32 and 1.36.

[APV21] Tarik Aougab, Priyam Patel, and Nicholas G. Vlamis, *Isometry groups of infinite-genus hyperbolic surfaces*, Math. Ann. **381** (2021), no. 1-2, 459–498, DOI 10.1007/s00208-021-02164-z. MR4322618 Cited in Problems 2.3, 2.33, 2.39, and 2.41.

[AR25] Dave Auckly and Daniel Ruberman. Families of diffeomorphisms, embeddings, and positive scalar curvature metrics via Seiberg-Witten theory, 2025. `arXiv:2501.11892`. Cited in Problems 4.68, 4.70, 4.71, and 4.75.

[Ara71] S. Ju. Arakelov, *Families of algebraic curves with fixed degeneracies* (Russian), Izv. Akad. Nauk SSSR Ser. Mat. **35** (1971), 1269–1293. MR321933 Cited in Problem 2.20.

[Art47] E. Artin, *Braids and permutations*, Ann. of Math. (2) **48** (1947), 643–649, DOI 10.2307/1969131. MR20989 Cited in Problem 2.7.

[AS05] Colin Adams and Eric Schoenfeld, *Totally geodesic Seifert surfaces in hyperbolic knot and link complements. I*, Geom. Dedicata **116** (2005), 237–247, DOI 10.1007/s10711-005-9018-z. MR2195448 Cited in Problem 3.14.

[AS12] Javier Aramayona and Juan Souto, *Homomorphisms between mapping class groups*, Geom. Topol. **16** (2012), no. 4, 2285–2341, DOI 10.2140/gt.2012.16.2285. MR3033518 Cited in Problem 2.4.

[AS16] Javier Aramayona and Juan Souto, *Rigidity phenomena in the mapping class group*, Handbook of Teichmüller theory. Vol. VI, IRMA Lect. Math. Theor. Phys., vol. 27, Eur. Math. Soc., Zürich, 2016, pp. 131–165. MR3618188 Cited in Problems 2.3 and 2.7.

[AS19] Mohammed Abouzaid and Ivan Smith, *Khovanov homology from Floer cohomology*, J. Amer. Math. Soc. **32** (2019), no. 1, 1–79, DOI 10.1090/jams/902. MR3867999 Cited in Problem 1.35.

[AS23] Ian Agol and Matthew Stover, *Congruence RFRS towers* (English, with English and French summaries), Ann. Inst. Fourier (Grenoble) **73** (2023), no. 1, 307–333, DOI 10.5802/aif.3532. With an appendix by Mehmet Haluk Şengün. MR4588931 Cited in Problem 3.29.

[Asa01] Mamoru Asada, *The faithfulness of the monodromy representations associated with certain families of algebraic curves*, J. Pure Appl. Algebra **159** (2001), no. 2-3, 123–147, DOI 10.1016/S0022-4049(00)00056-6. MR1828935 Cited in Problem 2.2.

[Asa23] Nobutaka Asano, *Right-left equivalent maps of simplified (2, 0)-trisections with different configurations of vanishing cycles*, Kyushu J. Math. **77** (2023), no. 2, 299–317. MR4665183 Cited in Problem 4.114.

[AT16a] Tetsuya Abe and Keiji Tagami, *Fibered knots with the same 0-surgery and the slice-ribbon conjecture*, Math. Res. Lett. **23** (2016), no. 2, 303–323, DOI 10.4310/MRL.2016.v23.n2.a1. MR3512887 Cited in Problem 1.39.

[AT16b] Tetsuya Abe and Motoo Tange. A construction of slice knots via annulus twists. *Michigan Math. J.*, 65(3):573–597, 2016. doi:10.1307/mmj/1472066149. Cited in Problem 1.10.

[AT24] Ian Agol and Chi Cheuk Tsang, *Dynamics of veering triangulations: infinitesimal components of their flow graphs and applications*, Algebr. Geom. Topol. **24** (2024), no. 6, 3401–3453, DOI 10.2140/agt.2024.24.3401. MR4812222 Cited in Problem 3.21.

[Ati69] M. F. Atiyah, *The signature of fibre-bundles*, Global Analysis (Papers in Honor of K. Kodaira), Univ. Tokyo Press, Tokyo, 1969, pp. 73–84. MR254864 Cited in Problem 2.24.

[Ati88a] Michael Atiyah, *New invariants of* 3- *and* 4-*dimensional manifolds*, The mathematical heritage of Hermann Weyl (Durham, NC, 1987), Proc. Sympos. Pure Math., vol. 48, Amer. Math. Soc., Providence, RI, 1988, pp. 285–299, DOI 10.1090/pspum/048/974342. MR974342 Cited in Problem 3.47.

[Ati88b] Michael Atiyah, *Topological quantum field theories*, Inst. Hautes Études Sci. Publ. Math. **68** (1988), 175–186 (1989). MR1001453 Cited in Problem 3.78.

[Auc94] David R. Auckly. Topological methods to compute Chern-Simons invariants. *Math. Proc. Cambridge Philos. Soc.*, 115(2):229–251, 1994. doi:10.1017/S0305004100072066. Cited in Problem 3.64.

[Auc97] David Auckly, *Surgery numbers of* 3-*manifolds: a hyperbolic example*, Geometric topology (Athens, GA, 1993), AMS/IP Stud. Adv. Math., vol. 2, Amer. Math. Soc., Providence, RI, 1997, pp. 21–34, DOI 10.1090/amsip/002.1/02. MR1470719 Cited in Problems 1.15 and 3.70.

[Auc23] David Auckly. Smoothly knotted surfaces that remain distinct after many internal stabilizations, 2023. arXiv:2307.16266. Cited in Problem 4.33.

[Aur00] Denis Auroux, *Symplectic 4-manifolds as branched coverings of* $\mathbf{CP}^2$, Invent. Math. **139** (2000), no. 3, 551–602, DOI 10.1007/s002220050019. MR1738061 Cited in Problem 4.90.

[Aur05] Denis Auroux, *A stable classification of Lefschetz fibrations*, Geom. Topol. **9** (2005), 203–217, DOI 10.2140/gt.2005.9.203. MR2115673 Cited in Problem 2.22.

[Aur06a] Denis Auroux, *The canonical pencils on Horikawa surfaces*, Geom. Topol. **10** (2006), 2173–2217, DOI 10.2140/gt.2006.10.2173. MR2284054 Cited in Problem 4.94.

[Aur06b] Denis Auroux, *Symplectic 4-manifolds, singular plane curves, and isotopy problems*, Floer homology, gauge theory, and low-dimensional topology, Clay Math. Proc., vol. 5, Amer. Math. Soc., Providence, RI, 2006, pp. 263–276. MR2249258 Cited in Problems 4.90 and 4.101.

[Aur10] Denis Auroux, *Fukaya categories of symmetric products and bordered Heegaard-Floer homology*, J. Gökova Geom. Topol. GGT **4** (2010), 1–54. MR2755992 Cited in Problem 2.47.

[Avd23] Russell Avdek. An algebraic generalization of Giroux's criterion, 2023. arXiv:2206.14710. Cited in Problem 4.108.

[AY13] Selman Akbulut and Kouichi Yasui, *Gluck twisting 4-manifolds with odd intersection form*, Math. Res. Lett. **20** (2013), no. 2, 385–389, DOI 10.4310/MRL.2013.v20.n2.a13. MR3151654 Cited in Problem 4.11.

[Baa05] S. Baader, *Quasipositivity and homogeneity*, Math. Proc. Cambridge Philos. Soc. **139** (2005), no. 2, 287–290, DOI 10.1017/S0305004105008698. MR2168087 Cited in Problem 1.82.

[Bab16] László Babai, *Graph isomorphism in quasipolynomial time [extended abstract]*, STOC'16—Proceedings of the 48th Annual ACM SIGACT Symposium on Theory of Computing, ACM, New York, 2016, pp. 684–697, DOI 10.1145/2897518.2897542. MR3536606 Cited in Problem 3.27.

[Bai77] James Leonard Bailey, *ALEXANDER INVARIANTS OF LINKS*, ProQuest LLC, Ann Arbor, MI, 1977. Thesis (Ph.D.)–The University of British Columbia (Canada). MR2627593 Cited in Problem 1.8.

[Bak06] Kenneth L. Baker, *Small genus knots in lens spaces have small bridge number*, Algebr. Geom. Topol. **6** (2006), 1519–1621, DOI 10.2140/agt.2006.6.1519. MR2253458 Cited in Problem 1.9.

[Bak16] Kenneth L. Baker, *A note on the concordance of fibered knots*, J. Topol. **9** (2016), no. 1, 1–4, DOI 10.1112/jtopol/jtv024. MR3465837 Cited in Problems 1.39, 1.50, and 1.53.

[Bal20] William Ballinger. Concordance invariants from the $E(-1)$ spectral sequence on Khovanov homology, 2020. arXiv:2004.10807. Cited in Problem 1.47.

[Bam21a] Richard H Bamler, *On the fundamental group of non-collapsed ancient Ricci flows*, J. Math. Study **58** (2025), no. 4, 417–428. MR5013739 Cited in Problem 4.129.

[Bam21b] Richard H. Bamler, *Recent developments in Ricci flows*, Notices Amer. Math. Soc. **68** (2021), no. 9, 1486–1498, DOI 10.1090/noti2343. MR4323820 Cited in Problem 4.15.

[Ban22] Ian M. Banfield. Christoffel words and the strong Fox conjecture for two-bridge knots, 2022. arXiv:2212.04561. Cited in Problem 1.7.

[Bar63] D. Barden. *The structure of manifolds*. PhD thesis, Cambridge University, 1963. Cited in Problems 4.46 and 4.56.

[Bar99] Jean-François Barraud, *Nodal symplectic spheres in* $\mathbf{CP}^2$ *with positive self-intersection*, Internat. Math. Res. Notices **9** (1999), 495–508, DOI 10.1155/S1073792899000252. MR1692591 Cited in Problem 4.102.

[Bar00] Jean-François Barraud, *Courbes pseudo-holomorphes équisingulières en dimension 4* (French, with English and French summaries), Bull. Soc. Math. France **128** (2000), no. 2, 179–206. MR1772440 Cited in Problem 4.102.

[Bar19] David Baraglia, *Obstructions to smooth group actions on 4-manifolds from families Seiberg-Witten theory*, Adv. Math. **354** (2019), 106730, 32, DOI 10.1016/j.aim.2019.106730. MR3981995 Cited in Problem 4.86.

[Bar21] David Baraglia, *Constraints on families of smooth 4-manifolds from Bauer-Furuta invariants*, Algebr. Geom. Topol. **21** (2021), no. 1, 317–349, DOI 10.2140/agt.2021.21.317. MR4224743 Cited in Problems 4.70, 4.83, and 4.86.

[Bar23a] David Baraglia. The mod 2 Seiberg-Witten invariants of spin structures and spin families, 2023. arXiv:2303.06883. Cited in Problem 4.62.

[Bar23b] David Baraglia, *Non-trivial smooth families of* $K3$ *surfaces*, Math. Ann. **387** (2023), no. 3-4, 1719–1744, DOI 10.1007/s00208-022-02508-3. MR4657435 Cited in Problem 4.75.

[Bar23c] David Baraglia. On the mapping class groups of simply-connected smooth 4-manifolds, 2023. arXiv:2310.18819. Cited in Problems 4.75 and 4.84.

[Bar25a] Filippo Baroni, *Classification of genus-two surfaces in* S^3, Algebr. Geom. Topol. **25** (2025), no. 8, 4719–4785, DOI 10.2140/agt.2025.25.4719. MR4998418 Cited in Problem 3.32.

[Bar25b] Filippo Baroni, *Uniformly polynomial-time classification of surface homeomorphisms*, Math. Ann. **392** (2025), no. 3, 3247–3298, DOI 10.1007/s00208-025-03159-w. MR4939663 Cited in Problem 2.10.

[Bas97] Ara Basmajian, *Large parameter spaces of quasiconformally distinct hyperbolic structures*, J. Anal. Math. **71** (1997), 75–85, DOI 10.1007/BF02788023. MR1454244 Cited in Problem 2.40.

[Bat14] Joshua Batson, *Nonorientable slice genus can be arbitrarily large*, Math. Res. Lett. **21** (2014), no. 3, 423–436, DOI 10.4310/MRL.2014.v21.n3.a1. MR3272020 Cited in Problem 1.47.

[Bau04] Stefan Bauer, *A stable cohomotopy refinement of Seiberg-Witten invariants. II*, Invent. Math. **155** (2004), no. 1, 21–40, DOI 10.1007/s00222-003-0289-4. MR2025299 Cited in Problems 4.3 and 4.64.

[Bau08] Stefan Bauer, *Almost complex 4-manifolds with vanishing first Chern class*, J. Differential Geom. **79** (2008), no. 1, 25–32. MR2414748 Cited in Problem 4.92.

[Bau12] Stefan A. Bauer. Intersection forms of spin four-manifolds, 2012. arXiv:1211.7092. Cited in Problem 4.15.

[Bay06] R. İnanç Baykur, *Kähler decomposition of 4-manifolds*, Algebr. Geom. Topol. **6** (2006), 1239–1265, DOI 10.2140/agt.2006.6.1239. MR2253445 Cited in Problem 4.120.

[Bay09] Refik İnanç Baykur, *Topology of broken Lefschetz fibrations and near-symplectic four-manifolds*, Pacific J. Math. **240** (2009), no. 2, 201–230, DOI 10.2140/pjm.2009.240.201. MR2485463 Cited in Problem 4.113.

[Bay12a] R. İnanç Baykur, *Broken Lefschetz fibrations and smooth structures on 4-manifolds*, Proceedings of the Freedman Fest, Geom. Topol. Monogr., vol. 18, Geom. Topol. Publ., Coventry, 2012, pp. 9–34, DOI 10.2140/gtm.2012.18.9. MR3084230 Cited in Problem 4.113.

[Bay12b] R. İnanç Baykur, *Non-holomorphic surface bundles and Lefschetz fibrations*, Math. Res. Lett. **19** (2012), no. 3, 567–574, DOI 10.4310/MRL.2012.v19.n3.a5. MR2998140 Cited in Problem 4.97.

[Bay14] R. İnanç Baykur, *Virtual Betti numbers and the symplectic Kodaira dimension of fibered 4-manifolds*, Proc. Amer. Math. Soc. **142** (2014), no. 12, 4377–4384, DOI 10.1090/S0002-9939-2014-12151-4. MR3267005 Cited in Problem 4.92.

[Bay16] R. İnanç Baykur, *Inequivalent Lefschetz fibrations and surgery equivalence of symplectic 4-manifolds*, J. Symplectic Geom. **14** (2016), no. 3, 671–686, DOI 10.4310/JSG.2016.v14.n3.a2. MR3548482 Cited in Problem 4.98.

[Bay19] R. İnanç Baykur, *Inequivalent Lefschetz fibrations on rational and ruled surfaces*, Breadth in contemporary topology, Proc. Sympos. Pure Math., vol. 102, Amer. Math. Soc., Providence, RI, 2019, pp. 21–28, DOI 10.1090/pspum/102/02. MR3967357 Cited in Problem 4.98.

[Bay22] R. İnanç Baykur, *Small exotic 4-manifolds and symplectic Calabi-Yau surfaces via genus-3 pencils*, Gauge theory and low-dimensional topology—progress and interaction, Open Book Ser., vol. 5, Math. Sci. Publ., Berkeley, CA, 2022, pp. 185–221. MR4600724 Cited in Problems 2.21, 4.3, 4.4, and 4.92.

[BB97] Mladen Bestvina and Noel Brady, *Morse theory and finiteness properties of groups*, Invent. Math. **129** (1997), no. 3, 445–470, DOI 10.1007/s002220050168. MR1465330 Cited in Problem 5.11.

[BB01] Stephen J. Bigelow and Ryan D. Budney, *The mapping class group of a genus two surface is linear*, Algebr. Geom. Topol. **1** (2001), 699–708, DOI 10.2140/agt.2001.1.699. MR1875613 Cited in Problem 2.3.

[BB04] Stéphane Baseilhac and Riccardo Benedetti, *Quantum hyperbolic invariants of 3-manifolds with* $\mathrm{PSL}(2,\mathbb{C})$*-characters*, Topology **43** (2004), no. 6, 1373–1423, DOI 10.1016/j.top.2004.02.001. MR2081430 Cited in Problem 2.50.

[BB15] Michel Boileau and Steven Boyer, *Graph manifold* $\mathbb{Z}$*-homology 3-spheres and taut foliations*, J. Topol. **8** (2015), no. 2, 571–585, DOI 10.1112/jtopol/jtv006. MR3356771 Cited in Problem 3.48.

[BB16] Jeffrey F. Brock and Kenneth W. Bromberg, *Inflexibility, Weil-Peterson distance, and volumes of fibered 3-manifolds*, Math. Res. Lett. **23** (2016), no. 3, 649–674, DOI 10.4310/MRL.2016.v23.n3.a4. MR3533189 Cited in Problem 2.18.

[BB18] Manuel Bärenz and John Barrett, *Dichromatic state sum models for four-manifolds from pivotal functors*, Comm. Math. Phys. **360** (2018), no. 2, 663–714, DOI 10.1007/s00220-017-3012-9. MR3800795 Cited in Problem 3.78.

[BB22] Ryan Budney and Benjamin A. Burton, *Embeddings of 3-manifolds in* S^4 *from the point of view of the 11-tetrahedron census*, Exp. Math. **31** (2022), no. 3, 988–1013, DOI 10.1080/10586458.2020.1740836. MR4477418 Cited in Problem 4.25.

[BB25] Rachael Boyd and Corey Bregman, *Embedding spaces of split links*, Adv. Math. **470** (2025), Paper No. 110235, 41, DOI 10.1016/j.aim.2025.110235. MR4887219 Cited in Problem 1.89.

[BBB23] Martin Bridgeman, Jeffrey Brock, and Kenneth Bromberg, *The Weil-Petersson gradient flow of renormalized volume and 3-dimensional convex cores*, Geom. Topol. **27** (2023), no. 8, 3183–3228, DOI 10.2140/gt.2023.27.3183. MR4668096 Cited in Problem 2.18.

[BBG19a] Michel Boileau, Steven Boyer, and Cameron McA. Gordon, *Branched covers of quasi-positive links and L-spaces*, J. Topol. **12** (2019), no. 2, 536–576, DOI 10.1112/topo.12092. MR4072174 Cited in Problems 1.21, 1.22, and 2.47.

[BBG19b] Michel Boileau, Steven Boyer, and Cameron McA. Gordon, *On definite strongly quasipositive links and L-space branched covers*, Adv. Math. **357** (2019), 106828, 63, DOI 10.1016/j.aim.2019.106828. MR4016557 Cited in Problem 2.47.

[BBK15] Ryan Blair, John Burke, and Robin Koytcheff, *A prime decomposition theorem for the 2-string link monoid*, J. Knot Theory Ramifications **24** (2015), no. 2, 1550005, 24, DOI 10.1142/S0218216515500054. MR3334658 Cited in Problem 1.89.

[BBL20] S. Baader, I. Banfield, and L. Lewark, *Untwisting 3-strand torus knots*, Bull. Lond. Math. Soc. **52** (2020), no. 3, 429–436, DOI 10.1112/blms.12335. MR4171377 Cited in Problem 1.46.

[BBRW16] M. Boileau, S. Boyer, D. Rolfsen, and S. C. Wang, *One-domination of knots*, Illinois J. Math. **60** (2016), no. 1, 117–139. MR3665174 Cited in Problem 3.62.

[BC09] Ryan Budney and Fred Cohen, *On the homology of the space of knots*, Geom. Topol. **13** (2009), no. 1, 99–139, DOI 10.2140/gt.2009.13.99. MR2469515 Cited in Problem 1.89.

[BC12] Jason Behrstock and Ruth Charney, *Divergence and quasimorphisms of right-angled Artin groups*, Math. Ann. **352** (2012), no. 2, 339–356, DOI 10.1007/s00208-011-0641-8. MR2874959 Cited in Problem 2.28.

[BC17a] Nicolas Bergeron and Laurent Clozel, *Sur la cohomologie des variétés hyperboliques de dimension 7 trialitaires* (French, with French summary), Israel J. Math. **222** (2017), no. 1, 333–400, DOI 10.1007/s11856-017-1593-9. MR3736510 Cited in Problem 5.13.

[BC17b] Steven Boyer and Adam Clay, *Foliations, orders, representations, L-spaces and graph manifolds*, Adv. Math. **310** (2017), 159–234, DOI 10.1016/j.aim.2017.01.026. MR3620687 Cited in Problem 3.48.

[BC24a] A. Basmajian and Y. Chandran. A Bers type classification of big mapping class groups, 2024. `arXiv:2410.05606`. Cited in Problem 2.40.

[BC24b] Keegan Boyle and Wenzhao Chen, *Equivariant topological slice disks and negative amphichiral knots*, Indiana Univ. Math. J. **73** (2024), no. 5, 1623–1637. MR4845816 Cited in Problems 1.49 and 3.69.

[BC24c] Joseph Breen and Austin Christian. Bypass moves in convex hypersurface theory, 2024. `arXiv:2206.14710`. Cited in Problem 4.108.

[BCGPM16] Christian Blanchet, Francesco Costantino, Nathan Geer, and Bertrand Patureau-Mirand, *Non-semi-simple TQFTs, Reidemeister torsion and Kashaev's invariants*, Adv. Math. **301** (2016), 1–78, DOI 10.1016/j.aim.2016.06.003. MR3539369 Cited in Problem 3.78.

[BCK21] Hyungryul Baik, Inhyeok Choi, and Dongryul M. Kim. Simple length spectra as moduli for hyperbolic surfaces and rigidity of length identities, 2021. `arXiv:2012.05652`. Cited in Problem 2.16.

[BCL+24] Benjamin A. Burton, Hsien-Chih Chang, Maarten Löffler, Clément Maria, Arnaud de Mesmay, Saul Schleimer, Eric Sedgwick, and Jonathan Spreer, *Hard diagrams of the unknot*, Exp. Math. **33** (2024), no. 3, 482–500, DOI 10.1080/10586458.2022.2161676. MR4790980 Cited in Problems 1.76 and 1.78.

[BCV09] Sebastian Baader, Kai Cieliebak, and Thomas Vogel, *Legendrian ribbons in overtwisted contact structures*, J. Knot Theory Ramifications **18** (2009), no. 4, 523–529, DOI 10.1142/S0218216509006999. MR2514546 Cited in Problem 1.69.

[BcV16] Nicolas Bergeron, Mehmet Haluk Şengün, and Akshay Venkatesh, *Torsion homology growth and cycle complexity of arithmetic manifolds*, Duke Math. J. **165** (2016), no. 9, 1629–1693, DOI 10.1215/00127094-3450429. MR3513571 Cited in Problem 3.6.

[BD02] Jim Bryan and Ron Donagi, *Surface bundles over surfaces of small genus*, Geom. Topol. **6** (2002), 59–67, DOI 10.2140/gt.2002.6.59. MR1885589 Cited in Problem 4.97.

[BD15] Jeffrey F. Brock and Nathan M. Dunfield, *Injectivity radii of hyperbolic integer homology 3-spheres*, Geom. Topol. **19** (2015), no. 1, 497–523, DOI 10.2140/gt.2015.19.497. MR3318758 Cited in Problem 3.6.

[BD23] Etienne Batelier and Julien Ducoulombier, *Operadic actions on long knots and 2-string links*, Algebr. Geom. Topol. **23** (2023), no. 2, 833–882, DOI 10.2140/agt.2023.23.833. MR4587318 Cited in Problem 1.89.

[BDK07] Jeremy Brookman, James F. Davis, and Qayum Khan, *Manifolds homotopy equivalent to $P^n \# P^n$*, Math. Ann. **338** (2007), no. 4, 947–962, DOI 10.1007/s00208-007-0099-x. MR2317756 Cited in Problems 4.20 and 4.22.

[BDMS21] Francesco Bonsante, Jeffrey Danciger, Sara Maloni, and Jean-Marc Schlenker, *The induced metric on the boundary of the convex hull of a quasicircle in hyperbolic and anti-de Sitter geometry*, Geom. Topol. **25** (2021), no. 6, 2827–2911, DOI

10.2140/gt.2021.25.2827. With an appendix by Boubacar Diallo. MR4347307 Cited in Problem 3.25.

[BDR20] Juliette Bavard, Spencer Dowdall, and Kasra Rafi, *Isomorphisms between big mapping class groups*, Int. Math. Res. Not. IMRN **10** (2020), 3084–3099, DOI 10.1093/imrn/rny093. MR4098634 Cited in Problem 2.41.

[BDR23a] Anna Beliakova and Marco De Renzi, *Kerler-Lyubashenko functors on 4-dimensional 2-handlebodies*, Int. Math. Res. Not. IMRN **13** (2024), 10005–10080, DOI 10.1093/imrn/rnac039. MR4769129 Cited in Problem 3.78.

[BDR23b] Anna Beliakova and Marco De Renzi, *Refned Bobtcheva-Messia invariants of 4-dimensional 2-handlebodies*, Essays in geometry—dedicated to Norbert A'Campo, IRMA Lect. Math. Theor. Phys., vol. 34, EMS Press, Berlin, [2023] ©2023, pp. 387–431. MR4631276 Cited in Problem 3.78.

[BE78] Israel Berstein and Allan L. Edmonds, *The degree and branch set of a branced covering*, Invent. Math. **45** (1978), no. 3, 213–220, DOI 10.1007/BF01403169. MR474261 Cited in Problem 5.6.

[BE06] Nigel Boston and Jordan S. Ellenberg, *Pro-p groups and towers of rational homology spheres*, Geom. Topol. **10** (2006), 331–334, DOI 10.2140/gt.2006.10.331. MR2224459 Cited in Problem 3.10.

[BE13] John A. Baldwin and John B. Etnyre, *Admissible transverse surgery does not preserve tightness*, Math. Ann. **357** (2013), no. 2, 441–468, DOI 10.1007/s00208-013-0911-8. MR3096514 Cited in Problem 3.43.

[Bea79] Arnaud Beauville, *L'application canonique pour les surfaces de type général* (French), Invent. Math. **55** (1979), no. 2, 121–140, DOI 10.1007/BF01390086. MR553705 Cited in Problem 2.20.

[BEH$^+$15] R. İnanç Baykur, John Etnyre, Matthew Hedden, Keiko Kawamuro, and Jeremy Van Horn-Morris. Contact and symplectic geometry and the mapping class groups. American Institute of Mathematics, Official report of the 2nd SQuaRE meeting, 2015. Cited in Problem 1.69.

[Bel04] Mikhail Belolipetsky, *On volumes of arithmetic quotients of* $\mathrm{SO}(1,n)$, Ann. Sc. Norm. Super. Pisa Cl. Sci. (5) **3** (2004), no. 4, 749–770. MR2124587 Cited in Problem 4.124.

[Bel07] Mikhail Belolipetsky, *Addendum to: "On volumes of arithmetic quotients of* $\mathrm{SO}(1,n)$*" [Ann. Sc. Norm. Super. Pisa Cl. Sci. (5)* **3** *(2004), no. 4, 749–770; MR2124587]*, Ann. Sc. Norm. Super. Pisa Cl. Sci. (5) **6** (2007), no. 2, 263–268. MR2352518 Cited in Problem 4.124.

[BEM15] Matthew Strom Borman, Yakov Eliashberg, and Emmy Murphy, *Existence and classification of overtwisted contact structures in all dimensions*, Acta Math. **215** (2015), no. 2, 281–361, DOI 10.1007/s11511-016-0134-4. MR3455235 Cited in Problem 4.108.

[Ben83] Daniel Bennequin, *Entrelacements et équations de Pfaff* (French), Third Schnepfenried geometry conference, Vol. 1 (Schnepfenried, 1982), Astérisque, vol. 107, Soc. Math. France, Paris, 1983, pp. 87–161. MR753131 Cited in Problem 1.69.

[BER11] Kai-Uwe Bux, Mikhail V. Ershov, and Andrei S. Rapinchuk, *The congruence subgroup property for* $\mathrm{Aut}\, F_2$*: a group-theoretic proof of Asada's theorem*, Groups Geom. Dyn. **5** (2011), no. 2, 327–353, DOI 10.4171/GGD/130. MR2782176 Cited in Problem 2.2.

[Ber18] John Berge. Some knots with surgeries yielding lens spaces, 2018. `arXiv:1802.09722`. Cited in Problem 1.9.

[BF04] Stefan Bauer and Mikio Furuta, *A stable cohomotopy refinement of Seiberg-Witten invariants. I*, Invent. Math. **155** (2004), no. 1, 1–19, DOI 10.1007/s00222-003-0288-5. MR2025298 Cited in Problems 4.3, 4.61, and 4.64.

[BF15] R. İnanç Baykur and Stefan Friedl, *Virtually symplectic fibered 4-manifolds*, Indiana Univ. Math. J. **64** (2015), no. 4, 983–999, DOI 10.1512/iumj.2015.64.5591. MR3385783 Cited in Problem 4.92.

[BF18] Michel Boileau and Stefan Friedl, *Epimorphisms of 3-manifold groups*, Q. J. Math. **69** (2018), no. 3, 931–942, DOI 10.1093/qmath/hay007. MR3859215 Cited in Problem 1.18.

[BFH22] Aaron Brown, David Fisher, and Sebastian Hurtado, *Zimmer's conjecture: subexponential growth, measure rigidity, and strong property (T)*, Ann. of Math. (2) **196** (2022), no. 3, 891–940, DOI 10.4007/annals.2022.196.3.1. MR4502593 Cited in Problem 2.33.

[BFLL18] S. Baader, P. Feller, L. Lewark, and L. Liechti, *On the topological 4-genus of torus knots*, Trans. Amer. Math. Soc. **370** (2018), no. 4, 2639–2656, DOI 10.1090/tran/7051. MR3748580 Cited in Problem 1.46.

[BFS24] Ludovico Battista, Leonardo Ferrari, and Diego Santoro, *Dodecahedral L-spaces and hyperbolic 4-manifolds*, Comm. Anal. Geom. **32** (2024), no. 8, 2095–2134, DOI 10.4310/cag.241212004157. MR4841678 Cited in Problem 4.95.

[BFT23] Mladen Bestvina, Federica Fanoni, and Jing Tao. Towards Nielsen-Thurston classification for surfaces of infinite type, 2023. `arXiv:2303.12413`. Cited in Problem 2.37.

[BG95] Boris Botvinnik and Peter B. Gilkey, *The eta invariant and metrics of positive scalar curvature*, Math. Ann. **302** (1995), no. 3, 507–517, DOI 10.1007/BF01444505. MR1339924 Cited in Problem 4.128.

[BG19] Ryan Budney and David Gabai. Knotted 3-balls in S^4, 2019. `arXiv:1912.09029`. Cited in Problems 1.89, 4.23, 4.27, 4.67, 4.72, and 4.75.

[BG25] Ryan Budney and David Gabai, *On the automorphism groups of hyperbolic manifolds*, Int. Math. Res. Not. IMRN **7** (2025), Paper No. rnaf083, 28, DOI 10.1093/imrn/rnaf083. MR4889279 Cited in Problems 4.67, 4.72, and 4.75.

[BG24] Joe Boninger and Joshua Evan Greene, *Special alternating knots are band prime*, Int. Math. Res. Not. IMRN **10** (2024), 8758–8763, DOI 10.1093/imrn/rnae009. MR4749185 Cited in Problem 1.53.

[BGH08] Kenneth L. Baker, J. Elisenda Grigsby, and Matthew Hedden, *Grid diagrams for lens spaces and combinatorial knot Floer homology*, Int. Math. Res. Not. IMRN **10** (2008), Art. ID rnm024, 39, DOI 10.1093/imrn/rnn024. MR2429242 Cited in Problem 1.9.

[BGH25] Steven Boyer, Cameron McA. Gordon, and Ying Hu, *Cyclic branched covers of Seifert links and properties related to the ADE link conjecture*, J. Lond. Math. Soc. (2) **111** (2025), no. 6, Paper No. e70178, 53, DOI 10.1112/jlms.70178. MR4914619 Cited in Problem 1.22.

[BGPMR20] Christian Blanchet, Nathan Geer, Bertrand Patureau-Mirand, and Nicolai Reshetikhin, *Holonomy braidings, biquandles and quantum invariants of links with* $\mathrm{SL}_2(\mathbb{C})$ *flat connections*, Selecta Math. (N.S.) **26** (2020), no. 2, Paper No. 19, 58, DOI 10.1007/s00029-020-0545-0. MR4073970 Cited in Problem 3.78.

[BGR19] M. D. Baker, M. Goerner, and A. W. Reid, *All principal congruence link groups*, J. Algebra **528** (2019), 497–504, DOI 10.1016/j.jalgebra.2019.02.023. MR3957954 Cited in Problem 1.88.

[BGR22] M. D. Baker, M. Goerner, and A. W. Reid, *All principal congruence link groups*, J. Algebra **528** (2019), 497–504, DOI 10.1016/j.jalgebra.2019.02.023. MR3957954 Cited in Problem 1.88.

[BGW13] Steven Boyer, Cameron McA. Gordon, and Liam Watson, *On L-spaces and left-orderable fundamental groups*, Math. Ann. **356** (2013), no. 4, 1213–1245, DOI 10.1007/s00208-012-0852-7. MR3072799 Cited in Problem 3.48.

[BH73] Joan S. Birman and Hugh M. Hilden, *On isotopies of homeomorphisms of Riemann surfaces*, Ann. of Math. (2) **97** (1973), 424–439, DOI 10.2307/1970830. MR325959 Cited in Problem 2.29.

[BH13] Tara E. Brendle and Allen Hatcher, *Configuration spaces of rings and wickets*, Comment. Math. Helv. **88** (2013), no. 1, 131–162, DOI 10.4171/CMH/280. MR3008915 Cited in Problem 2.8.

[BH16a] R. Inanç Baykur and Kenta Hayano, *Hurwitz equivalence for Lefschetz fibrations and their multisections*, Real and complex singularities, Contemp. Math., vol. 675, Amer. Math. Soc., Providence, RI, 2016, pp. 1–24, DOI 10.1090/conm/675. MR3578715 Cited in Problem 2.22.

[BH16b] R. İnanç Baykur and Kenta Hayano, *Multisections of Lefschetz fibrations and topology of symplectic 4-manifolds*, Geom. Topol. **20** (2016), no. 4, 2335–2395, DOI 10.2140/gt.2016.20.2335. MR3548468 Cited in Problems 2.22, 4.92, and 4.98.

[BH23] R. Inanc Baykur and Noriyuki Hamada, Exotic 4-manifolds with signature zero, 2023. Selecta Math., to appear. `arXiv:2305.10908`. Cited in Problems 2.21, 4.3, and 4.30.

[BH24a] David Baraglia and Pedram Hekmati, *Equivariant Seiberg-Witten-Floer cohomology*, Algebr. Geom. Topol. **24** (2024), no. 1, 493–554, DOI 10.2140/agt.2024.24.493. MR4721374 Cited in Problems 3.75 and 3.76.

[BH24b] David Baraglia and Pedram Hekmati. New invariants of involutions from Seiberg-Witten-Floer theory, 2024. `arXiv:2403.00203`. Cited in Problem 1.47.

[BH24c] R. İnanç Baykur and Noriyuki Hamada, *Lefschetz fibrations with arbitrary signature*, J. Eur. Math. Soc. (JEMS) **26** (2024), no. 8, 2837–2895, DOI 10.4171/jems/1326. MR4756947 Cited in Problems 2.6, 2.21, 2.22, and 4.97.

[BH25] Mark Brittenham and Susan Hermiller. Unknotting number is not additive under connected sum, 2025. `arXiv:2506.24088`. Cited in Problem 1.3.

[Bha24] Deeparaj Bhat. Surgery exact triangles in instanton theory, 2024. `arXiv:2311.04242`. Cited in Problems 3.50 and 3.54.

[BHH23] Joseph Breen, Ko Honda, and Yang Huang. The Giroux correspondence in arbitrary dimensions, 2023. `arXiv:2307.02317`. Cited in Problem 4.108.

[BHH24] Joseph Breen, Ko Honda, and Yang Huang. The Giroux correspondence in arbitrary dimensions, 2024. URL: `https://arxiv.org/abs/2307.02317`, `arXiv:2307.02317`. Cited in Problem 3.39.

[BHM23] R. İnanç Baykur, Kenta Hayano, and Naoyuki Monden, *Unchaining surgery and topology of symplectic 4-manifolds*, Math. Z. **303** (2023), no. 3, Paper No. 77, 32, DOI 10.1007/s00209-023-03204-x. MR4552624 Cited in Problems 2.6, 4.92, and 4.98.

[BHS25] John A. Baldwin, Ying Hu, and Steven Sivek, *Khovanov homology and the cinquefoil*, J. Eur. Math. Soc. (JEMS) **27** (2025), no. 6, 2443–2465, DOI 10.4171/jems/1415. MR4889247 Cited in Problem 1.21.

[BHT01] Tara E. Brendle and Hessam Hamidi-Tehrani, *On the linearity problem for mapping class groups*, Algebr. Geom. Topol. **1** (2001), 445–468, DOI 10.2140/agt.2001.1.445. MR1852767 Cited in Problem 2.3.

[BHW22] Jonathan Bowden, Sebastian Wolfgang Hensel, and Richard Webb, *Quasi-morphisms on surface diffeomorphism groups*, J. Amer. Math. Soc. **35** (2022), no. 1, 211–231, DOI 10.1090/jams/981. MR4322392 Cited in Problem 4.85.

[BI09] Sebastian Baader and Masaharu Ishikawa, *Legendrian graphs and quasipositive diagrams* (English, with English and French summaries), Ann. Fac. Sci. Toulouse Math. (6) **18** (2009), no. 2, 285–305. MR2562830 Cited in Problem 1.69.

[Big01] Stephen J. Bigelow, *Braid groups are linear*, J. Amer. Math. Soc. **14** (2001), no. 2, 471–486, DOI 10.1090/S0894-0347-00-00361-1. MR1815219 Cited in Problem 2.3.

[Bin56] R. H. Bing, *A simple closed curve that pierces no disk*, J. Math. Pures Appl. (9) **35** (1956), 337–343. MR81461 Cited in Problem 1.81.

[Bin69] R. H. Bing. The elusive fixed point property. *Amer. Math. Monthly*, 76:119–132, 1969. `doi:10.2307/2317258`. Cited in Problem 5.1.

[BIP08] Dmitri Burago, Sergei Ivanov, and Leonid Polterovich, *Conjugation-invariant norms on groups of geometric origin*, Groups of diffeomorphisms, Adv. Stud. Pure Math., vol. 52, Math. Soc. Japan, Tokyo, 2008, pp. 221–250, DOI 10.2969/aspm/05210221. MR2509711 Cited in Problem 4.85.

[Bir69] Joan S. Birman, *Mapping class groups and their relationship to braid groups*, Comm. Pure Appl. Math. **22** (1969), 213–238, DOI 10.1002/cpa.3160220206. MR243519 Cited in Problem 4.67.

[Bir71] Joan S. Birman. On Siegel's modular group. *Math. Ann.*, 191:59–68, 1971. `doi:10.1007/BF01433472`. Cited in Problem 2.5.

[BJW18] Michel Boileau, Yeonhee Jang, and Richard Weidmann, *Meridional rank and bridge number for a class of links*, Pacific J. Math. **292** (2018), no. 1, 61–80, DOI 10.2140/pjm.2018.292.61. MR3708258 Cited in Problem 1.18.

[BK85] K. Burns and A. Katok. Manifolds with nonpositive curvature. *Ergodic Theory Dynam. Systems*, 5(2):307–317, 1985. `doi:10.1017/S0143385700002935`. Cited in Problem 2.16.

[BK02a] Mario Bonk and Bruce Kleiner, *Quasisymmetric parametrizations of two-dimensional metric spheres*, Invent. Math. **150** (2002), no. 1, 127–183, DOI 10.1007/s00222-002-0233-z. MR1930885 Cited in Problem 3.24.

[BK02b] Mario Bonk and Bruce Kleiner, *Rigidity for quasi-Möbius group actions*, J. Differential Geom. **61** (2002), no. 1, 81–106. MR1949785 Cited in Problem 3.24.

[BK03] V. Braungardt and D. Kotschick, *Clustering of critical points in Lefschetz fibrations and the symplectic Szpiro inequality*, Trans. Amer. Math. Soc. **355** (2003), no. 8, 3217–3226, DOI 10.1090/S0002-9947-03-03290-2. MR1974683 Cited in Problem 2.21.

[BK05] Mario Bonk and Bruce Kleiner, *Conformal dimension and Gromov hyperbolic groups with 2-sphere boundary*, Geom. Topol. **9** (2005), 219–246, DOI 10.2140/gt.2005.9.219. MR2116315 Cited in Problem 3.24.

[BK07] Scott Baldridge and Paul Kirk, *On symplectic 4-manifolds with prescribed fundamental group*, Comment. Math. Helv. **82** (2007), no. 4, 845–875, DOI 10.4171/CMH/112. MR2341842 Cited in Problem 4.91.

[BK08] Ara Basmajian and Youngju Kim, *Geometrically infinite surfaces with discrete length spectra*, Geom. Dedicata **137** (2008), 219–240, DOI 10.1007/s10711-008-9294-5. MR2449153 Cited in Problem 2.40.

[BK15a] R. İnanç Baykur and Seiichi Kamada, *Classification of broken Lefschetz fibrations with small fiber genera*, J. Math. Soc. Japan **67** (2015), no. 3, 877–901, DOI 10.2969/jmsj/06730877. MR3376572 Cited in Problem 4.113.

[BK15b] John Burke and Robin Koytcheff, *A colored operad for string link infection*, Algebr. Geom. Topol. **15** (2015), no. 6, 3371–3408, DOI 10.2140/agt.2015.15.3371. MR3450765 Cited in Problem 1.89.

[BK17] R. İnanç Baykur and Mustafa Korkmaz, *Small Lefschetz fibrations and exotic 4-manifolds*, Math. Ann. **367** (2017), no. 3-4, 1333–1361, DOI 10.1007/s00208-016-1466-2. MR3623227 Cited in Problem 2.21.

[BK20] David Baraglia and Hokuto Konno, *A gluing formula for families Seiberg-Witten invariants*, Geom. Topol. **24** (2020), no. 3, 1381–1456, DOI 10.2140/gt.2020.24.1381. MR4157556 Cited in Problems 4.68, 4.70, and 4.80.

[BK22] David Baraglia and Hokuto Konno, *On the Bauer-Furuta and Seiberg-Witten invariants of families of 4-manifolds*, J. Topol. **15** (2022), no. 2, 505–586, DOI 10.1112/topo.12229. MR4441598 Cited in Problems 4.61 and 4.70.

[BK23] David Baraglia and Hokuto Konno, *A note on the Nielsen realization problem for $K3$ surfaces*, Proc. Amer. Math. Soc. **151** (2023), no. 9, 4079–4087, DOI 10.1090/proc/15544. MR4607650 Cited in Problems 4.70 and 4.84.

[BK24a] David Baraglia and Hokuto Konno. Irreducible 4-manifolds can admit exotic diffeomorphisms, 2024. `arXiv:2412.14398`. Cited in Problem 4.68.

[BK24b] R. İnanç Baykur and Mustafa Korkmaz, *Geography of surface bundles over surfaces*, Math. Ann. **390** (2024), no. 4, 5793–5817, DOI 10.1007/s00208-024-02899-5. MR4816121 Cited in Problem 4.97.

[BKK16] Hyungryul Baik, Sang-hyun Kim, and Thomas Koberda, *Right-angled Artin groups in the C^∞ diffeomorphism group of the real line*, Israel J. Math. **213** (2016), no. 1, 175–182, DOI 10.1007/s11856-016-1307-8. MR3509472 Cited in Problem 2.27.

[BKK+21] Stefan Behrens, Boldizsár Kalmár, Min Hoon Kim, Mark Powell, and Arunima Ray (eds.), *The disc embedding theorem*, Oxford University Press, Oxford, 2021. MR4519498 Cited in Problems 1.46, 3.73, and 4.54.

[BKK24] Mauricio Bustamante, Manuel Krannich, and Alexander Kupers, *Finiteness properties of automorphism spaces of manifolds with finite fundamental group*, Math. Ann. **388** (2024), no. 4, 3321–3371, DOI 10.1007/s00208-023-02594-x. MR4721772 Cited in Problems 4.71 and 4.75.

[BKKT20] Christian Bonatti, Sang-hyun Kim, Thomas Koberda, and Michele Triestino, *Small C^1 actions of semidirect products on compact manifolds*, Algebr. Geom. Topol. **20** (2020), no. 6, 3183–3203, DOI 10.2140/agt.2020.20.3183. MR4185939 Cited in Problem 3.22.

[BKL98] Joan Birman, Ki Hyoung Ko, and Sang Jin Lee, *A new approach to the word and conjugacy problems in the braid groups*, Adv. Math. **139** (1998), no. 2, 322–353, DOI 10.1006/aima.1998.1761. MR1654165 Cited in Problem 2.9.

[BKST25] Fraser Binns, Sungkyung Kang, Jonathan Simone, and Paula Truöl, *On the nonorientable four-ball genus of torus knots*, Algebr. Geom. Topol. **25** (2025), no. 4, 2209–2251, DOI 10.2140/agt.2025.25.2209. MR4950903 Cited in Problem 1.47.

[BL74] Dan Burghelea and Richard Lashof, *The homotopy type of the space of diffeomorphisms. I, II*, Trans. Amer. Math. Soc. **196** (1974), 1–36; ibid. 196 (1974), 37–50, DOI 10.2307/1997010. MR356103 Cited in Problem 4.73.

[BL89] Steven A. Bleiler and Richard A. Litherland, *Lens spaces and Dehn surgery*, Proc. Amer. Math. Soc. **107** (1989), no. 4, 1127–1131, DOI 10.2307/2047677. MR984783 Cited in Problem 1.9.

[BL02] Christian Bohr and Ronnie Lee. Homology cobordism and classical knot invariants. *Comment. Math. Helv.*, 77(2):363–382, 2002. doi:10.1007/s00014-002-8344-0. Cited in Problem 3.71.

[BL23] Henk Bruin and Olga Lukina, *Rotated odometers*, J. Lond. Math. Soc. (2) **107** (2023), no. 6, 1983–2024, DOI 10.1112/jlms.12731. MR4598177 Cited in Problem 2.42.

[BLL65] W. Browder, J. Levine, and G. R. Livesay, *Finding a boundary for an open manifold*, Amer. J. Math. **87** (1965), 1017–1028, DOI 10.2307/2373259. MR189046 Cited in Problem 4.49.

[BLL+21] Sarah Blackwell, Noémie Legout, Caitlin Leverson, Maÿlis Limouzineau, Ziva Myer, Yu Pan, Samantha Pezzimenti, Lara Simone Suárez, and Lisa Traynor, *Constructions of Lagrangian cobordisms*, Research directions in symplectic and contact geometry and topology, Assoc. Women Math. Ser., vol. 27, Springer, Cham, [2021] ©2021, pp. 245–272, DOI 10.1007/978-3-030-80979-9_5. MR4417718 Cited in Problem 1.65.

[BLM+83] Joan S. Birman, Alex Lubotzky, and John McCarthy, *Abelian and solvable subgroups of the mapping class groups*, Duke Math. J. **50** (1983), no. 4, 1107–1120, DOI 10.1215/S0012-7094-83-05046-9. MR726319 Cited in Problem 2.41.

[BLM23] Richard H. Bamler, Chao Li, and Christos Mantoulidis, *Decomposing 4-manifolds with positive scalar curvature*, Adv. Math. **430** (2023), Paper No. 109231, 17, DOI 10.1016/j.aim.2023.109231. MR4621960 Cited in Problems 4.128 and 4.129.

[BLSY24] John A. Baldwin, Zhenkun Li, Steven Sivek, and Fan Ye, *Small Dehn surgery and* SU(2), Geom. Topol. **28** (2024), no. 4, 1891–1922, DOI 10.2140/gt.2024.28.1891. MR4777704 Cited in Problem 1.14.

[BLW10] Arthur Bartels, Wolfgang Lück, and Shmuel Weinberger, *On hyperbolic groups with spheres as boundary*, J. Differential Geom. **86** (2010), no. 1, 1–16. MR2772544 Cited in Problem 3.24.

[BLY23] John A. Baldwin, Zhenkun Li, and Fan Ye, *Sutured instanton homology and Heegaard diagrams*, Compos. Math. **159** (2023), no. 9, 1898–1915, DOI 10.1112/s0010437x23007303. MR4622392 Cited in Problems 3.53 and 3.59.

[BLZ24] Maciej Borodzik, Beibei Liu, and Ian Zemke. Lattice homology, formality, and plumbed l-space links, 2024. arXiv:2210.15792. Cited in Problem 3.57.

[BM81] Robert Brooks and J. Peter Matelski, *The dynamics of 2-generator subgroups of* PSL(2, **C**), Riemann surfaces and related topics: Proceedings of the 1978 Stony Brook Conference (State Univ. New York, Stony Brook, N.Y., 1978), Ann. of Math. Stud., No. 97, Princeton Univ. Press, Princeton, NJ, 1981, pp. 65–71. MR624805 Cited in Problem 5.1.

[BM91] Mladen Bestvina and Geoffrey Mess, *The boundary of negatively curved groups*, J. Amer. Math. Soc. **4** (1991), no. 3, 469–481, DOI 10.2307/2939264. MR1096169 Cited in Problem 3.24.

[BM06] Robert W. Bell and Dan Margalit, *Braid groups and the co-Hopfian property*, J. Algebra **303** (2006), no. 1, 275–294, DOI 10.1016/j.jalgebra.2005.10.038. MR2253663 Cited in Problem 2.7.

[BM19] Tara E. Brendle and Dan Margalit, *Normal subgroups of mapping class groups and the metaconjecture of Ivanov*, J. Amer. Math. Soc. **32** (2019), no. 4, 1009–1070, DOI 10.1090/jams/927. MR4013739 Cited in Problem 2.4.

[BM23] Keegan Boyle and Jeffrey Musyt, *Equivariant cobordisms between freely periodic knots*, Canad. Math. Bull. **66** (2023), no. 2, 450–457, DOI 10.4153/S000843952200042X. MR4584473 Cited in Problem 1.49.

[BM24a] Kenneth L. Baker and Kimihiko Motegi. The stable crossing number of a twist family of knots and the satellite crossing number conjecture, 2024. `arXiv:2404.05308`. Cited in Problem 1.2.

[BM24b] Thomas Barthelmé and Kathryn Mann, *Orbit equivalences of $\mathbb{R}$-covered Anosov flows and hyperbolic-like actions on the line*, Geom. Topol. **28** (2024), no. 2, 867–899, DOI 10.2140/gt.2024.28.867. Appendix written jointly with Jonathan Bowden. MR4718129 Cited in Problem 3.21.

[BM25] R. İnanç Baykur and Porter Morgan. On nonorientable 4–manifolds, 2025. Math. Res. Lett., to appear. URL: `https://arxiv.org/abs/2506.20950`, `arXiv:2506.20950`. Cited in Problems 4.57 and 4.113.

[BMM16] Joan Birman, Dan Margalit, and William Menasco, *Efficient geodesics and an effective algorithm for distance in the complex of curves*, Math. Ann. **366** (2016), no. 3-4, 1253–1279, DOI 10.1007/s00208-015-1357-y. MR3563237 Cited in Problem 2.10.

[BMR95] B. H. Bowditch, C. Maclachlan, and A. W. Reid, *Arithmetic hyperbolic surface bundles*, Math. Ann. **302** (1995), no. 1, 31–60, DOI 10.1007/BF01444486. MR1329446 Cited in Problem 2.19.

[BMRS20] M. R. Bridson, D. B. McReynolds, A. W. Reid, and R. Spitler, *Absolute profinite rigidity and hyperbolic geometry*, Ann. of Math. (2) **192** (2020), no. 3, 679–719, DOI 10.4007/annals.2020.192.3.1. MR4172619 Cited in Problem 3.8.

[BMT20] Kenneth L. Baker, Kimihiko Motegi, and Toshie Takata, *The strong slope conjecture for twisted generalized Whitehead doubles*, Quantum Topol. **11** (2020), no. 3, 545–608, DOI 10.4171/qt/242. MR4164004 Cited in Problem 1.26.

[BMT21] Kenneth L. Baker, Kimihiko Motegi, and Toshie Takata, *The strong slope conjecture for cablings and connected sums*, New York J. Math. **27** (2021), 676–704. MR4250271 Cited in Problem 1.26.

[BMVHM17] R. İnanç Baykur, Naoyuki Monden, and Jeremy Van Horn-Morris, *Positive factorizations of mapping classes*, Algebr. Geom. Topol. **17** (2017), no. 3, 1527–1555, DOI 10.2140/agt.2017.17.1527. MR3677934 Cited in Problem 2.21.

[BN90] S. Boyer and A. Nicas, *Varieties of group representations and Casson's invariant for rational homology 3-spheres*, Trans. Amer. Math. Soc. **322** (1990), no. 2, 507–522, DOI 10.2307/2001712. MR972701 Cited in Problem 3.51.

[BN17] Hans U. Boden and Matthias Nagel, *Concordance group of virtual knots*, Proc. Amer. Math. Soc. **145** (2017), no. 12, 5451–5461, DOI 10.1090/proc/13667. MR3717970 Cited in Problem 1.84.

[BNS23] Elia Bruè, Aaron Naber, and Daniele Semola. Six dimensional counterexample to the Milnor Conjecture, 2023. `arXiv:2311.12155`. Cited in Problem 5.15.

[BNS25] Elia Bruè, Aaron Naber, and Daniele Semola, *Fundamental groups and the Milnor conjecture*, Ann. of Math. (2) **201** (2025), no. 1, 225–289, DOI 10.4007/annals.2025.201.1.4. MR4848671 Cited in Problem 5.15.

[BNV23] Assaf Bar-Natan and Yvon Verberne, *The grand arc graph*, Math. Z. **305** (2023), no. 2, Paper No. 20, 21, DOI 10.1007/s00209-023-03337-z. MR4641121 Cited in Problem 2.38.

[BO01] Michel Boileau and Stepan Orevkov, *Quasi-positivité d'une courbe analytique dans une boule pseudo-convexe* (French, with English and French summaries), C. R. Acad. Sci. Paris Sér. I Math. **332** (2001), no. 9, 825–830, DOI 10.1016/S0764-4442(01)01945-0. MR1836094 Cited in Problems 1.64, 1.68, 1.70, and 4.104.

[BO04] Francis Bonahon and Jean-Pierre Otal, *Laminations measurées de plissage des variétés hyperboliques de dimension 3* (French, with English summary), Ann. of Math. (2) **160** (2004), no. 3, 1013–1055, DOI 10.4007/annals.2004.160.1013. MR2144972 Cited in Problem 3.25.

[Bod19] Benjamin Bode, *Constructing links of isolated singularities of polynomials $\mathbb{R}^4 \to \mathbb{R}^2$*, J. Knot Theory Ramifications **28** (2019), no. 1, 1950009, 21, DOI 10.1142/S0218216519500093. MR3910952 Cited in Problem 1.87.

[Bod20] Benjamin Bode, *Real algebraic links in S^3 and braid group actions on the set of n-adic integers*, J. Knot Theory Ramifications **29** (2020), no. 6, 2050039, 44, DOI 10.1142/S021821652050039X. MR4125176 Cited in Problem 1.87.

[Bod23] Benjamin Bode, *Twisting and satellite operations on P-fibered braids*, Comm. Anal. Geom. **31** (2023), no. 8, 2013–2038, DOI 10.4310/cag.2023.v31.n8.a5. MR4785581 Cited in Problem 1.87.

[Bol09] Dmitry Bolotov, *About the macroscopic dimension of certain PSC-manifolds*, Algebr. Geom. Topol. **9** (2009), no. 1, 21–27, DOI 10.2140/agt.2009.9.21. MR2471130 Cited in Problem 4.129.

[Bon83] Francis Bonahon, *Cobordism of automorphisms of surfaces*, Ann. Sci. École Norm. Sup. (4) **16** (1983), no. 2, 237–270. MR732345 Cited in Problem 2.45.

[Bon86] Francis Bonahon, *Bouts des variétés hyperboliques de dimension* 3 (French), Ann. of Math. (2) **124** (1986), no. 1, 71–158, DOI 10.2307/1971388. MR847953 Cited in Problem 2.17.

[Bon88] Francis Bonahon, *The geometry of Teichmüller space via geodesic currents*, Invent. Math. **92** (1988), no. 1, 139–162, DOI 10.1007/BF01393996. MR931208 Cited in Problem 2.17.

[Bon96] Francis Bonahon, *Shearing hyperbolic surfaces, bending pleated surfaces and Thurston's symplectic form* (English, with English and French summaries), Ann. Fac. Sci. Toulouse Math. (6) **5** (1996), no. 2, 233–297. MR1413855 Cited in Problem 3.25.

[Bon05] Francis Bonahon, *Kleinian groups which are almost Fuchsian*, J. Reine Angew. Math. **587** (2005), 1–15, DOI 10.1515/crll.2005.2005.587.1. MR2186972 Cited in Problem 3.25.

[Boo23] David Boozer, *Computer bounds for Kronheimer-Mrowka foam evaluation*, Exp. Math. **32** (2023), no. 4, 615–630, DOI 10.1080/10586458.2021.1982078. MR4669283 Cited in Problem 1.23.

[Boo25] David Boozer. The combinatorial and gauge-theoretic foam evaluation functors are not the same. *Math. Ann.*, 392(1):47–56, 2025. `doi:10.1007/s00208-024-03064-8`. Cited in Problem 1.23.

[Bor81] A. Borel, *Commensurability classes and volumes of hyperbolic 3-manifolds*, Ann. Scuola Norm. Sup. Pisa Cl. Sci. (4) **8** (1981), no. 1, 1–33. MR616899 Cited in Problem 3.2.

[Bow09] Brian H. Bowditch, *Atoroidal surface bundles over surfaces*, Geom. Funct. Anal. **19** (2009), no. 4, 943–988, DOI 10.1007/s00039-009-0033-3. MR2570310 Cited in Problem 2.23.

[Bow11] Jonathan Bowden, *On closed leaves of foliations, multisections and stable commutator lengths*, J. Topol. Anal. **3** (2011), no. 4, 491–509, DOI 10.1142/S1793525311000696. MR2887673 Cited in Problem 4.123.

[Bow12] Jonathan Bowden, *Exactly fillable contact structures without Stein fillings*, Algebr. Geom. Topol. **12** (2012), no. 3, 1803–1810, DOI 10.2140/agt.2012.12.1803. MR2979998 Cited in Problem 4.110.

[Bow16] Jonathan Bowden, *Approximating C^0-foliations by contact structures*, Geom. Funct. Anal. **26** (2016), no. 5, 1255–1296, DOI 10.1007/s00039-016-0387-2. MR3568032 Cited in Problems 3.15, 3.40, and 3.48.

[Boy88] Philip Boyland, *An analog of Sharkovski's theorem for twist maps*, Hamiltonian dynamical systems (Boulder, CO, 1987), Contemp. Math., vol. 81, Amer. Math. Soc., Providence, RI, 1988, pp. 119–133, DOI 10.1090/conm/081/986261. MR986261 Cited in Problem 2.43.

[Boy93] Steven Boyer, *Realization of simply-connected 4-manifolds with a given boundary*, Comment. Math. Helv. **68** (1993), no. 1, 20–47, DOI 10.1007/BF02565808. MR1201200 Cited in Problem 4.45.

[BP92] Riccardo Benedetti and Carlo Petronio, *Lectures on hyperbolic geometry*, Universitext, Springer-Verlag, Berlin, 1992, DOI 10.1007/978-3-642-58158-8. MR1219310 Cited in Problem 3.3.

[BPR21] Stefan Behrens, Mark Powell, and Arunima Ray, *Context for the disc embedding theorem*, The disc embedding theorem, Oxford Univ. Press, Oxford, 2021, pp. 1–26. MR4519499 Cited in Problem 1.62.

[BPVdV84] W. Barth, C. Peters, and A. Van de Ven. *Compact complex surfaces*, volume 4 of *Ergebnisse der Mathematik und ihrer Grenzgebiete (3) [Results in Mathematics and Related Areas (3)]*. Springer-Verlag, Berlin, 1984. doi:10.1007/978-3-642-96754-2. Cited in Problem 4.90.

[BR23a] Mark D. Baker and Alan W. Reid, *Infinitely many arithmetic alternating links*, Algebr. Geom. Topol. **23** (2023), no. 6, 2857–2866, DOI 10.2140/agt.2023.23.2857. MR4640142 Cited in Problem 1.88.

[BR23b] Keegan Boyle and Nicholas Rouse. Obstructions to free periodicity and symmetric L-space knots, 2023. arXiv:2310.01705. Cited in Problem 1.49.

[Bre97] Glen E. Bredon, *Topology and geometry*, Graduate Texts in Mathematics, vol. 139, Springer-Verlag, New York, 1997. Corrected third printing of the 1993 original. MR1700700 Cited in Problem 3.73.

[Bri66] Egbert Brieskorn, *Beispiele zur Differentialtopologie von Singularitäten* (German), Invent. Math. **2** (1966), 1–14, DOI 10.1007/BF01403388. MR206972 Cited in Problem 4.32.

[Bri83] M. Brin, *Curves isotopic to tame curves*, Continua, decompositions, manifolds (Austin, Tex., 1980), Univ. Texas Press, Austin, TX, 1983, pp. 163–166. MR711988 Cited in Problem 1.81.

[Bri10] Martin R. Bridson, *Semisimple actions of mapping class groups on* CAT(0) *spaces*, Geometry of Riemann surfaces, London Math. Soc. Lecture Note Ser., vol. 368, Cambridge Univ. Press, Cambridge, 2010, pp. 1–14. MR2665003 Cited in Problem 2.3.

[BRM15] Khalid Bou-Rabee and D. B. McReynolds, *Extremal behavior of divisibility functions*, Geom. Dedicata **175** (2015), 407–415, DOI 10.1007/s10711-014-9955-5. MR3323650 Cited in Problem 2.3.

[Bro62] Morton Brown. Locally flat imbeddings of topological manifolds. *Ann. of Math. (2)*, 75:331–341, 1962. doi:10.2307/1970177. Cited in Problem 5.5.

[Bro72] William Browder, *Surgery on simply-connected manifolds*, Ergebnisse der Mathematik und ihrer Grenzgebiete [Results in Mathematics and Related Areas], Band 65, Springer-Verlag, New York-Heidelberg, 1972. MR358813 Cited in Problem 4.46.

[Bro77] Morton Brown, *A short short proof of the Cartwright-Littlewood theorem*, Proc. Amer. Math. Soc. **65** (1977), no. 2, 372, DOI 10.2307/2041926. MR461491 Cited in Problem 5.2.

[Bro84] Morton Brown, *A new proof of Brouwer's lemma on translation arcs*, Houston J. Math. **10** (1984), no. 1, 35–41. MR736573 Cited in Problem 5.2.

[Bro03] Jeffrey F. Brock, *Weil-Petersson translation distance and volumes of mapping tori*, Comm. Anal. Geom. **11** (2003), no. 5, 987–999, DOI 10.4310/CAG.2003.v11.n5.a6. MR2032506 Cited in Problem 2.18.

[BRW05] Steven Boyer, Dale Rolfsen, and Bert Wiest, *Orderable 3-manifold groups* (English, with English and French summaries), Ann. Inst. Fourier (Grenoble) **55** (2005), no. 1, 243–288, DOI 10.5802/aif.2098. MR2141698 Cited in Problems 3.20 and 3.48.

[Bry98] Jim Bryan, *Seiberg-Witten theory and* $\mathbf{Z}/2^p$ *actions on spin* 4*-manifolds*, Math. Res. Lett. **5** (1998), no. 1-2, 165–183, DOI 10.4310/MRL.1998.v5.n2.a3. MR1617929 Cited in Problems 4.44 and 4.86.

[BS88a] Steven Bleiler and Martin Scharlemann, *A projective plane in* $\mathbf{R}^4$ *with three critical points is standard. Strongly invertible knots have property P*, Topology **27** (1988), no. 4, 519–540, DOI 10.1016/0040-9383(88)90030-4. MR976593 Cited in Problem 4.29.

[BS88b] P. Buser and K.-D. Semmler, *The geometry and spectrum of the one-holed torus*, Comment. Math. Helv. **63** (1988), no. 2, 259–274, DOI 10.1007/BF02566766. MR948781 Cited in Problem 2.16.

[BS98] R. Benedetti and M. Shiota, *On real algebraic links in* S^3 (English, with Italian summary), Boll. Unione Mat. Ital. Sez. B Artic. Ric. Mat. (8) **1** (1998), no. 3, 585–609. MR1662333 Cited in Problem 1.87.

[BS13] R. İnanç Baykur and Nathan Sunukjian, *Round handles, logarithmic transforms and smooth 4-manifolds*, J. Topol. **6** (2013), no. 1, 49–63, DOI 10.1112/jtopol/jts027. MR3029421 Cited in Problems 4.7, 4.34, and 4.57.

[BS16] R. İnanç Baykur and Nathan Sunukjian, *Knotted surfaces in 4-manifolds and stabilizations*, J. Topol. **9** (2016), no. 1, 215–231, DOI 10.1112/jtopol/jtv039. MR3465848 Cited in Problem 4.33.

[BS18a] John A. Baldwin and Steven Sivek, *Invariants of Legendrian and transverse knots in monopole knot homology*, J. Symplectic Geom. **16** (2018), no. 4, 959–1000, DOI 10.4310/JSG.2018.v16.n4.a3. MR3917725 Cited in Problem 1.66.

[BS18b] John A. Baldwin and Steven Sivek, *Stein fillings and* SU(2) *representations*, Geom. Topol. **22** (2018), no. 7, 4307–4380, DOI 10.2140/gt.2018.22.4307. MR3890778 Cited in Problems 3.51 and 3.52.

[BS18c] R. İnanç Baykur and Osamu Saeki, *Simplified broken Lefschetz fibrations and trisections of 4-manifolds*, Proc. Natl. Acad. Sci. USA **115** (2018), no. 43, 10894–10900, DOI 10.1073/pnas.1717175115. MR3871793 Cited in Problems 4.113 and 4.114.

[BS21] John A. Baldwin and Steven Sivek, *Framed instanton homology and concordance*, J. Topol. **14** (2021), no. 4, 1113–1175, DOI 10.1112/topo.12207. MR4332488 Cited in Problem 3.61.

[BS22] John A. Baldwin and Steven Sivek, *Instanton L-spaces and splicing* (English, with English and French summaries), Ann. H. Lebesgue **5** (2022), 1213–1233, DOI 10.5802/ahl.148. MR4526251 Cited in Problem 3.52.

[BS23a] John A. Baldwin and Steven Sivek, *Instantons and L-space surgeries*, J. Eur. Math. Soc. (JEMS) **25** (2023), no. 10, 4033–4122, DOI 10.4171/jems/1280. MR4634689 Cited in Problem 1.14.

[BS23b] R. İnanç Baykur and Osamu Saeki, *Simplifying indefinite fibrations on 4-manifolds*, Trans. Amer. Math. Soc. **376** (2023), no. 5, 3011–3062, DOI 10.1090/tran/8325. MR4577327 Cited in Problems 4.113 and 4.114.

[BSS24] R. Inanc Baykur, Andras I. Stipsicz, and Zoltan Szabo. Smooth structures on four-manifolds with finite cyclic fundamental groups, 2024. `arXiv:2406.09007`. Cited in Problems 4.2, 4.3, and 4.91.

[BSZ25] John A. Baldwin, Steven Sivek, and Jonathan Zung. Pseudo-Anosov flows on hyperbolic l-spaces, 2025. `arXiv:2505.21113`. Cited in Problem 3.21.

[BT12] Carl-Friedrich Bödigheimer and Ulrike Tillmann, *Embeddings of braid groups into mapping class groups and their homology*, Configuration spaces, CRM Series, vol. 14, Ed. Norm., Pisa, 2012, pp. 173–191, DOI 10.1007/978-88-7642-431-1_7. MR3203638 Cited in Problem 2.29.

[Bud07] Ryan Budney, *Little cubes and long knots*, Topology **46** (2007), no. 1, 1–27, DOI 10.1016/j.top.2006.09.001. MR2288724 Cited in Problem 1.89.

[Bud10] Ryan Budney, *Topology of knot spaces in dimension 3*, Proc. Lond. Math. Soc. (3) **101** (2010), no. 2, 477–496, DOI 10.1112/plms/pdp058. MR2679699 Cited in Problem 1.89.

[Bud12] Ryan Budney, *An operad for splicing*, J. Topol. **5** (2012), no. 4, 945–976, DOI 10.1112/jtopol/jts024. MR3001316 Cited in Problem 1.89.

[Bud16] Ryan Budney. A gorgeous but incomplete proof of "The Smale Conjecture", 2016. `https://ldtopology.wordpress.com/2016/10/02/a-gorgeous-but-incomplete-proof-of-the-smale-conjecture/`. Accessed: 2024-01-25. Cited in Problem 1.75.

[Bul97] Doug Bullock. Rings of $SL_2(\mathbf{C})$-characters and the Kauffman bracket skein module. *Comment. Math. Helv.*, 72(4):521–542, 1997. `doi:10.1007/s000140050032`. Cited in Problem 3.65.

[Bur11] Benjamin A. Burton, *The Pachner graph and the simplification of 3-sphere triangulations*, Computational geometry (SCG'11), ACM, New York, 2011, pp. 153–162, DOI 10.1145/1998196.1998220. MR2919606 Cited in Problem 3.28.

[Bur14] Benjamin A. Burton. The cusped hyperbolic census is complete, 2014. `arXiv:1405.2695`. Cited in Problem 1.12.

[Bur18] Stephan D. Burton, *The determinant and volume of 2-bridge links and alternating 3-braids*, New York J. Math. **24** (2018), 293–316. MR3778507 Cited in Problem 1.29.

[But14] J. O. Button, *A 3-manifold group which is not four dimensional linear*, J. Pure Appl. Algebra **218** (2014), no. 9, 1604–1619, DOI 10.1016/j.jpaa.2014.01.001. MR3188859 Cited in Problem 3.37.

[But19] Jack Oliver Button, *Aspects of non positive curvature for linear groups with no infinite order unipotents*, Groups Geom. Dyn. **13** (2019), no. 1, 277–292, DOI 10.4171/GGD/484. MR3900771 Cited in Problem 2.3.

[BV13] Nicolas Bergeron and Akshay Venkatesh, *The asymptotic growth of torsion homology for arithmetic groups*, J. Inst. Math. Jussieu **12** (2013), no. 2, 391–447, DOI 10.1017/S1474748012000667. MR3028790 Cited in Problem 3.6.

[BVHM15] R. İnanç Baykur and Jeremy Van Horn-Morris, *Families of contact 3-manifolds with arbitrarily large Stein fillings*, J. Differential Geom. **101** (2015), no. 3, 423–465. With an appendix by Samuel Lisi and Chris Wendl. MR3415768 Cited in Problem 2.21.

[BVHM16] R. İnanç Baykur and Jeremy Van Horn-Morris, *Topological complexity of symplectic 4-manifolds and Stein fillings*, J. Symplectic Geom. **14** (2016), no. 1, 171–202, DOI 10.4310/JSG.2016.v14.n1.a7. MR3523254 Cited in Problem 2.21.

[BVHM18] R. İnanç Baykur and Jeremy Van Horn-Morris, *Fillings of genus-1 open books and 4-braids*, Int. Math. Res. Not. IMRN **5** (2018), 1329–1346, DOI 10.1093/imrn/rnw281. MR3801464 Cited in Problems 4.104 and 4.105.

[BVV18] John Baldwin and David Shea Vela-Vick, *A note on the knot Floer homology of fibered knots*, Algebr. Geom. Topol. **18** (2018), no. 6, 3669–3690, DOI 10.2140/agt.2018.18.3669. MR3868231 Cited in Problem 3.55.

[BW84] Michel Boileau and Claude Weber, *Le problème de J. Milnor sur le nombre gordien des nœuds algébriques* (French), Enseign. Math. (2) **30** (1984), no. 3-4, 173–222. MR767901 Cited in Problem 1.5.

[BW01] Mark Brittenham and Ying-Qing Wu, *The classification of exceptional Dehn surgeries on 2-bridge knots*, Comm. Anal. Geom. **9** (2001), no. 1, 97–113, DOI 10.4310/CAG.2001.v9.n1.a4. MR1807953 Cited in Problem 1.15.

[BW12] Nicolas Bergeron and Daniel T. Wise, *A boundary criterion for cubulation*, Amer. J. Math. **134** (2012), no. 3, 843–859, DOI 10.1353/ajm.2012.0020. MR2931226 Cited in Problem 5.13.

[BW16a] Mark C. Bell and Richard C. H. Webb. Polynomial-time algorithms for the curve graph, 2016. arXiv:1609.09392. Cited in Problem 2.10.

[BW16b] Francis Bonahon and Helen Wong, *Representations of the Kauffman bracket skein algebra I: invariants and miraculous cancellations*, Invent. Math. **204** (2016), no. 1, 195–243, DOI 10.1007/s00222-015-0611-y. MR3480556 Cited in Problem 3.78.

[BW16c] Francis Bonahon and Helen Wong, *The Witten-Reshetikhin-Turaev representation of the Kauffman bracket skein algebra*, Proc. Amer. Math. Soc. **144** (2016), no. 6, 2711–2724, DOI 10.1090/proc/12927. MR3477089 Cited in Problem 3.78.

[BW17] Francis Bonahon and Helen Wong, *Representations of the Kauffman bracket skein algebra II: Punctured surfaces*, Algebr. Geom. Topol. **17** (2017), no. 6, 3399–3434, DOI 10.2140/agt.2017.17.3399. MR3709650 Cited in Problems 2.50 and 3.78.

[BW19] Francis Bonahon and Helen Wong, *Representations of the Kauffman bracket skein algebra III: closed surfaces and naturality*, Quantum Topol. **10** (2019), no. 2, 325–398, DOI 10.4171/QT/125. MR3950651 Cited in Problem 3.78.

[BWY21] Francis Bonahon, Helen Wong, and Tian Yang. Asymptotics of quantum invariants of surface diffeomorphisms i: conjecture and algebraic computations, 2021. arXiv:2112.12852. Cited in Problem 2.50.

[BWY22] Francis Bonahon, Helen Wong, and Tian Yang. Asymptotics of quantum invariants of surface diffeomorphisms II: The figure-eight knot complement, 2022. arXiv:2203.05730. Cited in Problem 2.50.

[BZ84] M. Boileau and H. Zieschang, *Heegaard genus of closed orientable Seifert 3-manifolds*, Invent. Math. **76** (1984), no. 3, 455–468, DOI 10.1007/BF01388469. MR746538 Cited in Problems 3.35 and 4.117.

[BZH14] Gerhard Burde, Heiner Zieschang, and Michael Heusener, *Knots*, Third, fully revised and extended edition, De Gruyter Studies in Mathematics, vol. 5, De Gruyter, Berlin, 2014. MR3156509 Cited in Problem 1.0.

[Cad98] Carlos Alberto Cadavid, *On a remarkable set of words in the mapping class group*, ProQuest LLC, Ann Arbor, MI, 1998. Thesis (Ph.D.)–The University of Texas at Austin. MR2699379 Cited in Problem 2.21.

[Cai35] S. S. Cairns, *Triangulation of the manifold of class one*, Bull. Amer. Math. Soc. **41** (1935), no. 8, 549–552, DOI 10.1090/S0002-9904-1935-06140-3. MR1563139 Cited in Problem 4.119.

[Cal03] Danny Calegari, *Problems in foliations and laminations of 3-manifolds*, Topology and geometry of manifolds (Athens, GA, 2001), Proc. Sympos. Pure Math., vol. 71, Amer. Math. Soc., Providence, RI, 2003, pp. 297–335, DOI 10.1090/pspum/071/2024640. MR2024640 Cited in Problem 3.31.

[Cal06a] Danny Calegari, *Dynamical forcing of circular groups*, Trans. Amer. Math. Soc. **358** (2006), no. 8, 3473–3491, DOI 10.1090/S0002-9947-05-03754-2. MR2218985 Cited in Problem 2.34.

[Cal06b] Danny Calegari, *Universal circles for quasigeodesic flows*, Geom. Topol. **10** (2006), 2271–2298, DOI 10.2140/gt.2006.10.2271. MR2284058 Cited in Problem 3.19.

[Cal09a] Danny Calegari. Big mapping class groups and dynamics. Geometry and the imagination, `https://lamington.wordpress.com/2009/06/22/big-mapping-class-groups-and-dynamics/`, 2009. Cited in Problem 2.38.

[Cal09b] Danny Calegari, *scl*, MSJ Memoirs, vol. 20, Mathematical Society of Japan, Tokyo, 2009, DOI 10.1142/e018. MR2527432 Cited in Problem 2.31.

[CAL14] Jose-Ignacio Cogolludo-Agustín and Anatoly Libgober, *Mordell-Weil groups of elliptic threefolds and the Alexander module of plane curves*, J. Reine Angew. Math. **697** (2014), 15–55, DOI 10.1515/crelle-2012-0096. MR3281651 Cited in Problem 4.103.

[Can79] J. W. Cannon, *Shrinking cell-like decompositions of manifolds. Codimension three*, Ann. of Math. (2) **110** (1979), no. 1, 83–112, DOI 10.2307/1971245. MR541330 Cited in Problem 5.8.

[Can94] James W. Cannon, *The combinatorial Riemann mapping theorem*, Acta Math. **173** (1994), no. 2, 155–234, DOI 10.1007/BF02398434. MR1301392 Cited in Problem 3.24.

[Cap74] Sylvain E. Cappell, *Unitary nilpotent groups and Hermitian K-theory. I*, Bull. Amer. Math. Soc. **80** (1974), 1117–1122, DOI 10.1090/S0002-9904-1974-13636-0. MR358815 Cited in Problem 4.22.

[Car91] J. Scott Carter, *Extending immersions of curves to properly immersed surfaces*, Topology Appl. **40** (1991), no. 3, 287–306, DOI 10.1016/0166-8641(91)90111-X. MR1124843 Cited in Problem 1.83.

[Cas16] Fabrice Castel, *Geometric representations of the braid groups* (English, with English and French summaries), Astérisque **378** (2016), vi+175. MR3499033 Cited in Problem 2.7.

[Cas22] Roger Casals, *Lagrangian skeleta and plane curve singularities*, J. Fixed Point Theory Appl. **24** (2022), no. 2, Paper No. 34, 43, DOI 10.1007/s11784-022-00939-8. MR4405603 Cited in Problem 1.67.

[Cat09] Fabrizio Catanese. Canonical symplectic structures and deformations of algebraic surfaces. *Commun. Contemp. Math.*, 11(3):481–493, 2009. `doi:10.1142/S0219199709003478`. Cited in Problem 4.94.

[Cat17] Fabrizio Catanese, *Kodaira fibrations and beyond: methods for moduli theory*, Jpn. J. Math. **12** (2017), no. 2, 91–174, DOI 10.1007/s11537-017-1569-x. MR3694930 Cited in Problem 2.24.

[Cau17] Sabin Cautis, *Remarks on coloured triply graded link invariants*, Algebr. Geom. Topol. **17** (2017), no. 6, 3811–3836, DOI 10.2140/agt.2017.17.3811. MR3709661 Cited in Problem 1.34.

[CC22] Danny Calegari and Lvzhou Chen, *Normal subgroups of big mapping class groups*, Trans. Amer. Math. Soc. Ser. B **9** (2022), 957–976, DOI 10.1090/btran/108. MR4498366 Cited in Problem 2.43.

[CCF21] John Cantwell, Lawrence Conlon, and Sergio R. Fenley, *Endperiodic automorphisms of surfaces and foliations*, Ergodic Theory Dynam. Systems **41** (2021), no. 1, 66–212, DOI 10.1017/etds.2019.56. MR4190052 Cited in Problems 2.37 and 2.45.

[CCPR+24] Linyi Chen, Grant Crider-Phillips, Braeden Reinoso, Joshua Sabloff, and Leyu Yao, *Non-orientable Lagrangian fillings of Legendrian knots*, Math. Proc. Cambridge Philos. Soc. **176** (2024), no. 1, 123–153, DOI 10.1017/s0305004123000440. MR4680483 Cited in Problem 1.68.

[CD03] Danny Calegari and Nathan M. Dunfield, *Laminations and groups of homeomorphisms of the circle*, Invent. Math. **152** (2003), no. 1, 149–204, DOI 10.1007/s00222-002-0271-6. MR1965363 Cited in Problems 3.5, 3.19, 3.22, and 3.48.

[CD06] Frank Calegari and Nathan M. Dunfield, *Automorphic forms and rational homology 3-spheres*, Geom. Topol. **10** (2006), 295–329, DOI 10.2140/gt.2006.10.295. MR2224458 Cited in Problems 3.5 and 3.10.

[CDGW] Marc Culler, Nathan M. Dunfield, Matthias Goerner, and Jeffrey R. Weeks. SnapPy, a computer program for studying the geometry and topology of 3-manifolds. Available at `http://snappy.computop.org`. Cited in Problem 3.30.

[CDR03] Robert Connelly, Erik D. Demaine, and Günter Rote, *Straightening polygonal arcs and convexifying polygonal cycles*, Discrete Comput. Geom. **30** (2003), no. 2, 205–239, DOI 10.1007/s00454-003-0006-7. U.S.-Hungarian Workshops on Discrete Geometry and Convexity (Budapest, 1999/Auburn, AL, 2000). MR2007962 Cited in Problem 1.72.

[CDR22] Vincent Colin, Pierre Dehornoy, and Ana Rechtman, *On the existence of supporting broken book decompositions for contact forms in dimension 3*, Invent. Math. **231** (2023), no. 3, 1489–1539, DOI 10.1007/s00222-022-01160-7. MR4549092 Cited in Problem 3.46.

[CDT24] Frank Calegari, Vesselin Dimitrov, and Yunqing Tang. The linear independence of 1, $\zeta(2)$, and $l(2, \chi_{-3})$, 2024. `arXiv:2408.15403`. Cited in Problem 3.2.

[CE12] Kai Cieliebak and Yakov Eliashberg, *From Stein to Weinstein and back*, American Mathematical Society Colloquium Publications, vol. 59, American Mathematical Society, Providence, RI, 2012. Symplectic geometry of affine complex manifolds, DOI 10.1090/coll/059. MR3012475 Cited in Problem 5.17.

[Cer68] Jean Cerf, *Sur les difféomorphismes de la sphère de dimension trois* ($\Gamma_4 = 0$) (French), Lecture Notes in Mathematics, No. 53, Springer-Verlag, Berlin-New York, 1968. MR229250 Cited in Problem 4.23.

[Čer69] A. V. Černavskiĭ, *Local contractibility of the group of homeomorphisms of a manifold* (Russian), Mat. Sb. (N.S.) **79(121)** (1969), 307–356. MR259925 Cited in Problem 5.4.

[CET21] James Conway, John B. Etnyre, and Bülent Tosun, *Symplectic fillings, contact surgeries, and Lagrangian disks*, Int. Math. Res. Not. IMRN **8** (2021), 6020–6050, DOI 10.1093/imrn/rny291. MR4251270 Cited in Problem 1.65.

[CF94] Louis Crane and Igor B. Frenkel, *Four-dimensional topological quantum field theory, Hopf categories, and the canonical bases*, J. Math. Phys. **35** (1994), no. 10, 5136–5154, DOI 10.1063/1.530746. Topology and physics. MR1295461 Cited in Problem 3.67.

[CF17] A. Costa and M. Farber, *Large random simplicial complexes, II; the fundamental group*, J. Topol. Anal. **9** (2017), no. 3, 441–483, DOI 10.1142/S1793525317500170. MR3661651 Cited in Problem 5.11.

[CFF92] C. Croke, A. Fathi, and J. Feldman, *The marked length-spectrum of a surface of nonpositive curvature*, Topology **31** (1992), no. 4, 847–855, DOI 10.1016/0040-9383(92)90013-8. MR1191384 Cited in Problem 2.16.

[CFH16] Anthony Conway, Stefan Friedl, and Gerrit Herrmann, *Linking forms revisited*, Pure Appl. Math. Q. **12** (2016), no. 4, 493–515, DOI 10.4310/PAMQ.2016.v12.n4.a3. MR3842851 Cited in Problem 3.73.

[CFHH13] Tim D. Cochran, Bridget D. Franklin, Matthew Hedden, and Peter D. Horn, *Knot concordance and homology cobordism*, Proc. Amer. Math. Soc. **141** (2013), no. 6, 2193–2208, DOI 10.1090/S0002-9939-2013-11471-1. MR3034445 Cited in Problem 1.58.

[CG78] A. J. Casson and C. McA. Gordon, *On slice knots in dimension three*, Algebraic and geometric topology (Proc. Sympos. Pure Math., Stanford Univ., Stanford, Calif., 1976), Proc. Sympos. Pure Math., XXXII, Amer. Math. Soc., Providence, RI, 1978, pp. 39–53. MR520521 Cited in Problem 3.73.

[CG83a] A. J. Casson and C. McA. Gordon, *A loop theorem for duality spaces and fibred ribbon knots*, Invent. Math. **74** (1983), no. 1, 119–137, DOI 10.1007/BF01388533. MR722728 Cited in Problems 1.10, 1.50, 1.52, 1.55, and 1.59.

[CG83b] J. H. Conway and C. McA. Gordon, *Knots and links in spatial graphs*, J. Graph Theory **7** (1983), no. 4, 445–453, DOI 10.1002/jgt.3190070410. MR722061 Cited in Problem 1.86.

[CG86] A. J. Casson and C. McA. Gordon, *Cobordism of classical knots*, À la recherche de la topologie perdue, Progr. Math., vol. 62, Birkhäuser Boston, Boston, MA, 1986, pp. 181–199. With an appendix by P. M. Gilmer. MR900252 Cited in Problems 1.46 and 3.73.

[CG22] Roger Casals and Honghao Gao, *Infinitely many Lagrangian fillings*, Ann. of Math. (2) **195** (2022), no. 1, 207–249, DOI 10.4007/annals.2022.195.1.3. MR4358415 Cited in Problems 1.67 and 4.104.

[CGH09] Vincent Colin, Emmanuel Giroux, and Ko Honda, *Finitude homotopique et isotopique des structures de contact tendues* (French, with French summary), Publ. Math. Inst. Hautes Études Sci. **109** (2009), 245–293, DOI 10.1007/s10240-009-0022-y. MR2511589 Cited in Problem 1.63.

[CGH16] Daniel Cristofaro-Gardiner and Michael Hutchings, *From one Reeb orbit to two*, J. Differential Geom. **102** (2016), no. 1, 25–36. MR3447085 Cited in Problem 3.46.

[CGH17a] Vincent Colin, Paolo Ghiggini, and Ko Honda. The equivalence of Heegaard Floer homology and embedded contact homology via open book decompositions I, 2017. `arXiv:1208.1074`. Cited in Problem 3.47.

[CGH17b] Vincent Colin, Paolo Ghiggini, and Ko Honda. The equivalence of Heegaard Floer homology and embedded contact homology via open book decompositions II, 2017. `arXiv:1208.1077`. Cited in Problem 3.47.

[CGH20] Vincent Colin, Paolo Ghiggini, and Ko Honda. An exposition of the equivalence of Heegaard Floer homology and embedded contact homology. In *Characters in low-dimensional topology*, volume 760 of *Contemp. Math.*, pages 45–101. Amer. Math. Soc., [Providence], RI, [2020] ©2020. `doi:10.1090/conm/760/15286`. Cited in Problem 5.18.

[CGHHL23] Daniel Cristofaro-Gardiner, Umberto Hryniewicz, Michael Hutchings, and Hui Liu, *Contact three-manifolds with exactly two simple Reeb orbits*, Geom. Topol. **27** (2023), no. 9, 3801–3831, DOI 10.2140/gt.2023.27.3801. MR4674840 Cited in Problem 3.46.

[CGHHL24] Dan Cristofaro-Gardiner, Umberto Hryniewicz, Michael Hutchings, and Hui Liu. Proof of Hofer-Wysocki-Zehnder's two or infinity conjecture, 2024. `arXiv:2310.07636`. Cited in Problem 3.46.

[CGHN00] David Coulson, Oliver A. Goodman, Craig D. Hodgson, and Walter D. Neumann, *Computing arithmetic invariants of 3-manifolds*, Experiment. Math. **9** (2000), no. 1, 127–152. MR1758805 Cited in Problem 3.63.

[CGHP19] Dan Cristofaro-Gardiner, Michael Hutchings, and Daniel Pomerleano, *Torsion contact forms in three dimensions have two or infinitely many Reeb orbits*, Geom. Topol. **23** (2019), no. 7, 3601–3645, DOI 10.2140/gt.2019.23.3601. MR4047649 Cited in Problem 3.46.

[CGHPM23] Francesco Costantino, Nathan Geer, Benjamin Haïoun, and Bertrand Patureau-Mirand. Skein (3+ 1)-TQFTs from non-semisimple ribbon categories, 2023. `arXiv:2306.03225`. Cited in Problem 3.78.

[CGL06] R. Chamanara, F. P. Gardiner, and N. Lakic, *A hyperelliptic realization of the horseshoe and baker maps*, Ergodic Theory Dynam. Systems **26** (2006), no. 6, 1749–1768, DOI 10.1017/S0143385706000484. MR2279264 Cited in Problem 2.42.

[CGLS87] Marc Culler, C. McA. Gordon, J. Luecke, and Peter B. Shalen, *Dehn surgery on knots*, Ann. of Math. (2) **125** (1987), no. 2, 237–300, DOI 10.2307/1971311. MR881270 Cited in Problem 1.9.

[CGPM14] Francesco Costantino, Nathan Geer, and Bertrand Patureau-Mirand, *Quantum invariants of 3-manifolds via link surgery presentations and non-semi-simple categories*, J. Topol. **7** (2014), no. 4, 1005–1053, DOI 10.1112/jtopol/jtu006. MR3286896 Cited in Problem 3.78.

[CGPS20] Sungbong Chun, Sergei Gukov, Sunghyuk Park, and Nikita Sopenko, *3d-3d correspondence for mapping tori*, J. High Energy Phys. **9** (2020), 152, 59, DOI 10.1007/jhep09(2020)152. MR4190272 Cited in Problem 3.68.

[CH81] Andrew J. Casson and John L. Harer, *Some homology lens spaces which bound rational homology balls*, Pacific J. Math. **96** (1981), no. 1, 23–36. MR634760 Cited in Problem 4.58.

[CH15] Alexander Coward and Joel Hass, *Topological and physical link theory are distinct*, Pacific J. Math. **276** (2015), no. 2, 387–400, DOI 10.2140/pjm.2015.276.387. MR3374064 Cited in Problem 1.71.

[Cha04] R. Chamanara, *Affine automorphism groups of surfaces of infinite type*, In the tradition of Ahlfors and Bers, III, Contemp. Math., vol. 355, Amer. Math. Soc., Providence, RI, 2004, pp. 123–145, DOI 10.1090/conm/355/06449. MR2145060 Cited in Problem 2.42.

[Cha08] Jae Choon Cha, *Topological minimal genus and L^2-signatures*, Algebr. Geom. Topol. **8** (2008), no. 2, 885–909, DOI 10.2140/agt.2008.8.885. MR2443100 Cited in Problem 1.45.

[Cha10] Baptiste Chantraine, *Lagrangian concordance of Legendrian knots*, Algebr. Geom. Topol. **10** (2010), no. 1, 63–85, DOI 10.2140/agt.2010.10.63. MR2580429 Cited in Problem 1.68.

[Cha15] Baptiste Chantraine, *Lagrangian concordance is not a symmetric relation*, Quantum Topol. **6** (2015), no. 3, 451–474, DOI 10.4171/QT/68. MR3392961 Cited in Problem 1.66.

[Che55] Shiing-shen Chern, *On curvature and characteristic classes of a Riemann manifold*, Abh. Math. Sem. Univ. Hamburg **20** (1955), 117–126, DOI 10.1007/BF02960745. MR75647 Cited in Problem 4.126.

[Che02] Yuri Chekanov, *Differential algebra of Legendrian links*, Invent. Math. **150** (2002), no. 3, 441–483, DOI 10.1007/s002220200212. MR1946550 Cited in Problem 5.16.

[Che18] Lei Chen, *The number of fiberings of a surface bundle over a surface*, Algebr. Geom. Topol. **18** (2018), no. 4, 2245–2263, DOI 10.2140/agt.2018.18.2245. MR3797066 Cited in Problem 2.24.

[Che20a] Weimin Chen, *Finite group actions on symplectic Calabi-Yau 4-manifolds with $b_1 > 0$*, J. Gökova Geom. Topol. GGT **14** (2020), 1–54. MR4206325 Cited in Problem 4.92.

[Che20b] Weimin Chen, *On a class of symplectic 4-orbifolds with vanishing canonical class*, J. Gökova Geom. Topol. GGT **14** (2020), 55–90. MR4206326 Cited in Problem 4.92.

[Che23] Jie Chen. *Flat knots and invariants.* PhD thesis, McMaster University, 2023. URL: `https://www.math.mcmaster.ca/%7Eboden/students/Chen-PhD.pdf`. Cited in Problem 1.83.

[Che25] Jacopo G. Chen. Non-cobordant hyperbolic manifolds, 2025. `arXiv:2501.11610`. Cited in Problem 4.126.

[CHH13] Tim D. Cochran, Shelly Harvey, and Peter Horn, *Filtering smooth concordance classes of topologically slice knots*, Geom. Topol. **17** (2013), no. 4, 2103–2162, DOI 10.2140/gt.2013.17.2103. MR3109864 Cited in Problem 1.44.

[CHLR08] T. Chinburg, E. Hamilton, D. D. Long, and A. W. Reid, *Geodesics and commensurability classes of arithmetic hyperbolic 3-manifolds*, Duke Math. J. **145** (2008), no. 1, 25–44, DOI 10.1215/00127094-2008-045. MR2451288 Cited in Problem 3.17.

[Cho04] Young-Eun Choi, *Positively oriented ideal triangulations on hyperbolic three-manifolds*, Topology **43** (2004), no. 6, 1345–1371, DOI 10.1016/j.top.2004.02.002. MR2081429 Cited in Problem 3.3.

[Chr22] Micah Chrisman, *Milnor's concordance invariants for knots on surfaces*, Algebr. Geom. Topol. **22** (2022), no. 5, 2293–2353, DOI 10.2140/agt.2022.22.2293. MR4503338 Cited in Problem 1.83.

[CHT24] Vincent Colin, Ko Honda, and Yin Tian, *Applications of higher-dimensional Heegaard Floer homology to contact topology*, J. Topol. **17** (2024), no. 3, Paper No. e12349, 77, DOI 10.1112/topo.12349. MR4775502 Cited in Problem 5.18.

[Cie98] Kai Cieliebak, *A geometric obstruction to the contact type property*, Math. Z. **228** (1998), no. 3, 451–487, DOI 10.1007/PL00004626. MR1637792 Cited in Problem 4.107.

[CJ94] Andrew Casson and Douglas Jungreis, *Convergence groups and Seifert fibered 3-manifolds*, Invent. Math. **118** (1994), no. 3, 441–456, DOI 10.1007/BF01231540. MR1296353 Cited in Problem 3.24.

[CJN14] Gonzalo Castro, Eduardo Jorquera, and Andrés Navas, *Sharp regularity for certain nilpotent group actions on the interval*, Math. Ann. **359** (2014), no. 1-2, 101–152, DOI 10.1007/s00208-013-0995-1. MR3201895 Cited in Problem 2.34.

[CJS95] R. L. Cohen, J. D. S. Jones, and G. B. Segal, *Floer's infinite-dimensional Morse theory and homotopy theory*, The Floer memorial volume, Progr. Math., vol. 133, Birkhäuser, Basel, 1995, pp. 297–325. MR1362832 Cited in Problem 3.47.

[CK14] M. B. Can and Ç. Karakurt, *Calculating Heegaard-Floer homology by counting lattice points in tetrahedra*, Acta Math. Hungar. **144** (2014), no. 1, 43–75, DOI 10.1007/s10474-014-0432-2. MR3267169 Cited in Problem 3.57.

[CK25] Abhijit Champanerkar and Ilya Kofman. Geometric bounds for spanning tree entropy of planar lattice graphs, 2025. `arXiv:2505.05688`. Cited in Problem 1.29.

[CKL19] Abhijit Champanerkar, Ilya Kofman, and Matilde Lalín, *Mahler measure and the vol-det conjecture*, J. Lond. Math. Soc. (2) **99** (2019), no. 3, 872–900, DOI 10.1112/jlms.12200. MR3977894 Cited in Problem 1.29.

[CKM19] Lei Chen, Kevin Kordek, and Dan Margalit. Homomorphisms between braid groups, 2019. `arXiv:1910.00712`. Cited in Problems 2.4 and 2.7.

[CKP16] Abhijit Champanerkar, Ilya Kofman, and Jessica S. Purcell, *Geometrically and diagrammatically maximal knots*, J. Lond. Math. Soc. (2) **94** (2016), no. 3, 883–908, DOI 10.1112/jlms/jdw062. MR3614933 Cited in Problem 1.29.

[CKP20] Jae Choon Cha, Min Hoon Kim, and Mark Powell, *A family of freely slice good boundary links*, Math. Ann. **376** (2020), no. 3-4, 1009–1030, DOI 10.1007/s00208-019-01907-3. MR4081108 Cited in Problem 1.62.

[CKP23] Anthony Conway, Min Hoon Kim, and Wojciech Politarczyk, *Nonslice linear combinations of iterated torus knots*, Algebr. Geom. Topol. **23** (2023), no. 2, 765–802, DOI 10.2140/agt.2023.23.765. MR4587316 Cited in Problem 1.39.

[CKS98] Jason Cantarella, Robert B. Kusner, and John M. Sullivan. Tight knot values deviate from linear relations. *Nature*, 392:237, 1998. URL: `http://torus.math.uiuc.edu/jms/Papers/scicor/scicor.ps.gzpapers3://publication/uuid/A1CD2FF8-BAEE-4CCC-A98C-BB5ECC1DC83A`. Cited in Problem 1.72.

[CKS02] Jason Cantarella, Robert B. Kusner, and John M. Sullivan. On the minimum ropelength of knots and links. *Inventiones Mathematicae*, 150:257–286, 2002. URL: `http://www.springerlink.com/index/10.1007/s00222-002-0234-y`, `doi:10.1007/s00222-002-0234-y`. Cited in Problems 1.71 and 1.74.

[CKY97] Louis Crane, Louis H. Kauffman, and David N. Yetter, *State-sum invariants of 4-manifolds*, J. Knot Theory Ramifications **6** (1997), no. 2, 177–234, DOI 10.1142/S0218216597000145. MR1452438 Cited in Problem 3.78.

[CL85] A. J. Casson and D. D. Long, *Algorithmic compression of surface automorphisms*, Invent. Math. **81** (1985), no. 2, 295–303, DOI 10.1007/BF01389054. MR799268 Cited in Problem 1.52.

[CL98] D. Cooper and D. D. Long, *Representation theory and the A-polynomial of a knot*, Chaos Solitons Fractals **9** (1998), no. 4-5, 749–763, DOI 10.1016/S0960-0779(97)00102-1. Knot theory and its applications. MR1628754 Cited in Problem 1.25.

[CL99] D. Cooper and W. B. R. Lickorish, *Mutations of links in genus 2 handlebodies*, Proc. Amer. Math. Soc. **127** (1999), no. 1, 309–314, DOI 10.1090/S0002-9939-99-04871-6. MR1605940 Cited in Problem 3.61.

[CL01] D. Cooper and D. D. Long, *Some surface subgroups survive surgery*, Geom. Topol. **5** (2001), 347–367, DOI 10.2140/gt.2001.5.347. MR1825666 Cited in Problem 3.13.

[CL14] Alexander Coward and Marc Lackenby, *An upper bound on Reidemeister moves*, Amer. J. Math. **136** (2014), no. 4, 1023–1066, DOI 10.1353/ajm.2014.0027. MR3245186 Cited in Problem 1.77.

[CL24] Otis Chodosh and Chao Li, *Generalized soap bubbles and the topology of manifolds with positive scalar curvature*, Ann. of Math. (2) **199** (2024), no. 2, 707–740, DOI 10.4007/annals.2024.199.2.3. MR4713021 Cited in Problem 4.129.

[CL25] Lvzhou Chen and Yash Lodha, *The Wiegold problem and free products of left-orderable groups*, Preprint, 2025, `arXiv:2510.26073`, math.GR. Cited in Problem 3.39.

[Cla17] Corrin Clarkson, *Three-manifold mutations detected by Heegaard Floer homology*, Algebr. Geom. Topol. **17** (2017), no. 1, 1–16, DOI 10.2140/agt.2017.17.1. MR3604370 Cited in Problems 2.48 and 3.61.

[CLL23] Otis Chodosh, Chao Li, and Yevgeny Liokumovich, *Classifying sufficiently connected PSC manifolds in 4 and 5 dimensions*, Geom. Topol. **27** (2023), no. 4, 1635–1655, DOI 10.2140/gt.2023.27.1635. MR4602422 Cited in Problem 4.129.

[CLM12] Matt T. Clay, Christopher J. Leininger, and Johanna Mangahas, *The geometry of right-angled Artin subgroups of mapping class groups*, Groups Geom. Dyn. **6** (2012), no. 2, 249–278, DOI 10.4171/GGD/157. MR2914860 Cited in Problem 2.28.

[CM01] Chun Cao and G. Robert Meyerhoff, *The orientable cusped hyperbolic 3-manifolds of minimum volume*, Invent. Math. **146** (2001), no. 3, 451–478, DOI 10.1007/s002220100167. MR1869847 Cited in Problem 3.1.

[CM05] Marston Conder and Colin Maclachlan, *Compact hyperbolic 4-manifolds of small volume*, Proc. Amer. Math. Soc. **133** (2005), no. 8, 2469–2476, DOI 10.1090/S0002-9939-05-07634-3. MR2138890 Cited in Problems 4.19 and 4.124.

[CM19] Austin Christian and Michael Menke. Splitting symplectic fillings, 2019. `arXiv:1909.00420`. Cited in Problem 4.110.

[CM24] Gonzalo Contreras and Marco Mazzucchelli, *Proof of the C^2-stability conjecture for geodesic flows of closed surfaces*, Duke Math. J. **173** (2024), no. 2, 347–390, DOI 10.1215/00127094-2023-0010. MR4728177 Cited in Problem 3.47.

[CMM21] Matt Clay, Johanna Mangahas, and Dan Margalit, *Right-angled Artin groups as normal subgroups of mapping class groups*, Compos. Math. **157** (2021), no. 8, 1807–1852, DOI 10.1112/S0010437X21007417. MR4292178 Cited in Problem 2.29.

[CMM24] Otis Chodosh, Davi Maximo, and Anubhav Mukherjee. Complete riemannian 4-manifolds with uniformly positive scalar curvature, 2024. `arXiv:2407.05574`. Cited in Problem 4.128.

[CMP15] Tobias Holck Colding, William P. Minicozzi II, and Erik Kjær Pedersen, *Mean curvature flow*, Bull. Amer. Math. Soc. (N.S.) **52** (2015), no. 2, 297–333, DOI 10.1090/S0273-0979-2015-01468-0. MR3312634 Cited in Problem 4.1.

[CMP21] Jae Choon Cha, Allison N. Miller, and Mark Powell, *Two-solvable and two-bipolar knots with large four-genera*, Math. Res. Lett. **28** (2021), no. 2, 331–382, DOI 10.4310/MRL.2021.v28.n2.a2. MR4258974 Cited in Problem 1.45.

[CNPP06] Clément Caubel, András Némethi, and Patrick Popescu-Pampu, *Milnor open books and Milnor fillable contact 3-manifolds*, Topology **45** (2006), no. 3, 673–689, DOI 10.1016/j.top.2006.01.002. MR2218761 Cited in Problem 5.18.

[CNS16] Christopher Cornwell, Lenhard Ng, and Steven Sivek, *Obstructions to Lagrangian concordance*, Algebr. Geom. Topol. **16** (2016), no. 2, 797–824, DOI 10.2140/agt.2016.16.797. MR3493408 Cited in Problems 1.65, 1.66, 1.68, and 3.52.

[Coc54] W. H. Cockcroft, *On two-dimensional aspherical complexes*, Proc. London Math. Soc. (3) **4** (1954), 375–384, DOI 10.1112/plms/s3-4.1.375. MR63042 Cited in Problem 5.11.

[Coc84] Tim Cochran, *Slice links in S^4*, Trans. Amer. Math. Soc. **285** (1984), no. 1, 389–401, DOI 10.2307/1999487. MR748845 Cited in Problem 4.40.

[Col01] Vincent Colin, *Sur la torsion des structures de contact tendues* (French, with English and French summaries), Ann. Sci. École Norm. Sup. (4) **34** (2001), no. 2, 267–286, DOI 10.1016/S0012-9593(00)01061-2. MR1841879 Cited in Problems 3.40 and 3.42.

[Con59] Paul Conrad, *Right-ordered groups*, Michigan Math. J. **6** (1959), 267–275. MR106954 Cited in Problem 3.20.

[Con70] J. H. Conway, *An enumeration of knots and links, and some of their algebraic properties*, Computational Problems in Abstract Algebra (Proc. Conf., Oxford, 1967), Pergamon, Oxford-New York-Toronto, Ont., 1970, pp. 329–358. MR258014 Cited in Problem 1.0.

[COP23] Anthony Conway, Patrick Orson, and Mark Powell. Unknotting nonorientable surfaces, 2023. `arXiv:2306.12305`. Cited in Problem 4.29.

[Cor18] James Cornish, *Growth Rate of 3-Manifold Homologies under Branched Covers*, ProQuest LLC, Ann Arbor, MI, 2018. Thesis (Ph.D.)–Columbia University. MR3818831 Cited in Problem 2.47.

[COT03] Tim D. Cochran, Kent E. Orr, and Peter Teichner, *Knot concordance, Whitney towers and L^2-signatures*, Ann. of Math. (2) **157** (2003), no. 2, 433–519, DOI 10.4007/annals.2003.157.433. MR1973052 Cited in Problems 1.44 and 1.61.

[CP01] John Crisp and Luis Paris. The solution to a conjecture of Tits on the subgroup generated by the squares of the generators of an Artin group. *Invent. Math.*, 145(1):19–36, 2001. doi:10.1007/s002220100138. Cited in Problems 2.28 and 2.29.

[CP23] Anthony Conway and Mark Powell, *Embedded surfaces with infinite cyclic knot group*, Geom. Topol. **27** (2023), no. 2, 739–821, DOI 10.2140/gt.2023.27.739. MR4589564 Cited in Problems 4.29, 4.30, and 4.45.

[CPV18] Vincent Colin, Francisco Presas, and Thomas Vogel, *Notes on open book decompositions for Engel structures*, Algebr. Geom. Topol. **18** (2018), no. 7, 4275–4303, DOI 10.2140/agt.2018.18.4275. MR3892245 Cited in Problem 4.121.

[CPV21] Yassin Chandran, Priyam Patel, and Nicholas G. Vlamis. Infinite-type surfaces and mapping class groups: Open problems. Available at https://https://www.patelp.com/uploads/2/5/7/9/25792573/inftypeproblems.pdf, 2021. Cited in Problems 2.39 and 2.40.

[CR72] P. E. Conner and Frank Raymond, *Manifolds with few periodic homeomorphisms*, Proceedings of the Second Conference on Compact Transformation Groups (Univ. Massachusetts, Amherst, Mass., 1971), Lecture Notes in Math., Vol. 299, Springer, Berlin-New York, 1972, pp. 1–75. MR358835 Cited in Problem 4.84.

[CR88] Sylvain Cappell and Daniel Ruberman, *Imbeddings and homology cobordisms of lens spaces*, Comment. Math. Helv. **63** (1988), no. 1, 75–88, DOI 10.1007/BF02566753. MR928028 Cited in Problem 4.59.

[CR09] Fabrizio Catanese and Sönke Rollenske, *Double Kodaira fibrations*, J. Reine Angew. Math. **628** (2009), 205–233, DOI 10.1515/CRELLE.2009.024. MR2503241 Cited in Problem 4.97.

[CR16] Adam Clay and Dale Rolfsen, *Ordered groups and topology*, Graduate Studies in Mathematics, vol. 176, American Mathematical Society, Providence, RI, 2016, DOI 10.1090/gsm/176. MR3560661 Cited in Problems 3.20 and 3.22.

[Cro59] Richard Crowell, *Genus of alternating link types*, Ann. of Math. (2) **69** (1959), 258–275, DOI 10.2307/1970181. MR99665 Cited in Problem 1.8.

[Cro90] Christopher B. Croke. Rigidity for surfaces of nonpositive curvature. *Comment. Math. Helv.*, 65(1):150–169, 1990. doi:10.1007/BF02566599. Cited in Problem 2.16.

[CRR00] Sylvain Cappell, Andrew Ranicki, and Jonathan Rosenberg (eds.), *Surveys on surgery theory. Vol. 1*, Annals of Mathematics Studies, vol. 145, Princeton University Press, Princeton, NJ, 2000. Papers dedicated to C. T. C. Wall. MR1746325 Cited in the Introduction.

[CRR01] Sylvain Cappell, Andrew Ranicki, and Jonathan Rosenberg (eds.), *Surveys on surgery theory. Vol. 2*, Annals of Mathematics Studies, vol. 149, Princeton University Press, Princeton, NJ, 2001. Papers dedicated to C. T. C. Wall. MR1818769 Cited in the Introduction.

[CRS22] Ted Chinburg, Alan W. Reid, and Matthew Stover, *Azumaya algebras and canonical components*, Int. Math. Res. Not. IMRN **7** (2022), 4969–5036, DOI 10.1093/imrn/rnaa209. MR4403955 Cited in Problem 1.85.

[CS71] Sylvain E. Cappell and Julius L. Shaneson, *On four dimensional surgery and applications*, Comment. Math. Helv. **46** (1971), 500–528, DOI 10.1007/BF02566862. MR301750 Cited in Problems 4.18 and 4.22.

[CS74] Sylvain E. Cappell and Julius L. Shaneson, *The codimension two placement problem and homology equivalent manifolds*, Ann. of Math. (2) **99** (1974), 277–348, DOI 10.2307/1970901. MR339216 Cited in Problems 4.59 and 4.88.

[CS76] Sylvain E. Cappell and Julius L. Shaneson, *Some new four-manifolds*, Ann. of Math. (2) **104** (1976), no. 1, 61–72, DOI 10.2307/1971056. MR418125 Cited in Problems 4.1, 4.87, and 4.88.

[CS79] Sylvanin E. Cappell and Julius L. Shaneson, *Embeddings and immersions of four-dimensional manifolds in* $\mathbf{R}^6$, Geometric topology (Proc. Georgia Topology Conf., Athens, Ga., 1977), Academic Press, New York-London, 1979, pp. 301–303. MR537736 Cited in Problem 4.60.

[CS80] Sylvain E. Cappell and Julius L. Shaneson, *Link cobordism*, Comment. Math. Helv. **55** (1980), no. 1, 20–49, DOI 10.1007/BF02566673. MR569244 Cited in Problem 1.62.

[CS87a] Sylvain E. Cappell and Julius L. Shaneson, *Corrigendum to: "Smooth nontrivial 4-dimensional s-cobordisms"*, Bull. Amer. Math. Soc. (N.S.) **17** (1987), no. 2, 401, DOI 10.1090/S0273-0979-1987-15616-3. MR903747 Cited in Problem 4.56.

[CS87b] Sylvain E. Cappell and Julius L. Shaneson, *Smooth nontrivial 4-dimensional s-cobordisms*, Bull. Amer. Math. Soc. (N.S.) **17** (1987), no. 1, 141–143, DOI 10.1090/S0273-0979-1987-15542-X. MR888891 Cited in Problem 4.56.

[CS92] Marc Culler and Peter B. Shalen, *Paradoxical decompositions, 2-generator Kleinian groups, and volumes of hyperbolic 3-manifolds*, J. Amer. Math. Soc. **5** (1992), no. 2, 231–288, DOI 10.2307/2152768. MR1135928 Cited in Problem 3.23.

[CS98] J. W. Cannon and E. L. Swenson, *Recognizing constant curvature discrete groups in dimension* 3, Trans. Amer. Math. Soc. **350** (1998), no. 2, 809–849, DOI 10.1090/S0002-9947-98-02107-2. MR1458317 Cited in Problem 3.24.

[CS10] Donald I. Cartwright and Tim Steger, *Enumeration of the 50 fake projective planes* (English, with English and French summaries), C. R. Math. Acad. Sci. Paris **348** (2010), no. 1-2, 11–13, DOI 10.1016/j.crma.2009.11.016. MR2586735 Cited in Problem 4.91.

[CS12b] Marc Culler and Peter B. Shalen, *Margulis numbers for Haken manifolds*, Israel J. Math. **190** (2012), 445–475, DOI 10.1007/s11856-011-0189-z. MR2956250 Cited in Problem 3.23.

[CS21] Aaron Calderon and Nick Salter, *Higher spin mapping class groups and strata of abelian differentials over Teichmüller space*, Adv. Math. **389** (2021), Paper No. 107926, 56, DOI 10.1016/j.aim.2021.107926. MR4289049 Cited in Problem 2.6.

[CS23] Aaron Calderon and Nick Salter, *Framed mapping class groups and the monodromy of strata of abelian differentials*, J. Eur. Math. Soc. (JEMS) **25** (2023), no. 12, 4719–4790, DOI 10.4171/jems/1290. MR4662301 Cited in Problems 2.6 and 2.26.

[CSS08] John Crisp, Michah Sageev, and Mark Sapir, *Surface subgroups of right-angled Artin groups*, Internat. J. Algebra Comput. **18** (2008), no. 3, 443–491, DOI 10.1142/S0218196708004536. MR2422070 Cited in Problem 2.30.

[CT07] James W. Cannon and William P. Thurston, *Group invariant Peano curves*, Geom. Topol. **11** (2007), 1315–1355, DOI 10.2140/gt.2007.11.1315. MR2326947 Cited in Problem 5.13.

[CT19] Michelle Chu and Stephan Tillmann, *Reflections on trisection genus*, Rev. Roumaine Math. Pures Appl. **64** (2019), no. 4, 395–402. MR4049694 Cited in Problem 4.117.

[CV35] Stefan Cohn-Vossen, *Kürzeste Wege und Totalkrümmung auf Flächen* (German), Compositio Math. **2** (1935), 69–133. MR1556908 Cited in Problem 5.15.

[CW04] John Crisp and Bert Wiest, *Embeddings of graph braid and surface groups in right-angled Artin groups and braid groups*, Algebr. Geom. Topol. **4** (2004), 439–472, DOI 10.2140/agt.2004.4.439. MR2077673 Cited in Problem 2.30.

[CW07] Fabrizio Catanese and Bronislaw Wajnryb, *Diffeomorphism of simply connected algebraic surfaces*, J. Differential Geom. **76** (2007), no. 2, 177–213, DOI 10.4310/jdg/1180135677. MR2330412 Cited in Problem 4.94.

[CWLRS25] Tamunonye Cheetham-West, Alexander Lubotzky, Alan W. Reid, and Ryan Spitler, *Property FA is not a profinite property*, Groups Geom. Dyn. **19** (2025), no. 3, 1081–1087, DOI 10.4171/ggd/802. MR4945540 Cited in Problem 3.9.

[CY93] Louis Crane and David Yetter, *A categorical construction of 4D topological quantum field theories*, Quantum topology, Ser. Knots Everything, vol. 3, World Sci. Publ., River Edge, NJ, 1993, pp. 120–130, DOI 10.1142/9789812796387_0005. MR1273569 Cited in Problem 3.78.

[CY18] Qingtao Chen and Tian Yang, *Volume conjectures for the Reshetikhin-Turaev and the Turaev-Viro invariants*, Quantum Topol. **9** (2018), no. 3, 419–460, DOI 10.4171/QT/111. MR3827806 Cited in Problem 3.4.

[CY24] Tommaso Cremaschi and Andrew Yarmola. Knots in circle bundles are determined by their complements, 2024. `arXiv:2401.02895`. Cited in Problem 1.12.

[Dav85] Michael W. Davis, *A hyperbolic 4-manifold*, Proc. Amer. Math. Soc. **93** (1985), no. 2, 325–328, DOI 10.2307/2044771. MR770546 Cited in Problem 4.126.

[Dav98] Michael W. Davis, *The cohomology of a Coxeter group with group ring coefficients*, Duke Math. J. **91** (1998), no. 2, 297–314, DOI 10.1215/S0012-7094-98-09113-X. MR1600586 Cited in Problem 3.38.

[Dav05] James F. Davis, *The Borel/Novikov conjectures and stable diffeomorphisms of 4-manifolds*, Geometry and topology of manifolds, Fields Inst. Commun., vol. 47, Amer. Math. Soc., Providence, RI, 2005, pp. 63–76, DOI 10.1090/fic/047/05. MR2189926 Cited in Problem 4.52.

[Dav20] Christopher W. Davis, *Topological concordance of knots in homology spheres and the solvable filtration*, J. Topol. **13** (2020), no. 1, 343–355, DOI 10.1112/topo.12126. MR4138741 Cited in Problem 1.61.

[Dav25] Christopher W. Davis, *Whitney tower concordance and knots in homology spheres*, Algebr. Geom. Topol. **25** (2025), no. 6, 3503–3521, DOI 10.2140/agt.2025.25.3503. MR4972718 Cited in Problem 1.61.

[DDH89] Steven Diaz, Ron Donagi, and David Harbater, *Every curve is a Hurwitz space*, Duke Math. J. **59** (1989), no. 3, 737–746, DOI 10.1215/S0012-7094-89-05933-4. MR1046746 Cited in Problem 2.2.

[DeB06] Jason DeBlois, *Totally geodesic surfaces and homology*, Algebr. Geom. Topol. **6** (2006), 1413–1428, DOI 10.2140/agt.2006.6.1413. MR2253453 Cited in Problem 3.14.

[DFL14] Michael W. Davis, Jim Fowler, and Jean-François Lafont, *Aspherical manifolds that cannot be triangulated*, Algebr. Geom. Topol. **14** (2014), no. 2, 795–803, DOI 10.2140/agt.2014.14.795. MR3159970 Cited in Problem 5.8.

[DFPR82] John D. Dixon, Edward W. Formanek, John C. Poland, and Luis Ribes, *Profinite completions and isomorphic finite quotients*, J. Pure Appl. Algebra **23** (1982), no. 3, 227–231, DOI 10.1016/0022-4049(82)90098-6. MR644274 Cited in Problem 3.0.

[DG81] Joan L. Dyer and Edna K. Grossman, *The automorphism groups of the braid groups*, Amer. J. Math. **103** (1981), no. 6, 1151–1169, DOI 10.2307/2374228. MR636956 Cited in Problem 2.7.

[DG05] Thomas Delzant and Misha Gromov, *Cuts in Kähler groups*, Infinite groups: geometric, combinatorial and dynamical aspects, Progr. Math., vol. 248, Birkhäuser, Basel, 2005, pp. 31–55, DOI 10.1007/3-7643-7447-0_3. MR2195452 Cited in Problem 5.13.

[DG10] François Dahmani and Vincent Guirardel, *Foliations for solving equations in groups: free, virtually free, and hyperbolic groups*, J. Topol. **3** (2010), no. 2, 343–404, DOI 10.1112/jtopol/jtq010. MR2651364 Cited in Problem 2.36.

[DG20] Renaud Detcherry and Stavros Garoufalidis, *A diagrammatic approach to the AJ conjecture*, Math. Ann. **378** (2020), no. 1-2, 447–484, DOI 10.1007/s00208-020-02028-y. MR4150924 Cited in Problem 1.25.

[DG25] Nathan M. Dunfield and Sherry Gong, *Ribbon concordances and slice obstructions: experiments and examples*, 2025, arXiv:2512.21825, math.GT, Cited in Problem 1.52.

[DGST10] Nathan M. Dunfield, Stavros Garoufalidis, Alexander Shumakovitch, and Morwen Thistlethwaite, *Behavior of knot invariants under genus 2 mutation*, New York J. Math. **16** (2010), 99–123. MR2657370 Cited in Problem 3.61.

[DHKK14] G. Dimitrov, F. Haiden, L. Katzarkov, and M. Kontsevich, *Dynamical systems and categories*, The influence of Solomon Lefschetz in geometry and topology, Contemp. Math., vol. 621, Amer. Math. Soc., Providence, RI, 2014, pp. 133–170, DOI 10.1090/conm/621/12421. MR3289326 Cited in Problem 2.47.

[DHM15] Kelly Delp, Diane Hoffoss, and Jason Fox Manning. Problems in groups, geometry, and three-manifolds, 2015. arXiv:1512.04620. Cited in Problem 3.24.

[DHM23] Irving Dai, Matthew Hedden, and Abhishek Mallick, *Corks, involutions, and Heegaard Floer homology*, J. Eur. Math. Soc. (JEMS) **25** (2023), no. 6, 2319–2389, DOI 10.4171/jems/1239. MR4592871 Cited in Problems 3.74, 3.76, and 4.58.

[DHMS24] Irving Dai, Matthew Hedden, Abhishek Mallick, and Matthew Stoffregen, *Rank-expanding satellites, Whitehead doubles, and Heegaard Floer homology*, J. Topol.

17 (2024), no. 4, Paper No. e70008, 40, DOI 10.1112/topo.70008. MR4832710 Cited in Problem 1.40.

[DHST21] Irving Dai, Jennifer Hom, Matthew Stoffregen, and Linh Truong, *More concordance homomorphisms from knot Floer homology*, Geom. Topol. **25** (2021), no. 1, 275–338, DOI 10.2140/gt.2021.25.275. MR4226231 Cited in Problem 1.38.

[DHST23] Irving Dai, Jennifer Hom, Matthew Stoffregen, and Linh Truong, *An infinite-rank summand of the homology cobordism group*, Duke Math. J. **172** (2023), no. 12, 2365–2432, DOI 10.1215/00127094-2022-0082. MR4654053 Cited in Problems 3.69 and 4.25.

[DHV24] V. Delecroix, P. Hubert, and F. Valdez. Infinite translation surfaces in the wild, 2024. arXiv:2403.05424. Cited in Problem 2.42.

[Dia93] Yuanan Diao. Minimal knotted polygons on the cubic lattice. *Journal of Knot Theory and Its Ramifications*, 02:413–425, 1993. doi:10.1142/S0218216593000234. Cited in Problem 1.73.

[Dia24] Yuanan Diao, *The ropelength conjecture of alternating knots*, Math. Proc. Cambridge Philos. Soc. **177** (2024), no. 2, 367–369, DOI 10.1017/S0305004124000288. MR4834873 Cited in Problem 1.74.

[DIS+25] Aliakbar Daemi, Hayato Imori, Kouki Sato, Christopher Scaduto, and Masaki Taniguchi, *Instantons, special cycles and knot concordance*, Geom. Topol. **29** (2025), no. 8, 4189–4298, DOI 10.2140/gt.2025.29.4189. MR5000788 Cited in Problem 1.44.

[DIS24] Aliakbar Daemi, Nobuo Iida, and Christopher Scaduto. Rank three instantons, representations and sutures, 2024. arXiv:2402.10448. Cited in Problem 3.51.

[DJ91] Michael W. Davis and Tadeusz Januszkiewicz, *Hyperbolization of polyhedra*, J. Differential Geom. **34** (1991), no. 2, 347–388. MR1131435 Cited in Problem 5.8.

[DK90] S. K. Donaldson and P. B. Kronheimer, *The geometry of four-manifolds*, Oxford Mathematical Monographs, The Clarendon Press, Oxford University Press, New York, 1990. Oxford Science Publications. MR1079726 Cited in Problem 4.12.

[DK22] Renaud Detcherry and Efstratia Kalfagianni, *Cosets of monodromies and quantum representations*, Indiana Univ. Math. J. **71** (2022), no. 3, 1101–1129. MR4448580 Cited in Problem 2.49.

[DKdlNG20] Valentina Disarlo, Thomas Koberda, and J. de la Nuez González. The model theory of the curve graph, 2020. arXiv:2008.10490. Cited in Problem 2.4.

[DKK17] Heather A. Dye, Aaron Kaestner, and Louis H. Kauffman, *Khovanov homology, Lee homology and a Rasmussen invariant for virtual knots*, J. Knot Theory Ramifications **26** (2017), no. 3, 1741001, 57, DOI 10.1142/S0218216517410012. MR3627701 Cited in Problem 1.84.

[DKM+24] Irving Dai, Sungkyung Kang, Abhishek Mallick, JungHwan Park, and Matthew Stoffregen, *The* $(2,1)$*-cable of the figure-eight knot is not smoothly slice*, Invent. Math. **238** (2024), no. 2, 371–390, DOI 10.1007/s00222-024-01286-w. MR4809438 Cited in Problem 1.50.

[DKM25] Renaud Detcherry, Efstratia Kalfagianni, and Shashini Marasinghe. Seifert cobordisms and the Chen-Yang volume conjecture, 2025. arXiv:2505.01546. Cited in Problem 3.4.

[DKS25] Renaud Detcherry, Efstratia Kalfagianni, and Adam S. Sikora, *Kauffman bracket skein modules of small 3-manifolds*, Adv. Math. **467** (2025), Paper No. 110169, 45, DOI 10.1016/j.aim.2025.110169. MR4871425 Cited in Problems 3.65 and 3.66.

[dlNG22] Javier de la Nuez Gonzalez. Non-Roelcke precompactness of groups of surface homeomorphisms, 2022. arXiv:2202.06527. Cited in Problem 2.33.

[DLW18] Josef G. Dorfmeister, Tian-Jun Li, and Weiwei Wu, *Stability and existence of surfaces in symplectic 4-manifolds with* $b^+ = 1$, J. Reine Angew. Math. **742** (2018), 115–155, DOI 10.1515/crelle-2015-0083. MR3849624 Cited in Problem 4.100.

[DLME24] Aliakbar Daemi, Tye Lidman, and Mike Miller Eismeier, Filtered instanton homology and cosmetic surgery, 2024. arXiv:2410.21248. Cited in Problem 1.12.

[DM19] Irving Dai and Ciprian Manolescu, *Involutive Heegaard Floer homology and plumbed three-manifolds*, J. Inst. Math. Jussieu **18** (2019), no. 6, 1115–1155, DOI 10.1017/s1474748017000329. MR4021102 Cited in Problem 3.57.

[DM25a] Raquel Díaz and Pedro M. G. Manchón. New torsion patterns in Khovanov homology, 2025. arXiv:2508.00606. Cited in Problem 1.30.

[DM25b] Raquel Díaz and Pedro M. G. Manchón, *A pattern for torsion in Khovanov homology*, Fund. Math. **270** (2025), no. 1, 75–97, DOI 10.4064/fm240810-12-3. MR4930307 Cited in Problem 1.30.

[DME22] Aliakbar Daemi and Mike Miller Eismeier. Instantons and rational homology spheres, 2022. arXiv:2210.14071. Cited in Problem 3.59.

[dMRST21] Arnaud de Mesmay, Yo'av Rieck, Eric Sedgwick, and Martin Tancer, *The unbearable hardness of unknotting*, Adv. Math. **381** (2021), Paper No. 107648, 36, DOI 10.1016/j.aim.2021.107648. MR4215750 Cited in Problem 1.80.

[Dom22] George Domat, *Big pure mapping class groups are never perfect*, Math. Res. Lett. **29** (2022), no. 3, 691–726, DOI 10.4310/mrl.2022.v29.n3.a4. Appendix with Ryan Dickmann. MR4516036 Cited in Problem 2.39.

[Don78] Harold Donnelly, *Eta invariants for G-spaces*, Indiana Univ. Math. J. **27** (1978), no. 6, 889–918, DOI 10.1512/iumj.1978.27.27060. MR511246 Cited in Problem 4.89.

[Don83] S. K. Donaldson, *An application of gauge theory to four-dimensional topology*, J. Differential Geom. **18** (1983), no. 2, 279–315. MR710056 Cited in Problem 4.51.

[Don87a] S. K. Donaldson, *Irrationality and the h-cobordism conjecture*, J. Differential Geom. **26** (1987), no. 1, 141–168. MR892034 Cited in Problems 4.30, 4.46, and 4.56.

[Don87b] S. K. Donaldson, *The orientation of Yang-Mills moduli spaces and 4-manifold topology*, J. Differential Geom. **26** (1987), no. 3, 397–428. MR910015 Cited in Problems 4.15, 4.30, and 4.70.

[Don90] S. K. Donaldson, *Polynomial invariants for smooth four-manifolds*, Topology **29** (1990), no. 3, 257–315, DOI 10.1016/0040-9383(90)90001-Z. MR1066174 Cited in Problems 4.70 and 4.83.

[Don96] S. K. Donaldson, *Symplectic submanifolds and almost-complex geometry*, J. Differential Geom. **44** (1996), no. 4, 666–705. MR1438190 Cited in Problem 4.109.

[Don99] S. K. Donaldson, *Lefschetz pencils on symplectic manifolds*, J. Differential Geom. **53** (1999), no. 2, 205–236. MR1802722 Cited in Problems 4.95 and 4.100.

[Don00] S. K. Donaldson, *Moment maps and diffeomorphisms [MR1701920 (2001a:53122)]*, Surveys in differential geometry, Surv. Differ. Geom., vol. 7, Int. Press, Somerville, MA, 2000, pp. 107–127, DOI 10.4310/SDG.2002.v7.n1.a5. MR1919424 Cited in Problem 4.93.

[Don06] S. K. Donaldson. Two-forms on four-manifolds and elliptic equations. In *Inspired by S. S. Chern*, volume 11 of *Nankai Tracts Math.*, pages 153–172. World Sci. Publ., Hackensack, NJ, 2006. URL: https://doi.org/10.1142/9789812772688_0007, doi:10.1142/9789812772688_0007. Cited in Problem 4.93.

[DP13] I. A. Dynnikov and M. V. Prasolov, *Bypasses for rectangular diagrams. A proof of the Jones conjecture and related questions*, Trans. Moscow Math. Soc., posted on 2013, 97–144, DOI 10.1090/s0077-1554-2014-00210-7. MR3235791 Cited in Problem 1.63.

[DP23a] Michelle Daher and Mark Powell. Smoothing 3-manifolds in 5-manifolds, 2023. arXiv:2309.15962. Cited in Problem 4.40.

[DP23b] Alessio Di Prisa, *The equivariant concordance group is not abelian*, Bull. Lond. Math. Soc. **55** (2023), no. 1, 502–507, DOI 10.1112/blms.12741. MR4568356 Cited in Problem 1.49.

[Dre69] Andreas Dress, *Newman's theorems on transformation groups*, Topology **8** (1969), 203–207, DOI 10.1016/0040-9383(69)90010-X. MR238353 Cited in Problem 5.2.

[DRG24] Georgios Dimitroglou Rizell and Roman Golovko. Instability of Legendrian knottedness, and non-regular Lagrangian concordances of knots, 2024. arXiv:2409.00290. Cited in Problem 1.65.

[DRGG+22] Marco De Renzi, Azat M. Gainutdinov, Nathan Geer, Bertrand Patureau-Mirand, and Ingo Runkel. 3-Dimensional TQFTs from non-semisimple modular categories. *Selecta Math. (N.S.)*, 28(2):Paper No. 42, 60, 2022. doi:10.1007/s00029-021-00737-z. Cited in Problem 3.78.

[DS89] S. K. Donaldson and D. P. Sullivan, *Quasiconformal 4-manifolds*, Acta Math. **163** (1989), no. 3-4, 181–252, DOI 10.1007/BF02392736. MR1032074 Cited in Problems 4.61 and 4.131.

[DS03] Simon Donaldson and Ivan Smith, *Lefschetz pencils and the canonical class for symplectic four-manifolds*, Topology **42** (2003), no. 4, 743–785, DOI 10.1016/S0040-9383(02)00024-1. MR1958528 Cited in Problems 2.22 and 4.100.

[DS24a] Aliakbar Daemi and Christopher Scaduto, *Chern-Simons functional, singular instantons, and the four-dimensional clasp number*, J. Eur. Math. Soc. (JEMS) **26** (2024), no. 6, 2127–2190, DOI 10.4171/jems/1320. MR4742808 Cited in Problem 1.47.

[DS24b] Bruno Dular and Jean-Marc Schlenker. Convex co-compact hyperbolic manifolds are determined by their pleating lamination, 2024. `arXiv:2403.10090`. Cited in Problem 3.25.

[DSS23] Hirofumi Sasahira and Matthew Stoffregen, *Seiberg-Witten Floer spectra for $b_1 > 0$*, Memoirs of the European Mathematical Society, vol. 17, EMS Press, Berlin, 2025, DOI 10.4171/mems/17. MR4867561 Cited in Problems 3.56 and 3.57.

[Dun20] Nathan M. Dunfield, *A census of exceptional Dehn fillings*, Characters in low-dimensional topology, Contemp. Math., vol. 760, Amer. Math. Soc., [Providence], RI, [2020] ©2020, pp. 143–155, DOI 10.1090/conm/760/15289. MR4193924 Cited in Problem 1.11.

[DV09] Robert J. Daverman and Gerard A. Venema, *Embeddings in manifolds*, Graduate Studies in Mathematics, vol. 106, American Mathematical Society, Providence, RI, 2009, DOI 10.1090/gsm/106. MR2561389 Cited in Problems 1.61 and 1.81.

[DW15] J. Elisenda Grigsby, Anthony M. Licata, and Stephan M. Wehrli, *Annular Khovanov-Lee homology, braids, and cobordisms*, Pure Appl. Math. Q. **13** (2017), no. 3, 389–436, DOI 10.4310/pamq.2017.v13.n3.a2. MR3882203 Cited in Problem 4.59.

[DW21] Renaud Detcherry and Maxime Wolff, *A basis for the Kauffman skein module of the product of a surface and a circle*, Algebr. Geom. Topol. **21** (2021), no. 6, 2959–2993, DOI 10.2140/agt.2021.21.2959. MR4344875 Cited in Problem 3.66.

[Dyn03] I. A. Dynnikov, *Recognition algorithms in knot theory* (Russian, with Russian summary), Uspekhi Mat. Nauk **58** (2003), no. 6(354), 45–92, DOI 10.1070/RM2003v058n06ABEH000675; English transl., Russian Math. Surveys **58** (2003), no. 6, 1093–1139. MR2054090 Cited in Problem 1.79.

[Dyn06] I. A. Dynnikov, *Arc-presentations of links: monotonic simplification*, Fund. Math. **190** (2006), 29–76, DOI 10.4064/fm190-0-3. MR2232855 Cited in Problems 1.63 and 1.76.

[Dys51] F. J. Dyson. Continuous functions defined on spheres. *Ann. of Math. (2)*, 54:534–536, 1951. `doi:10.2307/1969487`. Cited in Problem 2.14.

[Edm87] Allan L. Edmonds. Construction of group actions on four-manifolds. *Trans. Amer. Math. Soc.*, 299(1):155–170, 1987. `doi:10.2307/2000487`. Cited in Problem 3.75.

[Edm05] Allan L. Edmonds, *Homology lens spaces in topological 4-manifolds*, Illinois J. Math. **49** (2005), no. 3, 827–837. MR2210261 Cited in Problem 4.28.

[Edm15] Allan L. Edmonds, *The Euler characteristic of a Haken 4-manifold*, Geometry, groups and dynamics, Contemp. Math., vol. 639, Amer. Math. Soc., Providence, RI, 2015, pp. 217–234, DOI 10.1090/conm/639/12796. MR3379830 Cited in Problem 4.19.

[Edw06] Robert D. Edwards. Suspensions of homology spheres, 2006. URL: `https://arxiv.org/abs/math/0610573`, `arXiv:math/0610573`. Cited in Problem 5.8.

[EE82] A. L. Edmonds and J. H. Ewing, *Remarks on the cobordism group of surface diffeomorphisms*, Math. Ann. **259** (1982), no. 4, 497–504, DOI 10.1007/BF01466055. MR660044 Cited in Problem 2.45.

[EF06] John B. Etnyre and Terry Fuller, *Realizing 4-manifolds as achiral Lefschetz fibrations*, Int. Math. Res. Not., posted on 2006, Art. ID 70272, 21, DOI 10.1155/IMRN/2006/70272. MR2219214 Cited in Problem 4.120.

[EF09] Yakov Eliashberg and Maia Fraser, *Topologically trivial Legendrian knots*, J. Symplectic Geom. **7** (2009), no. 2, 77–127, DOI 10.4310/jsg.2009.v7.n2.a4. MR2496415 Cited in Problem 1.63.

[Eft09] Eaman Eftekhary. Seifert fibered homology spheres with trivial Heegaard Floer homology, 2009. `arXiv:0909.3975`. Cited in Problem 3.50.

[Eft18] Eaman Eftekhary, *Bordered Floer homology and existence of incompressible tori in homology spheres*, Compos. Math. **154** (2018), no. 6, 1222–1268, DOI 10.1112/s0010437x18007054. MR3826457 Cited in Problem 3.50.

[EG57] Samuel Eilenberg and Tudor Ganea, *On the Lusternik-Schnirelmann category of abstract groups*, Ann. of Math. (2) **65** (1957), 517–518, DOI 10.2307/1970062. MR85510 Cited in Problem 5.11.

[EG92] Yakov Eliashberg and Mikhael Gromov, *Embeddings of Stein manifolds of dimension n into the affine space of dimension $3n/2+1$*, Ann. of Math. (2) **136** (1992), no. 1, 123–135, DOI 10.2307/2946547. MR1173927 Cited in Problem 4.60.

[EGG+22] Tobias Ekholm, Angus Gruen, Sergei Gukov, Piotr Kucharski, Sunghyuk Park, Marko Stošić, and Piotr Sułkowski, *Branches, quivers, and ideals for knot complements*, J. Geom. Phys. **177** (2022), Paper No. 104520, 75, DOI 10.1016/j.geomphys.2022.104520. MR4408378 Cited in Problem 3.68.

[EGL20] Yakov Eliashberg, Sheel Ganatra, and Oleg Lazarev, *Flexible Lagrangians*, Int. Math. Res. Not. IMRN **8** (2020), 2408–2435, DOI 10.1093/imrn/rny078. MR4090744 Cited in Problem 1.65.

[EH01a] John B. Etnyre and Ko Honda, *Knots and contact geometry. I. Torus knots and the figure eight knot*, J. Symplectic Geom. **1** (2001), no. 1, 63–120, DOI 10.4310/jsg.2001.v1.n1.a3. MR1959579 Cited in Problem 1.63.

[EH01b] John B. Etnyre and Ko Honda, *On the nonexistence of tight contact structures*, Ann. of Math. (2) **153** (2001), no. 3, 749–766, DOI 10.2307/2661367. MR1836287 Cited in Problems 3.40 and 4.108.

[EH03] John B. Etnyre and Ko Honda, *On connected sums and Legendrian knots*, Adv. Math. **179** (2003), no. 1, 59–74, DOI 10.1016/S0001-8708(02)00027-0. MR2004728 Cited in Problem 1.63.

[EH19] Ben Elias and Matthew Hogancamp. On the computation of torus link homology. *Compos. Math.*, 155(1):164–205, 2019. `doi:10.1112/s0010437x18007571`. Cited in Problem 1.33.

[EHK16] Tobias Ekholm, Ko Honda, and Tamás Kálmán, *Legendrian knots and exact Lagrangian cobordisms*, J. Eur. Math. Soc. (JEMS) **18** (2016), no. 11, 2627–2689, DOI 10.4171/JEMS/650. MR3562353 Cited in Problems 1.65 and 1.67.

[Eis09] Michael Eisermann, *The Jones polynomial of ribbon links*, Geom. Topol. **13** (2009), no. 2, 623–660, DOI 10.2140/gt.2009.13.623. MR2469525 Cited in Problem 1.50.

[EJ16] Jens Kristian Egsgaard and Søren Fuglede Jørgensen, *The homological content of the Jones representations at $q=-1$*, J. Knot Theory Ramifications **25** (2016), no. 11, 1650062, 25, DOI 10.1142/S0218216516500620. MR3558421 Cited in Problem 2.49.

[EK71] Robert D. Edwards and Robion C. Kirby, *Deformations of spaces of imbeddings*, Ann. of Math. (2) **93** (1971), 63–88, DOI 10.2307/1970753. MR283802 Cited in Problem 5.4.

[EK10] Ben Elias and Dan Krasner, *Rouquier complexes are functorial over braid cobordisms*, Homology Homotopy Appl. **12** (2010), no. 2, 109–146, DOI 10.4310/hha.2010.v12.n2.a4. MR2721032 Cited in Problem 1.34.

[EKM12] John B. Etnyre, Rafal Komendarczyk, and Patrick Massot, *Tightness in contact metric 3-manifolds*, Invent. Math. **188** (2012), no. 3, 621–657, DOI 10.1007/s00222-011-0355-2. MR2917179 Cited in Problem 3.40.

[EKM16] John B. Etnyre, Rafal Komendarczyk, and Patrick Massot, *Quantitative Darboux theorems in contact geometry*, Trans. Amer. Math. Soc. **368** (2016), no. 11, 7845–7881, DOI 10.1090/tran/6821. MR3546786 Cited in Problem 3.40.

[EKS16] Tobias Ekholm, Thomas Kragh, and Ivan Smith, *Lagrangian exotic spheres*, J. Topol. Anal. **8** (2016), no. 3, 375–397, DOI 10.1142/S1793525316500199. MR3509566 Cited in Problem 4.106.

[EKT03] Shalom Eliahou, Louis H. Kauffman, and Morwen B. Thistlethwaite, *Infinite families of links with trivial Jones polynomial*, Topology **42** (2003), no. 1, 155–169, DOI 10.1016/S0040-9383(02)00012-5. MR1928648 Cited in Problem 1.24.

[EL83] Beno Eckmann and Peter Linnell, *Poincaré duality groups of dimension two. II*, Comment. Math. Helv. **58** (1983), no. 1, 111–114, DOI 10.1007/BF02564628. MR699010 Cited in Problem 3.38.

[EL09] Michael Eisermann and Christoph Lamm. For which triangles is Pick's formula almost correct? *Experiment. Math.*, 18(2):187–191, 2009. http://projecteuclid.org/euclid.em/1259158428. Cited in Problem 1.46.

[EL16] Alexander P. Ellis and Aaron D. Lauda, *An odd categorification of* $U_q(\mathfrak{sl}_2)$, Quantum Topol. **7** (2016), no. 2, 329–433, DOI 10.4171/QT/78. MR3459963 Cited in Problem 1.37.

[Eli89] Y. Eliashberg, *Classification of overtwisted contact structures on* 3*-manifolds*, Invent. Math. **98** (1989), no. 3, 623–637, DOI 10.1007/BF01393840. MR1022310 Cited in Problem 3.39.

[Eli90] Yakov Eliashberg, *Filling by holomorphic discs and its applications*, Geometry of low-dimensional manifolds, 2 (Durham, 1989), London Math. Soc. Lecture Note Ser., vol. 151, Cambridge Univ. Press, Cambridge, 1990, pp. 45–67. MR1171908 Cited in Problems 3.39 and 4.110.

[Eli92] Yakov Eliashberg, *Contact* 3*-manifolds twenty years since J. Martinet's work* (English, with French summary), Ann. Inst. Fourier (Grenoble) **42** (1992), no. 1-2, 165–192, DOI 10.5802/aif.1288. MR1162559 Cited in Problems 1.62, 1.69, and 3.15.

[ELT12] John B. Etnyre, Douglas J. LaFountain, and Bülent Tosun, *Legendrian and transverse cables of positive torus knots*, Geom. Topol. **16** (2012), no. 3, 1639–1689, DOI 10.2140/gt.2012.16.1639. MR2967060 Cited in Problem 1.63.

[EM80] Beno Eckmann and Heinz Müller, *Poincaré duality groups of dimension two*, Comment. Math. Helv. **55** (1980), no. 4, 510–520, DOI 10.1007/BF02566702. MR604709 Cited in Problem 3.38.

[EMM22] John B. Etnyre, Hyunki Min, and Anubhav Mukherjee, *On 3-manifolds that are boundaries of exotic 4-manifolds*, Trans. Amer. Math. Soc. **375** (2022), no. 6, 4307–4332, DOI 10.1090/tran/8586. MR4419060 Cited in Problem 4.6.

[EMn92] Mario Eudave Muñoz, *Band sums of links which yield composite links. The cabling conjecture for strongly invertible knots*, Trans. Amer. Math. Soc. **330** (1992), no. 2, 463–501, DOI 10.2307/2153918. MR1112545 Cited in Problem 1.11.

[EN85] David Eisenbud and Walter Neumann, *Three-dimensional link theory and invariants of plane curve singularities*, Annals of Mathematics Studies, vol. 110, Princeton University Press, Princeton, NJ, 1985. MR817982 Cited in Problem 1.8.

[EN03] John B. Etnyre and Lenhard L. Ng, *Problems in low dimensional contact topology*, Topology and geometry of manifolds (Athens, GA, 2001), Proc. Sympos. Pure Math., vol. 71, Amer. Math. Soc., Providence, RI, 2003, pp. 337–357, DOI 10.1090/pspum/071/2024641. MR2024641 Cited in Problem 1.63.

[EN22] John B. Etnyre and Lenhard L. Ng, *Legendrian contact homology in* $\mathbb{R}^3$, Surveys in differential geometry 2020. Surveys in 3-manifold topology and geometry, Surv. Differ. Geom., vol. 25, Int. Press, Boston, MA, [2022] ©2022, pp. 103–161. MR4479751 Cited in Problem 5.16.

[End95] Hisaaki Endo, *Linear independence of topologically slice knots in the smooth cobordism group*, Topology Appl. **63** (1995), no. 3, 257–262, DOI 10.1016/0166-8641(94)00062-8. MR1334309 Cited in Problem 1.38.

[End21] Hisaaki Endo, *Lefschetz fibrations*, Sugaku Expositions **34** (2021), no. 2, 175–204, DOI 10.1090/suga/462. Translation of [3675915]. MR4327688 Cited in Problem 2.22.

[Eng70] Renate Engmann, *Nicht-homöomorphe Heegaard-Zerlegungen vom Geschlecht* 2 *der zusammenhängenden Summe zweier Linsenräume* (German), Abh. Math. Sem. Univ. Hamburg **35** (1970), 33–38, DOI 10.1007/BF02992472. MR283803 Cited in Problem 4.115.

[ENS02] John B. Etnyre, Lenhard L. Ng, and Joshua M. Sabloff, *Invariants of Legendrian knots and coherent orientations*, J. Symplectic Geom. **1** (2002), no. 2, 321–367, DOI 10.4310/jsg.2001.v1.n2.a5. MR1959585 Cited in Problem 5.16.

[ENS18] Tobias Ekholm, Lenhard Ng, and Vivek Shende, *A complete knot invariant from contact homology*, Invent. Math. **211** (2018), no. 3, 1149–1200, DOI 10.1007/s00222-017-0761-1. MR3763406 Cited in Problems 4.106 and 5.16.

[ENV13] John B. Etnyre, Lenhard L. Ng, and Vera Vértesi, *Legendrian and transverse twist knots*, J. Eur. Math. Soc. (JEMS) **15** (2013), no. 3, 969–995, DOI 10.4171/JEMS/383. MR3085098 Cited in Problem 1.63.

[EO08] John B. Etnyre and Burak Ozbagci, *Invariants of contact structures from open books*, Trans. Amer. Math. Soc. **360** (2008), no. 6, 3133–3151, DOI 10.1090/S0002-9947-08-04459-0. MR2379791 Cited in Problem 3.45.

[EP88] D. B. A. Epstein and R. C. Penner, *Euclidean decompositions of noncompact hyperbolic manifolds*, J. Differential Geom. **27** (1988), no. 1, 67–80. MR918457 Cited in Problem 3.3.

[EP96] Y. Eliashberg and L. Polterovich, *Local Lagrangian 2-knots are trivial*, Ann. of Math. (2) **144** (1996), no. 1, 61–76, DOI 10.2307/2118583. MR1405943 Cited in Problem 1.67.

[EP23] Yakov Eliashberg and Dishant M. Pancholi, *Honda-Huang's work on contact convexity revisited*, Essays in geometry—dedicated to Norbert A'Campo, IRMA Lect. Math. Theor. Phys., vol. 34, EMS Press, Berlin, [2023] ©2023, pp. 453–492. MR4631278 Cited in Problem 4.108.

[EQ16] Ben Elias and You Qi, *A categorification of quantum* $\mathfrak{sl}(2)$ *at prime roots of unity*, Adv. Math. **299** (2016), 863–930, DOI 10.1016/j.aim.2016.06.002. MR3519482 Cited in Problem 3.67.

[ES22] Viveka Erlandsson and Juan Souto, *Mirzakhani's curve counting and geodesic currents*, Progress in Mathematics, vol. 345, Birkhäuser/Springer, Cham, [2022] ©2022, DOI 10.1007/978-3-031-08705-9. MR4501245 Cited in Problem 2.17.

[ET98] Yakov M. Eliashberg and William P. Thurston, *Confoliations*, University Lecture Series, vol. 13, American Mathematical Society, Providence, RI, 1998, DOI 10.1090/ulect/013. MR1483314 Cited in Problems 3.15, 3.39, and 3.40.

[Etn04a] John B. Etnyre, *On symplectic fillings*, Algebr. Geom. Topol. **4** (2004), 73–80, DOI 10.2140/agt.2004.4.73. MR2023278 Cited in Problem 3.39.

[Etn04b] John B. Etnyre, *Planar open book decompositions and contact structures*, Int. Math. Res. Not. **79** (2004), 4255–4267, DOI 10.1155/S1073792804142207. MR2126827 Cited in Problem 3.45.

[Etn05] John B. Etnyre, *Legendrian and transversal knots*, Handbook of knot theory, Elsevier B. V., Amsterdam, 2005, pp. 105–185, DOI 10.1016/B978-044451452-3/50004-6. MR2179261 Cited in Problems 1.62 and 1.69.

[ETW18] Michael Ehrig, Daniel Tubbenhauer, and Paul Wedrich, *Functoriality of colored link homologies*, Proc. Lond. Math. Soc. (3) **117** (2018), no. 5, 996–1040, DOI 10.1112/plms.12154. MR3877770 Cited in Problem 1.37.

[EVHM15] John B. Etnyre and Jeremy Van Horn-Morris, *Monoids in the mapping class group*, Interactions between low-dimensional topology and mapping class groups, Geom. Topol. Monogr., vol. 19, Geom. Topol. Publ., Coventry, 2015, pp. 319–365, DOI 10.2140/gtm.2015.19.319. MR3609913 Cited in Problem 1.64.

[Far72] F. T. Farrell, *The obstruction to fibering a manifold over a circle*, Indiana Univ. Math. J. **21** (1971/72), 315–346, DOI 10.1512/iumj.1971.21.21024. MR290397 Cited in Problem 4.127.

[FC99] V. V. Fok and L. O. Chekhov, *Quantum Teichmüller spaces* (Russian, with Russian summary), Teoret. Mat. Fiz. **120** (1999), no. 3, 511–528, DOI 10.1007/BF02557246; English transl., Theoret. and Math. Phys. **120** (1999), no. 3, 1245–1259. MR1737362 Cited in Problem 3.78.

[Fed59] Herbert Federer, *Curvature measures*, Trans. Amer. Math. Soc. **93** (1959), 418–491, DOI 10.2307/1993504. MR110078 Cited in Problem 1.71.

[Fee22] Paul M. N. Feehan. Bialynicki-Birula theory, Morse-Bott theory, and resolution of singularities for analytic spaces, 2022. `arXiv:2206.14710`. Cited in Problem 4.90.

[Fen16] Sérgio R. Fenley, *Quasigeodesic pseudo-Anosov flows in hyperbolic 3-manifolds and connections with large scale geometry*, Adv. Math. **303** (2016), 192–278, DOI 10.1016/j.aim.2016.05.015. MR3552525 Cited in Problem 3.21.

[FF03] Benson Farb and John Franks, *Groups of homeomorphisms of one-manifolds. III. Nilpotent subgroups*, Ergodic Theory Dynam. Systems **23** (2003), no. 5, 1467–1484, DOI 10.1017/S0143385702001712. MR2018608 Cited in Problems 2.34 and 3.22.

[FF20] Benson Farb and John Franks, *Groups of homeomorphisms of one-manifolds, I: Actions of nonlinear groups*, What's next?—the mathematical legacy of William P. Thurston, Ann. of Math. Stud., vol. 205, Princeton Univ. Press, Princeton, NJ, 2020, pp. 116–140, DOI 10.2307/j.ctvthhdvv.9. MR4205638 Cited in Problem 2.27.

[FGL02] Charles Frohman, Răzvan Gelca, and Walter Lofaro, *The A-polynomial from the noncommutative viewpoint*, Trans. Amer. Math. Soc. **354** (2002), no. 2, 735–747, DOI 10.1090/S0002-9947-01-02889-6. MR1862565 Cited in Problem 1.25.

[FGM21] Federica Fanoni, Tyrone Ghaswala, and Alan McLeay, *Homeomorphic subsurfaces and the omnipresent arcs* (English, with English and French summaries), Ann. H. Lebesgue **4** (2021), 1565–1593, DOI 10.5802/ahl.110. MR4353971 Cited in Problem 2.38.

[FGMW10] Michael Freedman, Robert Gompf, Scott Morrison, and Kevin Walker, *Man and machine thinking about the smooth 4-dimensional Poincaré conjecture*, Quantum Topol. **1** (2010), no. 2, 171–208, DOI 10.4171/QT/5. MR2657647 Cited in Problems 1.28, 1.60, and 4.43.

[FHMT07] Stefan Friedl, Ian Hambleton, Paul Melvin, and Peter Teichner, *Non-smoothable four-manifolds with infinite cyclic fundamental group*, Int. Math. Res. Not. IMRN **11** (2007), Art. ID rnm031, 20, DOI 10.1093/imrn/rnm031. MR2344268 Cited in Problem 4.51.

[FHW94] Michael H. Freedman, Zheng-Xu He, and Zhenghan Wang, *Möbius energy of knots and unknots*, Ann. of Math. (2) **139** (1994), no. 1, 1–50, DOI 10.2307/2946626. MR1259363 Cited in Problem 1.75.

[Fic84] Henry Clay Fickle, *Knots, $\mathbf{Z}$-homology 3-spheres and contractible 4-manifolds*, Houston J. Math. **10** (1984), no. 4, 467–493. MR774711 Cited in Problem 4.58.

[Fin77] Ronald Fintushel, *Circle actions on simply connected 4-manifolds*, Trans. Amer. Math. Soc. **230** (1977), 147–171, DOI 10.2307/1997715. MR458456 Cited in Problem 4.89.

[Fin78] Ronald Fintushel, *Classification of circle actions on 4-manifolds*, Trans. Amer. Math. Soc. **242** (1978), 377–390, DOI 10.2307/1997745. MR496815 Cited in Problem 4.89.

[Fin02] Sergey Finashin, *Knotting of algebraic curves in* $\mathbb{C}\mathrm{P}^2$, Topology **41** (2002), no. 1, 47–55, DOI 10.1016/S0040-9383(00)00023-9. MR1871240 Cited in Problems 4.8 and 4.101.

[Fin09] Sergey Finashin, *Exotic embeddings of* $\#6\mathbb{R}\mathrm{P}^2$ *in the 4-sphere*, Proceedings of Gökova Geometry-Topology Conference 2008, Gökova Geometry/Topology Conference (GGT), Gökova, 2009, pp. 151–169. MR2528682 Cited in Problem 4.29.

[FK65] Robert Fricke and Felix Klein, *Vorlesungen über die Theorie der automorphen Funktionen. Band 1: Die gruppentheoretischen Grundlagen. Band II: Die funktionentheoretischen Ausführungen und die Andwendungen* (German), Bibliotheca Mathematica Teubneriana, Bande 3, vol. 4, Johnson Reprint Corp., New York; B. G. Teubner Verlagsgesellschaft, Stuttgart, 1965. MR183872 Cited in Problem 2.16.

[FK16] Michael Freedman and Vyacheslav Krushkal, *Engel relations in 4-manifold topology*, Forum Math. Sigma **4** (2016), Paper No. e22, 57, DOI 10.1017/fms.2016.20. MR3536924 Cited in Problem 4.47.

[FKBL19] Charles Frohman, Joanna Kania-Bartoszynska, and Thang Lê. Unicity for representations of the Kauffman bracket skein algebra. *Invent. Math.*, 215(2):609–650, 2019. `doi:10.1007/s00222-018-0833-x`. Cited in Problems 2.50 and 3.78.

[FKBL25] Charles D. Frohman, Joanna Kania-Bartoszynska, and Thang T. Q. Lê, *Sliced skein algebras and geometric Kauffman bracket*, Adv. Math. **463** (2025), Paper No. 110118, 65, DOI 10.1016/j.aim.2025.110118. MR4857567 Cited in Problem 2.50.

[FKLL23] Elizabeth Field, Heejoung Kim, Christopher Leininger, and Marissa Loving, *End-periodic homeomorphisms and volumes of mapping tori*, J. Topol. **16** (2023), no. 1, 57–105, DOI 10.1112/topo.12277. MR4532490 Cited in Problems 2.18, 2.37, and 2.44.

[FKLL25] Elizabeth Field, Autumn Kent, Christopher Leininger, and Marissa Loving, *A lower bound on volumes of end-periodic mapping tori*, J. Topol. **18** (2025), no. 3, Paper No. e70037, 36, DOI 10.1112/topo.70037. MR4954690 Cited in Problem 2.18.

[FKP11] David Futer, Efstratia Kalfagianni, and Jessica S. Purcell, *Slopes and colored Jones polynomials of adequate knots*, Proc. Amer. Math. Soc. **139** (2011), no. 5, 1889–1896, DOI 10.1090/S0002-9939-2010-10617-2. MR2763776 Cited in Problem 1.26.

[FKR23] L. Ferrari, A. Kolpakov, and A. W. Reid, *Infinitely many arithmetic hyperbolic rational homology 3-spheres that bound geometrically*, Trans. Amer. Math. Soc.

376 (2023), no. 3, 1979–1997, DOI 10.1090/tran/8816. MR4549697 Cited in Problem 3.77.

[FKV87] S. M. Finashin, M. Kreck, and O. Ya. Viro, *Exotic knottings of surfaces in the* 4*-sphere*, Bull. Amer. Math. Soc. (N.S.) **17** (1987), no. 2, 287–290, DOI 10.1090/S0273-0979-1987-15562-5. MR903734 Cited in Problems 4.29 and 4.30.

[FL15] Paul M. N. Feehan and Thomas G. Leness, *Witten's conjecture for many four-manifolds of simple type*, J. Eur. Math. Soc. (JEMS) **17** (2015), no. 4, 899–923, DOI 10.4171/JEMS/521. MR3349302 Cited in Problem 4.62.

[FL18] Paul M. N. Feehan and Thomas G. Leness, *An* SO(3)*-monopole cobordism formula relating Donaldson and Seiberg-Witten invariants*, Mem. Amer. Math. Soc. **256** (2018), no. 1226, xiv+234, DOI 10.1090/memo/1226. MR3897982 Cited in Problems 3.47 and 4.62.

[FL23] Paul M. N. Feehan and Thomas G. Leness. Virtual Morse-Bott index, moduli spaces of pairs, and applications to topology of smooth four-manifolds, 2023. `arXiv:2010.15789`. Cited in Problem 4.90.

[FL24a] Benson Farb and Eduard Looijenga. The Nielsen realization problem for K3 surfaces. *J. Differential Geom.*, 127(2):505–549, 2024. `doi:10.4310/jdg/1717772420`. Cited in Problem 4.84.

[FL24b] Paul M. N. Feehan and Thomas G. Leness, SO(3) *monopoles, level-one Seiberg-Witten moduli spaces, and Witten's conjecture in low degrees*, Proceedings of the 1999 Georgia Topology Conference (Athens, GA), Topology Appl. **124** (2002), no. 2, 221–326, DOI 10.1016/S0166-8641(01)00233-4. MR1936209 Cited in Problem 4.90.

[FLL23] Peter Feller, Lukas Lewark, and Andrew Lobb, *Almost positive links are strongly quasipositive*, Math. Ann. **385** (2023), no. 1-2, 481–510, DOI 10.1007/s00208-021-02328-x. MR4542723 Cited in Problem 1.69.

[Flo88] Andreas Floer, *An instanton-invariant for* 3*-manifolds*, Comm. Math. Phys. **118** (1988), no. 2, 215–240. MR956166 Cited in Problem 3.47.

[Flo89] Andreas Floer, *Witten's complex and infinite-dimensional Morse theory*, J. Differential Geom. **30** (1989), no. 1, 207–221. MR1001276 Cited in Problem 3.47.

[FM57] Ralph H. Fox and John W. Milnor. Singularities of 2-spheres in 4-space and equivalence of knots. *Bull. Amer. Math. Soc.*, 63:406, 1957. Cited in Problem 1.37.

[FM66] Ralph H. Fox and John W. Milnor, *Singularities of* 2*-spheres in* 4*-space and cobordism of knots*, Osaka Math. J. **3** (1966), 257–267. MR211392 Cited in Problems 1.37, 1.38, and 1.43.

[FM88] Robert Friedman and John W. Morgan, *On the diffeomorphism types of certain algebraic surfaces. I*, J. Differential Geom. **27** (1988), no. 2, 297–369. MR925124 Cited in Problems 4.70, 4.75, and 4.83.

[FM12] Benson Farb and Dan Margalit, *A primer on mapping class groups*, Princeton Mathematical Series, vol. 49, Princeton University Press, Princeton, NJ, 2012. MR2850125 Cited in Problem 2.0.

[FM16] Peter Feller and Duncan McCoy, *On 2-bridge knots with differing smooth and topological slice genera*, Proc. Amer. Math. Soc. **144** (2016), no. 12, 5435–5442, DOI 10.1090/proc/13147. MR3556284 Cited in Problem 1.46.

[FM17] David Futer and Christian Millichap, *Spectrally similar incommensurable 3-manifolds*, Proc. Lond. Math. Soc. (3) **115** (2017), no. 2, 411–447, DOI 10.1112/plms.12045. MR3684110 Cited in Problem 3.17.

[FO93] Mikio Furuta and Hiroshi Ohta, *Differentiable structures on punctured* 4*-manifolds*, Topology Appl. **51** (1993), no. 3, 291–301, DOI 10.1016/0166-8641(93)90083-P. MR1237394 Cited in Problems 4.22 and 4.65.

[Fox61] R. H. Fox, *A quick trip through knot theory*, Topology of 3-manifolds and related topics (Proc. The Univ. of Georgia Institute, 1961), Prentice-Hall, Inc., Englewood Cliffs, NJ, 1961, pp. 120–167. MR140099 Cited in Problem 4.35.

[Fox62] R. H. Fox, *Some problems in knot theory*, Topology of 3-manifolds and related topics (Proc. The Univ. of Georgia Institute, 1961), Prentice-Hall, Inc., Englewood Cliffs, NJ, 1961, pp. 168–176. MR140100 Cited in Problems 1.7 and 1.50.

[FP77] Ronald Fintushel and Peter Sie Pao, *Identification of certain* 4*-manifolds with group actions*, Proc. Amer. Math. Soc. **67** (1977), no. 2, 344–350, DOI 10.2307/2041299. MR501042 Cited in Problem 4.89.

[FP92] Edward Formanek and Claudio Procesi, *The automorphism group of a free group is not linear*, J. Algebra **149** (1992), no. 2, 494–499, DOI 10.1016/0021-8693(92)90029-L. MR1172442 Cited in Problem 2.3.

[FP15] Federica Fanoni and Hugo Parlier, *Systoles and kissing numbers of finite area hyperbolic surfaces*, Algebr. Geom. Topol. **15** (2015), no. 6, 3409–3433, DOI 10.2140/agt.2015.15.3409. MR3450766 Cited in Problem 2.12.

[FP20] Stefan Friedl and Mark Powell, *Homotopy ribbon concordance and Alexander polynomials*, Arch. Math. (Basel) **115** (2020), no. 6, 717–725, DOI 10.1007/s00013-020-01517-5. MR4179701 Cited in Problem 1.53.

[FPS07] Ronald Fintushel, B. Doug Park, and Ronald J. Stern, *Reverse engineering small 4-manifolds*, Algebr. Geom. Topol. **7** (2007), 2103–2116, DOI 10.2140/agt.2007.7.2103. MR2366188 Cited in Problem 4.3.

[FPS22] David Futer, Jessica S. Purcell, and Saul Schleimer, *Effective bilipschitz bounds on drilling and filling*, Geom. Topol. **26** (2022), no. 3, 1077–1188, DOI 10.2140/gt.2022.26.1077. MR4466646 Cited in Problems 1.12 and 3.23.

[FPS24] David Futer, Jessica S. Purcell, and Saul Schleimer, Code for verifying the cosmetic surgery conjecture on a given 3-manifold, 2024. `https://github.com/saulsch/Cosmetic`. Cited in Problem 1.12.

[FPS25] David Futer, Jessica S. Purcell, and Saul Schleimer, *Excluding cosmetic surgeries on hyperbolic 3-manifolds*, J. Comput. Geom. **16** (2025), no. 1, 694–736, `doi:doi.org/10.20382/jocg.v16i1a19`. Cited in Problem 1.12.

[FQ90] Michael H. Freedman and Frank Quinn, *Topology of 4-manifolds*, Princeton Mathematical Series, vol. 39, Princeton University Press, Princeton, NJ, 1990. MR1201584 Cited in Problems 1.46, 1.61, 1.62, 3.73, 3.74, 4.4, 4.5, 4.12, 4.20, 4.26, 4.29, 4.30, 4.35, 4.41, 4.46, 4.49, 4.51, 4.52, 4.54, and 4.56.

[FR11] Bell Foozwell and Hyam Rubinstein, *Introduction to the theory of Haken n-manifolds*, Topology and geometry in dimension three, Contemp. Math., vol. 560, Amer. Math. Soc., Providence, RI, 2011, pp. 71–84, DOI 10.1090/conm/560/11092. MR2866924 Cited in Problem 4.19.

[Fre82] Michael Hartley Freedman, *The topology of four-dimensional manifolds*, J. Differential Geometry **17** (1982), no. 3, 357–453. MR679066 Cited in Problems 1.61, 3.73, 4.1, 4.9, 4.20, 4.21, 4.25, 4.28, 4.50, 4.54, 4.76, and 4.80.

[Fre84] Michael H. Freedman, *The disk theorem for four-dimensional manifolds*, Proceedings of the International Congress of Mathematicians, Vol. 1, 2 (Warsaw, 1983), PWN, Warsaw, 1984, pp. 647–663. MR804721 Cited in Problems 1.62 and 4.46.

[Fre22] Michael Freedman. The 2-width of embedded 3-manifolds. *Peking Math. J.*, 5(1):21–35, 2022. `doi:10.1007/s42543-021-00035-9`. Cited in Problem 2.15.

[Frø96] Kim A. Frøyshov, *The Seiberg-Witten equations and four-manifolds with boundary*, Math. Res. Lett. **3** (1996), no. 3, 373–390, DOI 10.4310/MRL.1996.v3.n3.a7. MR1397685 Cited in Problem 4.61.

[Frø02] Kim A. Frøyshov, *Equivariant aspects of Yang-Mills Floer theory*, Topology **41** (2002), no. 3, 525–552, DOI 10.1016/S0040-9383(01)00018-0. MR1910040 Cited in Problem 4.61.

[FRW24] Ethan Farber, Braeden Reinoso, and Luya Wang, *Fixed-point-free pseudo-Anosov homeomorphisms, knot Floer homology and the cinquefoil*, Geom. Topol. **28** (2024), no. 9, 4337–4381, DOI 10.2140/gt.2024.28.4337. MR4845456 Cited in Problems 1.14, 1.21, and 1.22.

[FS81] Ronald Fintushel and Ronald J. Stern, *An exotic free involution on S^4*, Ann. of Math. (2) **113** (1981), no. 2, 357–365, DOI 10.2307/2006987. MR607896 Cited in Problems 4.87 and 4.88.

[FS84] Ronald Fintushel and Ronald J. Stern, *A μ-invariant one homology* 3*-sphere that bounds an orientable rational ball*, Four-manifold theory (Durham, N.H., 1982), Contemp. Math., vol. 35, Amer. Math. Soc., Providence, RI, 1984, pp. 265–268, DOI 10.1090/conm/035/780582. MR780582 Cited in Problems 1.60 and 3.72.

[FS87a] Ronald Fintushel and Ronald Stern, *Rational homology cobordisms of spherical space forms*, Topology **26** (1987), no. 3, 385–393, DOI 10.1016/0040-9383(87)90008-5. MR899056 Cited in Problem 4.59.

[FS87b] Ronald Fintushel and Ronald J. Stern, O(2) *actions on the 5-sphere*, Invent. Math. **87** (1987), no. 3, 457–476, DOI 10.1007/BF01389237. MR874031 Cited in Problem 4.58.

[FS90] Ronald Fintushel and Ronald J. Stern, *Instanton homology of Seifert fibred homology three spheres*, Proc. London Math. Soc. (3) **61** (1990), no. 1, 109–137, DOI 10.1112/plms/s3-61.1.109. MR1051101 Cited in Problems 3.52 and 3.64.

[FS94] Ronald Fintushel and Ronald J. Stern, *A fake 4-manifold with* $\pi_1 = \mathbf{Z}$ *and* $b^+ = 4$, Turkish J. Math. **18** (1994), no. 1, 1–6. MR1270433 Cited in Problem 4.51.

[FS98] Ronald Fintushel and Ronald J. Stern. Knots, links, and 4-manifolds. *Invent. Math.*, 134(2):363–400, 1998. doi:10.1007/s002220050268. Cited in Problems 4.2, 4.63, 4.99, and 4.130.

[FS99a] Ronald Fintushel and Ronald J. Stern, *Nondiffeomorphic symplectic 4-manifolds with the same Seiberg-Witten invariants*, Proceedings of the Kirbyfest (Berkeley, CA, 1998), Geom. Topol. Monogr., vol. 2, Geom. Topol. Publ., Coventry, 1999, pp. 103–111, DOI 10.2140/gtm.1999.2.103. MR1734403 Cited in Problem 4.8.

[FS99b] Ronald Fintushel and Ronald J. Stern, *Symplectic surfaces in a fixed homology class*, J. Differential Geom. **52** (1999), no. 2, 203–222. MR1758295 Cited in Problem 4.101.

[FS04] Ronald Fintushel and Ronald J. Stern, *Families of simply connected 4-manifolds with the same Seiberg-Witten invariants*, Topology **43** (2004), no. 6, 1449–1467, DOI 10.1016/j.top.2004.03.002. MR2081432 Cited in Problem 4.113.

[FS11] Ronald Fintushel and Ronald J. Stern, *Pinwheels and nullhomologous surgery on 4-manifolds with* $b^+ = 1$, Algebr. Geom. Topol. **11** (2011), no. 3, 1649–1699, DOI 10.2140/agt.2011.11.1649. MR2821436 Cited in Problems 4.2, 4.57, and 4.111.

[FS18] Michael Freedman and Martin Scharlemann. Powell moves and the Goeritz group, 2018. arXiv:1804.05909. Cited in Problem 2.15.

[FSS22] Steven Frankel, Saul Schleimer, and Henry Segerman. From veering triangulations to link spaces and back again, 2022. arXiv:1911.00006. Cited in Problem 3.21.

[FT95a] Michael H. Freedman and Peter Teichner, 4*-manifold topology. I. Subexponential groups*, Invent. Math. **122** (1995), no. 3, 509–529, DOI 10.1007/BF01231454. MR1359602 Cited in Problem 4.46.

[FT95b] Michael H. Freedman and Peter Teichner, 4*-manifold topology. II. Dwyer's filtration and surgery kernels*, Invent. Math. **122** (1995), no. 3, 531–557, DOI 10.1007/BF01231455. MR1359603 Cited in Problem 1.62.

[FT05] Stefan Friedl and Peter Teichner, *New topologically slice knots*, Geom. Topol. **9** (2005), 2129–2158, DOI 10.2140/gt.2005.9.2129. MR2209368 Cited in Problem 1.46.

[FU14] Krzysztof Frączek and Corinna Ulcigrai, *Non-ergodic* $\mathbb{Z}$*-periodic billiards and infinite translation surfaces*, Invent. Math. **197** (2014), no. 2, 241–298, DOI 10.1007/s00222-013-0482-z. MR3232007 Cited in Problem 2.42.

[Fuj90] Michihiko Fujii, *Hyperbolic* 3*-manifolds with totally geodesic boundary*, Osaka J. Math. **27** (1990), no. 3, 539–553. MR1075164 Cited in Problem 4.126.

[Ful98] Terry Fuller. Diffeomorphism types of genus 2 Lefschetz fibrations. *Math. Ann.*, 311(1):163–176, 1998. doi:10.1007/s002080050182. Cited in Problem 4.94.

[Ful23] Terry Fuller, *Unchaining surgery, branched covers, and pencils on elliptic surfaces*, Algebr. Geom. Topol. **23** (2023), no. 6, 2867–2893, DOI 10.2140/agt.2023.23.2867. MR4640143 Cited in Problem 4.92.

[Fun99] Louis Funar, *On the TQFT representations of the mapping class groups*, Pacific J. Math. **188** (1999), no. 2, 251–274, DOI 10.2140/pjm.1999.188.251. MR1684208 Cited in Problem 2.6.

[Fun13] Louis Funar, *Torus bundles not distinguished by TQFT invariants*, Geom. Topol. **17** (2013), no. 4, 2289–2344, DOI 10.2140/gt.2013.17.2289. With an appendix by Funar and Andrei Rapinchuk. MR3109869 Cited in Problem 3.8.

[Fur90] Mikio Furuta, *Homology cobordism group of homology* 3*-spheres*, Invent. Math. **100** (1990), no. 2, 339–355, DOI 10.1007/BF01231190. MR1047138 Cited in Problem 3.64.

[Fur01] M. Furuta, *Monopole equation and the $\frac{11}{8}$-conjecture*, Math. Res. Lett. **8** (2001), no. 3, 279–291, DOI 10.4310/MRL.2001.v8.n3.a5. MR1839478 Cited in Problems 4.15, 4.64, and 4.131.

[FV11] Stefan Friedl and Stefano Vidussi, *Twisted Alexander polynomials detect fibered 3-manifolds*, Ann. of Math. (2) **173** (2011), no. 3, 1587–1643, DOI 10.4007/annals.2011.173.3.8. MR2800721 Cited in Problem 4.99.

[FV13] Stefan Friedl and Stefano Vidussi, *On the topology of symplectic Calabi-Yau 4-manifolds*, J. Topol. **6** (2013), no. 4, 945–954, DOI 10.1112/jtopol/jtt020. MR3145145 Cited in Problem 4.92.

[FW86] John L. Friedman and Donald M. Witt, *Homotopy is not isotopy for homeomorphisms of* 3*-manifolds*, Topology **25** (1986), no. 1, 35–44, DOI 10.1016/0040-9383(86)90003-0. MR836722 Cited in Problem 4.69.

[FY19] Joel Fine and Chengjian Yao, *A report on the hypersymplectic flow*, Pure Appl. Math. Q. **15** (2019), no. 4, 1219–1260, DOI 10.4310/PAMQ.2019.v15.n4.a7. MR4085670 Cited in Problem 4.93.

[GÏ5] Matthias Görner, *Regular tessellation link complements*, Exp. Math. **24** (2015), no. 2, 225–246, DOI 10.1080/10586458.2014.986310. MR3350527 Cited in Problem 3.11.

[GA70] Francisco Javier Gonzalez Acuna, *ON HOMOLOGY SPHERES*, ProQuest LLC, Ann Arbor, MI, 1970. Thesis (Ph.D.)–Princeton University. MR2619599 Cited in Problem 3.69.

[Gab83] David Gabai, *Foliations and the topology of* 3*-manifolds*, J. Differential Geom. **18** (1983), no. 3, 445–503. MR723813 Cited in Problem 3.40.

[Gab85] David Gabai, *The simple loop conjecture*, J. Differential Geom. **21** (1985), no. 1, 143–149. MR806708 Cited in Problem 3.36.

[Gab87a] David Gabai, *Foliations and the topology of* 3*-manifolds. II*, J. Differential Geom. **26** (1987), no. 3, 461–478. MR910017 Cited in Problems 1.10 and 1.13.

[Gab87b] David Gabai, *Foliations and the topology of* 3*-manifolds. III*, J. Differential Geom. **26** (1987), no. 3, 479–536. MR910018 Cited in Problems 1.13 and 4.23.

[Gab92] David Gabai, *Convergence groups are Fuchsian groups*, Ann. of Math. (2) **136** (1992), no. 3, 447–510, DOI 10.2307/2946597. MR1189862 Cited in Problem 3.24.

[Gab97] David Gabai, *Problems in foliations and laminations*, Geometric topology (Athens, GA, 1993), AMS/IP Stud. Adv. Math., vol. 2, Amer. Math. Soc., Providence, RI, 1997, pp. 1–33, DOI 10.1090/amsip/002.2/01. MR1470750 Cited in Problem 3.15.

[Gab00] Damien Gaboriau. Coût des relations d'équivalence et des groupes. *Invent. Math.*, 139(1):41–98, 2000. `doi:10.1007/s002229900019`. Cited in Problem 3.35.

[Gab13] Boštjan Gabrovšek, *The categorification of the Kauffman bracket Skein module of* $\mathbb{RP}^3$, Bull. Aust. Math. Soc. **88** (2013), no. 3, 407–422, DOI 10.1017/S0004972713000105. MR3189291 Cited in Problem 1.32.

[Gab20] David Gabai, *The 4-dimensional light bulb theorem*, J. Amer. Math. Soc. **33** (2020), no. 3, 609–652, DOI 10.1090/jams/920. MR4127900 Cited in Problem 4.34.

[Gab22] David Gabai. 3-spheres in the 4-sphere and pseudo-isotopies of $S^1 \times S^3$, 2022. `arXiv:2212.02004`. Cited in Problems 4.23, 4.68, 4.74, 4.75, and 4.77.

[Gad07] Siddhartha Gadgil, *Degree-one maps, surgery and four-manifolds*, Bull. Lond. Math. Soc. **39** (2007), no. 3, 419–424, DOI 10.1112/blms/bdm019. MR2331569 Cited in Problem 3.62.

[Gad10] Siddhartha Gadgil, *Open manifolds, Ozsváth-Szabó invariants and exotic* $\mathbb{R}^4$*'s*, Expo. Math. **28** (2010), no. 3, 254–261, DOI 10.1016/j.exmath.2009.09.002. MR2671000 Cited in Problem 4.61.

[Gai18] Fyodor Gainullin, *Heegaard Floer homology and knots determined by their complements*, Algebr. Geom. Topol. **18** (2018), no. 1, 69–109, DOI 10.2140/agt.2018.18.69. MR3748239 Cited in Problem 1.12.

[Gal24] Daniel A. P. Galvin. The Casson-Sullivan invariant for homeomorphisms of 4-manifolds, 2024. `arXiv:2405.07928`. Cited in Problems 4.33 and 4.45.

[GAn75] F. González-Acuña, *Homomorphs of knot groups*, Ann. of Math. (2) **102** (1975), no. 2, 373–377, DOI 10.2307/1971036. MR379671 Cited in Problem 1.18.

[GAn94] F. González-Acuña, *A characterization of 2-knot groups*, Rev. Mat. Iberoamericana **10** (1994), no. 2, 221–228, DOI 10.4171/RMI/151. MR1286475 Cited in Problem 4.35.

[GAnS86] Francisco González-Acuña and Hamish Short, *Knot surgery and primeness*, Math. Proc. Cambridge Philos. Soc. **99** (1986), no. 1, 89–102, DOI 10.1017/S0305004100063969. MR809502 Cited in Problem 1.11.

[Gar69] F. A. Garside, *The braid group and other groups*, Quart. J. Math. Oxford Ser. (2) **20** (1969), 235–254, DOI 10.1093/qmath/20.1.235. MR248801 Cited in Problem 2.9.

[Gar04] Stavros Garoufalidis, *On the characteristic and deformation varieties of a knot*, Proceedings of the Casson Fest, Geom. Topol. Monogr., vol. 7, Geom. Topol. Publ., Coventry, 2004, pp. 291–309, DOI 10.2140/gtm.2004.7.291. MR2172488 Cited in Problem 1.25.

[Gar11] Stavros Garoufalidis, *The Jones slopes of a knot*, Quantum Topol. **2** (2011), no. 1, 43–69, DOI 10.4171/QT/13. MR2763086 Cited in Problem 1.26.

[Gas16] Jonah Gaster, *Lifting curves simply*, Int. Math. Res. Not. IMRN **18** (2016), 5559–5568, DOI 10.1093/imrn/rnv316. MR3567251 Cited in Problem 2.11.

[Gay06] David T. Gay, *Four-dimensional symplectic cobordisms containing three-handles*, Geom. Topol. **10** (2006), 1749–1759, DOI 10.2140/gt.2006.10.1749. MR2284049 Cited in Problems 3.42 and 3.43.

[Gay25] David T. Gay, *Diffeomorphisms of the 4-sphere, Cerf theory and Montesinos twins*, Algebr. Geom. Topol. **25** (2025), no. 5, 2817–2849, DOI 10.2140/agt.2025.25.2817. MR4996240 Cited in Problem 4.67.

[Geh67] F. W. Gehring. Extension theorems for quasiconformal mappings in n-space. *J. Analyse Math.*, 19:149–169, 1967. doi:10.1007/BF02788713. Cited in Problem 4.23.

[Gei08] Hansjörg Geiges, *An introduction to contact topology*, Cambridge Studies in Advanced Mathematics, vol. 109, Cambridge University Press, Cambridge, 2008, DOI 10.1017/CBO9780511611438. MR2397738 Cited in Problem 1.62.

[Gel02] Răzvan Gelca, *Non-commutative trigonometry and the A-polynomial of the trefoil knot*, Math. Proc. Cambridge Philos. Soc. **133** (2002), no. 2, 311–323, DOI 10.1017/S0305004102006047. MR1912404 Cited in Problem 1.25.

[Ger96] Sylvain Gervais, *Presentation and central extensions of mapping class groups*, Trans. Amer. Math. Soc. **348** (1996), no. 8, 3097–3132, DOI 10.1090/S0002-9947-96-01509-7. MR1327256 Cited in Problem 2.9.

[Ger21] Chris Gerig, *No homotopy 4-sphere invariants using* ECH = SWF, Algebr. Geom. Topol. **21** (2021), no. 5, 2543–2569, DOI 10.2140/agt.2021.21.2543. MR4334519 Cited in Problem 4.1.

[GG20] Shijie Gu and Craig R. Guilbault, *Compactifications of manifolds with boundary*, J. Topol. Anal. **12** (2020), no. 4, 1073–1101, DOI 10.1142/S1793525319500754. MR4146574 Cited in Problem 4.49.

[GGH+23] David Gabai, David T. Gay, Daniel Hartman, Vyacheslav Krushkal, and Mark Powell. Pseudo-isotopies of simply connected 4-manifolds, 2023. arXiv:2311.11196. Cited in Problems 4.67, 4.68, 4.72, 4.76, and 4.82.

[GGH25] David Gabai, David T. Gay, and Daniel Hartman. Pseudo-Isotopy and Diffeomorphisms of the 4-Sphere I: Loops of Spheres, 2025. arXiv:2505.12088. Cited in Problem 4.67.

[Gha22] Nakisa Ghanbarian, *The non-orientable 4-genus for knots with 10 crossings*, J. Knot Theory Ramifications **31** (2022), no. 5, Paper No. 2250034, 46, DOI 10.1142/S0218216522500341. MR4450931 Cited in Problem 1.47.

[Ghi08] Paolo Ghiggini, *Knot Floer homology detects genus-one fibred knots*, Amer. J. Math. **130** (2008), no. 5, 1151–1169, DOI 10.1353/ajm.0.0016. MR2450204 Cited in Problems 1.19 and 1.21.

[GHM+21] David Gabai, Robert Haraway, Robert Meyerhoff, Nathaniel Thurston, and Andrew Yarmola. Hyperbolic 3-manifolds of low cusp volume, 2021. arXiv:2109.14570. Cited in Problem 3.1.

[GHMR23] Sergei Gukov, James Halverson, Ciprian Manolescu, and Fabian Ruehle. Searching for ribbons with machine learning, 2023. arXiv:2304.09304. Cited in Problem 1.52.

[GHSR20] Eugene Gorsky, Graham Hawkes, Anne Schilling, and Julianne Rainbolt. Generalized q, t-Catalan numbers. *Algebr. Comb.*, 3(4):855–886, 2020. doi:10.5802/alco.120. Cited in Problem 1.33.

[Gif66] Charles H. Giffen. The generalized Smith conjecture. *Amer. J. Math.*, 88:187–198, 1966. doi:10.2307/2373054. Cited in Problem 4.87.

[Gil82] Patrick M. Gilmer, *On the slice genus of knots*, Invent. Math. **66** (1982), no. 2, 191–197, DOI 10.1007/BF01389390. MR656619 Cited in Problem 3.73.

[Gil84] Patrick M. Gilmer, *Ribbon concordance and a partial order on S-equivalence classes*, Topology Appl. **18** (1984), no. 2-3, 313–324, DOI 10.1016/0166-8641(84)90016-6. MR769297 Cited in Problems 1.43 and 1.53.

[Gir02] Emmanuel Giroux, *Géométrie de contact: de la dimension trois vers les dimensions supérieures* (French, with French summary), Proceedings of the International Congress of Mathematicians, Vol. II (Beijing, 2002), Higher Ed. Press, Beijing, 2002, pp. 405–414. MR1957051 Cited in Problems 3.39, 3.44, 4.108, and 4.109.

[Gir17] Emmanuel Giroux, *Remarks on Donaldson's symplectic submanifolds*, Pure Appl. Math. Q. **13** (2017), no. 3, 369–388, DOI 10.4310/pamq.2017.v13.n3.a1. MR3882202 Cited in Problem 4.109.

[GJ11] Joshua Greene and Stanislav Jabuka, *The slice-ribbon conjecture for 3-stranded pretzel knots*, Amer. J. Math. **133** (2011), no. 3, 555–580, DOI 10.1353/ajm.2011.0022. MR2808326 Cited in Problem 1.50.

[GJS25] Iordan Ganev, David Jordan, and Pavel Safronov, *The quantum Frobenius for character varieties and multiplicative quiver varieties*, J. Eur. Math. Soc. (JEMS) **27** (2025), no. 7, 3023–3084, DOI 10.4171/jems/1427. MR4910600 Cited in Problem 2.50.

[GJS23] Sam Gunningham, David Jordan, and Pavel Safronov. The finiteness conjecture for skein modules. *Invent. Math.*, 232(1):301–363, 2023. doi:10.1007/s00222-022-01167-0. Cited in Problems 3.65 and 3.66.

[GK16] David Gay and Robion Kirby, *Trisecting 4-manifolds*, Geom. Topol. **20** (2016), no. 6, 3097–3132, DOI 10.2140/gt.2016.20.3097. MR3590351 Cited in Problems 4.110, 4.111, 4.113, and 4.114.

[GK23] Marco Golla and Fabien Kütle, *Symplectic isotopy of rational cuspidal sextics and septics*, Int. Math. Res. Not. IMRN **8** (2023), 6504–6578, DOI 10.1093/imrn/rnab364. MR4574379 Cited in Problem 4.102.

[GL80] Mikhael Gromov and H. Blaine Lawson, Jr. The classification of simply connected manifolds of positive scalar curvature. *Ann. of Math. (2)*, 111(3):423–434, 1980. doi:10.2307/1971103. Cited in Problem 4.129.

[GL83] Patrick M. Gilmer and Charles Livingston, *On embedding 3-manifolds in 4-space*, Topology **22** (1983), no. 3, 241–252, DOI 10.1016/0040-9383(83)90011-3. MR710099 Cited in Problems 1.48, 4.25, and 4.59.

[GL87] C. McA. Gordon and J. Luecke, *Only integral Dehn surgeries can yield reducible manifolds*, Math. Proc. Cambridge Philos. Soc. **102** (1987), no. 1, 97–101, DOI 10.1017/S0305004100067086. MR886439 Cited in Problems 1.11 and 1.15.

[GL89] C. McA. Gordon and J. Luecke, *Knots are determined by their complements*, J. Amer. Math. Soc. **2** (1989), no. 2, 371–415, DOI 10.2307/1990979. MR965210 Cited in Problems 1.12, 3.27, 3.70, and 4.118.

[GL03] O. Gonzalez and R. de la Llave, *Existence of ideal knots*, J. Knot Theory Ramifications **12** (2003), no. 1, 123–133, DOI 10.1142/S0218216503002354. MR1953628 Cited in Problem 1.74.

[GL05] Stavros Garoufalidis and Thang T. Q. Lê, *The colored Jones function is q-holonomic*, Geom. Topol. **9** (2005), 1253–1293, DOI 10.2140/gt.2005.9.1253. MR2174266 Cited in Problem 1.25.

[GL06] C. McA. Gordon and John Luecke, *Knots with unknotting number 1 and essential Conway spheres*, Algebr. Geom. Topol. **6** (2006), 2051–2116, DOI 10.2140/agt.2006.6.2051. MR2263059 Cited in Problem 1.3.

[GL11] Patrick M. Gilmer and Charles Livingston, *The nonorientable 4-genus of knots*, J. Lond. Math. Soc. (2) **84** (2011), no. 3, 559–577, DOI 10.1112/jlms/jdr024. MR2855790 Cited in Problem 1.47.

[GL15] Eugene Gorsky and Lukas Lewark. On stable $\mathfrak{sl}_3$-homology of torus knots. *Exp. Math.*, 24(2):162–174, 2015. doi:10.1080/10586458.2014.963746. Cited in Problem 1.31.

[GL16] Joshua Evan Greene and Adam Simon Levine, *Strong Heegaard diagrams and strong L-spaces*, Algebr. Geom. Topol. **16** (2016), no. 6, 3167–3208, DOI 10.2140/agt.2016.16.3167. MR3584256 Cited in Problem 3.53.

[GL23] Joshua Evan Greene and Andrew Lobb, *Cyclic quadrilaterals and smooth Jordan curves*, Invent. Math. **234** (2023), no. 3, 931–935, DOI 10.1007/s00222-023-01212-6. MR4665774 Cited in Problem 2.14.

[GL24] Joshua Evan Greene and Andrew Lobb. Polynomial inscriptions, 2024. arXiv:2206.14710. Cited in Problem 2.14.

[GL25] Daniel Galvin and Roberto Ladu, *Non-smoothable homeomorphisms of 4-manifolds with boundary*, Adv. Math. **467** (2025), Paper No. 110191, 22, DOI 10.1016/j.aim.2025.110191. MR4875371 Cited in Problems 4.70 and 4.83.

[Glu62] Herman Gluck, *The embedding of two-spheres in the four-sphere*, Trans. Amer. Math. Soc. **104** (1962), 308–333, DOI 10.2307/1993581. MR146807 Cited in Problem 4.9.

[GLVDV20] Stavros Garoufalidis, Christine Ruey Shan Lee, and Roland van der Veen, *The slope conjecture for Montesinos knots*, Internat. J. Math. **31** (2020), no. 7, 2050056, 66, DOI 10.1142/S0129167X20500561. MR4123945 Cited in Problem 1.26.

[GLW18] J. Elisenda Grigsby, Anthony M. Licata, and Stephan M. Wehrli. Annular Khovanov homology and knotted Schur-Weyl representations. *Compos. Math.*, 154(3):459–502, 2018. doi:10.1112/S0010437X17007540. Cited in Problem 1.36.

[GM93] Robert E. Gompf and Tomasz S. Mrowka, *Irreducible 4-manifolds need not be complex*, Ann. of Math. (2) **138** (1993), no. 1, 61–111, DOI 10.2307/2946635. MR1230927 Cited in Problem 4.63.

[GM99] Oscar Gonzalez and John H. Maddocks, *Global curvature, thickness, and the ideal shapes of knots*, Proc. Natl. Acad. Sci. USA **96** (1999), no. 9, 4769–4773, DOI 10.1073/pnas.96.9.4769. MR1692638 Cited in Problem 1.71.

[GM03] E Giroux and JP Mohsen. Structures de contact et fibrations symplectiques au-dessus du cercle. *Lecture notes*, 2003. Cited in Problem 4.108.

[GM09] Frederick W. Gehring and Gaven J. Martin, *Minimal co-volume hyperbolic lattices. I. The spherical points of a Kleinian group*, Ann. of Math. (2) **170** (2009), no. 1, 123–161, DOI 10.4007/annals.2009.170.123. MR2521113 Cited in Problem 3.1.

[GM19] Patrick M. Gilmer and Gregor Masbaum, *On the skein module of the product of a surface and a circle*, Proc. Amer. Math. Soc. **147** (2019), no. 9, 4091–4106, DOI 10.1090/proc/14553. MR3993800 Cited in Problem 3.66.

[GM21] Sergei Gukov and Ciprian Manolescu, *A two-variable series for knot complements*, Quantum Topol. **12** (2021), no. 1, 1–109, DOI 10.4171/qt/145. MR4233201 Cited in Problem 3.68.

[GMM09] David Gabai, Robert Meyerhoff, and Peter Milley, *Minimum volume cusped hyperbolic three-manifolds*, J. Amer. Math. Soc. **22** (2009), no. 4, 1157–1215, DOI 10.1090/S0894-0347-09-00639-0. MR2525782 Cited in Problem 3.1.

[GMM10] David Gabai, Robert Meyerhoff, and Peter Milley, *Mom technology and hyperbolic 3-manifolds*, In the tradition of Ahlfors-Bers. V, Contemp. Math., vol. 510, Amer. Math. Soc., Providence, RI, 2010, pp. 84–107. MR2581832 Cited in Problem 3.2.

[GMSvdM02] O. Gonzalez, J. H. Maddocks, F. Schuricht, and H. von der Mosel, *Global curvature and self-contact of nonlinearly elastic curves and rods*, Calc. Var. Partial Differential Equations **14** (2002), no. 1, 29–68, DOI 10.1007/s005260100089. MR1883599 Cited in Problem 1.74.

[GN] Oliver Goodman and Walter Neumann. Snap for hyperbolic 3-manifolds. Available at https://sourceforge.net/projects/snap-pari/. Cited in Problem 3.63.

[GN14] J. Elisenda Grigsby and Yi Ni, *Sutured Khovanov homology distinguishes braids from other tangles*, Math. Res. Lett. **21** (2014), no. 6, 1263–1275, DOI 10.4310/MRL.2014.v21.n6.a4. MR3335847 Cited in Problem 1.28.

[GN25] Daniel Galvin and Isacco Nonino. Pseudo-isotopy versus isotopy for homeomorphisms of 4-manifolds, 2025. arXiv:2506.11905. Cited in Problem 4.77.

[GNS25] David Gabai, Patrick Naylor, and Hannah Schwartz, *Doubles of Gluck twists: a five-dimensional approach.* part A, Adv. Math. **480** (2025), no. part A, Paper No. 110455, 29, DOI 10.1016/j.aim.2025.110455. MR4939424 Cited in Problems 4.9 and 4.23.

[GO23] Joshua Evan Greene and Brendan Owens. Alternating links, rational balls, and cube tilings, 2023. J. Eur. Math. Soc., to appear. arXiv:2212.06248. Cited in Problem 1.53.

[Goe33] Lebrecht Goeritz, *Die abbildungen der brezelfläche und der vollbrezel vom geschlecht 2* (German), Abh. Math. Sem. Univ. Hamburg **9** (1933), no. 1, 244–259, DOI 10.1007/BF02940650. MR3069602 Cited in Problem 2.15.

[Goe12] Sebastian Goette. Computations and applications of η invariants. In *Global differential geometry*, volume 17 of *Springer Proc. Math.*, pages 401–433. Springer, Heidelberg, 2012. URL: https://doi.org/10.1007/978-3-642-22842-1_13, doi:10.1007/978-3-642-22842-1_13. Cited in Problem 3.63.

[Gol15] Marco Golla, *Ozsváth-Szabó invariants of contact surgeries*, Geom. Topol. **19** (2015), no. 1, 171–235, DOI 10.2140/gt.2015.19.171. MR3318750 Cited in Problem 3.41.

[Gom84] Robert E. Gompf, *Stable diffeomorphism of compact* 4*-manifolds*, Topology Appl. **18** (1984), no. 2-3, 115–120, DOI 10.1016/0166-8641(84)90004-X. MR769285 Cited in Problem 4.7.

[Gom85] Robert E. Gompf, *An infinite set of exotic* $\mathbf{R}^4$*'s*, J. Differential Geom. **21** (1985), no. 2, 283–300. MR816673 Cited in Problem 4.13.

[Gom91a] Robert E. Gompf, *Killing the Akbulut-Kirby* 4*-sphere, with relevance to the Andrews-Curtis and Schoenflies problems*, Topology **30** (1991), no. 1, 97–115, DOI 10.1016/0040-9383(91)90036-4. MR1081936 Cited in Problem 4.9.

[Gom91b] Robert E. Gompf, *On Cappell-Shaneson* 4*-spheres*, Topology Appl. **38** (1991), no. 2, 123–136, DOI 10.1016/0166-8641(91)90079-2. MR1094545 Cited in Problem 4.87.

[Gom93] Robert E. Gompf, *An exotic menagerie*, J. Differential Geom. **37** (1993), no. 1, 199–223. MR1198606 Cited in Problem 4.5.

[Gom10] Robert E. Gompf, *More Cappell-Shaneson spheres are standard*, Algebr. Geom. Topol. **10** (2010), no. 3, 1665–1681, DOI 10.2140/agt.2010.10.1665. MR2683748 Cited in Problem 4.1.

[Gom13] Robert E. Gompf, *Smooth embeddings with Stein surface images*, J. Topol. **6** (2013), no. 4, 915–944, DOI 10.1112/jtopol/jtt017. MR3145144 Cited in Problem 4.107.

[Gom18] Robert E. Gompf, *Group actions, corks and exotic smoothings of* $\mathbb{R}^4$, Invent. Math. **214** (2018), no. 3, 1131–1168, DOI 10.1007/s00222-018-0819-8. MR3878728 Cited in Problem 4.71.

[Gom25] Robert E. Gompf. On sections of maps from 4-manifolds to the 2-sphere, 2025. URL: https://arxiv.org/abs/2506.18066, arXiv:2506.18066. Cited in Problem 2.22.

[Gon21] Sherry Gong, *On the Kronheimer-Mrowka concordance invariant*, J. Topol. **14** (2021), no. 1, 1–28, DOI 10.1112/topo.12175. MR4186131 Cited in Problems 1.44 and 4.43.

[Gor74] C. McA. Gordon, *On the higher-dimensional Smith conjecture*, Proc. London Math. Soc. (3) **29** (1974), 98–110, DOI 10.1112/plms/s3-29.1.98. MR356073 Cited in Problem 4.87.

[Gor75] C. McA. Gordon, *Knots, homology spheres, and contractible* 4*-manifolds*, Topology **14** (1975), 151–172, DOI 10.1016/0040-9383(75)90024-5. MR402762 Cited in Problem 1.40.

[Gor76] C. McA. Gordon, *Knots in the* 4*-sphere*, Comment. Math. Helv. **51** (1976), no. 4, 585–596, DOI 10.1007/BF02568175. MR440561 Cited in Problems 4.9 and 4.36.

[Gor78] C. McA. Gordon, *Some aspects of classical knot theory*, Knot theory (Proc. Sem., Plans-sur-Bex, 1977), Lecture Notes in Math., vol. 685, Springer, Berlin-New York, 1978, pp. 1–60. MR521730 Cited in Problem 4.35.

[Gor81] C. McA. Gordon, *Ribbon concordance of knots in the* 3*-sphere*, Math. Ann. **257** (1981), no. 2, 157–170, DOI 10.1007/BF01458281. MR634459 Cited in Problems 1.50, 1.53, 1.54, 1.56, 1.85, and 4.38.

[Gor91] Cameron McA. Gordon, *Dehn surgery on knots*, Proceedings of the International Congress of Mathematicians, Vol. I, II (Kyoto, 1990), Math. Soc. Japan, Tokyo, 1991, pp. 631–642. MR1159250 Cited in Problem 1.12.

[Gor09] Cameron Gordon, *Dehn surgery and 3-manifolds*, Low dimensional topology, IAS/Park City Math. Ser., vol. 15, Amer. Math. Soc., Providence, RI, 2009, pp. 21–71, DOI 10.1090/pcms/015/03. MR2503492 Cited in Problems 1.8 and 1.9.

[GOR13] Eugene Gorsky, Alexei Oblomkov, and Jacob Rasmussen, *On stable Khovanov homology of torus knots*, Exp. Math. **22** (2013), no. 3, 265–281, DOI 10.1080/10586458.2013.798553. MR3171092 Cited in Problem 1.31.

[Gor22] Cameron McA. Gordon, *On the homeomorphism problem for 4-manifolds*, New Zealand J. Math. **52** (2021 [2021), 821–826, DOI 10.53733/205. MR4387995 Cited in Problem 4.53.

[GPPV20] Sergei Gukov, Du Pei, Pavel Putrov, and Cumrun Vafa, *BPS spectra and 3-manifold invariants*, J. Knot Theory Ramifications **29** (2020), no. 2, 2040003, 85, DOI 10.1142/S0218216520400039. MR4089709 Cited in Problems 1.32 and 3.68.

[GPV17] Sergei Gukov, Pavel Putrov, and Cumrun Vafa, *Fivebranes and 3-manifold homology*, J. High Energy Phys. **7** (2017), 071, front matter+80, DOI 10.1007/JHEP07(2017)071. MR3686727 Cited in Problems 1.32 and 3.68.

[GR62] Murray Gerstenhaber and Oscar S. Rothaus, *The solution of sets of equations in groups*, Proc. Nat. Acad. Sci. U.S.A. **48** (1962), 1531–1533, DOI 10.1073/pnas.48.9.1531. MR166296 Cited in Problem 4.118.

[GR83] D. Gillman and D. Rolfsen, *The Zeeman conjecture for standard spines is equivalent to the Poincaré conjecture*, Topology **22** (1983), no. 3, 315–323, DOI 10.1016/0040-9383(83)90017-4. MR710105 Cited in Problem 5.12.

[Gre13a] Joshua Evan Greene, *Lattices, graphs, and Conway mutation*, Invent. Math. **192** (2013), no. 3, 717–750, DOI 10.1007/s00222-012-0421-4. MR3049933 Cited in Problem 1.17.

[Gre13b] Joshua Evan Greene, *The lens space realization problem*, Ann. of Math. (2) **177** (2013), no. 2, 449–511, DOI 10.4007/annals.2013.177.2.3. MR3010805 Cited in Problem 1.9.

[Gre13c] Joshua Evan Greene, *A spanning tree model for the Heegaard Floer homology of a branched double-cover*, J. Topol. **6** (2013), no. 2, 525–567, DOI 10.1112/jtopol/jtt007. MR3065184 Cited in Problem 3.53.

[Gre13d] Joshua Evan Greene, *A surgery triangle for lattice cohomology*, Algebr. Geom. Topol. **13** (2013), no. 1, 441–451, DOI 10.2140/agt.2013.13.441. MR3031647 Cited in Problem 3.57.

[Gre15] Joshua Evan Greene, *L-space surgeries, genus bounds, and the cabling conjecture*, J. Differential Geom. **100** (2015), no. 3, 491–506. MR3352796 Cited in Problem 1.11.

[Gre17] Joshua Evan Greene, *Alternating links and definite surfaces*, Duke Math. J. **166** (2017), no. 11, 2133–2151, DOI 10.1215/00127094-2017-0004. With an appendix by András Juhász and Marc Lackenby. MR3694566 Cited in Problem 1.82.

[Gre19] Joshua Evan Greene, *On loops intersecting at most once*, Geom. Funct. Anal. **29** (2019), no. 6, 1828–1843, DOI 10.1007/s00039-019-00517-0. MR4034921 Cited in Problem 2.12.

[Gri91] H. B. Griffiths, *The topology of square pegs in round holes*, Proc. London Math. Soc. (3) **62** (1991), no. 3, 647–672, DOI 10.1112/plms/s3-62.3.647. MR1095236 Cited in Problem 2.14.

[Gro85] M. Gromov, *Pseudo holomorphic curves in symplectic manifolds*, Invent. Math. **82** (1985), no. 2, 307–347, DOI 10.1007/BF01388806. MR809718 Cited in Problems 3.39, 4.101, and 4.107.

[Gro88] M. Gromov, *Width and related invariants of Riemannian manifolds* (English, with French summary), Astérisque **163-164** (1988), 6, 93–109, 282 (1989). On the geometry of differentiable manifolds (Rome, 1986). MR999972 Cited in Problem 4.129.

[Gru72] Branko Grünbaum, *Arrangements and spreads*, Conference Board of the Mathematical Sciences Regional Conference Series in Mathematics, No. 10, American Mathematical Society, Providence, RI, 1972. MR307027 Cited in Problem 2.14.

[GRW17] Søren Galatius and Oscar Randal-Williams, *Homological stability for moduli spaces of high dimensional manifolds. II*, Ann. of Math. (2) **186** (2017), no. 1, 127–204, DOI 10.4007/annals.2017.186.1.4. MR3665002 Cited in Problem 4.75.

[GRW18] Søren Galatius and Oscar Randal-Williams, *Homological stability for moduli spaces of high dimensional manifolds. I*, J. Amer. Math. Soc. **31** (2018), no. 1, 215–264, DOI 10.1090/jams/884. MR3718454 Cited in Problem 4.78.

[GRW24] Søren Galatius and Oscar Randal-Williams, *The Alexander trick for homology spheres*, Int. Math. Res. Not. IMRN **24** (2024), 14689–14703, DOI 10.1093/imrn/rnae255. MR4843757 Cited in Problem 4.48.

[GS80] David E. Galewski and Ronald J. Stern, *Classification of simplicial triangulations of topological manifolds*, Ann. of Math. (2) **111** (1980), no. 1, 1–34, DOI 10.2307/1971215. MR558395 Cited in Problems 3.69 and 5.8.

[GS99] Robert E. Gompf and András I. Stipsicz, *4-manifolds and Kirby calculus*, Graduate Studies in Mathematics, vol. 20, American Mathematical Society, Providence, RI, 1999, DOI 10.1090/gsm/020. MR1707327 Cited in Problems 1.10, 4.11, 4.95, and 4.120.

[GS03] Răzvan Gelca and Jeremy Sain, *The noncommutative A-ideal of a* $(2, 2p+1)$*-torus knot determines its Jones polynomial*, J. Knot Theory Ramifications **12** (2003), no. 2, 187–201, DOI 10.1142/S021821650300238X. MR1967240 Cited in Problem 1.25.

[GS04] Răzvan Gelca and Jeremy Sain, *The computation of the non-commutative generalization of the A-polynomial of the figure-eight knot*, J. Knot Theory Ramifications **13** (2004), no. 6, 785–808, DOI 10.1142/S0218216504003482. MR2088746 Cited in Problem 1.25.

[GS22a] Paolo Ghiggini and Gilberto Spano. Knot Floer homology of fibred knots and Floer homology of surface diffeomorphisms, 2022. `arXiv:2201.12411`. Cited in Problem 1.21.

[GS22b] Marco Golla and Laura Starkston, *The symplectic isotopy problem for rational cuspidal curves*, Compos. Math. **158** (2022), no. 7, 1595–1682, DOI 10.1112/s0010437x2200762x. MR4483879 Cited in Problems 4.102 and 4.104.

[GST10] Robert E. Gompf, Martin Scharlemann, and Abigail Thompson, *Fibered knots and potential counterexamples to the property 2R and slice-ribbon conjectures*, Geom. Topol. **14** (2010), no. 4, 2305–2347, DOI 10.2140/gt.2010.14.2305. MR2740649 Cited in Problems 1.10 and 1.50.

[GSY22] Roberta Guadagni, Joshua M. Sabloff, and Matthew Yacavone, *Legendrian satellites and decomposable cobordisms*, J. Knot Theory Ramifications **31** (2022), no. 13, Paper No. 2250071, 33, DOI 10.1142/s0218216522500717. MR4523297 Cited in Problem 1.65.

[GT04] Stavros Garoufalidis and Peter Teichner, *On knots with trivial Alexander polynomial*, J. Differential Geom. **67** (2004), no. 1, 167–193. MR2153483 Cited in Problem 1.62.

[Gué09] François Guéritaud, *Triangulated cores of punctured-torus groups*, J. Differential Geom. **81** (2009), no. 1, 91–142. MR2477892 Cited in Problem 3.25.

[Guk05] Sergei Gukov. Three-dimensional quantum gravity, Chern-Simons theory, and the A-polynomial. *Comm. Math. Phys.*, 255(3):577–627, 2005. `doi:10.1007/s00220-005-1312-y`. Cited in Problem 1.25.

[Gut22] Gary Guth. For exotic surfaces with boundary, one stabilization is not enough, 2022. J. Eur. Math. Soc., to appear. `https://doi.org/10.4171/jems/1541`. `arXiv:2207.11847`. Cited in Problem 4.33.

[GvdV16] Stavros Garoufalidis and Roland van der Veen, *Quadratic integer programming and the slope conjecture*, New York J. Math. **22** (2016), 907–932. MR3548130 Cited in Problem 1.26.

[GW10] J. Elisenda Grigsby and Stephan M. Wehrli, *Khovanov homology, sutured Floer homology and annular links*, Algebr. Geom. Topol. **10** (2010), no. 4, 2009–2039, DOI 10.2140/agt.2010.10.2009. MR2728482 Cited in Problems 1.28 and 1.36.

[GW12] Davide Gaiotto and Edward Witten, *Knot invariants from four-dimensional gauge theory*, Adv. Theor. Math. Phys. **16** (2012), no. 3, 935–1086, DOI 10.4310/atmp.2012.v16.n3.a5. MR3024278 Cited in Problem 1.32.

[GW25] Onkar Singh Gujral and Joshua Wang, *A minimality property for knots without Khovanov 2-torsion*, Algebr. Geom. Topol. **25** (2025), no. 7, 4073–4075, DOI 10.2140/agt.2025.25.4073. MR4986964 Cited in Problem 1.30.

[GY20] David Gabai and Mehdi Yazdi, *The fully marked surface theorem*, Acta Math. **225** (2020), no. 2, 369–413, DOI 10.4310/acta.2020.v225.n2.a4. MR4205410 Cited in Problem 3.15.

[Haa85] Andrew Haas, *Length spectra as moduli for hyperbolic surfaces*, Duke Math. J. **52** (1985), no. 4, 923–934, DOI 10.1215/S0012-7094-85-05249-4. MR816393 Cited in Problem 2.16.

[Haï15] Peter Haïssinsky, *Hyperbolic groups with planar boundaries*, Invent. Math. **201** (2015), no. 1, 239–307, DOI 10.1007/s00222-014-0552-x. MR3359053 Cited in Problem 3.24.

[Hak61] Wolfgang Haken, *Theorie der Normalflächen* (German), Acta Math. **105** (1961), 245–375, DOI 10.1007/BF02559591. MR141106 Cited in Problem 1.76.

[Hak66] Wolfgang Haken, *On homotopy 3-spheres*, Illinois J. Math. **10** (1966), 159–178. MR219072 Cited in Problem 3.33.

[Hak68] Wolfgang Haken, *Some results on surfaces in 3-manifolds*, Studies in Modern Topology, Studies in Mathematics, Vol. 5, Math. Assoc. America, 1968, pp. 39–98. MR224071 Cited in Problem 3.32.

[Hak70] Wolfgang Haken, *Various aspects of the three-dimensional Poincaré problem*, Topology of Manifolds (Proc. Inst., Univ. of Georgia, Athens, Ga., 1969), Markham Publishing Co., Chicago, IL, 1970, pp. 140–152. MR273624 Cited in Problem 4.117.

[Ham14] Noriyuki Hamada. Upper bounds for the minimal number of singular fibers in a Lefschetz fibration over the torus. *Michigan Math. J.*, 63(2):275–291, 2014. doi:10.1307/mmj/1401973051. Cited in Problem 2.21.

[Ham15] Ursula Hamenstädt, *Incompressible surfaces in rank one locally symmetric spaces*, Geom. Funct. Anal. **25** (2015), no. 3, 815–859, DOI 10.1007/s00039-015-0330-y. MR3361773 Cited in Problem 5.13.

[Ham17] Noriyuki Hamada. Sections of the Matsumoto-Cadavid-Korkmaz Lefschetz fibration, 2017. arXiv:1610.08458. Cited in Problems 2.21, 2.22, and 4.98.

[Ham20] Ursula Hamenstädt. Signature of surface bundles and bounded cohomology, 2020. arXiv:2011.05792. Cited in Problem 4.97.

[HAMS93] C. Hog-Angeloni, W. Metzler, and A. J. Sieradski. *Two-dimensional Homotopy and Combinatorial Group Theory*, volume 197 of *London Math. Soc. Lect. Note Ser.* Cambridge Univ. Press, 1993. Cited in Problems 5.0, 5.10, 5.11, and 5.12.

[Han38] W. Hantzsche, *Einlagerung von Mannigfaltigkeiten in euklidische Räume* (German), Math. Z. **43** (1938), no. 1, 38–58, DOI 10.1007/BF01181085. MR1545714 Cited in Problem 4.25.

[Han23] Jonathan Hanselman, *Heegaard Floer homology and cosmetic surgeries in S^3*, J. Eur. Math. Soc. (JEMS) **25** (2023), no. 5, 1627–1669, DOI 10.4171/jems/1218. MR4592856 Cited in Problem 1.12.

[Har81] Richard Hartley, *Knots with free period*, Canadian J. Math. **33** (1981), no. 1, 91–102, DOI 10.4153/CJM-1981-009-7. MR608857 Cited in Problem 1.49.

[Har85] John L. Harer, *Stability of the homology of the mapping class groups of orientable surfaces*, Ann. of Math. (2) **121** (1985), no. 2, 215–249, DOI 10.2307/1971172. MR786348 Cited in Problem 4.78.

[Har86] John L. Harer, *The virtual cohomological dimension of the mapping class group of an orientable surface*, Invent. Math. **84** (1986), no. 1, 157–176, DOI 10.1007/BF01388737. MR830043 Cited in Problem 2.41.

[Has86] Joel Hass. Minimal surfaces in foliated manifolds. *Comment. Math. Helv.*, 61(1):1–32, 1986. doi:10.1007/BF02621899. Cited in Problem 3.18.

[Has87] Joel Hass, *Minimal surfaces in manifolds with S^1 actions and the simple loop conjecture for Seifert fibered spaces*, Proc. Amer. Math. Soc. **99** (1987), no. 2, 383–388, DOI 10.2307/2046646. MR870806 Cited in Problem 3.36.

[Has95] Joel Hass, *Acylindrical surfaces in 3-manifolds*, Michigan Math. J. **42** (1995), no. 2, 357–365, DOI 10.1307/mmj/1029005233. MR1342495 Cited in Problem 3.12.

[Has15] Joel Hass. Minimal fibrations of hyperbolic 3-manifolds, 2015. arXiv:1512.04145. Cited in Problem 3.18.

[Hat76] Allen Hatcher, *Homeomorphisms of sufficiently large P^2-irreducible 3-manifolds*, Topology **15** (1976), no. 4, 343–347, DOI 10.1016/0040-9383(76)90027-6. MR420620 Cited in Problem 1.89.

[Hat83] Allen E. Hatcher, *A proof of the Smale conjecture,* $\mathrm{Diff}(S^3) \simeq \mathrm{O}(4)$, Ann. of Math. (2) **117** (1983), no. 3, 553–607, DOI 10.2307/2007035. MR701256 Cited in Problems 1.75 and 1.89.

[Hat02] Allen Hatcher. Topological moduli spaces of knots. `https://pi.math.cornell.edu/~hatcher/Papers/knotspaces.pdf`, 2002. Cited in Problem 1.89.

[Hay11] Kenta Hayano, *On genus-1 simplified broken Lefschetz fibrations*, Algebr. Geom. Topol. **11** (2011), no. 3, 1267–1322, DOI 10.2140/agt.2011.11.1267. MR2801419 Cited in Problem 4.113.

[Hay14] Kenta Hayano, *Complete classification of genus-1 simplified broken Lefschetz fibrations*, Hiroshima Math. J. **44** (2014), no. 2, 223–234. MR3251824 Cited in Problem 4.113.

[Hay15] Andriy Haydys, *Fukaya-Seidel category and gauge theory*, J. Symplectic Geom. **13** (2015), no. 1, 151–207, DOI 10.4310/JSG.2015.v13.n1.a5. MR3338833 Cited in Problem 1.32.

[Hay20] Kenta Hayano, *On diagrams of simplified trisections and mapping class groups*, Osaka J. Math. **57** (2020), no. 1, 17–37. MR4052626 Cited in Problems 4.113 and 4.114.

[Hay21a] Kyle Hayden. Exotically knotted disks and complex curves, 2021. `arXiv:2003.13681`. Cited in Problems 4.31 and 4.104.

[Hay21b] Kyle Hayden, *Quasipositive links and Stein surfaces*, Geom. Topol. **25** (2021), no. 3, 1441–1477, DOI 10.2140/gt.2021.25.1441. MR4268165 Cited in Problem 1.69.

[Hay22] Kyle Hayden, *Legendrian ribbons and strongly quasipositive links in an open book* (English, with English and French summaries), J. Math. Pures Appl. (9) **165** (2022), 42–57, DOI 10.1016/j.matpur.2022.07.002. MR4470109 Cited in Problem 1.69.

[Hay23] Kyle Hayden. An atomic approach to Wall-type stabilization problems, 2023. `arXiv:2302.10127`. Cited in Problem 4.33.

[He19] Siqi He, *A gluing theorem for the Kapustin-Witten equations with a Nahm pole*, J. Topol. **12** (2019), no. 3, 855–915, DOI 10.1112/topo.12102. MR4072160 Cited in Problem 1.32.

[Hed10] Matthew Hedden. Notions of positivity and the Ozsváth-Szabó concordance invariant. *J. Knot Theory Ramifications*, 19(5):617–629, 2010. `doi:10.1142/S0218216510008017`. Cited in Problems 1.9, 1.21, 1.69, and 1.70.

[Hed11] Matthew Hedden, *On Floer homology and the Berge conjecture on knots admitting lens space surgeries*, Trans. Amer. Math. Soc. **363** (2011), no. 2, 949–968, DOI 10.1090/S0002-9947-2010-05117-7. MR2728591 Cited in Problems 1.9 and 1.20.

[Hem79] Geoffrey Hemion, *On the classification of homeomorphisms of* 2*-manifolds and the classification of* 3*-manifolds*, Acta Math. **142** (1979), no. 1-2, 123–155, DOI 10.1007/BF02395059. MR512214 Cited in Problem 3.32.

[Hem87] John Hempel, *Residual finiteness for* 3*-manifolds*, Combinatorial group theory and topology (Alta, Utah, 1984), Ann. of Math. Stud., vol. 111, Princeton Univ. Press, Princeton, NJ, 1987, pp. 379–396. MR895623 Cited in Problems 3.37 and 4.118.

[Hem01] John Hempel, *3-manifolds as viewed from the curve complex*, Topology **40** (2001), no. 3, 631–657, DOI 10.1016/S0040-9383(00)00033-1. MR1838999 Cited in Problem 2.10.

[Hem14] John Hempel. Some 3-manifold groups with the same finite quotients, 2014. `arXiv:1409.3509`. Cited in Problem 3.8.

[Hen12] Kristen Hendricks, *A rank inequality for the knot Floer homology of double branched covers*, Algebr. Geom. Topol. **12** (2012), no. 4, 2127–2178, DOI 10.2140/agt.2012.12.2127. MR3020203 Cited in Problem 2.48.

[Hen15] Kristen Hendricks, *Localization of the link Floer homology of doubly-periodic knots*, J. Symplectic Geom. **13** (2015), no. 3, 545–608, DOI 10.4310/JSG.2015.v13.n3.a2. MR3412087 Cited in Problem 1.43.

[HH18a] Noriyuki Hamada and Kenta Hayano, *Topology of holomorphic Lefschetz pencils on the four-torus*, Algebr. Geom. Topol. **18** (2018), no. 3, 1515–1572, DOI 10.2140/agt.2018.18.1515. MR3784012 Cited in Problems 2.21 and 4.92.

[HH18b] Ko Honda and Yang Huang. Bypass attachments in higher-dimensional contact topology, 2018. arXiv:1803.09142. Cited in Problem 4.108.

[HH19] Ko Honda and Yang Huang. Convex hypersurface theory in contact topology, 2019. arXiv:1907.06025. Cited in Problem 4.108.

[HH23] Ian Hambleton and Jonathan A. Hillman, *Quotients of $S^2 \times S^2$*, J. Lond. Math. Soc. (2) **108** (2023), no. 4, 1393–1416, DOI 10.1112/jlms.12783. MR4655268 Cited in Problem 4.50.

[HHAMR20] Jens Harlander, Cynthia Hog-Angeloni, Wolfgang Metzler, and Stephan Rosebrock. Low-dimensional topology, problems in. Encyclopedia of Mathematics. Website, 2020. http://encyclopediaofmath.org/index.php?title=Low-dimensional_topology,_problems_in&oldid=50797. Revision as of 17:46, 1 July 2020. Cited in Problem 5.0.

[HHL21] Kristen Hendricks, Jennifer Hom, and Tye Lidman, *Applications of involutive Heegaard Floer homology*, J. Inst. Math. Jussieu **20** (2021), no. 1, 187–224, DOI 10.1017/S147474801900015X. MR4205781 Cited in Problems 3.69 and 3.70.

[HIH22] Robert Haraway III and Neil R. Hoffman. On the complexity of cusped non-hyperbolicity, 2022. arXiv:1907.01675. Cited in Problem 3.30.

[HIKK22] Susumu Hirose, Daiki Iguchi, Eiko Kin, and Yuya Koda, *Goeritz groups of bridge decompositions*, Int. Math. Res. Not. IMRN **12** (2022), 9308–9356, DOI 10.1093/imrn/rnab001. MR4436208 Cited in Problem 2.8.

[Hil74] Hugh M. Hilden, *Every closed orientable 3-manifold is a 3-fold branched covering space of S^3*, Bull. Amer. Math. Soc. **80** (1974), 1243–1244, DOI 10.1090/S0002-9904-1974-13699-2. MR350719 Cited in Problem 5.6.

[Hil81] Jonathan A. Hillman, *The Torres conditions are insufficient*, Math. Proc. Cambridge Philos. Soc. **89** (1981), no. 1, 19–22, DOI 10.1017/S0305004100057893. MR591967 Cited in Problem 1.8.

[Hil00] Jonathan A. Hillman, *Complex surfaces which are fibre bundles*, Topology Appl. **100** (2000), no. 2-3, 187–191, DOI 10.1016/S0166-8641(98)00085-6. MR1733043 Cited in Problem 4.97.

[Hil02] J. A. Hillman, *Four-manifolds, geometries and knots*, Geometry & Topology Monographs, vol. 5, Geometry & Topology Publications, Coventry, 2002. MR1943724 Cited in Problems 4.4 and 4.35.

[Hil12] Jonathan Hillman, *Algebraic invariants of links*, 2nd ed., Series on Knots and Everything, vol. 52, World Scientific Publishing Co. Pte. Ltd., Hackensack, NJ, 2012, DOI 10.1142/8493. MR2931688 Cited in Problem 1.8.

[Hil15] Jonathan A. Hillman, *Sections of surface bundles*, Interactions between low-dimensional topology and mapping class groups, Geom. Topol. Monogr., vol. 19, Geom. Topol. Publ., Coventry, 2015, pp. 1–19, DOI 10.2140/gtm.2015.19.1. MR3609902 Cited in Problem 2.25.

[Hil20] Jonathan A. Hillman, *Poincaré duality in dimension 3*, The Open Book Series, vol. 3, Mathematical Sciences Publishers, Berkeley, CA, 2020. MR4235127 Cited in Problem 3.38.

[Hil23] Jonathan A. Hillman. Homotopy types of 4-manifolds with 3-manifold fundamental groups, 2023. arXiv:2307.15292. Cited in Problems 4.50 and 4.54.

[Hil24] J. A. Hillman. Locally flat embeddings of 3-manifolds in S^4, 2024. arXiv:2408.10535. Cited in Problem 4.25.

[Hil25] Thomas Hill, *Large-scale geometry of pure mapping class groups of infinite-type surfaces*, Proc. Amer. Math. Soc. **153** (2025), no. 6, 2667–2680, DOI 10.1090/proc/17181. MR4892635 Cited in Problem 2.46.

[Hir74] Ulrich Hirsch, *Über offene Abbildungen auf die 3-Sphäre* (German), Math. Z. **140** (1974), 203–230, DOI 10.1007/BF01214163. MR362313 Cited in Problem 5.6.

[Hir93] Susumu Hirose, *On diffeomorphisms over T^2-knots*, Proc. Amer. Math. Soc. **119** (1993), no. 3, 1009–1018, DOI 10.2307/2160546. MR1155598 Cited in Problem 4.39.

[Hir02] Susumu Hirose, *On diffeomorphisms over surfaces trivially embedded in the 4-sphere*, Algebr. Geom. Topol. **2** (2002), 791–824, DOI 10.2140/agt.2002.2.791. MR1928177 Cited in Problem 4.39.

[Hir05] Susumu Hirose, *Surfaces in the complex projective plane and their mapping class groups*, Algebr. Geom. Topol. **5** (2005), 577–613, DOI 10.2140/agt.2005.5.577. MR2153115 Cited in Problem 4.39.

[Hir12] Susumu Hirose, *On diffeomorphisms over nonorientable surfaces standardly embedded in the 4-sphere*, Algebr. Geom. Topol. **12** (2012), no. 1, 109–130, DOI 10.2140/agt.2012.12.109. MR2889548 Cited in Problem 4.39.

[Hit87] N. J. Hitchin, *The self-duality equations on a Riemann surface*, Proc. London Math. Soc. (3) **55** (1987), no. 1, 59–126, DOI 10.1112/plms/s3-55.1.59. MR887284 Cited in Problem 4.90.

[Hit16] Nigel Hitchin, *Higgs bundles and diffeomorphism groups*, Surveys in differential geometry 2016. Advances in geometry and mathematical physics, Surv. Differ. Geom., vol. 21, Int. Press, Somerville, MA, 2016, pp. 139–163. MR3525096 Cited in Problem 4.120.

[HK88] Ian Hambleton and Matthias Kreck, *On the classification of topological 4-manifolds with finite fundamental group*, Math. Ann. **280** (1988), no. 1, 85–104, DOI 10.1007/BF01474183. MR928299 Cited in Problems 4.50, 4.52, 4.54, and 4.55.

[HK93a] Ian Hambleton and Matthias Kreck, *Cancellation, elliptic surfaces and the topology of certain four-manifolds*, J. Reine Angew. Math. **444** (1993), 79–100, DOI 10.1515/crll.1993.444.79. MR1241794 Cited in Problems 4.50, 4.52, and 4.54.

[HK93b] Ian Hambleton and Matthias Kreck, *Cancellation of hyperbolic forms and topological four-manifolds*, J. Reine Angew. Math. **443** (1993), 21–47, DOI 10.1515/crll.1993.443.21. MR1241127 Cited in Problems 4.45, 4.50, and 4.52.

[HK93c] Ian Hambleton and Matthias Kreck, *Cancellation of lattices and finite two-complexes*, J. Reine Angew. Math. **442** (1993), 91–109, DOI 10.1515/crll.1993.442.91. MR1234837 Cited in Problems 4.50 and 4.52.

[HK98] Craig D. Hodgson and Steven P. Kerckhoff, *Rigidity of hyperbolic cone-manifolds and hyperbolic Dehn surgery*, J. Differential Geom. **48** (1998), no. 1, 1–59. MR1622600 Cited in Problem 3.25.

[HK12a] Joel Hass and Greg Kuperberg. The complexity of recognizing the 3-sphere. *Oberwolfach Reports*, 9(2), 2012. `https://ems.press/content/serial-article-files/46393`. Cited in Problem 3.30.

[HK12b] Matthew Hedden and Paul Kirk, *Instantons, concordance, and Whitehead doubling*, J. Differential Geom. **91** (2012), no. 2, 281–319. MR2971290 Cited in Problem 1.40.

[HK21] Andrew Havens and Robin Koytcheff, *Spaces of knots in the solid torus, knots in the thickened torus, and links in the 3-sphere*, Geom. Dedicata **214** (2021), 671–737, DOI 10.1007/s10711-021-00633-y. MR4308297 Cited in Problem 1.89.

[HKL12] Matthew Hedden, Paul Kirk, and Charles Livingston, *Non-slice linear combinations of algebraic knots*, J. Eur. Math. Soc. (JEMS) **14** (2012), no. 4, 1181–1208, DOI 10.4171/JEMS/330. MR2928848 Cited in Problems 1.39 and 1.50.

[HKL16a] Matthew Hedden, Se-Goo Kim, and Charles Livingston, *Topologically slice knots of smooth concordance order two*, J. Differential Geom. **102** (2016), no. 3, 353–393. MR3466802 Cited in Problems 1.38 and 1.48.

[HKL16b] Jennifer Hom, Çağrı Karakurt, and Tye Lidman, *Surgery obstructions and Heegaard Floer homology*, Geom. Topol. **20** (2016), no. 4, 2219–2251, DOI 10.2140/gt.2016.20.2219. MR3548466 Cited in Problems 1.15 and 3.70.

[HKL19] Jonathan Hanselman, Çağatay Kutluhan, and Tye Lidman, *A remark on the geography problem in Heegaard Floer homology*, Breadth in contemporary topology, Proc. Sympos. Pure Math., vol. 102, Amer. Math. Soc., Providence, RI, 2019, pp. 103–111, DOI 10.1090/pspum/102/08. MR3967363 Cited in Problem 3.55.

[HKM02] Ko Honda, William H. Kazez, and Gordana Matić, *Convex decomposition theory*, Int. Math. Res. Not. **2** (2002), 55–88, DOI 10.1155/S1073792802101140. MR1874319 Cited in Problems 3.40 and 3.42.

[HKM07] Ko Honda, William H. Kazez, and Gordana Matić, *Right-veering diffeomorphisms of compact surfaces with boundary*, Invent. Math. **169** (2007), no. 2, 427–449, DOI 10.1007/s00222-007-0051-4. MR2318562 Cited in Problem 3.44.

[HKM20] Mark C. Hughes, Seungwon Kim, and Maggie Miller, *Isotopies of surfaces in 4-manifolds via banded unlink diagrams*, Geom. Topol. **24** (2020), no. 3, 1519–1569, DOI 10.2140/gt.2020.24.1519. MR4157558 Cited in Problem 4.9.

[HKM23] Kyle Hayden, Sungkyung Kang, and Anubhav Mukherjee. One stabilization is not enough for closed knotted surfaces, 2023. arXiv:2304.01504. Cited in Problems 4.7 and 4.33.

[HKT94] Ian Hambleton, Matthias Kreck, and Peter Teichner, *Nonorientable 4-manifolds with fundamental group of order* 2, Trans. Amer. Math. Soc. **344** (1994), no. 2, 649–665, DOI 10.2307/2154500. MR1234481 Cited in Problems 4.20 and 4.50.

[HKT09] Ian Hambleton, Matthias Kreck, and Peter Teichner, *Topological 4-manifolds with geometrically two-dimensional fundamental groups*, J. Topol. Anal. **1** (2009), no. 2, 123–151, DOI 10.1142/S1793525309000084. MR2541758 Cited in Problem 4.4.

[HL95] Ian Hambleton and Ronnie Lee, *Smooth group actions on definite 4-manifolds and moduli spaces*, Duke Math. J. **78** (1995), no. 3, 715–732, DOI 10.1215/S0012-7094-95-07826-0. MR1334207 Cited in Problems 4.61 and 4.86.

[HL19] Jennifer Hom and Tye Lidman, *A note on positive-definite, symplectic four-manifolds*, J. Eur. Math. Soc. (JEMS) **21** (2019), no. 1, 257–270, DOI 10.4171/JEMS/835. MR3880209 Cited in Problem 4.118.

[HL22] Jennifer Hom and Tye Lidman, *Dehn surgery and nonseparating two-spheres*, Gauge theory and low-dimensional topology—progress and interaction, Open Book Ser., vol. 5, Math. Sci. Publ., Berkeley, CA, 2022, pp. 145–153. MR4600722 Cited in Problem 1.13.

[HLL22a] Kristen Hendricks, Tye Lidman, and Robert Lipshitz, *Rank inequalities for the Heegaard Floer homology of branched covers*, Doc. Math. **27** (2022), 581–612. MR4424030 Cited in Problems 2.48 and 3.62.

[HLL22b] Jennifer Hom, Adam Simon Levine, and Tye Lidman, *Knot concordance in homology cobordisms*, Duke Math. J. **171** (2022), no. 15, 3089–3131, DOI 10.1215/00127094-2021-0110. MR4497224 Cited in Problem 1.61.

[HLM83a] Hugh M. Hilden, M. T. Lozano, and José María Montesinos, *Universal knots*, Bull. Amer. Math. Soc. (N.S.) **8** (1983), no. 3, 449–450, DOI 10.1090/S0273-0979-1983-15114-5. MR693959 Cited in Problem 4.122.

[HLM83b] Hugh M. Hilden, María Teresa Lozano, and José María Montesinos, *The Whitehead link, the Borromean rings and the knot* 9_{46} *are universal*, Collect. Math. **34** (1983), no. 1, 19–28. MR747855 Cited in Problem 4.122.

[HLM85] Hugh M. Hilden, María Teresa Lozano, and José María Montesinos, *On knots that are universal*, Topology **24** (1985), no. 4, 499–504, DOI 10.1016/0040-9383(85)90019-9. MR816529 Cited in Problem 4.122.

[HLP99] Joel Hass, Jeffrey C. Lagarias, and Nicholas Pippenger, *The computational complexity of knot and link problems*, J. ACM **46** (1999), no. 2, 185–211, DOI 10.1145/301970.301971. MR1693203 Cited in Problem 1.76.

[HLP22] Jennifer Hom, Tye Lidman, and JungHwan Park. Unknotting number and cabling, 2022. arXiv:2206.04196. Cited in Problem 1.4.

[HLSX22] Michael J. Hopkins, Jianfeng Lin, XiaoLin Danny Shi, and Zhouli Xu, *Intersection forms of spin 4-manifolds and the* Pin(2)*-equivariant Mahowald invariant*, Commun. Am. Math. Soc. **2** (2022), 22–132, DOI 10.1090/cams/4. MR4385297 Cited in Problems 4.15 and 4.131.

[HM97a] Allen Hatcher and Darryl McCullough, *Finiteness of classifying spaces of relative diffeomorphism groups of 3-manifolds*, Geom. Topol. **1** (1997), 91–109, DOI 10.2140/gt.1997.1.91. MR1486644 Cited in Problem 1.89.

[HM97b] Chuichiro Hayashi and Kimihiko Motegi, *Dehn surgery on knots in solid tori creating essential annuli*, Trans. Amer. Math. Soc. **349** (1997), no. 12, 4897–4930, DOI 10.1090/S0002-9947-97-01723-6. MR1373637 Cited in Problem 1.11.

[HM18] Matthew Hedden and Thomas E. Mark, *Floer homology and fractional Dehn twists*, Adv. Math. **324** (2018), 1–39, DOI 10.1016/j.aim.2017.11.008. MR3733880 Cited in Problem 2.47.

[HM19a] Siqi He and Rafe Mazzeo, *The extended Bogomolny equations and generalized Nahm pole boundary condition*, Geom. Topol. **23** (2019), no. 5, 2475–2517, DOI 10.2140/gt.2019.23.2475. MR4019897 Cited in Problem 1.32.

[HM19b] Matthew Hogancamp and Anton Mellit. Torus link homology, 2019. arXiv:1909.00418. Cited in Problem 1.33.

[HMV24] Elena S. Hafner, Karola Mészáros, and Alexander Vidinas, *Log-concavity of the Alexander polynomial*, Int. Math. Res. Not. IMRN **13** (2024), 10273–10284, DOI 10.1093/imrn/rnae058. MR4770369 Cited in Problem 1.7.

[HMY00] Kazuo Habiro, Yoshihiko Marumoto, and Yuichi Yamada, *Gluck surgery and framed links in 4-manifolds*, Knots in Hellas '98 (Delphi), Ser. Knots Everything, vol. 24, World Sci. Publ., River Edge, NJ, 2000, pp. 80–93, DOI 10.1142/9789812792679_0007. MR1865702 Cited in Problems 4.9 and 4.10.

[HN10] Joel Hass and Tahl Nowik, *Unknot diagrams requiring a quadratic number of Reidemeister moves to untangle*, Discrete Comput. Geom. **44** (2010), no. 1, 91–95, DOI 10.1007/s00454-009-9156-4. MR2639820 Cited in Problems 1.76 and 1.77.

[Hog18] Matthew Hogancamp, *Categorified Young symmetrizers and stable homology of torus links*, Geom. Topol. **22** (2018), no. 5, 2943–3002, DOI 10.2140/gt.2018.22.2943. MR3811775 Cited in Problem 1.31.

[Hom21] Jennifer Hom, *Lecture notes on Heegaard Floer homology*, Quantum field theory and manifold invariants, IAS/Park City Math. Ser., vol. 28, Amer. Math. Soc., Providence, RI, [2021] ©2021, pp. 171–200. MR4380679 Cited in Problem 1.19.

[Hor76a] Eiji Horikawa. Algebraic surfaces of general type with small c_1^2. II. *Invent. Math.*, 37(2):121–155, 1976. doi:10.1007/BF01418966. Cited in Problem 4.94.

[Hor76b] Eiji Horikawa, *Algebraic surfaces of general type with small C_1^2. I*, Ann. of Math. (2) **104** (1976), no. 2, 357–387, DOI 10.2307/1971050. MR424831 Cited in Problem 4.94.

[Hor78] Eiji Horikawa, *Algebraic surfaces of general type with small c_1^2. III*, Invent. Math. **47** (1978), no. 3, 209–248, DOI 10.1007/BF01579212. MR501370 Cited in Problem 4.94.

[Hor79] Eiji Horikawa. Algebraic surfaces of general type with small c_1^2. IV. *Invent. Math.*, 50(2):103–128, 1978/79. doi:10.1007/BF01390285. Cited in Problem 4.94.

[Hos20] Jim Hoste, *A note on Alexander polynomials of 2-bridge links*, J. Knot Theory Ramifications **29** (2020), no. 8, 1971003, 7, DOI 10.1142/S0218216519710032. MR4143690 Cited in Problem 1.8.

[How79] James Howie, *Aspherical and acyclic 2-complexes*, J. London Math. Soc. (2) **20** (1979), no. 3, 549–558, DOI 10.1112/jlms/s2-20.3.549. MR561147 Cited in Problem 5.11.

[How83] James Howie, *Some remarks on a problem of J. H. C. Whitehead*, Topology **22** (1983), no. 4, 475–485, DOI 10.1016/0040-9383(83)90038-1. MR715251 Cited in Problem 5.11.

[How85] James Howie, *On the asphericity of ribbon disc complements*, Trans. Amer. Math. Soc. **289** (1985), no. 1, 281–302, DOI 10.2307/1999700. MR779064 Cited in Problem 4.38.

[How02] James Howie, *A proof of the Scott-Wiegold conjecture on free products of cyclic groups*, J. Pure Appl. Algebra **173** (2002), no. 2, 167–176, DOI 10.1016/S0022-4049(02)00042-7. MR1915093 Cited in Problem 1.11.

[How17] Joshua A. Howie, *A characterisation of alternating knot exteriors*, Geom. Topol. **21** (2017), no. 4, 2353–2371, DOI 10.2140/gt.2017.21.2353. MR3654110 Cited in Problem 1.82.

[HP70] L. S. Husch and T. M. Price, *Finding a boundary for a 3-manifold*, Ann. of Math. (2) **91** (1970), 223–235, DOI 10.2307/1970605. MR264678 Cited in Problem 4.49.

[HP95a] Jim Hoste and Józef H. Przytycki, *The $(2,\infty)$-skein module of Whitehead manifolds*, J. Knot Theory Ramifications **4** (1995), no. 3, 411–427, DOI 10.1142/S021821659500020X. MR1347362 Cited in Problem 1.28.

[HP95b] Jim Hoste and Józef H. Przytycki, *The Kauffman bracket skein module of $S^1 \times S^2$*, Math. Z. **220** (1995), no. 1, 65–73, DOI 10.1007/BF02572603. MR1347158 Cited in Problem 3.65.

[HP97] Sa'ar Hersonsky and Frédéric Paulin, *On the rigidity of discrete isometry groups of negatively curved spaces*, Comment. Math. Helv. **72** (1997), no. 3, 349–388, DOI 10.1007/s000140050022. MR1476054 Cited in Problem 2.16.

[HPC21] Matthew Hedden and Juanita Pinzón-Caicedo, *Satellites of infinite rank in the smooth concordance group*, Invent. Math. **225** (2021), no. 1, 131–157, DOI 10.1007/s00222-020-01026-w. MR4270665 Cited in Problem 1.40.

[HR85] Craig Hodgson and J. H. Rubinstein, *Involutions and isotopies of lens spaces*, Knot theory and manifolds (Vancouver, B.C., 1983), Lecture Notes in Math., vol. 1144, Springer, Berlin, 1985, pp. 60–96, DOI 10.1007/BFb0075012. MR823282 Cited in Problem 1.16.

[HR17] Jens Harlander and Stephan Rosebrock, *Injective labeled oriented trees are aspherical*, Math. Z. **287** (2017), no. 1-2, 199–214, DOI 10.1007/s00209-016-1823-6. MR3694674 Cited in Problem 5.11.

[HRRW20] Jonathan Hanselman, Jacob Rasmussen, Sarah Dean Rasmussen, and Liam Watson, *L-spaces, taut foliations, and graph manifolds*, Compos. Math. **156** (2020), no. 3, 604–612, DOI 10.1112/s0010437x19007814. MR4055962 Cited in Problem 3.48.

[HRW22] Jonathan Hanselman, Jacob Rasmussen, and Liam Watson, *Heegaard Floer homology for manifolds with torus boundary: properties and examples*, Proc. Lond. Math. Soc. (3) **125** (2022), no. 4, 879–967, DOI 10.1112/plms.12473. MR4500201 Cited in Problem 3.61.

[HRW24] Jonathan Hanselman, Jacob Rasmussen, and Liam Watson, *Bordered Floer homology for manifolds with torus boundary via immersed curves*, J. Amer. Math. Soc. **37** (2024), no. 2, 391–498, DOI 10.1090/jams/1029. MR4695506 Cited in Problem 3.62.

[HS98] Chuichiro Hayashi and Koya Shimokawa, *Symmetric knots satisfy the cabling conjecture*, Math. Proc. Cambridge Philos. Soc. **123** (1998), no. 3, 501–529, DOI 10.1017/S0305004197002399. MR1607989 Cited in Problem 1.11.

[HS15] Kyle Hayden and Joshua M. Sabloff, *Positive knots and Lagrangian fillability*, Proc. Amer. Math. Soc. **143** (2015), no. 4, 1813–1821, DOI 10.1090/S0002-9939-2014-12365-3. MR3314092 Cited in Problem 1.68.

[HS20] Neil R. Hoffman and Nathan S. Sunukjian, *Null-homologous exotic surfaces in 4-manifolds*, Algebr. Geom. Topol. **20** (2020), no. 5, 2677–2685, DOI 10.2140/agt.2020.20.2677. MR4171577 Cited in Problem 4.30.

[HT97] Ian Hambleton and Peter Teichner, *A non-extended Hermitian form over* **Z**[**Z**], Manuscripta Math. **93** (1997), no. 4, 435–442, DOI 10.1007/BF02677483. MR1465889 Cited in Problem 4.51.

[HT04] Ian Hambleton and Mihail Tanase, *Permutations, isotropy and smooth cyclic group actions on definite 4-manifolds*, Geom. Topol. **8** (2004), 475–509, DOI 10.2140/gt.2004.8.475. MR2033485 Cited in Problem 4.86.

[HT07] Michael Hutchings and Clifford Henry Taubes, *Gluing pseudoholomorphic curves along branched covered cylinders. I*, J. Symplectic Geom. **5** (2007), no. 1, 43–137, DOI 10.4310/jsg.2007.v5.n1.a5. MR2371184 Cited in Problem 3.47.

[HT09] Michael Hutchings and Clifford Henry Taubes, *Gluing pseudoholomorphic curves along branched covered cylinders. II*, J. Symplectic Geom. **7** (2009), no. 1, 29–133, DOI 10.4310/jsg.2009.v7.n1.a2. MR2491716 Cited in Problem 3.47.

[HTT09] Joel Hass, Abigail Thompson, and William Thurston, *Stabilization of Heegaard splittings*, Geom. Topol. **13** (2009), no. 4, 2029–2050, DOI 10.2140/gt.2009.13.2029. MR2507114 Cited in Problem 3.34.

[Hut02] Michael Hutchings, *An index inequality for embedded pseudoholomorphic curves in symplectizations*, J. Eur. Math. Soc. (JEMS) **4** (2002), no. 4, 313–361, DOI 10.1007/s100970100041. MR1941088 Cited in Problem 3.47.

[Hut10] Michael Hutchings, *Embedded contact homology and its applications*, Proceedings of the International Congress of Mathematicians. Volume II, Hindustan Book Agency, New Delhi, 2010, pp. 1022–1041. MR2827830 Cited in Problem 5.18.

[Hut14] Michael Hutchings, *Lecture notes on embedded contact homology*, Contact and symplectic topology, Bolyai Soc. Math. Stud., vol. 26, János Bolyai Math. Soc., Budapest, 2014, pp. 389–484, DOI 10.1007/978-3-319-02036-5_9. MR3220947 Cited in Problem 5.18.

[HW08] Frédéric Haglund and Daniel T. Wise, *Special cube complexes*, Geom. Funct. Anal. **17** (2008), no. 5, 1551–1620, DOI 10.1007/s00039-007-0629-4. MR2377497 Cited in Problem 2.27.

[HW18] Matthew Hedden and Liam Watson, *On the geography and botany of knot Floer homology*, Selecta Math. (N.S.) **24** (2018), no. 2, 997–1037, DOI

10.1007/s00029-017-0351-5. MR3782416 Cited in Problems 1.21, 1.49, 1.56, and 3.55.

[HW19] Zheng Huang and Biao Wang, *Complex length of short curves and minimal fibrations of hyperbolic three-manifolds fibering over the circle*, Proc. Lond. Math. Soc. (3) **118** (2019), no. 6, 1305–1327, DOI 10.1112/plms.12216. MR3957822 Cited in Problem 3.18.

[HY08] Susumu Hirose and Akira Yasuhara, *Surfaces in 4-manifolds and their mapping class groups*, Topology **47** (2008), no. 1, 41–50, DOI 10.1016/j.top.2007.05.001. MR2415773 Cited in Problem 4.39.

[Hyd19] James Hyde, *The group of boundary fixing homeomorphisms of the disc is not left-orderable*, Ann. of Math. (2) **190** (2019), no. 2, 657–661, DOI 10.4007/annals.2019.190.2.5. MR3997131 Cited in Problem 2.33.

[HZ86] J. Harer and D. Zagier, *The Euler characteristic of the moduli space of curves*, Invent. Math. **85** (1986), no. 3, 457–485, DOI 10.1007/BF01390325. MR848681 Cited in Problem 2.13.

[IIS22] Kazuhiro Ichihara, Tetsuya Ito, and Toshio Saito, *On constraints for knots to admit chirally cosmetic surgeries and their calculations*, Pacific J. Math. **321** (2022), no. 1, 167–191, DOI 10.2140/pjm.2022.321.167. MR4555154 Cited in Problem 1.12.

[IJ18] Kazuhiro Ichihara and In Dae Jong, *Cosmetic banding on knots and links*, Osaka J. Math. **55** (2018), no. 4, 731–745. With an appendix by Hidetoshi Masai. MR3862782 Cited in Problem 1.12.

[IK98] Masa-Nori Ishida and Fumiharu Kato. The strong rigidity theorem for non-Archimedean uniformization. *Tohoku Math. J. (2)*, 50(4):537–555, 1998. `doi:10.2748/tmj/1178224897`. Cited in Problem 4.91.

[IK19] Tetsuya Ito and Keiko Kawamuro, *The defect of the Bennequin-Eliashberg inequality and Bennequin surfaces*, Indiana Univ. Math. J. **68** (2019), no. 3, 799–833, DOI 10.1512/iumj.2019.68.7662. MR3978257 Cited in Problem 1.69.

[IK20] Daiki Iguchi and Yuya Koda, *Twisted book decompositions and the Goeritz groups*, Topology Appl. **272** (2020), 107064, 15, DOI 10.1016/j.topol.2020.107064. MR4057555 Cited in Problem 2.8.

[IKMT25] Nobuo Iida, Hokuto Konno, Anubhav Mukherjee, and Masaki Taniguchi, *Diffeomorphisms of 4-manifolds with boundary and exotic embeddings*, Math. Ann. **391** (2025), no. 2, 1845–1897, DOI 10.1007/s00208-024-02974-x. MR4853007 Cited in Problem 4.70.

[Ilt22] Damian Iltgen, *A lower bound on the stable 4-genus of knots*, Algebr. Geom. Topol. **22** (2022), no. 5, 2239–2265, DOI 10.2140/agt.2022.22.2239. MR4503336 Cited in Problem 1.41.

[IMM23] Giovanni Italiano, Bruno Martelli, and Matteo Migliorini, *Hyperbolic 5-manifolds that fiber over S^1*, Invent. Math. **231** (2023), no. 1, 1–38, DOI 10.1007/s00222-022-01141-w. MR4526820 Cited in Problems 4.127 and 5.13.

[IMT25] Nobuo Iida, Anubhav Mukherjee, and Masaki Taniguchi, *An adjunction inequality for the Bauer-Furuta type invariants, with applications to sliceness and 4-manifold topology*, Adv. Math. **466** (2025), Paper No. 110134, 38, DOI 10.1016/j.aim.2025.110134. MR4866433 Cited in Problem 4.6.

[IO00] Kazuhiro Ichihara and Makoto Ozawa, *Hyperbolic knot complements without closed embedded totally geodesic surfaces*, J. Austral. Math. Soc. Ser. A **68** (2000), no. 3, 379–386. MR1753367 Cited in Problem 3.14.

[IP02] Massimiliano Iori and Riccardo Piergallini, *4-manifolds as covers of the 4-sphere branched over non-singular surfaces*, Geom. Topol. **6** (2002), 393–401, DOI 10.2140/gt.2002.6.393. MR1914574 Cited in Problems 4.122 and 5.6.

[Ish88] Masa-Nori Ishida. An elliptic surface covered by Mumford's fake projective plane. *Tohoku Math. J. (2)*, 40(3):367–396, 1988. `doi:10.2748/tmj/1178227980`. Cited in Problem 4.91.

[Isl21] Gabriel Islambouli, *Nielsen equivalence and trisections*, Geom. Dedicata **214** (2021), 303–317, DOI 10.1007/s10711-021-00617-y. MR4308281 Cited in Problem 4.115.

[Ito22] Tetsuya Ito. Cosmetic surgery on knots. In Tomotada Ohtsuki, editor, *Problems in low-dimensiontal topology, 2022*. Research Institute for Mathematical Sciences,

Kyoto University, Kyoto, Japan, 2022. URL: `https://www.kurims.kyoto-u.ac.jp/~kyodo/kokyuroku/contents/pdf/2227-12.pdf`. Cited in Problem 1.12.

[Iva76] N. V. Ivanov. Diffeomorphism groups of Waldhausen manifolds. *J Math Sci.*, 12:115–118, 1976. `doi:10.1007/BF01098421`. Cited in Problem 1.89.

[Iva97] Nikolai V. Ivanov, *Automorphism of complexes of curves and of Teichmüller spaces*, Internat. Math. Res. Notices **14** (1997), 651–666, DOI 10.1155/S1073792897000433. MR1460387 Cited in Problem 2.4.

[Iva08] S. V. Ivanov, *The computational complexity of basic decision problems in 3-dimensional topology*, Geom. Dedicata **131** (2008), 1–26, DOI 10.1007/s10711-007-9210-4. MR2369189 Cited in Problem 3.27.

[Iwa90] Zjuñici Iwase, *Dehn surgery along a torus T^2-knot. II*, Japan. J. Math. (N.S.) **16** (1990), no. 2, 171–196, DOI 10.4099/math1924.16.171. MR1091159 Cited in Problems 4.34 and 4.57.

[Jin97] Gyo Taek Jin, *Polygon indices and superbridge indices of torus knots and links*, J. Knot Theory Ramifications **6** (1997), no. 2, 281–289, DOI 10.1142/S0218216597000170. MR1452441 Cited in Problem 1.73.

[JK06] Bjørn Jahren and Sławomir Kwasik, *Manifolds homotopy equivalent to $\mathbf{R}\mathrm{P}^4\#\mathbf{R}\mathrm{P}^4$*, Math. Proc. Cambridge Philos. Soc. **140** (2006), no. 2, 245–252, DOI 10.1017/S0305004105008893. MR2212277 Cited in Problem 4.22.

[JM08] Stanislav Jabuka and Thomas E. Mark, *On the Heegaard Floer homology of a surface times a circle*, Adv. Math. **218** (2008), no. 3, 728–761, DOI 10.1016/j.aim.2008.01.009. MR2414320 Cited in Problem 3.54.

[JM13] Jesse Johnson and Darryl McCullough, *The space of Heegaard splittings*, J. Reine Angew. Math. **679** (2013), 155–179, DOI 10.1515/crelle.2012.016. MR3065157 Cited in Problem 2.15.

[JMM96] M. Juvan, A. Malnič, and B. Mohar, *Systems of curves on surfaces*, J. Combin. Theory Ser. B **68** (1996), no. 1, 7–22, DOI 10.1006/jctb.1996.0053. MR1405702 Cited in Problem 2.12.

[JMMZ22] Jason Joseph, Jeffrey Meier, Maggie Miller, and Alexander Zupan, *Bridge trisections and classical knotted surface theory*, Pacific J. Math. **319** (2022), no. 2, 343–369, DOI 10.2140/pjm.2022.319.343. MR4482720 Cited in Problem 1.18.

[JMZ21] András Juhász, Maggie Miller, and Ian Zemke, *Transverse invariants and exotic surfaces in the 4-ball*, Geom. Topol. **25** (2021), no. 6, 2963–3012, DOI 10.2140/gt.2021.25.2963. MR4347309 Cited in Problem 4.31.

[Joh83] Dennis Johnson, *The structure of the Torelli group. I. A finite set of generators for $\mathcal{I}$*, Ann. of Math. (2) **118** (1983), no. 3, 423–442, DOI 10.2307/2006977. MR727699 Cited in Problems 2.5 and 4.81.

[Joh86] F. E. A. Johnson, *A class of non-Kählerian manifolds*, Math. Proc. Cambridge Philos. Soc. **100** (1986), no. 3, 519–521, DOI 10.1017/S030500410006624X. MR857726 Cited in Problem 4.97.

[Joh95] Klaus Johannson. *Topology and combinatorics of 3-manifolds*, volume 1599 of *Lecture Notes in Mathematics*. Springer-Verlag, Berlin, 1995. `doi:10.1007/BFb0074005`. Cited in Problem 3.34.

[Joh99] F. E. A. Johnson. A rigidity theorem for group extensions. *Arch. Math. (Basel)*, 73(2):81–89, 1999. `doi:10.1007/s000130050371`. Cited in Problem 2.24.

[Jon85] Vaughan F. R. Jones, *A polynomial invariant for knots via von Neumann algebras*, Bull. Amer. Math. Soc. (N.S.) **12** (1985), no. 1, 103–111, DOI 10.1090/S0273-0979-1985-15304-2. MR766964 Cited in Problem 1.23.

[Jon87] V. F. R. Jones. Hecke algebra representations of braid groups and link polynomials. *Ann. of Math. (2)*, 126(2):335–388, 1987. `doi:10.2307/1971403`. Cited in Problem 2.3.

[Jor12] Eduardo Jorquera, *A universal nilpotent group of C^1 diffeomorphisms of the interval*, Topology Appl. **159** (2012), no. 8, 2115–2126, DOI 10.1016/j.topol.2012.02.003. MR2902746 Cited in Problem 3.22.

[JP25] Jason Joseph and Puttipong Pongtanapaisan, *Meridional rank and bridge number of knotted 2-spheres*, Canad. J. Math. **77** (2025), no. 1, 282–299, DOI 10.4153/S0008414X23000883. MR4861737 Cited in Problems 1.18 and 4.111.

[JRST20] William Jaco, J. Hyam Rubinstein, Jonathan Spreer, and Stephan Tillmann, $\mathbb{Z}_2$-*Thurston norm and complexity of 3-manifolds, II*, Algebr. Geom. Topol. **20** (2020), no. 1, 503–529, DOI 10.2140/agt.2020.20.503. MR4071382 Cited in Problem 3.28.

[Juh06] András Juhász, *Holomorphic discs and sutured manifolds*, Algebr. Geom. Topol. **6** (2006), 1429–1457, DOI 10.2140/agt.2006.6.1429. MR2253454 Cited in Problem 3.47.

[Juh15] András Juhász. A survey of Heegaard Floer homology. In *New ideas in low dimensional topology*, volume 56 of *Ser. Knots Everything*, pages 237–296. World Sci. Publ., Hackensack, NJ, 2015. URL: `https://doi.org/10.1142/9789814630627_0007`, `doi:10.1142/9789814630627\_0007`. Cited in Problem 3.48.

[Juh16] András Juhász, *Cobordisms of sutured manifolds and the functoriality of link Floer homology*, Adv. Math. **299** (2016), 940–1038, DOI 10.1016/j.aim.2016.06.005. MR3519484 Cited in Problem 1.19.

[JVC21] Stanislav Jabuka and Cornelia A. Van Cott, *On a nonorientable analogue of the Milnor conjecture*, Algebr. Geom. Topol. **21** (2021), no. 5, 2571–2625, DOI 10.2140/agt.2021.21.2571. MR4334520 Cited in Problem 1.47.

[JZ20a] Andrei Jaikin-Zapirain, *Recognition of being fibered for compact 3-manifolds*, Geom. Topol. **24** (2020), no. 1, 409–420, DOI 10.2140/gt.2020.24.409. MR4080486 Cited in Problem 3.8.

[JZ20b] András Juhász and Ian Zemke, *Distinguishing slice disks using knot Floer homology*, Selecta Math. (N.S.) **26** (2020), no. 1, Paper No. 5, 18, DOI 10.1007/s00029-019-0531-6. MR4045151 Cited in Problem 1.59.

[Kam92] Seiichi Kamada, *Projective planes in* 4*-sphere obtained by deform-spinnings*, Knots 90 (Osaka, 1990), de Gruyter, Berlin, 1992, pp. 125–132. MR1177417 Cited in Problem 4.37.

[Kan22a] Sungkyung Kang, *Link homology theories and ribbon concordances*, Quantum Topol. **13** (2022), no. 1, 183–205, DOI 10.4171/qt/162. MR4404800 Cited in Problem 1.56.

[Kan22b] Sungkyung Kang. One stabilization is not enough for contractible 4-manifolds, 2022. `arXiv:2210.07510`. Cited in Problems 4.7 and 4.33.

[Kap98] Michael Kapovich. On normal subgroups in the fundamental groups of complex surfaces, 1998. `arXiv:math/9808085`. Cited in Problem 2.23.

[Kar13] R. N. Karasëv. On two conjectures of Makeev. *Zap. Nauchn. Sem. S.-Peterburg. Otdel. Mat. Inst. Steklov. (POMI)*, 415:5–14, 2013. `doi:10.1007/s10958-016-2679-3`. Cited in Problem 2.14.

[Kas77] A. Kas, *On the deformation types of regular elliptic surfaces*, Complex analysis and algebraic geometry, Iwanami Shoten Publishers, Tokyo, 1977, pp. 107–111. MR569686 Cited in Problem 2.21.

[Kas97] R. M. Kashaev, *The hyperbolic volume of knots from the quantum dilogarithm*, Lett. Math. Phys. **39** (1997), no. 3, 269–275, DOI 10.1023/A:1007364912784. MR1434238 Cited in Problems 1.27, 1.29, and 2.50.

[Kas98] R. M. Kashaev, *Quantization of Teichmüller spaces and the quantum dilogarithm*, Lett. Math. Phys. **43** (1998), no. 2, 105–115, DOI 10.1023/A:1007460128279. MR1607296 Cited in Problem 3.78.

[Kas25] Thorben Kastenholz. Simplicial volume of open books in dimension 4, 2025. `arXiv:2504.10975`. Cited in Problem 4.121.

[Kat22] Yuya Kato, *Nonsmoothable actions of $\mathbb{Z}_2 \times \mathbb{Z}_2$ on spin four-manifolds*, Topology Appl. **307** (2022), Paper No. 107868, 13, DOI 10.1016/j.topol.2021.107868. MR4365044 Cited in Problem 4.86.

[Kau87] Louis H. Kauffman, *State models and the Jones polynomial*, Topology **26** (1987), no. 3, 395–407, DOI 10.1016/0040-9383(87)90009-7. MR899057 Cited in Problems 1.1 and 1.23.

[Kau99] Louis H. Kauffman, *Virtual knot theory*, European J. Combin. **20** (1999), no. 7, 663–690, DOI 10.1006/eujc.1999.0314. MR1721925 Cited in Problems 1.83 and 1.84.

[Kaw96] Akio Kawauchi, *A survey of knot theory*, Birkhäuser Verlag, Basel, 1996. Translated and revised from the 1990 Japanese original by the author. MR1417494 Cited in Problem 1.6.

[Kaw06] Akio Kawauchi, *Topological imitations and Reni-Mecchia-Zimmermann's conjecture*, Kyungpook Math. J. **46** (2006), no. 1, 1–9. MR2214795 Cited in Problem 1.16.

[Kaw09] Akio Kawauchi, *Rational-slice knots via strongly negative-amphicheiral knots*, Commun. Math. Res. **25** (2009), no. 2, 177–192. MR2554510 Cited in Problems 1.60 and 4.44.

[Kea89] C. Kearton, *Mutation of knots*, Proc. Amer. Math. Soc. **105** (1989), no. 1, 206–208, DOI 10.2307/2046757. MR929430 Cited in Problem 3.61.

[Kea04] C. Kearton. S-equivalence of knots. *J. Knot Theory Ramifications*, 13(6):709–717, 2004. doi:10.1142/S0218216504003408. Cited in Problem 1.6.

[Ken16] Autumn Kent, *Congruence kernels around affine curves*, J. Reine Angew. Math. **713** (2016), 1–20, DOI 10.1515/crelle-2014-0023. MR3483623 Cited in Problem 2.2.

[Ker65a] Michel A. Kervaire, *Les nœuds de dimensions supérieures* (French), Bull. Soc. Math. France **93** (1965), 225–271. MR189052 Cited in Problems 4.35 and 4.40.

[Ker65b] Michel A. Kervaire, *On higher dimensional knots*, Differential and Combinatorial Topology (A Symposium in Honor of Marston Morse), Princeton Univ. Press, Princeton, NJ, 1965, pp. 105–119. MR178475 Cited in Problem 4.35.

[Ker83] Steven P. Kerckhoff, *The Nielsen realization problem*, Ann. of Math. (2) **117** (1983), no. 2, 235–265, DOI 10.2307/2007076. MR690845 Cited in Problem 4.84.

[Keu06] JongHae Keum, *A fake projective plane with an order 7 automorphism*, Topology **45** (2006), no. 5, 919–927, DOI 10.1016/j.top.2006.06.006. MR2239523 Cited in Problem 4.91.

[Keu11] JongHae Keum. A fake projective plane constructed from an elliptic surface with multiplicities $(2, 4)$. *Sci. China Math.*, 54(8):1665–1678, 2011. doi:10.1007/s11425-011-4247-0. Cited in Problem 4.91.

[Kho00] Mikhail Khovanov, *A categorification of the Jones polynomial*, Duke Math. J. **101** (2000), no. 3, 359–426, DOI 10.1215/S0012-7094-00-10131-7. MR1740682 Cited in Problems 1.28, 1.32, 1.37, and 3.67.

[Kho05] Vu The Khoi. On the symplectic volume of the moduli space of spherical and Euclidean polygons. *Kodai Mathematical Journal*, 28:199–208, 2005. doi:10.2996/kmj/1111588046. Cited in Problem 1.72.

[Kho07] Mikhail Khovanov, *Triply-graded link homology and Hochschild homology of Soergel bimodules*, Internat. J. Math. **18** (2007), no. 8, 869–885, DOI 10.1142/S0129167X07004400. MR2339573 Cited in Problem 1.34.

[Kho16] Mikhail Khovanov, *Hopfological algebra and categorification at a root of unity: the first steps*, J. Knot Theory Ramifications **25** (2016), no. 3, 1640006, 26, DOI 10.1142/S021821651640006X. MR3475073 Cited in Problem 3.67.

[Kie20] Dawid Kielak, *Residually finite rationally solvable groups and virtual fibring*, J. Amer. Math. Soc. **33** (2020), no. 2, 451–486, DOI 10.1090/jams/936. MR4073866 Cited in Problem 5.13.

[Kim06] Hee Jung Kim, *Modifying surfaces in 4-manifolds by twist spinning*, Geom. Topol. **10** (2006), 27–56, DOI 10.2140/gt.2006.10.27. MR2207789 Cited in Problem 4.101.

[Kim10] Sang-hyun Kim, *On right-angled Artin groups without surface subgroups*, Groups Geom. Dyn. **4** (2010), no. 2, 275–307, DOI 10.4171/GGD/84. MR2595093 Cited in Problem 2.30.

[Kim20] Seungwon Kim. Gluck twist and unknotting of satellite 2-knots, 2020. arXiv:2009.07353. Cited in Problem 4.9.

[Kin53] Shin'ichi Kinoshita, *On some contractible continua without fixed point property*, Fund. Math. **40** (1953), 96–98, DOI 10.4064/fm-40-1-96-98. MR60225 Cited in Problem 5.1.

[Kin61] Shin'ichi Kinoshita, *On the Alexander polynomials of* 2*-spheres in a* 4*-sphere*, Ann. of Math. (2) **74** (1961), 518–531, DOI 10.2307/1970296. MR133126 Cited in Problem 4.37.

[Kin01] Simon A. King, *The size of triangulations supporting a given link*, Geom. Topol. **5** (2001), 369–398, DOI 10.2140/gt.2001.5.369. MR1825667 Cited in Problem 3.28.

[Kir78] Rob Kirby, *Problems in low dimensional manifold theory*, Algebraic and geometric topology (Proc. Sympos. Pure Math., Stanford Univ., Stanford, Calif., 1976), Proc. Sympos. Pure Math., XXXII, Amer. Math. Soc., Providence, RI, 1978, pp. 273–312. MR520548 Cited in the Introduction and Problems 2.5, 4.1, 4.9, 4.25, and 4.29.

[Kir97] Rob Kirby (ed.), *Problems in low-dimensional topology*, Geometric topology (Athens, GA, 1993), AMS/IP Stud. Adv. Math., vol. 2, Amer. Math. Soc., Providence, RI, 1997, pp. 35–473, DOI 10.1090/amsip/002.2/02. MR1470751 Cited in the Introduction and Problems 1.1, 1.2, 1.3, 1.9, 1.10, 1.12, 1.13, 1.15, 1.16, 1.17, 1.18, 1.20, 1.24, 1.38, 1.50, 1.52, 1.57, 1.58, 1.61, 1.62, 2.25, 3.1, 3.2, 3.5, 3.34, 3.36, 3.38, 3.50, 3.52, 3.61, 3.64, 4.1, 4.2, 4.15, 4.17, 4.19, 4.20, 4.21, 4.23, 4.35, 4.38, 4.46, 4.52, 4.54, 4.56, 4.58, 4.62, 4.67, 4.94, 4.117, 4.118, 4.124, 4.128, 5.0, 5.4, 5.7, 5.10, 5.11, and 5.12.

[Kir20] R. C. Kirby. Markov's theorem on the nonrecognizablility of 4-manifolds: an exposition, 2020. Celebratio Mathematica: Martin Scharlemann, `https://celebratio.org/Scharlemann_M/article/785/`. Cited in Problem 4.53.

[KK90] Paul A. Kirk and Eric P. Klassen, *Chern-Simons invariants of 3-manifolds and representation spaces of knot groups*, Math. Ann. **287** (1990), no. 2, 343–367, DOI 10.1007/BF01446898. MR1054574 Cited in Problem 3.64.

[KK93] Paul Kirk and Eric Klassen, *Chern-Simons invariants of 3-manifolds decomposed along tori and the circle bundle over the representation space of* T^2, Comm. Math. Phys. **153** (1993), no. 3, 521–557. MR1218931 Cited in Problem 3.64.

[KK00] Michael Kapovich and Bruce Kleiner, *Hyperbolic groups with low-dimensional boundary* (English, with English and French summaries), Ann. Sci. École Norm. Sup. (4) **33** (2000), no. 5, 647–669, DOI 10.1016/S0012-9593(00)01049-1. MR1834498 Cited in Problem 3.24.

[KK03] Vik. S. Kulikov and V. M. Kharlamov, *On braid monodromy factorizations* (Russian, with Russian summary), Izv. Ross. Akad. Nauk Ser. Mat. **67** (2003), no. 3, 79–118, DOI 10.1070/IM2003v067n03ABEH000436; English transl., Izv. Math. **67** (2003), no. 3, 499–534. MR1992194 Cited in Problem 4.102.

[KK14a] Sang-Hyun Kim and Thomas Koberda, *The geometry of the curve graph of a right-angled Artin group*, Internat. J. Algebra Comput. **24** (2014), no. 2, 121–169, DOI 10.1142/S021819671450009X. MR3192368 Cited in Problem 2.28.

[KK14b] Sang-Hyun Kim and Thomas Koberda, *An obstruction to embedding right-angled Artin groups in mapping class groups*, Int. Math. Res. Not. IMRN **14** (2014), 3912–3918, DOI 10.1093/imrn/rnt064. MR3239092 Cited in Problem 2.28.

[KK20] Sang-hyun Kim and Thomas Koberda, *Diffeomorphism groups of critical regularity*, Invent. Math. **221** (2020), no. 2, 421–501, DOI 10.1007/s00222-020-00953-y. MR4121156 Cited in Problem 2.34.

[KK21] Sang-hyun Kim and Thomas Koberda, *Structure and regularity of group actions on one-manifolds*, Springer Monographs in Mathematics, Springer, Cham, [2021] ©2021, DOI 10.1007/978-3-030-89006-3. MR4381312 Cited in Problem 2.34.

[KK22] Hiroaki Karuo and Julien Korinman. Azumaya loci of skein algebras, 2022. `arXiv:2211.13700`. Cited in Problems 2.50 and 3.78.

[KK23] Rob Kusner and Wöden Kusner, *A Gordian pair of links*, Geom. Dedicata **217** (2023), no. 3, Paper No. 47, 2, DOI 10.1007/s10711-023-00783-1. MR4557931 Cited in Problem 1.71.

[KK25] Rob Kusner and Wöden Kusner. *The Xarax unlinks are physically Gordian*, Circle Packings, Minimal Surfaces, and Discrete Differential Geometry Poster Session Abstracts, ICERM (2025), `https://app.icerm.brown.edu/assets/521/8574/8574_4906_021120251500_Slides.pdf` Cited in Problem 1.71.

[KKdlNG25] Sang-hyun Kim, Thomas Koberda, and J. de la Nuez González, *First order rigidity of homeomorphism groups of manifolds*, Commun. Am. Math. Soc. **5** (2025), 144–194, DOI 10.1090/cams/47. MR4904300 Cited in Problem 2.35.

[KKN21a] Marek Kaluba, Dawid Kielak, and Piotr W. Nowak, *On property (T) for* $\mathrm{Aut}(F_n)$ *and* $\mathrm{SL}_n(\mathbb{Z})$, Ann. of Math. (2) **193** (2021), no. 2, 539–562, DOI 10.4007/annals.2021.193.2.3. MR4224715 Cited in Problem 2.1.

[KKN21b] Tsuyoshi Kato, Hokuto Konno, and Nobuhiro Nakamura, *Rigidity of the mod 2 families Seiberg-Witten invariants and topology of families of spin 4-manifolds*, Compos. Math. **157** (2021), no. 4, 770–808, DOI 10.1112/s0010437x2000771x. MR4247572 Cited in Problem 4.70.

[KKR92] Myung Ho Kim, Sadayoshi Kojima, and Frank Raymond, *Homotopy invariants of nonorientable 4-manifolds*, Trans. Amer. Math. Soc. **333** (1992), no. 1, 71–81, DOI 10.2307/2154099. MR1028758 Cited in Problem 4.55.

[KL88] Sławomir Kwasik and Kyung Bai Lee, *Locally linear actions on 3-manifolds*, Math. Proc. Cambridge Philos. Soc. **104** (1988), no. 2, 253–260, DOI 10.1017/S0305004100065427. MR948910 Cited in Problem 4.86.

[KL93] Sławomir Kwasik and Terry Lawson, *Nonsmoothable Z_p actions on contractible 4-manifolds*, J. Reine Angew. Math. **437** (1993), 29–54, DOI 10.1515/crll.1993.437.29. MR1212252 Cited in Problems 3.75 and 4.86.

[KL99] Paul Kirk and Charles Livingston, *Twisted knot polynomials: inversion, mutation and concordance*, Topology **38** (1999), no. 3, 663–671, DOI 10.1016/S0040-9383(98)00040-8. MR1670424 Cited in Problem 3.61.

[KL09] Paul Kirk and Charles Livingston, *The geography problem for 4-manifolds with specified fundamental group*, Trans. Amer. Math. Soc. **361** (2009), no. 8, 4091–4124, DOI 10.1090/S0002-9947-09-04649-2. MR2500880 Cited in Problem 4.54.

[KL15] Çağrı Karakurt and Tye Lidman, *Rank inequalities for the Heegaard Floer homology of Seifert homology spheres*, Trans. Amer. Math. Soc. **367** (2015), no. 10, 7291–7322, DOI 10.1090/S0002-9947-2014-06451-9. MR3378830 Cited in Problem 3.62.

[KL23a] Efstratia Kalfagianni and Christine Ruey Shan Lee, *Jones diameter and crossing number of knots*, Adv. Math. **417** (2023), Paper No. 108937, 35, DOI 10.1016/j.aim.2023.108937. MR4554670 Cited in Problem 1.2.

[KL23b] Mikhail Khovanov and Robert Lipshitz, *Categorical lifting of the Jones polynomial: a survey*, Bull. Amer. Math. Soc. (N.S.) **60** (2023), no. 4, 483–506, DOI 10.1090/bull/1772. MR4642116 Cited in Problem 1.28.

[KL23c] Hokuto Konno and Jianfeng Lin. Homological instability for moduli spaces of smooth 4-manifolds, 2023. `arXiv:2211.03043`. Cited in Problems 4.76 and 4.78.

[KL24] Autumn E. Kent and Christopher J. Leininger. Atoroidal surface bundles, 2024. `arXiv:2405.12067`. Cited in Problem 2.23.

[KLMME24] Hokuto Konno, Jianfeng Lin, Anubhav Mukherjee, and Juan Muñoz-Echániz. On four-dimensional Dehn twists and Milnor fibrations, 2024. `arXiv:2409.11961`. Cited in Problems 4.68 and 4.89.

[KLS18] Tirasan Khandhawit, Jianfeng Lin, and Hirofumi Sasahira, *Unfolded Seiberg-Witten Floer spectra, I: Definition and invariance*, Geom. Topol. **22** (2018), no. 4, 2027–2114, DOI 10.2140/gt.2018.22.2027. MR3784516 Cited in Problem 3.47.

[KLS19] Thomas Koberda, Feng Luo, and Hongbin Sun, *An effective Lie-Kolchin theorem for quasi-unipotent matrices*, Linear Algebra Appl. **581** (2019), 304–323, DOI 10.1016/j.laa.2019.07.023. MR3985526 Cited in Problem 2.3.

[KLT20a] Çağatay Kutluhan, Yi-Jen Lee, and Clifford Taubes, *HF=HM, IV: The Sieberg-Witten Floer homology and ech correspondence*, Geom. Topol. **24** (2020), no. 7, 3219–3469, DOI 10.2140/gt.2020.24.3219. MR4194308 Cited in Problem 3.47.

[KLT20b] Çağatay Kutluhan, Yi-Jen Lee, and Clifford Henry Taubes, *HF=HM, V: Seiberg-Witten Floer homology and handle additions*, Geom. Topol. **24** (2020), no. 7, 3471–3748, DOI 10.2140/gt.2020.24.3471. MR4194309 Cited in Problem 3.47.

[KLT20c] Çağatay Kutluhan, Yi-Jen Lee, and Clifford Henry Taubes, HF = HM*, I: Heegaard Floer homology and Seiberg-Witten Floer homology*, Geom. Topol. **24** (2020), no. 6, 2829–2854, DOI 10.2140/gt.2020.24.2829. MR4194305 Cited in Problem 3.47.

[KLT20d] Çağatay Kutluhan, Yi-Jen Lee, and Clifford Henry Taubes, HF = HM*, II: Reeb orbits and holomorphic curves for the ech/Heegaard Floer correspondence*, Geom. Topol. **24** (2020), no. 6, 2855–3012, DOI 10.2140/gt.2020.24.2855. MR4194306 Cited in Problem 3.47.

[KM63] Michel A. Kervaire and John W. Milnor, *Groups of homotopy spheres. I*, Ann. of Math. (2) **77** (1963), 504–537, DOI 10.2307/1970128. MR148075 Cited in Problem 3.69.

[KM78] Robion Kirby and Paul Melvin, *Slice knots and property* R, Invent. Math. **45** (1978), no. 1, 57–59, DOI 10.1007/BF01406223. MR467754 Cited in Problem 1.58.

[KM94] P. B. Kronheimer and T. S. Mrowka, *The genus of embedded surfaces in the projective plane*, Math. Res. Lett. **1** (1994), no. 6, 797–808, DOI 10.4310/MRL.1994.v1.n6.a14. MR1306022 Cited in Problems 1.46, 4.32, and 4.44.

[KM95] P. B. Kronheimer and T. S. Mrowka, *Embedded surfaces and the structure of Donaldson's polynomial invariants*, J. Differential Geom. **41** (1995), no. 3, 573–734. MR1338483 Cited in Problems 4.61 and 4.62.

[KM04] P. B. Kronheimer and T. S. Mrowka, *Dehn surgery, the fundamental group and $SU(2)$*, Math. Res. Lett. **11** (2004), no. 5-6, 741–754, DOI 10.4310/MRL.2004.v11.n6.a3. MR2106239 Cited in Problems 1.14, 3.37, and 3.52.

[KM05] D. Kotschick and S. Morita, *Signatures of foliated surface bundles and the symplectomorphism groups of surfaces*, Topology **44** (2005), no. 1, 131–149, DOI 10.1016/j.top.2004.05.002. MR2104005 Cited in Problem 2.32.

[KM06] Olga Kharlampovich and Alexei Myasnikov, *Elementary theory of free non-abelian groups*, J. Algebra **302** (2006), no. 2, 451–552, DOI 10.1016/j.jalgebra.2006.03.033. MR2293770 Cited in Problems 2.35 and 2.36.

[KM07] Peter Kronheimer and Tomasz Mrowka, *Monopoles and three-manifolds*, New Mathematical Monographs, vol. 10, Cambridge University Press, Cambridge, 2007, DOI 10.1017/CBO9780511543111. MR2388043 Cited in Problem 3.47.

[KM10] Peter Kronheimer and Tomasz Mrowka, *Knots, sutures, and excision*, J. Differential Geom. **84** (2010), no. 2, 301–364. MR2652464 Cited in Problems 3.47, 3.52, and 3.59.

[KM11] P. B. Kronheimer and T. S. Mrowka, *Khovanov homology is an unknot-detector*, Publ. Math. Inst. Hautes Études Sci. **113** (2011), 97–208, DOI 10.1007/s10240-010-0030-y. MR2805599 Cited in Problems 1.24, 1.28, and 1.76.

[KM12] Jeremy Kahn and Vladimir Markovic, *Immersing almost geodesic surfaces in a closed hyperbolic three manifold*, Ann. of Math. (2) **175** (2012), no. 3, 1127–1190, DOI 10.4007/annals.2012.175.3.4. MR2912704 Cited in Problems 3.7, 3.13, 3.24, and 5.13.

[KM13a] Alexander Kolpakov and Bruno Martelli, *Hyperbolic four-manifolds with one cusp*, Geom. Funct. Anal. **23** (2013), no. 6, 1903–1933, DOI 10.1007/s00039-013-0247-2. MR3132905 Cited in Problem 5.14.

[KM13b] P. B. Kronheimer and T. S. Mrowka, *Gauge theory and Rasmussen's invariant*, J. Topol. **6** (2013), no. 3, 659–674, DOI 10.1112/jtopol/jtt008. MR3100886 Cited in Problems 1.44 and 4.43.

[KM14a] V. D. Mazurov and E. I. Khukhro (eds.), *The Kourovka notebook*, Eighteenth edition, Russian Academy of Sciences Siberian Division, Institute of Mathematics, Novosibirsk, 2014. Unsolved problems in group theory, `arXiv:1401.0300`. MR3408705 Cited in Problem 4.26.

[KM14b] Peter B. Kronheimer and Ciprian Manolescu. Periodic Floer pro-spectra from the Seiberg-Witten equations, 2014. `arXiv:math/0203243`. Cited in Problem 3.47.

[KM18] Sadayoshi Kojima and Greg McShane, *Normalized entropy versus volume for pseudo-Anosovs*, Geom. Topol. **22** (2018), no. 4, 2403–2426, DOI 10.2140/gt.2018.22.2403. MR3784525 Cited in Problem 2.18.

[KM19a] P. B. Kronheimer and T. S. Mrowka, *Tait colorings, and an instanton homology for webs and foams*, J. Eur. Math. Soc. (JEMS) **21** (2019), no. 1, 55–119, DOI 10.4171/JEMS/831. MR3880205 Cited in Problem 1.23.

[KM19b] Peter B. Kronheimer and Tomasz S. Mrowka, *A deformation of instanton homology for webs*, Geom. Topol. **23** (2019), no. 3, 1491–1547, DOI 10.2140/gt.2019.23.1491. MR3956896 Cited in Problem 1.23.

[KM20] P. B. Kronheimer and T. S. Mrowka, *The Dehn twist on a sum of two $K3$ surfaces*, Math. Res. Lett. **27** (2020), no. 6, 1767–1783, DOI 10.4310/MRL.2020.v27.n6.a8. MR4216604 Cited in Problems 4.68, 4.70, 4.79, 4.80, and 4.82.

[KM21] Michael R. Klug and Maggie Miller, *Concordance of surfaces in 4-manifolds and the Freedman-Quinn invariant*, J. Topol. **14** (2021), no. 2, 560–586, DOI 10.1112/topo.12191. MR4286049 Cited in Problem 4.41.

[KM22] Michael Klug and Maggie Miller. Concordance of spheres in 4-manifolds with an immersed dual sphere, 2022. `arXiv:2211.07177`. Cited in Problem 4.41.

[KM24] Efstratia Kalfagianni and Rob Mcconkey, *Crossing numbers of cable knots*, Bull. Lond. Math. Soc. **56** (2024), no. 11, 3400–3411, DOI 10.1112/blms.13140. MR4828022 Cited in Problem 1.2.

[KMOS07] P. Kronheimer, T. Mrowka, P. Ozsváth, and Z. Szabó, *Monopoles and lens space surgeries*, Ann. of Math. (2) **165** (2007), no. 2, 457–546, DOI 10.4007/annals.2007.165.457. MR2299739 Cited in Problems 1.9 and 3.48.

[KMPW24] Vyacheslav Krushkal, Anubhav Mukherjee, Mark Powell, and Terrin Warren. Corks for exotic diffeomorphisms, 2024. `arXiv:2407.04696`. Cited in Problems 4.68 and 4.82.

[KMS86] Steven Kerckhoff, Howard Masur, and John Smillie, *Ergodicity of billiard flows and quadratic differentials*, Ann. of Math. (2) **124** (1986), no. 2, 293–311, DOI 10.2307/1971280. MR855297 Cited in Problem 2.42.

[KMT17] Thomas Koberda, Johanna Mangahas, and Samuel J. Taylor, *The geometry of purely loxodromic subgroups of right-angled Artin groups*, Trans. Amer. Math. Soc. **369** (2017), no. 11, 8179–8208, DOI 10.1090/tran/6933. MR3695858 Cited in Problem 2.28.

[KMT23a] Hokuto Konno, Abhishek Mallick, and Masaki Taniguchi. Exotic Dehn twists on 4-manifolds, 2023. `arXiv:2306.08607`. Cited in Problems 4.68, 4.70, and 4.89.

[KMT23b] Hokuto Konno, Jin Miyazawa, and Masaki Taniguchi. Involutions, links, and Floer cohomologies, 2023. `arXiv:2304.01115`. Cited in Problems 1.47 and 4.84.

[KN23] Hokuto Konno and Nobuhiro Nakamura, *Constraints on families of smooth 4-manifolds from* $\mathrm{Pin}^-(2)$*-monopole*, Algebr. Geom. Topol. **23** (2023), no. 1, 419–438, DOI 10.2140/agt.2023.23.419. MR4568007 Cited in Problem 4.70.

[KNO19] Marek Kaluba, Piotr W. Nowak, and Narutaka Ozawa, $\mathrm{Aut}(\mathbb{F}_5)$ *has property* (T), Math. Ann. **375** (2019), no. 3-4, 1169–1191, DOI 10.1007/s00208-019-01874-9. MR4023374 Cited in Problem 2.1.

[KNR22] Daniel Kasprowski, John Nicholson, and Benjamin Ruppik, *Homotopy classification of 4-manifolds whose fundamental group is dihedral*, Algebr. Geom. Topol. **22** (2022), no. 6, 2915–2949, DOI 10.2140/agt.2022.22.2915. MR4520681 Cited in Problems 4.50, 4.54, and 4.55.

[KNV24] Daniel Kasprowski, John Nicholson, and Simona Veselá. Stable equivalence relations on 4-manifolds, 2024. `arXiv:2405.06637`. Cited in Problem 4.18.

[KNY23] Tsuyoshi Kato, Nobuhiro Nakamura, and Kouichi Yasui, *The simple type conjecture for mod 2 Seiberg-Witten invariants*, J. Eur. Math. Soc. (JEMS) **25** (2023), no. 12, 4869–4877, DOI 10.4171/jems/1297. MR4662305 Cited in Problem 4.62.

[KO01] Mustafa Korkmaz and Burak Ozbagci, *Minimal number of singular fibers in a Lefschetz fibration*, Proc. Amer. Math. Soc. **129** (2001), no. 5, 1545–1549, DOI 10.1090/S0002-9939-00-05676-8. MR1713513 Cited in Problem 2.21.

[KO08] Mustafa Korkmaz and Burak Ozbagci, *On sections of elliptic fibrations*, Michigan Math. J. **56** (2008), no. 1, 77–87, DOI 10.1307/mmj/1213972398. MR2433657 Cited in Problem 2.22.

[Kob12] Thomas Koberda, *Right-angled Artin groups and a generalized isomorphism problem for finitely generated subgroups of mapping class groups*, Geom. Funct. Anal. **22** (2012), no. 6, 1541–1590, DOI 10.1007/s00039-012-0198-z. MR3000498 Cited in Problem 2.28.

[Koc06] Dessislava H. Kochloukova, *Some Novikov rings that are von Neumann finite and knot-like groups*, Comment. Math. Helv. **81** (2006), no. 4, 931–943, DOI 10.4171/CMH/81. MR2271229 Cited in Problem 1.53.

[Kod67] K. Kodaira, *A certain type of irregular algebraic surfaces*, J. Analyse Math. **19** (1967), 207–215, DOI 10.1007/BF02788717. MR216521 Cited in Problem 2.24.

[Kol08] János Kollár, *Is there a topological Bogomolov-Miyaoka-Yau inequality?*, Pure Appl. Math. Q. **4** (2008), no. 2, Special Issue: In honor of Fedor Bogomolov., 203–236, DOI 10.4310/PAMQ.2008.v4.n2.a1. MR2400877 Cited in Problems 4.58 and 5.7.

[Kon19] Hokuto Konno. Positive scalar curvature and higher-dimensional families of Seiberg-Witten equations. *Journal of Topology*, 12(4):1246–1265, 2019. `https://londmathsoc.onlinelibrary.wiley.com/doi/abs/10.1112/topo.12117`. `doi:10.1112/topo.12117`. Cited in Problem 4.128.

[Kon21] Hokuto Konno, *Characteristic classes via 4-dimensional gauge theory*, Geom. Topol. **25** (2021), no. 2, 711–773, DOI 10.2140/gt.2021.25.711. MR4251435 Cited in Problem 4.61.

[Kon24a] Hokuto Konno, *Dehn twists and the Nielsen realization problem for spin 4-manifolds*, Algebr. Geom. Topol. **24** (2024), no. 3, 1739–1753, DOI 10.2140/agt.2024.24.1739. MR4767887 Cited in Problem 4.84.

[Kon24b] Hokuto Konno, *The homology of moduli spaces of 4-manifolds may be infinitely generated*, Forum Math. Pi **12** (2024), Paper No. e25, 18, DOI 10.1017/fmp.2024.26. MR4831140 Cited in Problems 4.71 and 4.75.

[KOP+97] Vsevolod Katritch, Wilma K. Olson, Piotr Pieranski, Jacques Dubochet, and Andrzej Stasiak. Properties of ideal composite knots. *Nature*, 388:148–151, 1997. doi:10.1038/40582. Cited in Problem 1.74.

[KOPR21a] Min Hoon Kim, Patrick Orson, JungHwan Park, and Arunima Ray, *Good groups*, The disc embedding theorem, Oxford Univ. Press, Oxford, 2021, pp. 273–282. MR4519517 Cited in Problems 1.62 and 4.46.

[KOPR21b] Min Hoon Kim, Patrick Orson, JungHwan Park, and Arunima Ray, *Open problems*, The disc embedding theorem, Oxford Univ. Press, Oxford, 2021, pp. 353–382. MR4519521 Cited in Problem 4.46.

[Kor00] Mustafa Korkmaz, *On the linearity of certain mapping class groups*, Turkish J. Math. **24** (2000), no. 4, 367–371. MR1803819 Cited in Problem 2.3.

[Kor01] Mustafa Korkmaz, *Noncomplex smooth 4-manifolds with Lefschetz fibrations*, Internat. Math. Res. Notices **3** (2001), 115–128, DOI 10.1155/S107379280100006X. MR1810689 Cited in Problem 2.21.

[Kor06] Mustafa Korkmaz, *Problems on homomorphisms of mapping class groups*, Problems on mapping class groups and related topics, Proc. Sympos. Pure Math., vol. 74, Amer. Math. Soc., Providence, RI, 2006, pp. 81–89, DOI 10.1090/pspum/074/2264533. MR2264533 Cited in Problem 2.22.

[Kor23] Mustafa Korkmaz, *Low-dimensional linear representations of mapping class groups*, J. Topol. **16** (2023), no. 3, 899–935, DOI 10.1112/topo.12305. MR4637997 Cited in Problem 2.3.

[Kos24] Danica Kosanović, *Embedding calculus and grope cobordism of knots*, Adv. Math. **451** (2024), Paper No. 109779, 118, DOI 10.1016/j.aim.2024.109779. MR4760102 Cited in Problem 1.89.

[Kot89] Dieter Kotschick, *On manifolds homeomorphic to* $\mathbf{CP}^2\#8\overline{\mathbf{CP}}^2$, Invent. Math. **95** (1989), no. 3, 591–600, DOI 10.1007/BF01393892. MR979367 Cited in Problem 4.131.

[Kot99] D. Kotschick, *On regularly fibered complex surfaces*, Proceedings of the Kirbyfest (Berkeley, CA, 1998), Geom. Topol. Monogr., vol. 2, Geom. Topol. Publ., Coventry, 1999, pp. 291–298, DOI 10.2140/gtm.1999.2.291. MR1734413 Cited in Problem 4.97.

[KP15] Pierre-Vincent Koseleff and Daniel Pecker, *On Alexander-Conway polynomials of two-bridge links*. part 2, J. Symbolic Comput. **68** (2015), no. part 2, 215–229, DOI 10.1016/j.jsc.2014.09.011. MR3283863 Cited in Problem 1.8.

[KPR12] James Kaiser, Jessica S. Purcell, and Clint Rollins, *Volumes of chain links*, J. Knot Theory Ramifications **21** (2012), no. 11, 1250115, 17, DOI 10.1142/S0218216512501155. MR2969643 Cited in Problem 3.1.

[KPR22] Daniel Kasprowski, Mark Powell, and Arunima Ray, *Counterexamples in 4-manifold topology*, EMS Surv. Math. Sci. **9** (2022), no. 1, 193–249, DOI 10.4171/emss/56. MR4551461 Cited in Problem 4.18.

[KPR23] Daniel Kasprowski, Mark Powell, and Arunima Ray, *Gluck twists on concordant or homotopic spheres*, Math. Res. Lett. **30** (2023), no. 6, 1787–1811, DOI 10.4310/mrl.2023.v30.n6.a6. MR4779154 Cited in Problem 4.11.

[KPR24] Daniel Kasprowski, Mark Powell, and Benjamin Ruppik, *Homotopy classification of 4-manifolds with finite abelian 2-generator fundamental groups*, Math. Proc. Cambridge Philos. Soc. **177** (2024), no. 2, 263–283, DOI 10.1017/S0305004124000227. MR4834868 Cited in Problems 4.50, 4.54, and 4.55.

[KPRT24] Daniel Kasprowski, Mark Powell, Arunima Ray, and Peter Teichner. Embedding surfaces in 4-manifolds. *Geom. Topol.*, 28(5):2399–2482, 2024. doi:10.2140/gt.2024.28.2399. Cited in Problems 4.42 and 4.44.

[KPT21] Min Hoon Kim, Mark Powell, and Peter Teichner, *Round handle problem*, Pure Appl. Math. Q. **17** (2021), no. 1, 237–247, DOI 10.4310/PAMQ.2021.v17.n1.a6. MR4257584 Cited in Problem 4.47.

[KPT26] Sungkyung Kang, JungHwan Park, and Masaki Taniguchi, *Exotic Dehn twists and homotopy coherent group actions*, Invent. Math. **243** (2026), no. 1, 209–241, DOI 10.1007/s00222-025-01378-1. MR5008150 Cited in Problems 4.68 and 4.89.

[KPV24] Mikhail Khovanov, Krzysztof Putyra, and Pedro Vaz, *Odd two-variable Soergel bimodules and Rouquier complexes*, Algebraic and topological aspects of representation theory, Contemp. Math., vol. 791, Amer. Math. Soc., [Providence], RI, [2024] ©2024, pp. 205–227, DOI 10.1090/conm/791/15876. MR4694172 Cited in Problem 1.37.

[KQ00] Vyacheslav S. Krushkal and Frank Quinn, *Subexponential groups in 4-manifold topology*, Geom. Topol. **4** (2000), 407–430, DOI 10.2140/gt.2000.4.407. MR1796498 Cited in Problem 4.46.

[KR88a] P. H. Kropholler and M. A. Roller, *Splittings of Poincaré duality groups*, Math. Z. **197** (1988), no. 3, 421–438, DOI 10.1007/BF01418340. MR926850 Cited in Problem 3.38.

[KR88b] P. H. Kropholler and M. A. Roller, *Splittings of Poincaré duality groups. II*, J. London Math. Soc. (2) **38** (1988), no. 3, 410–420, DOI 10.1112/jlms/s2-38.3.410. MR972126 Cited in Problem 3.38.

[KR89] P. H. Kropholler and M. A. Roller, *Splittings of Poincaré duality groups. III*, J. London Math. Soc. (2) **39** (1989), no. 2, 271–284, DOI 10.1112/jlms/s2-39.2.271. MR991661 Cited in Problem 3.38.

[KR08a] Mikhail Khovanov and Lev Rozansky, *Matrix factorizations and link homology*, Fund. Math. **199** (2008), no. 1, 1–91, DOI 10.4064/fm199-1-1. MR2391017 Cited in Problem 1.37.

[KR08b] Mikhail Khovanov and Lev Rozansky. Matrix factorizations and link homology. II. *Geom. Topol.*, 12(3):1387–1425, 2008. `doi:10.2140/gt.2008.12.1387`. Cited in Problems 1.28 and 1.34.

[KR17] William H. Kazez and Rachel Roberts, *C^0 approximations of foliations*, Geom. Topol. **21** (2017), no. 6, 3601–3657, DOI 10.2140/gt.2017.21.3601. MR3693573 Cited in Problems 3.15, 3.40, and 3.48.

[KR21] Mikhail Khovanov and Louis-Hadrien Robert, *Foam evaluation and Kronheimer-Mrowka theories*, Adv. Math. **376** (2021), Paper No. 107433, 59, DOI 10.1016/j.aim.2020.107433. MR4178907 Cited in Problem 1.23.

[Kra02] Daan Krammer, *Braid groups are linear*, Ann. of Math. (2) **155** (2002), no. 1, 131–156, DOI 10.2307/3062152. MR1888796 Cited in Problem 2.3.

[Kre79] M. Kreck, *Isotopy classes of diffeomorphisms of $(k-1)$-connected almost-parallelizable $2k$-manifolds*, Algebraic topology, Aarhus 1978 (Proc. Sympos., Univ. Aarhus, Aarhus, 1978), Lecture Notes in Math., vol. 763, Springer, Berlin, 1979, pp. 643–663. MR561244 Cited in Problems 4.68, 4.76, and 4.80.

[Kre84a] M. Kreck, *Some closed 4-manifolds with exotic differentiable structure*, Algebraic topology, Aarhus 1982 (Aarhus, 1982), Lecture Notes in Math., vol. 1051, Springer, Berlin, 1984, pp. 246–262, DOI 10.1007/BFb0075570. MR764582 Cited in Problem 4.6.

[Kre84b] Matthias Kreck, *Bordism of diffeomorphisms and related topics*, Lecture Notes in Mathematics, vol. 1069, Springer-Verlag, Berlin, 1984. With an appendix by Neal W. Stoltzfus, DOI 10.1007/BFb0101007. MR755877 Cited in Problem 2.45.

[Kre90] Matthias Kreck, *On the homeomorphism classification of smooth knotted surfaces in the 4-sphere*, Geometry of low-dimensional manifolds, 1 (Durham, 1989), London Math. Soc. Lecture Note Ser., vol. 150, Cambridge Univ. Press, Cambridge, 1990, pp. 63–72. MR1171891 Cited in Problem 4.29.

[Kro97] P.B. Kronheimer. Some non-trivial families of symplectic structures. Preprint, available from `www.math.harvard.edu/~kronheim/diffsymp.pdf`, 1997. Cited in Problem 4.61.

[Kro98] P. B. Kronheimer, *Embedded surfaces and gauge theory in three and four dimensions*, Surveys in differential geometry, Vol. III (Cambridge, MA, 1996), Int. Press, Boston, MA, 1998, pp. 243–298. MR1677890 Cited in Problem 4.123.

[KS77] Robion C. Kirby and Laurence C. Siebenmann, *Foundational essays on topological manifolds, smoothings, and triangulations*, Annals of Mathematics Studies, No. 88, Princeton University Press, Princeton, NJ; University of Tokyo Press, Tokyo, 1977. With notes by John Milnor and Michael Atiyah. MR645390 Cited in Problems 4.46, 4.49, 4.56, 4.73, 4.119, and 5.5.

[KS88] Sławomir Kwasik and Reinhard Schultz, *Desuspension of group actions and the ribbon theorem*, Topology **27** (1988), no. 4, 443–457, DOI 10.1016/0040-9383(88)90023-7. MR976586 Cited in Problem 4.49.

[KS92] Sławomir Kwasik and Reinhard Schultz, *Vanishing of Whitehead torsion in dimension four*, Topology **31** (1992), no. 4, 735–756, DOI 10.1016/0040-9383(92)90005-3. MR1191376 Cited in Problem 4.56.

[KS96] Sławomir Kwasik and Reinhard Schultz, *Pseudo-isotopies of* 3*-manifolds*, Topology **35** (1996), no. 2, 363–376, DOI 10.1016/0040-9383(95)00017-8. MR1380504 Cited in Problem 4.69.

[KS98] Atsuko Katanaga and Osamu Saeki, *Embeddings of quaternion space in* S^4, J. Austral. Math. Soc. Ser. A **65** (1998), no. 3, 313–325. MR1660419 Cited in Problem 4.37.

[KS09] Mustafa Korkmaz and András I. Stipsicz, *Lefschetz fibrations on 4-manifolds*, Handbook of Teichmüller theory. Vol. II, IRMA Lect. Math. Theor. Phys., vol. 13, Eur. Math. Soc., Zürich, 2009, pp. 271–296, DOI 10.4171/055-1/9. MR2497784 Cited in Problem 2.22.

[KS16] Thomas Koberda and Ramanujan Santharoubane, *Quotients of surface groups and homology of finite covers via quantum representations*, Invent. Math. **206** (2016), no. 2, 269–292, DOI 10.1007/s00222-016-0652-x. MR3570293 Cited in Problem 2.49.

[KS19] Robin S. Krom and Dietmar A. Salamon, *The Donaldson geometric flow for symplectic four-manifolds*, J. Symplectic Geom. **17** (2019), no. 2, 381–417, DOI 10.4310/JSG.2019.v17.n2.a3. MR3992720 Cited in Problem 4.93.

[KS23] Michael R. Klug and Luuk Stehouwer, *Some properties of* $\mathrm{Pin}^{\pm}$*-structures on compact surfaces.* part B, Topology Appl. **339** (2023), no. part B, Paper No. 108678, 9, DOI 10.1016/j.topol.2023.108678. MR4640270 Cited in Problem 4.39.

[KSTY99] Atsuko Katanaga, Osamu Saeki, Masakazu Teragaito, and Yuichi Yamada, *Gluck surgery along a 2-sphere in a 4-manifold is realized by surgery along a projective plane*, Michigan Math. J. **46** (1999), no. 3, 555–571, DOI 10.1307/mmj/1030132479. MR1721575 Cited in Problem 4.37.

[KT57] Shin'ichi Kinoshita and Hidetaka Terasaka, *On unions of knots*, Osaka Math. J. **9** (1957), 131–153. MR98386 Cited in Problem 1.51.

[KT00] R. M. Kashaev and O. Tirkkonen, *A proof of the volume conjecture on torus knots* (Russian, with English and Russian summaries), Zap. Nauchn. Sem. S.-Peterburg. Otdel. Mat. Inst. Steklov. (POMI) **269** (2000), no. Vopr. Kvant. Teor. Polya i Stat. Fiz. 16, 262–268, 370, DOI 10.1023/A:1022608131142; English transl., J. Math. Sci. (N.Y.) **115** (2003), no. 1, 2033–2036. MR1805865 Cited in Problem 1.27.

[KT15] Efstratia Kalfagianni and Anh T. Tran, *Knot cabling and the degree of the colored Jones polynomial*, New York J. Math. **21** (2015), 905–941. MR3425627 Cited in Problem 1.26.

[KT17] Amey Kaloti and Bülent Tosun, *Hyperbolic rational homology spheres not admitting fillable contact structures*, Math. Res. Lett. **24** (2017), no. 6, 1693–1705, DOI 10.4310/MRL.2017.v24.n6.a6. MR3762690 Cited in Problem 3.40.

[KT20] Hokuto Konno and Masaki Taniguchi, *Positive scalar curvature and 10/8-type inequalities on 4-manifolds with periodic ends*, Invent. Math. **222** (2020), no. 3, 833–880, DOI 10.1007/s00222-020-00979-2. MR4169052 Cited in Problem 4.128.

[KT21] Dale Koenig and Anastasiia Tsvietkova, *NP-hard problems naturally arising in knot theory*, Trans. Amer. Math. Soc. Ser. B **8** (2021), 420–441, DOI 10.1090/btran/71. MR4273193 Cited in Problem 1.80.

[KT22a] Robion Kirby and Abigail Thompson. Trisections and link surgeries. *New Zealand J. Math.*, 52:145–152, 2021 [2021–2022]. `doi:10.53733/94`. Cited in Problem 4.116.

[KT22b] Hokuto Konno and Masaki Taniguchi, *The groups of diffeomorphisms and homeomorphisms of 4-manifolds with boundary.* part A, Adv. Math. **409** (2022), no. part

A, Paper No. 108627, 58, DOI 10.1016/j.aim.2022.108627. MR4469073 Cited in Problems 4.70 and 4.83.

[KT23] Hokuto Konno and Masaki Taniguchi, *Positive scalar curvature and homology cobordism invariants*, J. Topol. **16** (2023), no. 2, 679–719, DOI 10.1112/topo.12299. MR4637974 Cited in Problem 4.128.

[Kuh06] Sally M. Kuhlmann, *Geodesic knots in cusped hyperbolic 3-manifolds*, Algebr. Geom. Topol. **6** (2006), 2151–2162, DOI 10.2140/agt.2006.6.2151. MR2263061 Cited in Problem 3.16.

[Kuh08] Sally Kuhlmann, *Geodesic knots in closed hyperbolic 3-manifolds*, Geom. Dedicata **131** (2008), 181–211, DOI 10.1007/s10711-007-9227-8. MR2369199 Cited in Problem 3.16.

[Kup14] Greg Kuperberg, *Knottedness is in* NP*, modulo GRH*, Adv. Math. **256** (2014), 493–506, DOI 10.1016/j.aim.2014.01.007. MR3177300 Cited in Problem 1.76.

[Kup19a] Greg Kuperberg, *Algorithmic homeomorphism of 3-manifolds as a corollary of geometrization*, Pacific J. Math. **301** (2019), no. 1, 189–241, DOI 10.2140/pjm.2019.301.189. MR4007377 Cited in Problems 3.27 and 3.30.

[Kup19b] Alexander Kupers, *Some finiteness results for groups of automorphisms of manifolds*, Geom. Topol. **23** (2019), no. 5, 2277–2333, DOI 10.2140/gt.2019.23.2277. MR4019894 Cited in Problem 4.71.

[Kup21] Alexander Kupers, *Zeeman's conjecture*, Grad. J. Math. **6** (2021), no. 1, 35–42. MR4287600 Cited in Problem 5.12.

[KW91] Victor Klee and Stan Wagon, *Old and new unsolved problems in plane geometry and number theory*, The Dolciani Mathematical Expositions, vol. 11, Mathematical Association of America, Washington, DC, 1991. MR1133201 Cited in Problem 2.14.

[KW07] Anton Kapustin and Edward Witten, *Electric-magnetic duality and the geometric Langlands program*, Commun. Number Theory Phys. **1** (2007), no. 1, 1–236, DOI 10.4310/CNTP.2007.v1.n1.a1. MR2306566 Cited in Problem 1.32.

[KY21] Vincent Koziarz and Sai-Kee Yeung, *Stability of the Albanese fibration on the Cartwright-Steger surface*, Taiwanese J. Math. **25** (2021), no. 2, 251–256, DOI 10.11650/tjm/201108. MR4255211 Cited in Problem 4.97.

[KZ97] M. Kontsevich and A. Zorich. Lyapunov exponents and Hodge theory, 1997. `arXiv:hep-th/9701164`. Cited in Problem 2.26.

[KZ03] Maxim Kontsevich and Anton Zorich. Connected components of the moduli spaces of Abelian differentials with prescribed singularities. *Invent. Math.*, 153(3):631–678, 2003. `doi:10.1007/s00222-003-0303-x`. Cited in Problem 2.26.

[KZ03] Maxim Kontsevich and Anton Zorich, *Connected components of the moduli spaces of Abelian differentials with prescribed singularities*, Invent. Math. **153** (2003), no. 3, 631–678, DOI 10.1007/s00222-003-0303-x. MR2000471 Cited in Problem 2.26.

[LÔ6] Thang T. Q. Lê, *The colored Jones polynomial and the A-polynomial of knots*, Adv. Math. **207** (2006), no. 2, 782–804, DOI 10.1016/j.aim.2006.01.006. MR2271986 Cited in Problem 1.25.

[LÔ9] Thang T. Q. Lê. Hyperbolic volume, Mahler measure, and homology growth. Talk at Columbia University, `http://www.math.columbia.edu/~volconf09/notes/leconf.pdf`, 2009. Cited in Problem 3.6.

[LÎ6] Thang T. Q. Lê, *The colored Jones polynomial and the AJ conjecture* (English, with English and French summaries), Lectures on quantum topology in dimension three, Panor. Synthèses, vol. 48, Soc. Math. France, Paris, 2016, pp. 33–90. MR3617632 Cited in Problem 1.23.

[Lab97] C. Labruere, *Generalized braid groups and mapping class groups*, J. Knot Theory Ramifications **6** (1997), no. 5, 715–726, DOI 10.1142/S021821659700039X. MR1468903 Cited in Problem 2.29.

[Lac97] Marc Lackenby, *Dehn surgery on knots in 3-manifolds*, J. Amer. Math. Soc. **10** (1997), no. 4, 835–864, DOI 10.1090/S0894-0347-97-00241-5. MR1443548 Cited in Problems 1.12 and 1.13.

[Lac00] Marc Lackenby, *Word hyperbolic Dehn surgery*, Invent. Math. **140** (2000), no. 2, 243–282, DOI 10.1007/s002220000047. MR1756996 Cited in Problem 3.11.

[Lac06] Marc Lackenby, *Heegaard splittings, the virtually Haken conjecture and property* (τ), Invent. Math. **164** (2006), no. 2, 317–359, DOI 10.1007/s00222-005-0480-x. MR2218779 Cited in Problem 3.35.

[Lac09] Marc Lackenby, *The crossing number of composite knots*, J. Topol. **2** (2009), no. 4, 747–768, DOI 10.1112/jtopol/jtp028. MR2574742 Cited in Problem 1.1.

[Lac14] Marc Lackenby, *The crossing number of satellite knots*, Algebr. Geom. Topol. **14** (2014), no. 4, 2379–2409, DOI 10.2140/agt.2014.14.2379. MR3331616 Cited in Problem 1.2.

[Lac15] Marc Lackenby, *A polynomial upper bound on Reidemeister moves*, Ann. of Math. (2) **182** (2015), no. 2, 491–564, DOI 10.4007/annals.2015.182.2.3. MR3418524 Cited in Problems 1.76 and 1.77.

[Lac17a] Marc Lackenby, *Elementary knot theory*, Lectures on geometry, Clay Lect. Notes, Oxford Univ. Press, Oxford, 2017, pp. 29–64. MR3676592 Cited in Problem 1.80.

[Lac17b] Marc Lackenby, *Some conditionally hard problems on links and 3-manifolds*, Discrete Comput. Geom. **58** (2017), no. 3, 580–595, DOI 10.1007/s00454-017-9905-8. MR3690662 Cited in Problem 3.27.

[Lac21a] Marc Lackenby, *The efficient certification of knottedness and Thurston norm*, Adv. Math. **387** (2021), Paper No. 107796, 142, DOI 10.1016/j.aim.2021.107796. MR4274879 Cited in Problems 1.76 and 3.27.

[Lac21b] Marc Lackenby, *Links with splitting number one*, Geom. Dedicata **214** (2021), 319–351, DOI 10.1007/s10711-021-00618-x. MR4308282 Cited in Problem 1.80.

[Lad25] Roberto Ladu, *The Akbulut cork is not universal*, Selecta Math. (N.S.) **31** (2025), no. 4, Paper No. 74, 14, DOI 10.1007/s00029-025-01061-6. MR4940360 Cited in Problem 4.14.

[LaF10] Douglas J. LaFountain, *Studying uniform thickness. I. Legendrian simple iterated torus knots*, Algebr. Geom. Topol. **10** (2010), no. 2, 891–916, DOI 10.2140/agt.2010.10.891. MR2629769 Cited in Problem 1.63.

[Lam21] Christoph Lamm, *The search for nonsymmetric ribbon knots*, Exp. Math. **30** (2021), no. 3, 349–363, DOI 10.1080/10586458.2018.1540313. MR4309311 Cited in Problem 1.51.

[Lar15] Kyle Larson. *Some Constructions Involving Surgery on Surfaces in 4-manifolds.* PhD thesis, University of Texas, Austin, 2015. URL: `https://repositories.lib.utexas.edu/server/api/core/bitstreams/32f94088-e19b-4b49-8dd8-7b2320ef72e3/content`. Cited in Problem 4.26.

[Lar18] Kyle Larson, *Surgery on tori in the 4-sphere*, Math. Proc. Cambridge Philos. Soc. **164** (2018), no. 1, 109–124, DOI 10.1017/S0305004116000876. MR3733241 Cited in Problem 4.34.

[Lar19] Tim Large. Equivariant Floer theory and double covers of three-manifolds, 2019. `arXiv:1910.12119`. Cited in Problem 3.62.

[Las71] Richard K. Lashof, *A nonsmoothable knot*, Bull. Amer. Math. Soc. **77** (1971), 613–614, DOI 10.1090/S0002-9904-1971-12773-8. MR279815 Cited in Problem 4.32.

[Law78] Terry Lawson, *Open book decompositions for odd dimensional manifolds*, Topology **17** (1978), no. 2, 189–192, DOI 10.1016/S0040-9383(78)90024-1. MR494132 Cited in Problem 4.121.

[Law84] Terry Lawson, *Detecting the standard embedding of* $\mathbf{R}P^2$ *in* S^4, Math. Ann. **267** (1984), no. 4, 439–448, DOI 10.1007/BF01455961. MR742889 Cited in Problem 4.29.

[LC21] Peter Lambert-Cole. Stein trisections and homotopy 4-balls, 2021. `arXiv:2104.02003`. Cited in Problems 4.23 and 4.107.

[LC23] Peter Lambert-Cole, *Symplectic surfaces and bridge position*, Geom. Dedicata **217** (2023), no. 1, Paper No. 8, 15, DOI 10.1007/s10711-022-00742-2. MR4505010 Cited in Problem 4.101.

[LCM22] Peter Lambert-Cole and Jeffrey Meier. Bridge trisections in rational surfaces. *J. Topol. Anal.*, 14(3):655–708, 2022. `doi:10.1142/S1793525321500047`. Cited in Problems 4.111, 4.113, and 4.115.

[Lea02] Jason Paige Leasure, *Geodesics in the complex of curves of a surface*, ProQuest LLC, Ann Arbor, MI, 2002. Thesis (Ph.D.)–The University of Texas at Austin. MR2705485 Cited in Problem 2.10.

[LeB95] Claude LeBrun, *Anti-self-dual Riemannian 4-manifolds*, Twistor theory (Plymouth), Lecture Notes in Pure and Appl. Math., vol. 169, Dekker, New York, 1995, pp. 81–94. MR1306958 Cited in Problem 4.130.

[LeB02] Claude LeBrun, *Hyperbolic manifolds, harmonic forms, and Seiberg-Witten invariants*, Proceedings of the Euroconference on Partial Differential Equations and their Applications to Geometry and Physics (Castelvecchio Pascoli, 2000), Geom. Dedicata **91** (2002), 137–154, DOI 10.1023/A:1016222709901. MR1919897 Cited in Problems 2.23 and 4.95.

[Lec15] Ana G. Lecuona, *On the slice-ribbon conjecture for pretzel knots*, Algebr. Geom. Topol. **15** (2015), no. 4, 2133–2173, DOI 10.2140/agt.2015.15.2133. MR3402337 Cited in Problem 1.50.

[Lee22] Christine Ruey Shan Lee, *Jones slopes and coarse volume of near-alternating knots*, Comm. Anal. Geom. **30** (2022), no. 4, 891–948. MR4550162 Cited in Problem 1.26.

[Lei06] Christopher J. Leininger, *Small curvature surfaces in hyperbolic 3-manifolds*, J. Knot Theory Ramifications **15** (2006), no. 3, 379–411, DOI 10.1142/S0218216506004531. MR2217503 Cited in Problem 3.14.

[Lev65] J. Levine, *A characterization of knot polynomials*, Topology **4** (1965), 135–141, DOI 10.1016/0040-9383(65)90061-3. MR180964 Cited in Problem 1.8.

[Lev12] Adam Simon Levine, *Slicing mixed Bing-Whitehead doubles*, J. Topol. **5** (2012), no. 3, 713–726, DOI 10.1112/jtopol/jts019. MR2971611 Cited in Problem 1.62.

[Lev16] Adam Simon Levine, *Nonsurjective satellite operators and piecewise-linear concordance*, Forum Math. Sigma **4** (2016), Paper No. e34, 47, DOI 10.1017/fms.2016.31. MR3589337 Cited in Problem 1.61.

[Lev23] Adam Simon Levine, *A note on rationally slice knots*, New York J. Math. **29** (2023), 1363–1372. MR4689111 Cited in Problem 1.60.

[Lew23a] Lukas Lewark. Foamho, an $\mathfrak{sl}(3)$-homology calculator. URL: https://people.math.ethz.ch/ llewark/foamho.php, November 2023. Cited in Problem 1.31.

[Lew23b] Lukas Lewark. Khoca, a knot homology calculator. URL: https://people.math.ethz.ch/ llewark/khoca.php, November 2023. Cited in Problem 1.31.

[Li00] Tian-Jun Li, *Symplectic Parshin-Arakelov inequality*, Internat. Math. Res. Notices **18** (2000), 941–954, DOI 10.1155/S1073792800000490. MR1792283 Cited in Problem 2.21.

[Li06a] Tian-Jun Li, *Quaternionic bundles and Betti numbers of symplectic 4-manifolds with Kodaira dimension zero*, Int. Math. Res. Not., posted on 2006, Art. ID 37385, 28, DOI 10.1155/IMRN/2006/37385. MR2264722 Cited in Problem 4.92.

[Li06b] Tian-Jun Li, *Symplectic 4-manifolds with Kodaira dimension zero*, J. Differential Geom. **74** (2006), no. 2, 321–352. MR2259057 Cited in Problem 4.92.

[Li08] Tian-Jun Li, *The space of symplectic structures on closed 4-manifolds*, Third International Congress of Chinese Mathematicians. Part 1, 2, AMS/IP Stud. Adv. Math., 42, pt. 1, vol. 2, Amer. Math. Soc., Providence, RI, 2008, pp. 259–277. MR2409637 Cited in Problems 4.93 and 4.100.

[Li10] Tian-Jun Li, *Symplectic Calabi-Yau surfaces*, Handbook of geometric analysis, No. 3, Adv. Lect. Math. (ALM), vol. 14, Int. Press, Somerville, MA, 2010, pp. 231–356. MR2743450 Cited in Problem 4.92.

[Li13] Tao Li, *Rank and genus of 3-manifolds*, J. Amer. Math. Soc. **26** (2013), no. 3, 777–829, DOI 10.1090/S0894-0347-2013-00767-5. MR3037787 Cited in Problems 3.35 and 4.117.

[Li19] Tian-Jun Li, *Kodaira dimension in low dimensional topology*, Tsinghua lectures in mathematics, Adv. Lect. Math. (ALM), vol. 45, Int. Press, Somerville, MA, [2019] ©2019, pp. 265–291. MR3971557 Cited in Problem 4.92.

[Li22] Tao Li, *Heegaard genus, degree-one maps, and amalgamation of 3-manifolds*, J. Topol. **15** (2022), no. 3, 1540–1579, DOI 10.1112/topo.12253. MR4461853 Cited in Problem 3.33.

[Li24] Tao Li, *Taut foliations of 3-manifolds with Heegaard genus 2*, Duke Math. J. **173** (2024), no. 8, 1427–1475, DOI 10.1215/00127094-2023-0038. MR4756955 Cited in Problem 3.48.

[Lib82] A. Libgober, *Alexander polynomial of plane algebraic curves and cyclic multiple planes*, Duke Math. J. **49** (1982), no. 4, 833–851. MR683005 Cited in Problem 4.103.

[Lib83] A. Libgober, *Alexander invariants of plane algebraic curves*, Singularities, Part 2 (Arcata, Calif., 1981), Proc. Sympos. Pure Math., vol. 40, Amer. Math. Soc., Providence, RI, 1983, pp. 135–143, DOI 10.1090/pspum/040.2/713242. MR713242 Cited in Problem 4.103.

[Lib21] Anatoly Libgober, *Complements to ample divisors and singularities*, Handbook of geometry and topology of singularities II, Springer, Cham, [2021] ©2021, pp. 501–567, DOI 10.1007/978-3-030-78024-1_10. MR4367444 Cited in Problem 4.103.

[Lic63] André Lichnerowicz, *Spineurs harmoniques* (French), C. R. Acad. Sci. Paris **257** (1963), 7–9. MR156292 Cited in Problem 4.128.

[Lic79] W. B. Raymond Lickorish, *Shake-slice knots*, Topology of low-dimensional manifolds (Proc. Second Sussex Conf., Chelwood Gate, 1977), Lecture Notes in Math., vol. 722, Springer, Berlin, 1979, pp. 67–70. MR547455 Cited in Problem 1.57.

[Lic97] W. B. Raymond Lickorish, *An introduction to knot theory*, Graduate Texts in Mathematics, vol. 175, Springer-Verlag, New York, 1997, DOI 10.1007/978-1-4612-0691-0. MR1472978 Cited in Problems 1.0 and 1.82.

[Lin04] Vladimir Lin. Braids and permutations, 2004. `arXiv:math/0404528`. Cited in Problem 2.7.

[Lin16] Francesco Lin, *Exact Lagrangian caps of Legendrian knots*, J. Symplectic Geom. **14** (2016), no. 1, 269–295, DOI 10.4310/JSG.2016.v14.n1.a10. MR3523257 Cited in Problem 1.65.

[Lin19] Jianfeng Lin, *The Seiberg-Witten equations on end-periodic manifolds and an obstruction to positive scalar curvature metrics*, J. Topol. **12** (2019), no. 2, 328–371, DOI 10.1112/topo.12090. MR3911569 Cited in Problem 4.128.

[Lin22] Jianfeng Lin. The family Seiberg-Witten invariant and nonsymplectic loops of diffeomorphisms, 2022. `arXiv:2208.12082`. Cited in Problem 4.75.

[Lin23] Jianfeng Lin, *Isotopy of the Dehn twist on $K3 \# K3$ after a single stabilization*, Geom. Topol. **27** (2023), no. 5, 1987–2012, DOI 10.2140/gt.2023.27.1987. MR4621924 Cited in Problems 4.68, 4.70, and 4.79.

[Lin24] Francesco Lin. A remark on taut foliations and Floer homology. *Math. Res. Lett.*, 31(6):1819–1825, 2024. `doi:10.4310/mrl.250211002052`. Cited in Problem 3.55.

[Lis07] Paolo Lisca, *Lens spaces, rational balls and the ribbon conjecture*, Geom. Topol. **11** (2007), 429–472, DOI 10.2140/gt.2007.11.429. MR2302495 Cited in Problems 1.46 and 1.50.

[Lis08] Paolo Lisca, *On symplectic fillings of lens spaces*, Trans. Amer. Math. Soc. **360** (2008), no. 2, 765–799, DOI 10.1090/S0002-9947-07-04228-6. MR2346471 Cited in Problem 4.105.

[Lit79] R. A. Litherland, *Signatures of iterated torus knots*, Topology of low-dimensional manifolds (Proc. Second Sussex Conf., Chelwood Gate, 1977), Lecture Notes in Math., vol. 722, Springer, Berlin, 1979, pp. 71–84. MR547456 Cited in Problem 1.39.

[Liu13] Gang Liu. 3-manifolds with nonnegative Ricci curvature. *Invent. Math.*, 193(2):367–375, 2013. `doi:10.1007/s00222-012-0428-x`. Cited in Problem 5.15.

[Liu23] Yi Liu, *Finite-volume hyperbolic 3-manifolds are almost determined by their finite quotient groups*, Invent. Math. **231** (2023), no. 2, 741–804, DOI 10.1007/s00222-022-01155-4. MR4542705 Cited in Problems 3.7 and 3.8.

[Liv54] George R. Livesay, *On a theorem of F. J. Dyson*, Ann. of Math. (2) **59** (1954), 227–229, DOI 10.2307/1969689. MR59546 Cited in Problem 2.14.

[Liv01] Charles Livingston. Examples in concordance, 2001. `arXiv:math/0101035v1`. Cited in Problem 1.42.

[Liv10] Charles Livingston, *The stable 4-genus of knots*, Algebr. Geom. Topol. **10** (2010), no. 4, 2191–2202, DOI 10.2140/agt.2010.10.2191. MR2745668 Cited in Problem 1.41.

[Liv24] Charles Livingston, *Connected sums of codimension two locally flat submanifolds*, Proc. Roy. Soc. Edinburgh Sect. A **154** (2024), no. 6, 1937–1944, DOI 10.1017/prm.2022.87. MR4883994 Cited in Problem 5.5.

[LL95] T. J. Li and A. Liu, *Symplectic structure on ruled surfaces and a generalized adjunction formula*, Math. Res. Lett. **2** (1995), no. 4, 453–471, DOI 10.4310/MRL.1995.v2.n4.a6. MR1355707 Cited in Problem 4.90.

[LL12] Adam Simon Levine and Sam Lewallen, *Strong L-spaces and left-orderability*, Math. Res. Lett. **19** (2012), no. 6, 1237–1244, DOI 10.4310/MRL.2012.v19.n6.a5. MR3091604 Cited in Problem 3.53.

[LL19] Youlin Li and Yajing Liu, *Hyperbolic 3-manifolds admitting no fillable contact structures*, Proc. Amer. Math. Soc. **147** (2019), no. 1, 351–360, DOI 10.1090/proc/13870. MR3876754 Cited in Problem 3.40.

[LL22] Francesco Lin and Michael Lipnowski, *The Seiberg-Witten equations and the length spectrum of hyperbolic three-manifolds*, J. Amer. Math. Soc. **35** (2022), no. 1, 233–293, DOI 10.1090/jams/982. MR4322393 Cited in Problem 3.53.

[LL23a] Aaron Landesman and Daniel Litt, *Applications of the algebraic geometry of the Putman-Wieland conjecture*, Proc. Lond. Math. Soc. (3) **127** (2023), no. 1, 116–133, DOI 10.1112/plms.12539. MR4611405 Cited in Problem 2.1.

[LL23b] Aaron Landesman and Daniel Litt, *An introduction to the algebraic geometry of the Putman-Wieland conjecture*, Eur. J. Math. **9** (2023), no. 2, Paper No. 40, 25, DOI 10.1007/s40879-023-00637-w. MR4593261 Cited in Problem 2.1.

[LLP23] Adam Simon Levine, Tye Lidman, and Lisa Piccirillo. New constructions and invariants of closed exotic 4-manifolds, 2023. arXiv:2307.08130. Cited in Problems 4.3 and 4.8.

[LLS25] Aaron Landesman, Daniel Litt, and Will Sawin, *Big monodromy for higher Prym representations*, Geom. Topol. **29** (2025), no. 5, 2733–2782, DOI 10.2140/gt.2025.29.2733. MR4950926 Cited in Problem 2.24.

[LLSS23] Wanlin Li, Daniel Litt, Nick Salter, and Padmavathi Srinivasan, *Surface bundles and the section conjecture*, Math. Ann. **386** (2023), no. 1-2, 877–942, DOI 10.1007/s00208-022-02421-9. MR4585164 Cited in Problem 2.25.

[LM72] R. Luke and W. K. Mason, *The space of homeomorphisms on a compact two-manifold is an absolute neighborhood retract*, Trans. Amer. Math. Soc. **164** (1972), 275–285, DOI 10.2307/1995974. MR301693 Cited in Problem 5.4.

[LM83] Charles Livingston and Paul Melvin, *Algebraic knots are algebraically dependent*, Proc. Amer. Math. Soc. **87** (1983), no. 1, 179–180, DOI 10.2307/2044377. MR677257 Cited in Problem 1.39.

[LM14] Eduard Looijenga and Gabriele Mondello, *The fine structure of the moduli space of abelian differentials in genus 3*, Geom. Dedicata **169** (2014), 109–128, DOI 10.1007/s10711-013-9845-2. MR3175239 Cited in Problem 2.26.

[LM15] Kyle Larson and Jeffrey Meier, *Fibered ribbon disks*, J. Knot Theory Ramifications **24** (2015), no. 14, 1550066, 22, DOI 10.1142/S0218216515500662. MR3434552 Cited in Problems 1.50 and 4.34.

[LM18a] Tye Lidman and Ciprian Manolescu, *The equivalence of two Seiberg-Witten Floer homologies* (English, with English and French summaries), Astérisque **399** (2018), vii+220. MR3818611 Cited in Problem 3.47.

[LM18b] Tye Lidman and Ciprian Manolescu, *Floer homology and covering spaces*, Geom. Topol. **22** (2018), no. 5, 2817–2838, DOI 10.2140/gt.2018.22.2817. MR3811772 Cited in Problem 3.62.

[LM25] Jianfeng Lin and Anubhav Mukherjee, *Family Bauer-Furuta invariant, exotic surfaces and Smale conjecture*, J. Assoc. Math. Res. **3** (2025), no. 2, 237–275, DOI 10.56994/JAMR.003.002.003. MR4958155 Cited in Problems 4.33 and 4.67.

[LM23a] Yevgeny Liokumovich and Davi Maximo, *Waist inequality for 3-manifolds with positive scalar curvature*, Perspectives in scalar curvature. Vol. 2, World Sci. Publ., Hackensack, NJ, [2023] ©2023, pp. 799–831. MR4577931 Cited in Problem 4.129.

[LM23b] Charles Livingston and Allison H. Moore. Knotinfo: Table of knot invariants, November 2023. https://knotinfo.org. Cited in Problems 1.5 and 1.31.

[LMP25] Tao Li, Yoav Moriah, and Tali Pinsky, *Tunnel number one knots satisfy the Berge conjecture*, Geom. Topol. **29** (2025), no. 6, 3271–3343, DOI 10.2140/gt.2025.29.3271. MR4965188 Cited in Problem 1.9.

[LMR06] D. D. Long, C. Maclachlan, and A. W. Reid, *Arithmetic Fuchsian groups of genus zero*, Pure Appl. Math. Q. **2** (2006), no. 2, Special Issue: In honor of John H. Coates., 569–599, DOI 10.4310/PAMQ.2006.v2.n2.a9. MR2251482 Cited in Problem 3.10.

[LMT23] Michael P. Landry, Yair N. Minsky, and Samuel J. Taylor, *Flows, growth rates, and the veering polynomial*, Ergodic Theory Dynam. Systems **43** (2023), no. 9, 3026–3107, DOI 10.1017/etds.2022.63. MR4624489 Cited in Problem 3.21.

[LMW08] Robert Lipshitz, Ciprian Manolescu, and Jiajun Wang, *Combinatorial cobordism maps in hat Heegaard Floer theory*, Duke Math. J. **145** (2008), no. 2, 207–247, DOI 10.1215/00127094-2008-050. MR2449946 Cited in Problem 3.60.

[LN14] Tian-Jun Li and Yi Ni, *Virtual Betti numbers and virtual symplecticity of 4-dimensional mapping tori*, Math. Z. **277** (2014), no. 1-2, 195–208, DOI 10.1007/s00209-013-1250-x. MR3205768 Cited in Problem 4.92.

[LN15] Eileen Li and Yi Ni, *Half-integral finite surgeries on knots in S^3* (English, with English and French summaries), Ann. Fac. Sci. Toulouse Math. (6) **24** (2015), no. 5, 1157–1178, DOI 10.5802/afst.1479. MR3485330 Cited in Problem 1.49.

[LN21] Boris Lishak and Alexander Nabutovsky, *Complexity of unknotting of trivial 2-knots*, J. Topol. Anal. **13** (2021), no. 3, 623–657, DOI 10.1142/S1793525320500272. MR4312449 Cited in Problem 1.71.

[Lob19] Andrew Lobb, *A counterexample to Batson's conjecture*, Math. Res. Lett. **26** (2019), no. 6, 1789, DOI 10.4310/MRL.2019.v26.n6.a8. MR4078695 Cited in Problem 1.47.

[Lom81] S. J. Lomonaco Jr., *The homotopy groups of knots. I. How to compute the algebraic 2-type*, Pacific J. Math. **95** (1981), no. 2, 349–390. MR632192 Cited in Problems 4.36 and 4.38.

[Lon86] D. D. Long, *Discs in compression bodies*, Pacific J. Math. **122** (1986), no. 1, 129–146. MR825228 Cited in Problem 1.52.

[Loo71] Eduard Looijenga, *A note on polynomial isolated singularities*, Indag. Math. **33** (1971), 418–421. Nederl. Akad. Wetensch. Proc. Ser. A **74**. MR303557 Cited in Problem 1.87.

[LOT13] Robert Lipshitz, Peter S. Ozsváth, and Dylan P. Thurston, *A faithful linear-categorical action of the mapping class group of a surface with boundary*, J. Eur. Math. Soc. (JEMS) **15** (2013), no. 4, 1279–1307, DOI 10.4171/JEMS/392. MR3055762 Cited in Problems 2.3 and 2.47.

[LOT15] Robert Lipshitz, Peter S. Ozsváth, and Dylan P. Thurston, *Bimodules in bordered Heegaard Floer homology*, Geom. Topol. **19** (2015), no. 2, 525–724, DOI 10.2140/gt.2015.19.525. MR3336273 Cited in Problem 2.47.

[LP01] Andrea Loi and Riccardo Piergallini, *Compact Stein surfaces with boundary as branched covers of B^4*, Invent. Math. **143** (2001), no. 2, 325–348, DOI 10.1007/s002220000106. MR1835390 Cited in Problem 3.44.

[LP25] Tye Lidman and Lisa Piccirillo. Distinguishing closed 4-manifolds by slicing, 2025. `arXiv:2505.14387`. Cited in Problem 4.42.

[LPCZ23] Tye Lidman, Juanita Pinzón-Caicedo, and Raphael Zentner, *Toroidal integer homology three-spheres have irreducible $SU(2)$-representations*, J. Topol. **16** (2023), no. 1, 344–367, DOI 10.1112/topo.12275. MR4575869 Cited in Problem 3.52.

[LPRT] Justin Lanier, Priyam Patel, Anja Randecker, and Jing Tao. AIM Problem List: Surfaces of infinite type. `available at http://aimpl.org/genusinfinity`. Cited in Problem 2.38.

[LR00] D. D. Long and A. W. Reid, *On the geometric boundaries of hyperbolic 4-manifolds*, Geom. Topol. **4** (2000), 171–178, DOI 10.2140/gt.2000.4.171. MR1769269 Cited in Problems 3.77, 4.124, and 4.126.

[LR01] D. D. Long and A. W. Reid, *Constructing hyperbolic manifolds which bound geometrically*, Math. Res. Lett. **8** (2001), no. 4, 443–455, DOI 10.4310/MRL.2001.v8.n4.a5. MR1849261 Cited in Problem 4.126.

[LR02] D. D. Long and A. W. Reid, *All flat manifolds are cusps of hyperbolic orbifolds*, Algebr. Geom. Topol. **2** (2002), 285–296, DOI 10.2140/agt.2002.2.285. MR1917053 Cited in Problem 5.14.

[LR11] Frédéric Le Roux, *Free planar actions of the Klein bottle group*, Geom. Topol. **15** (2011), no. 3, 1545–1567, DOI 10.2140/gt.2011.15.1545. MR2851071 Cited in Problem 2.33.

[LR20] D. D. Long and A. W. Reid, *Virtually spinning hyperbolic manifolds*, Proc. Edinb. Math. Soc. (2) **63** (2020), no. 2, 305–313, DOI 10.1017/s0013091519000324. MR4089376 Cited in Problem 4.125.

[LRS18] Jianfeng Lin, Daniel Ruberman, and Nikolai Saveliev, *A splitting theorem for the Seiberg-Witten invariant of a homology $S^1 \times S^3$*, Geom. Topol. **22** (2018), no. 5, 2865–2942, DOI 10.2140/gt.2018.22.2865. MR3811774 Cited in Problem 3.69.

[LRS21] Jianfeng Lin, Daniel Ruberman, and Nikolai Saveliev, *On the monopole Lefschetz number of finite-order diffeomorphisms*, Geom. Topol. **25** (2021), no. 7, 3591–3628, DOI 10.2140/gt.2021.25.3591. MR4372637 Cited in Problem 4.65.

[LRS23a] Jianfeng Lin, Daniel Ruberman, and Nikolai Saveliev, *On the Frøyshov invariant and monopole Lefschetz number*, J. Differential Geom. **123** (2023), no. 3, 523–593, DOI 10.4310/jdg/1683307008. MR4584860 Cited in Problem 3.74.

[LRS23b] Jianfeng Lin, Daniel Ruberman, and Nikolai Saveliev, *On the Frøyshov invariant and monopole Lefschetz number*, J. Differential Geom. **123** (2023), no. 3, 523–593, DOI 10.4310/jdg/1683307008. MR4584860 Cited in Problem 4.65.

[LS07] Paolo Lisca and András I. Stipsicz, *Ozsváth-Szabó invariants and tight contact 3-manifolds. III*, J. Symplectic Geom. **5** (2007), no. 4, 357–384, DOI 10.4310/jsg.2007.v5.n4.a1. MR2413308 Cited in Problem 3.48.

[LS09] Paolo Lisca and András I. Stipsicz, *On the existence of tight contact structures on Seifert fibered 3-manifolds*, Duke Math. J. **148** (2009), no. 2, 175–209, DOI 10.1215/00127094-2009-024. MR2524494 Cited in Problem 3.40.

[LS14] Robert Lipshitz and Sucharit Sarkar, *A refinement of Rasmussen's S-invariant*, Duke Math. J. **163** (2014), no. 5, 923–952, DOI 10.1215/00127094-2644466. MR3189434 Cited in Problems 3.57 and 4.43.

[LST08] Feng Luo, Saul Schleimer, and Stephan Tillmann, *Geodesic ideal triangulations exist virtually*, Proc. Amer. Math. Soc. **136** (2008), no. 7, 2625–2630, DOI 10.1090/S0002-9939-08-09387-8. MR2390535 Cited in Problem 3.3.

[LT88] W. B. R. Lickorish and M. B. Thistlethwaite, *Some links with nontrivial polynomials and their crossing-numbers*, Comment. Math. Helv. **63** (1988), no. 4, 527–539, DOI 10.1007/BF02566777. MR966948 Cited in Problem 1.1.

[LT10] Thang T. Q. Le and Anh T. Tran, *On the volume conjecture for cables of knots*, J. Knot Theory Ramifications **19** (2010), no. 12, 1673–1691, DOI 10.1142/S0218216510008534. MR2755495 Cited in Problem 1.27.

[LT15] Thang T. Q. Le and Anh T. Tran, *On the AJ conjecture for knots*, Indiana Univ. Math. J. **64** (2015), no. 4, 1103–1151, DOI 10.1512/iumj.2015.64.5602. With an appendix written jointly with Vu Q. Huynh. MR3385787 Cited in Problem 1.25.

[LT16] Robert Lipshitz and David Treumann, *Noncommutative Hodge-to-de Rham spectral sequence and the Heegaard Floer homology of double covers*, J. Eur. Math. Soc. (JEMS) **18** (2016), no. 2, 281–325, DOI 10.4171/JEMS/590. MR3459952 Cited in Problem 3.62.

[LT20] Naichung Conan Leung and Ryosuke Takahashi, *Energy bound for Kapustin-Witten solutions on $S^3 \times \mathbb{R}^+$*, Int. Math. Res. Not. IMRN **19** (2020), 6135–6148, DOI 10.1093/imrn/rny198. MR4165474 Cited in Problem 1.32.

[Luo88] Feng Luo. The existence of $K(\pi, 1)$ 4-manifolds which are rational homology 4-spheres. *Proc. Amer. Math. Soc.*, 104(4):1315–1321, 1988. doi:10.2307/2047635. Cited in Problem 4.124.

[LV77] J. Luukkainen and J. Väisälä, *Elements of Lipschitz topology*, Ann. Acad. Sci. Fenn. Ser. A I Math. **3** (1977), no. 1, 85–122, DOI 10.5186/aasfm.1977.0315. MR515647 Cited in Problem 4.23.

[LV24] Joan Licata and Vera Vértesi. The Giroux correspondence in dimension 3, 2024. URL: https://arxiv.org/abs/2408.01079, arXiv:2408.01079. Cited in Problem 3.39.

[LvdV16] Christine Ruey Shan Lee and Roland van der Veen, *Slopes for pretzel knots*, New York J. Math. **22** (2016), 1339–1364. MR3576292 Cited in Problem 1.26.

[LW90] Ronnie Lee and Dariusz M. Wilczyński, *Locally flat 2-spheres in simply connected 4-manifolds*, Comment. Math. Helv. **65** (1990), no. 3, 388–412, DOI 10.1007/BF02566615. MR1069816 Cited in Problem 4.45.

[LW93] Ronnie Lee and Dariusz M. Wilczyński. Representing homology classes by locally flat 2-spheres. *K-Theory*, 7(4):333–367, 1993. doi:10.1007/BF00962053. Cited in Problem 4.45.

[LW97] Ronnie Lee and Dariusz M. Wilczyński, *Representing homology classes by locally flat surfaces of minimum genus*, Amer. J. Math. **119** (1997), no. 5, 1119–1137. MR1473071 Cited in Problems 4.32, 4.44, and 4.45.

[LW00] Ronnie Lee and Dariusz M. Wilczyński, *Genus inequalities and four-dimensional surgery*, Topology **39** (2000), no. 2, 311–330, DOI 10.1016/S0040-9383(99)00017-8. MR1722016 Cited in Problem 4.44.

[LW01] Ronnie Lee and Dariusz M. Wilczyński. Erratum to: "Genus inequalities and four-dimensional surgery" [Topology **39** (2000), no. 2, 311–330; MR1722016 (2001j:57028)]. *Topology*, 40(5):1123, 2001. doi:10.1016/S0040-9383(00)00008-2. Cited in Problem 4.45.

[LW21] Samuel Lisi and Chris Wendl, *Spine removal surgery and the geography of symplectic fillings*, Michigan Math. J. **70** (2021), no. 2, 403–422, DOI 10.1307/mmj/1594260053. MR4278702 Cited in Problem 4.105.

[LX23] Jianfeng Lin and Yi Xie. Configuration space integrals and formal smooth structures, 2023. arXiv:2310.14156. Cited in Problems 4.70 and 4.73.

[LY24] Zhenkun Li and Fan Ye. 2-torsion in instanton Floer homology, 2024. arXiv:2405.16252. Cited in Problem 3.54.

[LYL19] Xudong Leng, Zhiqing Yang, and Ximin Liu, *The slope conjectures for 3-string Montesinos knots*, New York J. Math. **25** (2019), 45–70. MR3904877 Cited in Problem 1.26.

[LZ94] E. Luft and X. Zhang, *Symmetric knots and the cabling conjecture*, Math. Ann. **298** (1994), no. 3, 489–496, DOI 10.1007/BF01459747. MR1262772 Cited in Problem 1.11.

[LZ17] Thang T. Q. Lê and Xingru Zhang, *Character varieties, A-polynomials and the AJ conjecture*, Algebr. Geom. Topol. **17** (2017), no. 1, 157–188, DOI 10.2140/agt.2017.17.157. MR3604376 Cited in Problem 1.25.

[LZ19] Adam Simon Levine and Ian Zemke, *Khovanov homology and ribbon concordances*, Bull. Lond. Math. Soc. **51** (2019), no. 6, 1099–1103, DOI 10.1112/blms.12303. MR4041014 Cited in Problem 1.56.

[Mae07] Hiroshi Maehara, *On configurations of solid balls in 3-space: chromatic numbers and knotted cycles*, Graphs Combin. **23** (2007), no. suppl. 1, 307–320, DOI 10.1007/s00373-007-0702-7. MR2320637 Cited in Problem 1.73.

[Mak05] Sergey Maksymenko, *Path-components of Morse mappings spaces of surfaces*, Comment. Math. Helv. **80** (2005), no. 3, 655–690, DOI 10.4171/CMH/30. MR2165207 Cited in Problem 4.114.

[Man80] Benoit B. Mandelbrot. Fractal aspects of the iteration of $z \to \lambda z(1-z)$ for complex λ and z. *Annals of the New York Academy of Sciences*, 357(1):249–259, 1980. doi:10.1111/j.1749-6632.1980.tb29690.x. Cited in Problem 5.1.

[Man01] Marco Manetti. On the moduli space of diffeomorphic algebraic surfaces. *Invent. Math.*, 143(1):29–76, 2001. doi:10.1007/s002220000101. Cited in Problem 4.94.

[Man03] Ciprian Manolescu, *Seiberg-Witten-Floer stable homotopy type of three-manifolds with $b_1 = 0$*, Geom. Topol. **7** (2003), 889–932, DOI 10.2140/gt.2003.7.889. MR2026550 Cited in Problems 3.47 and 3.56.

[Man07] Vassily Olegovich Manturov, *Khovanov homology for virtual knots with arbitrary coefficients*, J. Knot Theory Ramifications **16** (2007), no. 3, 345–377, DOI 10.1142/S0218216507005336. MR2320160 Cited in Problem 1.32.

[Man13] Ciprian Manolescu, *The Conley index, gauge theory, and triangulations*, J. Fixed Point Theory Appl. **13** (2013), no. 2, 431–457, DOI 10.1007/s11784-013-0134-3. MR3122335 Cited in Problem 4.119.

[Man14] Ciprian Manolescu, *On the intersection forms of spin four-manifolds with boundary*, Math. Ann. **359** (2014), no. 3-4, 695–728, DOI 10.1007/s00208-014-1010-1. MR3231012 Cited in Problems 3.56 and 4.15.

[Man16a] Ciprian Manolescu, *An introduction to knot Floer homology*, Physics and mathematics of link homology, Contemp. Math., vol. 680, Amer. Math. Soc., Providence, RI, 2016, pp. 99–135, DOI 10.1090/conm/680. MR3591644 Cited in Problem 1.19.

[Man16b] Ciprian Manolescu, *Pin(2)-equivariant Seiberg-Witten Floer homology and the triangulation conjecture*, J. Amer. Math. Soc. **29** (2016), no. 1, 147–176, DOI 10.1090/jams829. MR3402697 Cited in Problems 3.47, 3.56, 3.69, and 5.8.

[Man18] Ciprian Manolescu, *Homology cobordism and triangulations*, Proceedings of the International Congress of Mathematicians—Rio de Janeiro 2018. Vol. II. Invited lectures, World Sci. Publ., Hackensack, NJ, 2018, pp. 1175–1191. MR3966804 Cited in Problem 3.68.

[Mar58] A. Markov, *The insolubility of the problem of homeomorphy* (Russian), Dokl. Akad. Nauk SSSR **121** (1958), 218–220. MR97793 Cited in Problem 4.53.

[Mar74] Albert Marden. The geometry of finitely generated kleinian groups. *Ann. of Math. (2)*, 99:383–462, 1974. doi:10.2307/1971059. Cited in Problem 3.0.

[Mar09] João Faria Martins, *The fundamental crossed module of the complement of a knotted surface*, Trans. Amer. Math. Soc. **361** (2009), no. 9, 4593–4630, DOI 10.1090/S0002-9947-09-04576-0. MR2506421 Cited in Problems 4.36 and 4.38.

[Mar13] Vladimir Markovic, *Criterion for Cannon's conjecture*, Geom. Funct. Anal. **23** (2013), no. 3, 1035–1061, DOI 10.1007/s00039-013-0228-5. MR3061779 Cited in Problem 3.24.

[Mar19] Dan Margalit, *Problems, questions, and conjectures about mapping class groups*, Breadth in contemporary topology, Proc. Sympos. Pure Math., vol. 102, Amer. Math. Soc., Providence, RI, 2019, pp. 157–186, DOI 10.1090/pspum/102/12. MR3967367 Cited in Problems 2.2, 2.3, and 2.27.

[Mar23] Gage Martin, *Annular Khovanov homology and meridional disks*, J. Knot Theory Ramifications **32** (2023), no. 2, Paper No. 2250088, 14, DOI 10.1142/S0218216522500882. MR4564613 Cited in Problem 1.28.

[Mas71] W. K. Mason, *The space of all self-homeomorphisms of a two-cell which fix the cell's boundary is an absolute retract*, Trans. Amer. Math. Soc. **161** (1971), 185–205, DOI 10.2307/1995936. MR286067 Cited in Problem 5.4.

[Mas99] Gregor Masbaum, *An element of infinite order in TQFT-representations of mapping class groups*, Low-dimensional topology (Funchal, 1998), Contemp. Math., vol. 233, Amer. Math. Soc., Providence, RI, 1999, pp. 137–139, DOI 10.1090/conm/233/03423. MR1701678 Cited in Problem 2.6.

[Mat71] John N. Mather, *The vanishing of the homology of certain groups of homeomorphisms*, Topology **10** (1971), 297–298, DOI 10.1016/0040-9383(71)90022-X. MR288777 Cited in Problem 4.85.

[Mat74] John N. Mather, *Commutators of diffeomorphisms*, Comment. Math. Helv. **49** (1974), 512–528, DOI 10.1007/BF02566746. MR356129 Cited in Problem 4.85.

[Mat78] Takao Matumoto, *Triangulation of manifolds*, Algebraic and geometric topology (Proc. Sympos. Pure Math., Stanford Univ., Stanford, Calif., 1976), Proc. Sympos. Pure Math., XXXII, Amer. Math. Soc., Providence, RI, 1978, pp. 3–6. MR520517 Cited in Problems 3.69 and 5.8.

[Mat82] Yukio Matsumoto, *On 4-manifolds fibered by tori*, Proc. Japan Acad. Ser. A Math. Sci. **58** (1982), no. 7, 298–301. MR682687 Cited in Problem 2.22.

[Mat87] S. V. Matveev, *The Zeeman conjecture for nonthickenable special polyhedra is equivalent to the Andrews-Curtis conjecture* (Russian), Sibirsk. Mat. Zh. **28** (1987), no. 6, 66–80, 218. MR928341 Cited in Problem 5.12.

[Mat88] Gordana Matić, SO(3)*-connections and rational homology cobordisms*, J. Differential Geom. **28** (1988), no. 2, 277–307. MR961516 Cited in Problem 4.59.

[Mat96] Yukio Matsumoto, *Lefschetz fibrations of genus two—a topological approach*, Topology and Teichmüller spaces (Katinkulta, 1995), World Sci. Publ., River Edge, NJ, 1996, pp. 123–148. MR1659687 Cited in Problem 2.21.

[Mat07] Sergei Matveev, *Algorithmic topology and classification of 3-manifolds*, 2nd ed., Algorithms and Computation in Mathematics, vol. 9, Springer, Berlin, 2007. MR2341532 Cited in Problem 3.32.

[Mat14] Benjamin Matschke, *A survey on the square peg problem*, Notices Amer. Math. Soc. **61** (2014), no. 4, 346–352, DOI 10.1090/noti1100. MR3184501 Cited in Problem 2.14.

[Mau13] Rasimate Maungchang, *The Sunada construction and the simple length spectrum*, Geom. Dedicata **163** (2013), 349–360, DOI 10.1007/s10711-012-9753-x. MR3032699 Cited in Problem 2.16.

[Maz59] Barry Mazur, *On embeddings of spheres*, Bull. Amer. Math. Soc. **65** (1959), 59–65, DOI 10.1090/S0002-9904-1959-10274-3. MR117693 Cited in Problem 4.23.

[Maz61] Barry Mazur, *A note on some contractible* 4*-manifolds*, Ann. of Math. (2) **73** (1961), 221–228, DOI 10.2307/1970288. MR125574 Cited in Problem 4.87.

[Maz62] Barry Mazur, *Symmetric homology spheres*, Illinois J. Math. **6** (1962), 245–250. MR140102 Cited in Problem 4.1.

[Maz63] Barry Mazur. Relative neighborhoods and the theorems of Smale. *Ann. of Math. (2)*, 77:232–249, 1963. `doi:10.2307/1970215`. Cited in Problems 4.46 and 4.56.

[McC21] Duncan McCoy, *Null-homologous twisting and the algebraic genus*, 2019–20 MATRIX annals, MATRIX Book Ser., vol. 4, Springer, Cham, [2021] ©2021, pp. 147–165. MR4294766 Cited in Problem 1.46.

[McD91] Dusa McDuff, *Symplectic manifolds with contact type boundaries*, Invent. Math. **103** (1991), no. 3, 651–671, DOI 10.1007/BF01239530. MR1091622 Cited in Problem 4.110.

[McD92] Dusa McDuff, *Singularities of J-holomorphic curves in almost complex* 4*-manifolds*, J. Geom. Anal. **2** (1992), no. 3, 249–266, DOI 10.1007/BF02921295. MR1164604 Cited in Problem 4.102.

[McD22] Clayton McDonald. Surface slices and homology spheres, 2022. `arXiv:2202.02696`. Cited in Problem 4.25.

[McM13] Curtis T. McMullen, *Entropy on Riemann surfaces and the Jacobians of finite covers*, Comment. Math. Helv. **88** (2013), no. 4, 953–964, DOI 10.4171/CMH/308. MR3134416 Cited in Problem 3.7.

[McM14] Curt McMullen. Dynamics, geometry, and moduli spaces seminar, 2014: Renormalized volume. *Seminar notes*, 2014. Cited in Problem 2.18.

[McR12] D. B. McReynolds, *The congruence subgroup problem for pure braid groups: Thurston's proof*, New York J. Math. **18** (2012), 925–942. MR3007206 Cited in Problem 2.2.

[ME23] Mike Miller Eismeier. Equivariant instanton homology, 2023. To appear in Memoirs of the AMS. `arXiv:1907.01091`. Cited in Problem 3.59.

[Mec01] Mattia Mecchia, *Hyperbolic 2-fold branched coverings*, Rend. Istit. Mat. Univ. Trieste **32** (2001), no. suppl. 1, 165–180 (2002). Dedicated to the memory of Marco Reni. MR1893397 Cited in Problem 1.16.

[Mei15] Jeffrey Meier, *Distinguishing topologically and smoothly doubly slice knots*, J. Topol. **8** (2015), no. 2, 315–351, DOI 10.1112/jtopol/jtu027. MR3356764 Cited in Problem 1.48.

[Mei18] Jeffrey Meier, *Trisections and spun four-manifolds*, Math. Res. Lett. **25** (2018), no. 5, 1497–1524, DOI 10.4310/MRL.2018.v25.n5.a7. MR3917737 Cited in Problem 4.113.

[Mel77] Paul Michael Melvin, *BLOWING UP AND DOWN IN 4-MANIFOLDS*, ProQuest LLC, Ann Arbor, MI, 1977. Thesis (Ph.D.)–University of California, Berkeley. MR2627246 Cited in Problems 4.9 and 4.10.

[Mel79] Paul Melvin, *Bordism of diffeomorphisms*, Topology **18** (1979), no. 2, 173–175, DOI 10.1016/0040-9383(79)90034-X. MR544157 Cited in Problems 2.45 and 3.74.

[Mel22] Anton Mellit, *Homology of torus knots*, Geom. Topol. **26** (2022), no. 1, 47–70, DOI 10.2140/gt.2022.26.47. MR4404874 Cited in Problems 1.31 and 1.33.

[Men84] W. Menasco, *Closed incompressible surfaces in alternating knot and link complements*, Topology **23** (1984), no. 1, 37–44, DOI 10.1016/0040-9383(84)90023-5. MR721450 Cited in Problem 1.17.

[Mes92] Geoffrey Mess, *The Torelli groups for genus* 2 *and* 3 *surfaces*, Topology **31** (1992), no. 4, 775–790, DOI 10.1016/0040-9383(92)90008-6. MR1191379 Cited in Problem 2.5.

[Mey73] Werner Meyer, *Die Signatur von Flächenbündeln* (German), Math. Ann. **201** (1973), 239–264, DOI 10.1007/BF01427946. MR331382 Cited in Problem 4.97.

[Mey81] Mark D. Meyerson, *Balancing acts*, Topology Proc. **6** (1981), no. 1, 59–75 (1982). MR650481 Cited in Problem 2.14.

[Mey87] Robert Meyerhoff, *A lower bound for the volume of hyperbolic* 3*-manifolds*, Canad. J. Math. **39** (1987), no. 5, 1038–1056, DOI 10.4153/CJM-1987-053-6. MR918586 Cited in Problem 3.23.

[Mij03] Aleksandar Mijatović, *Simplifying triangulations of* S^3, Pacific J. Math. **208** (2003), no. 2, 291–324, DOI 10.2140/pjm.2003.208.291. MR1971667 Cited in Problem 3.28.

[Mij05a] Aleksandar Mijatović, *Simplical structures of knot complements*, Math. Res. Lett. **12** (2005), no. 5-6, 843–856, DOI 10.4310/MRL.2005.v12.n6.a6. MR2189244 Cited in Problem 3.28.

[Mij05b] Aleksandar Mijatović, *Triangulations of fibre-free Haken 3-manifolds*, Pacific J. Math. **219** (2005), no. 1, 139–186, DOI 10.2140/pjm.2005.219.139. MR2174223 Cited in Problem 3.28.

[Mil58] John Milnor, *On the existence of a connection with curvature zero*, Comment. Math. Helv. **32** (1958), 215–223, DOI 10.1007/BF02564579. MR95518 Cited in Problem 3.15.

[Mil65] John Milnor, *Lectures on the h-cobordism theorem*, Princeton University Press, Princeton, NJ, 1965. Notes by L. Siebenmann and J. Sondow. MR190942 Cited in Problems 4.46 and 4.56.

[Mil68a] J. Milnor, *A note on curvature and fundamental group*, J. Differential Geometry **2** (1968), 1–7. MR232311 Cited in Problem 5.15.

[Mil68b] John Milnor, *Singular points of complex hypersurfaces*, Annals of Mathematics Studies, No. 61, Princeton University Press, Princeton, NJ; University of Tokyo Press, Tokyo, 1968. MR239612 Cited in Problem 1.87.

[Mil82] John Milnor, *Hyperbolic geometry: the first 150 years*, Bull. Amer. Math. Soc. (N.S.) **6** (1982), no. 1, 9–24, DOI 10.1090/S0273-0979-1982-14958-8. MR634431 Cited in Problem 3.2.

[Mil09] Peter Milley, *Minimum volume hyperbolic 3-manifolds*, J. Topol. **2** (2009), no. 1, 181–192, DOI 10.1112/jtopol/jtp006. MR2499442 Cited in Problem 3.1.

[Mil18] Allison N. Miller. A note on the topological sliceness of some 2-bridge knots. *Math. Proc. Cambridge Philos. Soc.*, 164(1):185–191, 2018. doi:10.1017/S0305004117000172. Cited in Problem 1.46.

[Mil21] Maggie Miller, *Extending fibrations of knot complements to ribbon disk complements*, Geom. Topol. **25** (2021), no. 3, 1479–1550, DOI 10.2140/gt.2021.25.1479. MR4268166 Cited in Problem 1.55.

[Min06] Yair N. Minsky, *Curve complexes, surfaces and 3-manifolds*, International Congress of Mathematicians. Vol. II, Eur. Math. Soc., Zürich, 2006, pp. 1001–1033. MR2275633 Cited in Problem 2.10.

[Mis21] Filip Misev, *On families of fibred knots with equal Seifert forms*, Comm. Anal. Geom. **29** (2021), no. 2, 465–482, DOI 10.4310/CAG.2021.v29.n2.a6. MR4250328 Cited in Problem 1.21.

[Miy94] Katura Miyazaki, *Nonsimple, ribbon fibered knots*, Trans. Amer. Math. Soc. **341** (1994), no. 1, 1–44, DOI 10.2307/2154613. MR1176509 Cited in Problems 1.39 and 1.50.

[Miy18] Katura Miyazaki, *A note on genera of band sums that are fibered*, J. Knot Theory Ramifications **27** (2018), no. 12, 1871002, 3, DOI 10.1142/S0218216518710025. MR3876350 Cited in Problem 1.53.

[Miy23] Jin Miyazawa. A gauge theoretic invariant of embedded surfaces in 4-manifolds and exotic P^2-knots, 2023. arXiv:2312.02041. Cited in Problems 4.1, 4.29, and 4.117.

[MK14] V. D. Mazurov and E. I. Khukhro (eds.), *The Kourovka notebook*, Eighteenth edition, Russian Academy of Sciences Siberian Division, Institute of Mathematics, Novosibirsk, 2014. Unsolved problems in group theory. MR3408705 Cited in Problem 3.39.

[MM86] Darryl McCullough and Andy Miller, *The genus* 2 *Torelli group is not finitely generated*, Topology Appl. **22** (1986), no. 1, 43–49, DOI 10.1016/0166-8641(86)90076-3. MR831180 Cited in Problem 4.81.

[MM99] Howard A. Masur and Yair N. Minsky, *Geometry of the complex of curves. I. Hyperbolicity*, Invent. Math. **138** (1999), no. 1, 103–149, DOI 10.1007/s002220050343. MR1714338 Cited in Problem 2.10.

[MM00] H. A. Masur and Y. N. Minsky, *Geometry of the complex of curves. II. Hierarchical structure*, Geom. Funct. Anal. **10** (2000), no. 4, 902–974, DOI 10.1007/PL00001643. MR1791145 Cited in Problem 2.38.

[MM01] Hitoshi Murakami and Jun Murakami, *The colored Jones polynomials and the simplicial volume of a knot*, Acta Math. **186** (2001), no. 1, 85–104, DOI 10.1007/BF02392716. MR1828373 Cited in Problems 1.24, 1.27, and 1.29.

[MM04] Howard A. Masur and Yair N. Minsky, *Quasiconvexity in the curve complex*, In the tradition of Ahlfors and Bers, III, Contemp. Math., vol. 355, Amer. Math. Soc., Providence, RI, 2004, pp. 309–320, DOI 10.1090/conm/355/06460. MR2145071 Cited in Problem 2.10.

[MM12] T. H. Marshall and G. J. Martin, *Minimal co-volume hyperbolic lattices, II: Simple torsion in a Kleinian group*, Ann. of Math. (2) **176** (2012), no. 1, 261–301, DOI 10.4007/annals.2012.176.1.4. MR2925384 Cited in Problem 3.1.

[MM25a] Marco Marengon and Stefan Mihajlović, *Unknotting number 21 knots are slice in* $K3$, Math. Res. Lett. **32** (2025), no. 3, 939–955, DOI 10.4310/mrl.250731115553. MR4945106 Cited in Problem 4.42.

[MM25b] Jeffrey Meier and Allison N. Miller. Slice disks modulo local knotting, 2025. `arXiv:2503.09870`. Cited in Problem 1.59.

[MMO$^+$02] Hitoshi Murakami, Jun Murakami, Miyuki Okamoto, Toshie Takata, and Yoshiyuki Yokota, *Kashaev's conjecture and the Chern-Simons invariants of knots and links*, Experiment. Math. **11** (2002), no. 3, 427–435. MR1959752 Cited in Problem 1.27.

[MMP24] Ciprian Manolescu, Marco Marengon, and Lisa Piccirillo. Relative genus bounds in indefinite four-manifolds. *Math. Ann.*, 390(1):1481–1506, 2024. `doi:10.1007/s00208-023-02787-4`. Cited in Problem 4.42.

[MMRS24] Marco Marengon, Allison N. Miller, Arunima Ray, and András I. Stipsicz, *A note on surfaces in* $\mathbb{CP}^2$ *and* $\mathbb{CP}^2\#\mathbb{CP}^2$, Proc. Amer. Math. Soc. Ser. B **11** (2024), 187–199, DOI 10.1090/bproc/218. MR4762681 Cited in Problem 4.44.

[MMSW23] Ciprian Manolescu, Marco Marengon, Sucharit Sarkar, and Michael Willis, *A generalization of Rasmussen's invariant, with applications to surfaces in some four-manifolds*, Duke Math. J. **172** (2023), no. 2, 231–311, DOI 10.1215/00127094-2022-0039. MR4541332 Cited in Problem 4.43.

[MN92] Robert Meyerhoff and Walter D. Neumann, *An asymptotic formula for the eta invariants of hyperbolic 3-manifolds*, Comment. Math. Helv. **67** (1992), no. 1, 28–46, DOI 10.1007/BF02566487. MR1144612 Cited in Problem 3.63.

[MN22] Ciprian Manolescu and Ikshu Neithalath, *Skein lasagna modules for 2-handlebodies*, J. Reine Angew. Math. **788** (2022), 37–76, DOI 10.1515/crelle-2022-0021. MR4445546 Cited in Problem 4.66.

[MN24] Maggie Miller and Patrick Naylor, *Trisections of nonorientable 4-manifolds*, Michigan Math. J. **74** (2024), no. 2, 403–447, DOI 10.1307/mmj/20216127. MR4739421 Cited in Problem 4.113.

[MNU24] Vicente Monreal, Jaime Negrete, and Giancarlo Urzúa. Classification of horikawa surfaces with t-singularities, 2024. `arXiv:2410.02943`. Cited in Problem 4.94.

[MO99] Hiroshi Maehara and Ai Oshiro, *On knotted necklaces of pearls*, European J. Combin. **20** (1999), no. 5, 411–420, DOI 10.1006/eujc.1998.0279. MR1702378 Cited in Problem 1.73.

[Moi77a] Edwin E. Moise, *Geometric topology in dimensions* 2 *and* 3, Graduate Texts in Mathematics, Vol. 47, Springer-Verlag, New York-Heidelberg, 1977. MR488059 Cited in Problems 4.1 and 4.119.

[Moi77b] Boris Moishezon, *Complex surfaces and connected sums of complex projective planes*, Lecture Notes in Mathematics, Vol. 603, Springer-Verlag, Berlin-New York, 1977. With an appendix by R. Livne. MR491730 Cited in Problem 2.21.

[Moi94] B. Moishezon, *The arithmetic of braids and a statement of Chisini*, Geometric topology (Haifa, 1992), Contemp. Math., vol. 164, Amer. Math. Soc., Providence, RI, 1994, pp. 151–175, DOI 10.1090/conm/164/01591. MR1282761 Cited in Problem 4.102.

[Mon74] José M. Montesinos, *A representation of closed orientable 3-manifolds as 3-fold branched coverings of* S^3, Bull. Amer. Math. Soc. **80** (1974), 845–846, DOI 10.1090/S0002-9904-1974-13535-4. MR358784 Cited in Problem 5.6.

[Mon85] José María Montesinos, *Lectures on 3-fold simple coverings and 3-manifolds*, Combinatorial methods in topology and algebraic geometry (Rochester, N.Y., 1982), Contemp. Math., vol. 44, Amer. Math. Soc., Providence, RI, 1985, pp. 157–177, DOI 10.1090/conm/044/813111. MR813111 Cited in Problem 1.16.

[Mon12] Naoyuki Monden. On minimal number of singular fibers in a genus-2 Lefschetz fibration. *Tokyo J. Math.*, 35(2):483–490, 2012. `doi:10.3836/tjm/1358951332`. Cited in Problem 2.21.

[Mon23] Ian Montague. Non-smoothable $\mathbb{Z}/p$-actions on nuclei, 2023. `arXiv:2401.00244`. Cited in Problem 4.89.

[Mor87] Shigeyuki Morita. Characteristic classes of surface bundles. *Invent. Math.*, 90(3):551–577, 1987. `doi:10.1007/BF01389178`. Cited in Problem 2.32.

[Mor89] Shigeyuki Morita, *Families of Jacobian manifolds and characteristic classes of surface bundles. II*, Math. Proc. Cambridge Philos. Soc. **105** (1989), no. 1, 79–101, DOI 10.1017/S0305004100001389. MR966142 Cited in Problem 2.25.

[MOR+24] Gordana Matić, Ferit Oztürk, Javier Reyes, András I. Stipsicz, and Giancarlo Urzúa. An exotic $5\mathbb{R}P^2$ in the 4-sphere, 2024. `arXiv:2312.03617`. Cited in Problem 4.29.

[Mos68] G. D. Mostow, *Quasi-conformal mappings in n-space and the rigidity of hyperbolic space forms*, Inst. Hautes Études Sci. Publ. Math. **34** (1968), 53–104. MR236383 Cited in Problem 3.0.

[Mos71] Louise Moser, *Elementary surgery along a torus knot*, Pacific J. Math. **38** (1971), 737–745. MR383406 Cited in Problem 1.9.

[Mos92] Lee Mosher, *Dynamical systems and the homology norm of a 3-manifold. I. Efficient intersection of surfaces and flows*, Duke Math. J. **65** (1992), no. 3, 449–500, DOI 10.1215/S0012-7094-92-06518-5. MR1154179 Cited in Problem 3.21.

[MOS09] Ciprian Manolescu, Peter Ozsváth, and Sucharit Sarkar, *A combinatorial description of knot Floer homology*, Ann. of Math. (2) **169** (2009), no. 2, 633–660, DOI 10.4007/annals.2009.169.633. MR2480614 Cited in Problem 3.60.

[Mot88] Kimihiko Motegi, *Haken manifolds and representations of their fundamental groups in* $\mathrm{SL}(2, \mathbf{C})$, Topology Appl. **29** (1988), no. 3, 207–212, DOI 10.1016/0166-8641(88)90019-3. MR953952 Cited in Problem 3.52.

[MOT25] Ciprian Manolescu, Peter S. Ozsváth, and Dylan P. Thurston, *Grid diagrams and Heegaard Floer invariants*, Ann. of Math. (2) **201** (2025), no. 1, 1–78, DOI 10.4007/annals.2025.201.1.1. MR4848040 Cited in Problem 3.60.

[MOY97] Tomasz Mrowka, Peter Ozsváth, and Baozhen Yu, *Seiberg-Witten monopoles on Seifert fibered spaces*, Comm. Anal. Geom. **5** (1997), no. 4, 685–791, DOI 10.4310/CAG.1997.v5.n4.a3. MR1611061 Cited in Problem 3.56.

[MP89] Kunio Murasugi and Józef H. Przytycki, *The skein polynomial of a planar star product of two links*, Math. Proc. Cambridge Philos. Soc. **106** (1989), no. 2, 273–276, DOI 10.1017/S0305004100078099. MR1002540 Cited in Problem 1.5.

[MP18] Allison N. Miller and Lisa Piccirillo, *Knot traces and concordance*, J. Topol. **11** (2018), no. 1, 201–220, DOI 10.1112/topo.12054. MR3784230 Cited in Problem 1.58.

[MP19] Allison N. Miller and Mark Powell, *Stabilization distance between surfaces*, Enseign. Math. **65** (2019), no. 3-4, 397–440, DOI 10.4171/lem/65-3/4-4. MR4113047 Cited in Problem 1.59.

[MP23] Ciprian Manolescu and Lisa Piccirillo, *From zero surgeries to candidates for exotic definite 4-manifolds*, J. Lond. Math. Soc. (2) **108** (2023), no. 5, 2001–2036, DOI 10.1112/jlms.12800. MR4668522 Cited in Problems 4.1 and 4.3.

[MP25] Daniel Minahan and Andrew Putman. The second rational homology of the torelli group, 2025. `arXiv:2504.00211`. Cited in Problem 2.5.

[MPS+18] Sujoy Mukherjee, Józef H. Przytycki, Marithania Silvero, Xiao Wang, and Seung Yeop Yang. Search for torsion in Khovanov homology. *Exp. Math.*, 27(4):488–497, 2018. `doi:10.1080/10586458.2017.1320242`. Cited in Problem 1.31.

[MR90a] Robert Meyerhoff and Daniel Ruberman, *Mutation and the η-invariant*, J. Differential Geom. **31** (1990), no. 1, 101–130. MR1030667 Cited in Problem 3.63.

[MR90b] Robert Meyerhoff and Daniel Ruberman, *Cutting and pasting and the η-invariant*, Duke Math. J. **61** (1990), no. 3, 747–761, DOI 10.1215/S0012-7094-90-06127-7. MR1084457 Cited in Problem 3.63.

[MR92] William Menasco and Alan W. Reid, *Totally geodesic surfaces in hyperbolic link complements*, Topology '90 (Columbus, OH, 1990), Ohio State Univ. Math. Res. Inst. Publ., vol. 1, de Gruyter, Berlin, 1992, pp. 215–226. MR1184413 Cited in Problem 3.14.

[MR02] Mattia Mecchia and Marco Reni, *Hyperbolic 2-fold branched coverings of links and their quotients*, Pacific J. Math. **202** (2002), no. 2, 429–447, DOI 10.2140/pjm.2002.202.429. MR1888173 Cited in Problem 1.17.

[MR03] Colin Maclachlan and Alan W. Reid. *The arithmetic of hyperbolic 3-manifolds*, volume 219 of *Graduate Texts in Mathematics*. Springer-Verlag, New York, 2003. doi:10.1007/978-1-4757-6720-9. Cited in Problems 2.19 and 3.10.

[MR23] Kathryn Mann and Kasra Rafi, *Large-scale geometry of big mapping class groups*, Geom. Topol. **27** (2023), no. 6, 2237–2296, DOI 10.2140/gt.2023.27.2237. MR4634747 Cited in Problem 2.46.

[MRS11] Tomasz Mrowka, Daniel Ruberman, and Nikolai Saveliev, *Seiberg-Witten equations, end-periodic Dirac operators, and a lift of Rohlin's invariant*, J. Differential Geom. **88** (2011), no. 2, 333–377. MR2838269 Cited in Problems 4.22 and 4.65.

[MRS20] Bruno Martelli, Stefano Riolo, and Leone Slavich, *Compact hyperbolic manifolds without spin structures*, Geom. Topol. **24** (2020), no. 5, 2647–2674, DOI 10.2140/gt.2020.24.2647. MR4194300 Cited in Problem 4.125.

[MRT14] Justin Malestein, Igor Rivin, and Louis Theran. Topological designs. *Geom. Dedicata*, 168:221–233, 2014. doi:10.1007/s10711-012-9827-9. Cited in Problem 2.12.

[MS78] T. Matumoto and L. Siebenmann, *The topological s-cobordism theorem fails in dimension* 4 *or* 5, Math. Proc. Cambridge Philos. Soc. **84** (1978), no. 1, 85–87, DOI 10.1017/S0305004100054918. MR489573 Cited in Problem 4.56.

[MS88] John W. Morgan and Peter B. Shalen, *Degenerations of hyperbolic structures. III. Actions of* 3*-manifold groups on trees and Thurston's compactness theorem*, Ann. of Math. (2) **127** (1988), no. 3, 457–519, DOI 10.2307/2007003. MR942518 Cited in Problem 3.20.

[MS97] John W. Morgan and Zoltán Szabó, *Homotopy $K3$ surfaces and mod* 2 *Seiberg-Witten invariants*, Math. Res. Lett. **4** (1997), no. 1, 17–21, DOI 10.4310/MRL.1997.v4.n1.a2. MR1432806 Cited in Problems 4.70 and 4.92.

[MS13] Howard Masur and Saul Schleimer, *The geometry of the disk complex*, J. Amer. Math. Soc. **26** (2013), no. 1, 1–62, DOI 10.1090/S0894-0347-2012-00742-5. MR2983005 Cited in Problem 2.10.

[MS15] Allison H. Moore and Laura Starkston, *Genus-two mutant knots with the same dimension in knot Floer and Khovanov homologies*, Algebr. Geom. Topol. **15** (2015), no. 1, 43–63, DOI 10.2140/agt.2015.15.43. MR3325731 Cited in Problems 1.21 and 3.61.

[MS17a] Dusa McDuff and Dietmar Salamon, *Introduction to symplectic topology*, 3rd ed., Oxford Graduate Texts in Mathematics, Oxford University Press, Oxford, 2017, DOI 10.1093/oso/9780198794899.001.0001. MR3674984 Cited in Problem 2.32.

[MS17b] Hugh Morton and Peter Samuelson. The HOMFLYPT skein algebra of the torus and the elliptic Hall algebra. *Duke Math. J.*, 166(5):801–854, 2017. doi:10.1215/00127094-3718881. Cited in Problem 1.36.

[MS21] Julien Marché and Ramanujan Santharoubane, *Asymptotics of quantum representations of surface groups* (English, with English and French summaries), Ann. Sci. Éc. Norm. Supér. (4) **54** (2021), no. 5, 1275–1296, DOI 10.24033/asens.2481. MR4363245 Cited in Problem 2.49.

[MSZ16] Jeffrey Meier, Trent Schirmer, and Alexander Zupan, *Classification of trisections and the generalized property R conjecture*, Proc. Amer. Math. Soc. **144** (2016), no. 11, 4983–4997, DOI 10.1090/proc/13105. MR3544545 Cited in Problems 4.112 and 4.113.

[MT88] B. Moishezon and M. Teicher, *Braid group technique in complex geometry. I. Line arrangements in* $\mathbf{C}P^2$, Braids (Santa Cruz, CA, 1986), Contemp.

Math., vol. 78, Amer. Math. Soc., Providence, RI, 1988, pp. 425–555, DOI 10.1090/conm/078/975093. MR975093 Cited in Problems 4.101 and 4.102.

[MT91] Boris Moishezon and Mina Teicher, *Braid group technique in complex geometry. II. From arrangements of lines and conics to cuspidal curves*, Algebraic geometry (Chicago, IL, 1989), Lecture Notes in Math., vol. 1479, Springer, Berlin, 1991, pp. 131–180, DOI 10.1007/BFb0086269. MR1181212 Cited in Problem 4.101.

[MT92] William W. Menasco and Morwen B. Thistlethwaite, *Surfaces with boundary in alternating knot exteriors*, J. Reine Angew. Math. **426** (1992), 47–65, DOI 10.1515/crll.1992.426.47. MR1155746 Cited in Problem 1.11.

[MT93] William Menasco and Morwen Thistlethwaite, *The classification of alternating links*, Ann. of Math. (2) **138** (1993), no. 1, 113–171, DOI 10.2307/2946636. MR1230928 Cited in Problems 1.23 and 1.82.

[MT16] Johanna Mangahas and Samuel J. Taylor, *Convex cocompactness in mapping class groups via quasiconvexity in right-angled Artin groups*, Proc. Lond. Math. Soc. (3) **112** (2016), no. 5, 855–881, DOI 10.1112/plms/pdw009. MR3502022 Cited in Problem 2.28.

[MT17] Kimihiko Motegi and Toshie Takata, *The slope conjecture for graph knots*, Math. Proc. Cambridge Philos. Soc. **162** (2017), no. 3, 383–392, DOI 10.1017/S0305004116000566. MR3628196 Cited in Problem 1.26.

[MT18] Thomas E. Mark and Bülent Tosun. Naturality of Heegaard Floer invariants under positive rational contact surgery. *J. Differential Geom.*, 110(2):281–344, 2018. `doi:10.4310/jdg/1538791245`. Cited in Problem 3.41.

[MT21] Agnese Mantione and Rafael Torres, *Geography of 4-manifolds with positive scalar curvature*, Expo. Math. **39** (2021), no. 4, 566–582, DOI 10.1016/j.exmath.2021.05.003. MR4340763 Cited in Problem 4.128.

[MT22] Thomas E. Mark and Bülent Tosun, *On contact type hypersurfaces in 4-space*, Invent. Math. **228** (2022), no. 1, 493–534, DOI 10.1007/s00222-021-01083-9. MR4392461 Cited in Problems 4.58 and 4.107.

[MT24] Vladimir Marković and Ognjen Tošić, *The second variation of the Hodge norm and higher Prym representations*, J. Topol. **17** (2024), no. 1, Paper No. e12322, 23, DOI 10.1112/topo.12322. MR4701642 Cited in Problem 2.1.

[MT25] Hitoshi Murakami and Anh T. Tran, *The colored Jones polynomial of a cable of the figure-eight knot*, J. Knot Theory Ramifications **34** (2025), no. 3, Paper No. 2340019, 24, DOI 10.1142/S0218216523400199. MR4886163 Cited in Problem 1.27.

[MTV07] Marco Mackaay, Paul Turner, and Pedro Vaz, *A remark on Rasmussen's invariant of knots*, J. Knot Theory Ramifications **16** (2007), no. 3, 333–344, DOI 10.1142/S0218216507005312. MR2320159 Cited in Problem 4.43.

[Mum79] D. Mumford. An algebraic surface with K ample, $(K^2) = 9$, $p_g = q = 0$. *Amer. J. Math.*, 101(1):233–244, 1979. `doi:10.2307/2373947`. Cited in Problem 4.91.

[Mun60] James Munkres, *Obstructions to the smoothing of piecewise-differentiable homeomorphisms*, Ann. of Math. (2) **72** (1960), 521–554, DOI 10.2307/1970228. MR121804 Cited in Problem 4.1.

[Mur58] Kunio Murasugi, *On the genus of the alternating knot. I, II*, J. Math. Soc. Japan **10** (1958), 94–105, 235–248, DOI 10.2969/jmsj/01010094. MR99664 Cited in Problem 1.8.

[Mur71] Kunio Murasugi, *On periodic knots*, Comment. Math. Helv. **46** (1971), 162–174, DOI 10.1007/BF02566836. MR292060 Cited in Problems 1.43 and 1.49.

[Mur85] Kunio Murasugi, *On the Alexander polynomial of alternating algebraic knots*, J. Austral. Math. Soc. Ser. A **39** (1985), no. 3, 317–333. MR802722 Cited in Problem 1.7.

[Mur87] Kunio Murasugi, *Jones polynomials and classical conjectures in knot theory*, Topology **26** (1987), no. 2, 187–194, DOI 10.1016/0040-9383(87)90058-9. MR895570 Cited in Problems 1.1 and 1.23.

[Mur07] Hitoshi Murakami, *Various generalizations of the volume conjecture*, The interaction of analysis and geometry, Contemp. Math., vol. 424, Amer. Math. Soc., Providence, RI, 2007, pp. 165–186, DOI 10.1090/conm/424/08100. MR2316336 Cited in Problem 1.27.

[MV08] Yoshihiko Mitsumatsu and Elmar Vogt, *Foliations and compact leaves on 4-manifolds. I. Realization and self-intersection of compact leaves*, Groups of diffeomorphisms, Adv. Stud. Pure Math., vol. 52, Math. Soc. Japan, Tokyo, 2008, pp. 415–442, DOI 10.2969/aspm/05210415. MR2509719 Cited in Problem 4.123.

[MW50] Saunders MacLane and J. H. C. Whitehead, *On the 3-type of a complex*, Proc. Nat. Acad. Sci. U.S.A. **36** (1950), 41–48, DOI 10.1073/pnas.36.1.41. MR33519 Cited in Problem 4.36.

[MW86] José M. Montesinos and Wilbur Whitten, *Constructions of two-fold branched covering spaces*, Pacific J. Math. **125** (1986), no. 2, 415–446. MR863536 Cited in Problem 1.17.

[MW95] Mario J. Micallef and Brian White, *The structure of branch points in minimal surfaces and in pseudoholomorphic curves*, Ann. of Math. (2) **141** (1995), no. 1, 35–85, DOI 10.2307/2118627. MR1314031 Cited in Problem 4.102.

[MW07] Ib Madsen and Michael Weiss, *The stable moduli space of Riemann surfaces: Mumford's conjecture*, Ann. of Math. (2) **165** (2007), no. 3, 843–941, DOI 10.4007/annals.2007.165.843. MR2335797 Cited in Problem 4.78.

[MW14] Rafe Mazzeo and Edward Witten, *The Nahm pole boundary condition*, The influence of Solomon Lefschetz in geometry and topology, Contemp. Math., vol. 621, Amer. Math. Soc., Providence, RI, 2014, pp. 171–226, DOI 10.1090/conm/621/12422. MR3289327 Cited in Problem 1.32.

[MW15] Victor Mikhaylov and Edward Witten, *Branes and supergroups*, Comm. Math. Phys. **340** (2015), no. 2, 699–832, DOI 10.1007/s00220-015-2449-y. MR3397031 Cited in Problem 1.37.

[MW17] Françoise Michel and Claude Weber, *Higher-dimensional knots according to Michel Kervaire*, EMS Series of Lectures in Mathematics, European Mathematical Society (EMS), Zürich, 2017, DOI 10.4171/180. MR3699860 Cited in Problem 4.35.

[MW20] Rafe Mazzeo and Edward Witten, *The KW equations and the Nahm pole boundary condition with knots*, Comm. Anal. Geom. **28** (2020), no. 4, 871–942, DOI 10.4310/CAG.2020.v28.n4.a4. MR4165312 Cited in Problem 1.32.

[MW23] Kathryn Mann and Maxime Wolff, *Reconstructing maps out of groups* (English, with English and French summaries), Ann. Sci. Éc. Norm. Supér. (4) **56** (2023), no. 4, 1135–1154, DOI 10.24033/asens.2551. MR4650155 Cited in Problem 2.34.

[MW24] Jacob Migdail and Stephan Wehrli. Functoriality of odd and generalized Khovanov homology in $\mathbb{R}^3 \times I$, 2024. `arXiv:2206.14710`. Cited in Problem 1.37.

[MW25] Ciprian Manolescu and Michael Willis, *A Rasmussen invariant for links in* $\mathbb{RP}^3$, Trans. Amer. Math. Soc. Ser. B **12** (2025), 789–830, DOI 10.1090/btran/221. MR4904031 Cited in Problem 1.49.

[MWW22] Scott Morrison, Kevin Walker, and Paul Wedrich, *Invariants of 4-manifolds from Khovanov-Rozansky link homology*, Geom. Topol. **26** (2022), no. 8, 3367–3420, DOI 10.2140/gt.2022.26.3367. MR4562565 Cited in Problems 1.32 and 4.66.

[MWW23] Ciprian Manolescu, Kevin Walker, and Paul Wedrich, *Skein lasagna modules and handle decompositions*, Adv. Math. **425** (2023), Paper No. 109071, 40, DOI 10.1016/j.aim.2023.109071. MR4589588 Cited in Problem 4.66.

[MY72] Deane Montgomery and C. T. Yang, *Differentiable pseudo-free circle actions on homotopy seven spheres*, Proceedings of the Second Conference on Compact Transformation Groups (Univ. Massachusetts, Amherst, Mass., 1971), Lecture Notes in Math., Vol. 298, Springer, Berlin-New York, 1972, pp. 41–101. MR362383 Cited in Problem 5.7.

[MY00] Hitoshi Murakami and Akira Yasuhara, *Four-genus and four-dimensional clasp number of a knot*, Proc. Amer. Math. Soc. **128** (2000), no. 12, 3693–3699, DOI 10.1090/S0002-9939-00-05461-7. MR1690998 Cited in Problem 1.47.

[MZ04] Mattia Mecchia and Bruno Zimmermann, *The number of knots and links with the same 2-fold branched covering*, Q. J. Math. **55** (2004), no. 1, 69–76, DOI 10.1093/qjmath/55.1.69. MR2043014 Cited in Problems 1.16 and 1.17.

[MZ17a] Jeffrey Meier and Alexander Zupan, *Bridge trisections of knotted surfaces in* S^4, Trans. Amer. Math. Soc. **369** (2017), no. 10, 7343–7386, DOI 10.1090/tran/6934. MR3683111 Cited in Problems 1.18, 2.8, 4.111, 4.112, and 4.117.

[MZ17b] Jeffrey Meier and Alexander Zupan, *Genus-two trisections are standard*, Geom. Topol. **21** (2017), no. 3, 1583–1630, DOI 10.2140/gt.2017.21.1583. MR3650079 Cited in Problems 4.111, 4.113, and 4.117.

[MZ18] Jeffrey Meier and Alexander Zupan, *Characterizing Dehn surgeries on links via trisections*, Proc. Natl. Acad. Sci. USA **115** (2018), no. 43, 10887–10893, DOI 10.1073/pnas.1717187115. MR3871792 Cited in Problems 4.112 and 4.115.

[MZ22] Jeffrey Meier and Alexander Zupan. Generalized square knots and homotopy 4-spheres. *J. Differential Geom.*, 122(1):69–129, 2022. `doi:10.4310/jdg/1668186788`. Cited in Problems 1.10, 1.50, and 4.112.

[MZ23] Maggie Miller and Alexander Zupan, *Equivalent characterizations of handle-ribbon knots*, Comm. Anal. Geom. **31** (2023), no. 9, 2157–2193, DOI 10.4310/cag.2023.v31.n9.a1. MR4785585 Cited in Problem 1.50.

[MZ25] Jeffrey Meier and Alexander Zupan, *Knots bounding nonisotopic ribbon disks*, J. Topol. **18** (2025), no. 4, Paper No. e70047, 18, DOI 10.1112/topo.70047. MR4994819 Cited in Problems 1.10, 1.50, and 1.59.

[Nab95] Alexander Nabutovsky, *Non-recursive functions, knots "with thick ropes", and self-clenching "thick" hyperspheres*, Comm. Pure Appl. Math. **48** (1995), no. 4, 381–428, DOI 10.1002/cpa.3160480402. MR1324407 Cited in Problem 1.71.

[Nak09] Nobuhiro Nakamura, *Bauer-Furuta invariants under $\mathbb{Z}_2$-actions*, Math. Z. **262** (2009), no. 1, 219–233, DOI 10.1007/s00209-008-0370-1. MR2491607 Cited in Problem 4.86.

[Nan18] Ollie Nanyes. Wild and non-compact knot theory. WordPress, 2018. `https://wildandnoncompactknots.wordpress.com`. Cited in Problem 1.81.

[Neg22] Andrei Neguţ. Hecke correspondences for smooth moduli spaces of sheaves. *Publ. Math. Inst. Hautes Études Sci.*, 135:337–418, 2022. `doi:10.1007/s10240-022-00131-1`. Cited in Problem 1.36.

[Ném05] András Némethi, *On the Ozsváth-Szabó invariant of negative definite plumbed 3-manifolds*, Geom. Topol. **9** (2005), 991–1042, DOI 10.2140/gt.2005.9.991. MR2140997 Cited in Problems 3.57, 5.0, and 5.18.

[Ném08] András Némethi. Lattice cohomology of normal surface singularities. *Publ. Res. Inst. Math. Sci.*, 44(2):507–543, 2008. `doi:10.2977/prims/1210167336`. Cited in Problems 3.57, 5.0, and 5.18.

[Ném11] András Némethi, *Two exact sequences for lattice cohomology*, Noncommutative geometry and global analysis, Contemp. Math., vol. 546, Amer. Math. Soc., Providence, RI, 2011, pp. 249–269, DOI 10.1090/conm/546/10793. MR2815139 Cited in Problem 3.57.

[Neu80] Walter D. Neumann, *An invariant of plumbed homology spheres*, Topology Symposium, Siegen 1979 (Proc. Sympos., Univ. Siegen, Siegen, 1979), Lecture Notes in Math., vol. 788, Springer, Berlin, 1980, pp. 125–144. MR585657 Cited in Problem 3.72.

[Neu98] Walter D. Neumann, *Hilbert's 3rd problem and invariants of* 3*-manifolds*, The Epstein birthday schrift, Geom. Topol. Monogr., vol. 1, Geom. Topol. Publ., Coventry, 1998, pp. 383–411, DOI 10.2140/gtm.1998.1.383. MR1668316 Cited in Problem 3.2.

[Neu03] W. Neumann. Topology of hypersurface singularities. In *Erich Kähler Mathematische Werken*, pages 727–736. de Gruyter, 2003. See also `https://arxiv.org/abs/1706.04386`. Cited in Problem 1.87.

[New66] M. H. A. Newman, *The engulfing theorem for topological manifolds*, Ann. of Math. (2) **84** (1966), 555–571, DOI 10.2307/1970460. MR203708 Cited in Problem 4.1.

[New77] Sheldon E. Newhouse, *Quasi-elliptic periodic points in conservative dynamical systems*, Amer. J. Math. **99** (1977), no. 5, 1061–1087, DOI 10.2307/2374000. MR455049 Cited in Problem 3.47.

[Ng05] Lenhard Ng, *Knot and braid invariants from contact homology. I*, Geom. Topol. **9** (2005), 247–297, DOI 10.2140/gt.2005.9.247. MR2116316 Cited in Problems 4.106 and 5.16.

[Ni07] Yi Ni, *Knot Floer homology detects fibred knots*, Invent. Math. **170** (2007), no. 3, 577–608, DOI 10.1007/s00222-007-0075-9. MR2357503 Cited in Problems 1.9, 1.13, 1.19, and 1.21.

[Ni09] Yi Ni, *Dehn surgeries that yield fibred 3-manifolds*, Math. Ann. **344** (2009), no. 4, 863–876, DOI 10.1007/s00208-008-0331-3. MR2507628 Cited in Problem 1.13.

[Ni11] Yi Ni, *Thurston norm and cosmetic surgeries*, Low-dimensional and symplectic topology, Proc. Sympos. Pure Math., vol. 82, Amer. Math. Soc., Providence, RI, 2011, pp. 53–63, DOI 10.1090/pspum/082/2768653. MR2768653 Cited in Problem 1.12.

[Ni13] Yi Ni, *Nonseparating spheres and twisted Heegaard Floer homology*, Algebr. Geom. Topol. **13** (2013), no. 2, 1143–1159, DOI 10.2140/agt.2013.13.1143. MR3044606 Cited in Problems 1.12 and 1.13.

[Ni14] Yi Ni, *Some applications of Gabai's internal hierarchy*, Adv. Math. **250** (2014), 467–495, DOI 10.1016/j.aim.2013.10.001. MR3122174 Cited in Problem 1.20.

[Ni17a] Yi Ni, *Fintushel-Stern knot surgery in torus bundles*, J. Topol. **10** (2017), no. 1, 164–177, DOI 10.1112/topo.12002. MR3653065 Cited in Problem 4.99.

[Ni17b] Yi Ni, *Virtual Betti numbers and virtual symplecticity of 4-dimensional mapping tori, II*, Sci. China Math. **60** (2017), no. 9, 1591–1598, DOI 10.1007/s11425-016-9052-8. MR3689187 Cited in Problem 4.92.

[Ni22] Yi Ni, *The next-to-top term in knot Floer homology*, Quantum Topol. **13** (2022), no. 3, 579–591, DOI 10.4171/qt/174. MR4537314 Cited in Problem 3.55.

[Ni23a] Yi Ni. A note on knot Floer homology and fixed points of monodromy. *Peking Math. J.*, 6(2):635–643, 2023. doi:10.1007/s42543-022-00051-3. Cited in Problem 1.21.

[Ni23b] Yi Ni. Null-homotopic knots have property R. *Math. Proc. Cambridge Philos. Soc.*, 175(1):217–223, 2023. doi:10.1017/S0305004123000129. Cited in Problem 1.13.

[Ni24] Yi Ni, *Property G and the 4-genus*, Trans. Amer. Math. Soc. Ser. B **11** (2024), 120–143, DOI 10.1090/btran/153. MR4688532 Cited in Problem 1.13.

[Nic99] Liviu I. Nicolaescu, *Eta invariants of Dirac operators on circle bundles over Riemann surfaces and virtual dimensions of finite energy Seiberg-Witten moduli spaces*, Israel J. Math. **114** (1999), 61–123, DOI 10.1007/BF02785572. MR1738674 Cited in Problem 3.63.

[Nic00] Liviu I. Nicolaescu, *Finite energy Seiberg-Witten moduli spaces on 4-manifolds bounding Seifert fibrations*, Comm. Anal. Geom. **8** (2000), no. 5, 1027–1096, DOI 10.4310/CAG.2000.v8.n5.a3. MR1846125 Cited in Problem 3.63.

[Nie92] Mark J. Nielsen, *Triangles inscribed in simple closed curves*, Geom. Dedicata **43** (1992), no. 3, 291–297, DOI 10.1007/BF00151519. MR1181760 Cited in Problem 2.14.

[Nis98] Haruko Nishi. SU(n)-Chern-Simons invariants of Seifert fibered 3-manifolds. *Internat. J. Math.*, 9(3):295–330, 1998. doi:10.1142/S0129167X98000130. Cited in Problem 3.64.

[Nit22] Martin Nitsche. Computer proofs for Property (T), and SDP duality, 2022. arXiv:2009.05134. Cited in Problem 2.1.

[NNP23] Csaba Nagy, John Nicholson, and Mark Powell. Simple homotopy types of even dimensional manifolds, 2023. arXiv:arXiv:2312.00322. Cited in Problem 4.18.

[Nor69] R. A. Norman. Dehn's lemma for certain 4-manifolds. *Invent. Math.*, 7:143–147, 1969. doi:10.1007/BF01389797. Cited in Problem 4.42.

[Nou14] Mohamed Ait Nouh, *The minimal genus problem in* $\mathbb{CP}^2\#\mathbb{CP}^2$, Algebr. Geom. Topol. **14** (2014), no. 2, 671–686, DOI 10.2140/agt.2014.14.671. MR3159966 Cited in Problem 4.44.

[Nov64] S. P. Novikov, *Homotopically equivalent smooth manifolds. I* (Russian), Izv. Akad. Nauk SSSR Ser. Mat. **28** (1964), 365–474. MR162246 Cited in Problem 4.46.

[NR78] Walter D. Neumann and Frank Raymond, *Seifert manifolds, plumbing, μ-invariant and orientation reversing maps*, Algebraic and geometric topology (Proc. Sympos., Univ. California, Santa Barbara, Calif., 1977), Lecture Notes in Math., vol. 664, Springer, Berlin-New York, 1978, pp. 163–196. MR518415 Cited in Problem 3.72.

[NRS+20] Lenhard Ng, Dan Rutherford, Vivek Shende, Steven Sivek, and Eric Zaslow, *Augmentations are sheaves*, Geom. Topol. **24** (2020), no. 5, 2149–2286, DOI 10.2140/gt.2020.24.2149. MR4194293 Cited in Problem 5.16.

[NS03] Swatee Naik and Theodore Stanford, *A move on diagrams that generates S-equivalence of knots*, J. Knot Theory Ramifications **12** (2003), no. 5, 717–724, DOI 10.1142/S0218216503002639. MR1999640 Cited in Problem 1.6.

[NS12] Daniel Nash and András I. Stipsicz, *Gluck twist on a certain family of 2-knots*, Michigan Math. J. **61** (2012), no. 4, 703–713, DOI 10.1307/mmj/1353098509. MR3049286 Cited in Problem 4.9.

[NS22] Patrick Naylor and Hannah R. Schwartz, *Gluck twisting roll spun knots*, Algebr. Geom. Topol. **22** (2022), no. 2, 973–990, DOI 10.2140/agt.2022.22.973. MR4464469 Cited in Problem 4.9.

[NS24] Ali Naseri Sadr. A Table theorem for surfaces with odd Euler characteristic, 2024. `arXiv:2206.14710`. Cited in Problem 2.14.

[NST24] Yuta Nozaki, Kouki Sato, and Masaki Taniguchi, *Filtered instanton Floer homology and the homology cobordism group*, J. Eur. Math. Soc. (JEMS) **26** (2024), no. 12, 4699–4761, DOI 10.4171/jems/1371. MR4780492 Cited in Problems 3.64, 3.70, and 3.74.

[NW11] Klaus Niederkrüger and Chris Wendl, *Weak symplectic fillings and holomorphic curves* (English, with English and French summaries), Ann. Sci. Éc. Norm. Supér. (4) **44** (2011), no. 5, 801–853, DOI 10.24033/asens.2155. MR2931519 Cited in Problem 4.105.

[NW14] Yi Ni and Zhongtao Wu, *Heegaard Floer correction terms and rational genus bounds*, Adv. Math. **267** (2014), 360–380, DOI 10.1016/j.aim.2014.09.006. MR3269182 Cited in Problem 1.20.

[NW15] Yi Ni and Zhongtao Wu, *Cosmetic surgeries on knots in S^3*, J. Reine Angew. Math. **706** (2015), 1–17, DOI 10.1515/crelle-2013-0067. MR3393360 Cited in Problem 1.12.

[O'B83] Gary O'Brien, *The missing boundary problem for smooth manifolds of dimension greater than or equal to six*, Topology Appl. **16** (1983), no. 3, 303–324, DOI 10.1016/0166-8641(83)90027-5. MR722123 Cited in Problem 4.49.

[Oer84] Ulrich Oertel, *Closed incompressible surfaces in complements of star links*, Pacific J. Math. **111** (1984), no. 1, 209–230. MR732067 Cited in Problem 3.14.

[O'H91] Jun O'Hara, *Energy of a knot*, Topology **30** (1991), no. 2, 241–247, DOI 10.1016/0040-9383(91)90010-2. MR1098918 Cited in Problem 1.75.

[Oht16] Tomotada Ohtsuki, *On the asymptotic expansion of the Kashaev invariant of the* 5_2 *knot*, Quantum Topol. **7** (2016), no. 4, 669–735, DOI 10.4171/QT/83. MR3593566 Cited in Problem 1.27.

[Oht17] Tomotada Ohtsuki, *On the asymptotic expansions of the Kashaev invariant of hyperbolic knots with seven crossings*, Internat. J. Math. **28** (2017), no. 13, 1750096, 143, DOI 10.1142/S0129167X17500963. MR3737074 Cited in Problem 1.27.

[Ona10] Sinem Çelik Onaran, *On sections of genus two Lefschetz fibrations*, Pacific J. Math. **248** (2010), no. 1, 203–216, DOI 10.2140/pjm.2010.248.203. MR2734172 Cited in Problem 2.22.

[OO05] Hiroshi Ohta and Kaoru Ono. Simple singularities and symplectic fillings. *J. Differential Geom.*, 69(1):1–42, 2005. `doi:10.4310/jdg/1121540338`. Cited in Problems 3.49 and 4.102.

[OP25] Patrick Orson and Mark Powell, *Mapping class groups of simply connected 4-manifolds with boundary*, J. Differential Geom. **131** (2025), no. 1, 199–275, DOI 10.4310/jdg/1755544135. MR4947553 Cited in Problem 4.83.

[OPR21a] Patrick Orson, Mark Powell, and Arunima Ray, *The s-cobordism theorem, the sphere embedding theorem, and the Poincaré conjecture*, The disc embedding theorem, Oxford Univ. Press, Oxford, 2021, pp. 283–293. MR4519518 Cited in Problem 4.56.

[OPR21b] Patrick Orson, Mark Powell, and Arunima Ray, *Surgery theory and the classification of closed, simply connected 4-manifolds*, The disc embedding theorem, Oxford Univ. Press, Oxford, 2021, pp. 331–351. MR4519520 Cited in Problem 4.46.

[OR23] A. Oblomkov and L. Rozansky. Homfly-PT homology of Coxeter links. *Transform. Groups*, 28(3):1245–1275, 2023. `doi:10.1007/s00031-023-09816-1`. Cited in Problem 1.33.

[Ore21] S. Yu. Orevkov, *Multivariate signatures of iterated torus links* (Russian, with Russian summary), Funktsional. Anal. i Prilozhen. **55** (2021), no. 1, 73–92, DOI 10.1134/S001626632101007X; English transl., Funct. Anal. Appl. **55** (2021), no. 1, 59–74. MR4213102 Cited in Problem 1.8.

[ORS13] Peter S. Ozsváth, Jacob Rasmussen, and Zoltán Szabó, *Odd Khovanov homology*, Algebr. Geom. Topol. **13** (2013), no. 3, 1465–1488, DOI 10.2140/agt.2013.13.1465. MR3071132 Cited in Problem 1.37.

[OS00] Peter Ozsváth and Zoltán Szabó, *The symplectic Thom conjecture*, Ann. of Math. (2) **151** (2000), no. 1, 93–124, DOI 10.2307/121113. MR1745017 Cited in Problems 4.44, 4.100, and 4.123.

[OS03a] Peter Ozsváth and Zoltán Szabó, *Absolutely graded Floer homologies and intersection forms for four-manifolds with boundary*, Adv. Math. **173** (2003), no. 2, 179–261, DOI 10.1016/S0001-8708(02)00030-0. MR1957829 Cited in Problem 4.61.

[OS03b] Peter Ozsváth and Zoltán Szabó, *On the Floer homology of plumbed three-manifolds*, Geom. Topol. **7** (2003), 185–224, DOI 10.2140/gt.2003.7.185. MR1988284 Cited in Problems 3.50, 3.57, and 5.18.

[OS04a] Burak Ozbagci and András I. Stipsicz, *Contact 3-manifolds with infinitely many Stein fillings*, Proc. Amer. Math. Soc. **132** (2004), no. 5, 1549–1558, DOI 10.1090/S0002-9939-03-07328-3. MR2053364 Cited in Problem 4.105.

[OS04b] Peter Ozsváth and Zoltán Szabó, *Holomorphic disks and genus bounds*, Geom. Topol. **8** (2004), 311–334, DOI 10.2140/gt.2004.8.311. MR2023281 Cited in Problems 1.19, 1.76, 3.48, and 3.49.

[OS04c] Peter Ozsváth and Zoltán Szabó, *Holomorphic disks and knot invariants*, Adv. Math. **186** (2004), no. 1, 58–116, DOI 10.1016/j.aim.2003.05.001. MR2065507 Cited in Problems 1.19, 1.20, 2.47, and 3.47.

[OS04d] Peter Ozsváth and Zoltán Szabó, *Holomorphic disks and three-manifold invariants: properties and applications*, Ann. of Math. (2) **159** (2004), no. 3, 1159–1245, DOI 10.4007/annals.2004.159.1159. MR2113020 Cited in Problems 3.47 and 3.54.

[OS04e] Peter Ozsváth and Zoltán Szabó, *Holomorphic disks and topological invariants for closed three-manifolds*, Ann. of Math. (2) **159** (2004), no. 3, 1027–1158, DOI 10.4007/annals.2004.159.1027. MR2113019 Cited in Problems 2.47 and 3.47.

[OS05a] Peter Ozsváth and Zoltán Szabó, *Heegaard Floer homology and contact structures*, Duke Math. J. **129** (2005), no. 1, 39–61, DOI 10.1215/S0012-7094-04-12912-4. MR2153455 Cited in Problem 3.47.

[OS05b] Peter Ozsváth and Zoltán Szabó, *On knot Floer homology and lens space surgeries*, Topology **44** (2005), no. 6, 1281–1300, DOI 10.1016/j.top.2005.05.001. MR2168576 Cited in Problems 1.21, 1.31, and 1.49.

[OS05c] Peter Ozsváth and Zoltán Szabó, *On the Heegaard Floer homology of branched double-covers*, Adv. Math. **194** (2005), no. 1, 1–33, DOI 10.1016/j.aim.2004.05.008. MR2141852 Cited in Problem 1.37.

[OS08] Peter Ozsváth and Zoltán Szabó, *Holomorphic disks, link invariants and the multi-variable Alexander polynomial*, Algebr. Geom. Topol. **8** (2008), no. 2, 615–692, DOI 10.2140/agt.2008.8.615. MR2443092 Cited in Problem 3.47.

[OS11] Peter S. Ozsváth and Zoltán Szabó, *Knot Floer homology and rational surgeries*, Algebr. Geom. Topol. **11** (2011), no. 1, 1–68, DOI 10.2140/agt.2011.11.1. MR2764036 Cited in Problems 1.12 and 3.50.

[OS24] Brendan Owens and Frank Swenton, *An algorithm to find ribbon disks for alternating knots*, Exp. Math. **33** (2024), no. 3, 437–455, DOI 10.1080/10586458.2022.2158968. MR4790978 Cited in Problem 1.52.

[OSS05] Peter Ozsváth, András Stipsicz, and Zoltán Szabó, *Planar open books and Floer homology*, Int. Math. Res. Not. **54** (2005), 3385–3401, DOI 10.1155/IMRN.2005.3385. MR2200085 Cited in Problem 3.45.

[OSS14a] Peter Ozsváth, András I. Stipsicz, and Zoltán Szabó, *Knots in lattice homology*, Comment. Math. Helv. **89** (2014), no. 4, 783–818, DOI 10.4171/CMH/334. MR3284295 Cited in Problem 3.57.

[OSS14b] Peter Ozsváth, András I. Stipsicz, and Zoltán Szabó, *A spectral sequence on lattice homology*, Quantum Topol. **5** (2014), no. 4, 487–521, DOI 10.4171/QT/56. MR3317341 Cited in Problem 5.18.

[OSS17a] Peter S. Ozsváth, András I. Stipsicz, and Zoltán Szabó, *Unoriented knot Floer homology and the unoriented four-ball genus*, Int. Math. Res. Not. IMRN **17** (2017), 5137–5181, DOI 10.1093/imrn/rnw143. MR3694597 Cited in Problem 1.47.

[OSS17b] Peter S. Ozsváth, András I. Stipsicz, and Zoltán Szabó, *Concordance homomorphisms from knot Floer homology*, Adv. Math. **315** (2017), 366–426, DOI 10.1016/j.aim.2017.05.017. MR3667589 Cited in Problem 1.38.

[Ota82] Jean-Pierre Otal, *Présentations en ponts du nœud trivial* (French, with English summary), C. R. Acad. Sci. Paris Sér. I Math. **294** (1982), no. 16, 553–556. MR679942 Cited in Problem 4.112.

[Ota90] Jean-Pierre Otal, *Le spectre marqué des longueurs des surfaces à courbure négative* (French), Ann. of Math. (2) **131** (1990), no. 1, 151–162, DOI 10.2307/1971511. MR1038361 Cited in Problem 2.16.

[Ota01] Jean-Pierre Otal, *The hyperbolization theorem for fibered 3-manifolds*, SMF/AMS Texts and Monographs, vol. 7, American Mathematical Society, Providence, RI; Société Mathématique de France, Paris, 2001. Translated from the 1996 French original by Leslie D. Kay. MR1855976 Cited in Problems 2.14 and 4.127.

[Ouy97] Mingqing Ouyang, *A simplicial formula for the η-invariant of hyperbolic 3-manifolds*, Topology **36** (1997), no. 2, 411–421, DOI 10.1016/0040-9383(96)00013-4. MR1415596 Cited in Problem 3.63.

[OY18] Tomotada Ohtsuki and Yoshiyuki Yokota, *On the asymptotic expansions of the Kashaev invariant of the knots with 6 crossings*, Math. Proc. Cambridge Philos. Soc. **165** (2018), no. 2, 287–339, DOI 10.1017/S0305004117000494. MR3834003 Cited in Problem 1.27.

[Pac91] Udo Pachner, *P.L. homeomorphic manifolds are equivalent by elementary shellings*, European J. Combin. **12** (1991), no. 2, 129–145, DOI 10.1016/S0195-6698(13)80080-7. MR1095161 Cited in Problem 3.28.

[Pak08] Igor Pak. The discrete square peg problem, 2008. `arXiv:0804.0657`. Cited in Problem 2.14.

[Pan17a] Yu Pan, *The augmentation category map induced by exact Lagrangian cobordisms*, Algebr. Geom. Topol. **17** (2017), no. 3, 1813–1870, DOI 10.2140/agt.2017.17.1813. MR3677941 Cited in Problem 1.66.

[Pan17b] Yu Pan, *Exact Lagrangian fillings of Legendrian $(2, n)$ torus links*, Pacific J. Math. **289** (2017), no. 2, 417–441, DOI 10.2140/pjm.2017.289.417. MR3667178 Cited in Problem 1.67.

[Pan24] Tushar Pandey. The Bonahon-Wong-Yang volume conjecture for the four-puncture sphere, 2024. `arXiv:2311.13151`. Cited in Problem 2.50.

[Pao77] Peter Sie Pao, *The topological structure of* 4*-manifolds with effective torus actions. I*, Trans. Amer. Math. Soc. **227** (1977), 279–317, DOI 10.2307/1997462. MR431231 Cited in Problem 4.89.

[Pao78] Peter Sie Pao, *Nonlinear circle actions on the* 4*-sphere and twisting spun knots*, Topology **17** (1978), no. 3, 291–296, DOI 10.1016/0040-9383(78)90033-2. MR508892 Cited in Problem 4.9.

[Pao01] Luisa Paoluzzi, *On hyperbolic type involutions*, Rend. Istit. Mat. Univ. Trieste **32** (2001), no. suppl. 1, 221–256 (2002). Dedicated to the memory of Marco Reni. MR1893400 Cited in Problem 1.16.

[Pao05] Luisa Paoluzzi, *Hyperbolic knots and cyclic branched covers*, Publ. Mat. **49** (2005), no. 2, 257–284, DOI 10.5565/PUBLMAT_49205_01. MR2177068 Cited in Problem 1.16.

[Par02] Jongil Park. The geography of Spin symplectic 4-manifolds. *Math. Z.*, 240(2):405–421, 2002. `doi:10.1007/s002090100390`. Cited in Problem 4.3.

[Par13a] John Pardon, *The Hilbert-Smith conjecture for three-manifolds*, J. Amer. Math. Soc. **26** (2013), no. 3, 879–899, DOI 10.1090/S0894-0347-2013-00766-3. MR3037790 Cited in Problem 5.3.

[Par13b] Hugo Parlier, *Kissing numbers for surfaces*, J. Topol. **6** (2013), no. 3, 777–791, DOI 10.1112/jtopol/jtt012. MR3100890 Cited in Problem 2.12.

[Par19] John Pardon, *Totally disconnected groups (not) acting on two-manifolds*, Breadth in contemporary topology, Proc. Sympos. Pure Math., vol. 102, Amer. Math. Soc., Providence, RI, 2019, pp. 187–193, DOI 10.1090/pspum/102/13. MR3967368 Cited in Problem 5.3.

[Pat14] Priyam Patel. On a theorem of Peter Scott. *Proc. Amer. Math. Soc.*, 142(8):2891–2906, 2014. `doi:10.1090/S0002-9939-2014-12031-4`. Cited in Problem 2.11.

[Pen24] Mark Pencovitch, *Unknotting nonorientable surfaces of genus 4 and 5*, Linear Algebra Appl. **702** (2024), 195–217, DOI 10.1016/j.laa.2024.08.014. MR4792484 Cited in Problem 4.29.

[Per86] B. Perron, *Pseudo-isotopies et isotopies en dimension quatre dans la catégorie topologique* (French), Topology **25** (1986), no. 4, 381–397, DOI 10.1016/0040-9383(86)90018-2. MR862426 Cited in Problems 4.48, 4.72, 4.76, 4.77, 4.80, 4.81, and 4.82.

[Per02] Grisha Perelman. The entropy formula for the Ricci flow and its geometric applications, 2002. `arXiv:math/0211159`. Cited in Problems 3.0, 4.1, and 4.118.

[Per03a] Grisha Perelman. Finite extinction time for the solutions to the Ricci flow on certain three-manifolds, 2003. `arXiv:math/0307245`. Cited in Problems 3.0, 4.1, and 4.118.

[Per03b] Grisha Perelman. Ricci flow with surgery on three-manifolds, 2003. `arXiv:math/0303109`. Cited in Problems 3.0, 4.1, and 4.118.

[Per08a] Tim Perutz, *Lagrangian matching invariants for fibred four-manifolds. II*, Geom. Topol. **12** (2008), no. 3, 1461–1542, DOI 10.2140/gt.2008.12.1461. MR2421133 Cited in Problem 2.48.

[Per08b] Timothy Perutz, *Hamiltonian handleslides for Heegaard Floer homology*, Proceedings of Gökova Geometry-Topology Conference 2007, Gökova Geometry/Topology Conference (GGT), Gökova, 2008, pp. 15–35. MR2509747 Cited in Problem 2.48.

[Per82] B. Perron. Le nœud "huit" est algébrique réel. *Invent. Math.*, 65(3):441–451, 1981/82. `doi:10.1007/BF01396628`. Cited in Problem 1.87.

[Pet75] Ted Petrie, *Equivariant quasi-equivalence, transversality and normal cobordism*, Proceedings of the International Congress of Mathematicians (Vancouver, B.C., 1974), Canad. Math. Congr., Montreal, QC, 1975, pp. 537–541. MR431234 Cited in Problem 5.7.

[Pet10] Thomas David Peters, *Computations of Heegaard Floer homology: Torus bundles, L-spaces, and correction terms*, ProQuest LLC, Ann Arbor, MI, 2010. Thesis (Ph.D.)–Columbia University. MR2782352 Cited in Problem 3.48.

[Pic05] Anne Pichon. Real analytic germs $f\overline{g}$ and open-book decompositions of the 3-sphere. *Internat. J. Math.*, 16(1):1–12, 2005. `doi:10.1142/S0129167X05002710`. Cited in Problem 1.87.

[Pic19] Lisa Piccirillo, *Shake genus and slice genus*, Geom. Topol. **23** (2019), no. 5, 2665–2684, DOI 10.2140/gt.2019.23.2665. MR4019900 Cited in Problem 1.57.

[Pie91] R. Piergallini, *Covering moves*, Trans. Amer. Math. Soc. **325** (1991), no. 2, 903–920, DOI 10.2307/2001654. MR1010413 Cited in Problem 1.16.

[Pie95] R. Piergallini, *Four-manifolds as 4-fold branched covers of S^4*, Topology **34** (1995), no. 3, 497–508, DOI 10.1016/0040-9383(94)00034-I. MR1341805 Cited in Problem 5.6.

[Pla86] M. L. Platt, *Insufficiency of Torres' conditions for two-component classical links*, Trans. Amer. Math. Soc. **296** (1986), no. 1, 125–136, DOI 10.2307/2000564. MR837802 Cited in Problem 1.8.

[Pla06] Olga Plamenevskaya, *Transverse knots, branched double covers and Heegaard Floer contact invariants*, J. Symplectic Geom. **4** (2006), no. 2, 149–170, DOI 10.4310/jsg.2006.v4.n2.a2. MR2275002 Cited in Problem 1.64.

[Pla12] Olga Plamenevskaya, *On Legendrian surgeries between lens spaces*, J. Symplectic Geom. **10** (2012), no. 2, 165–181, DOI 10.4310/jsg.2012.v10.n2.a1. MR2926993 Cited in Problem 4.105.

[Poo86] Y. Sun Poon, *Compact self-dual manifolds with positive scalar curvature*, J. Differential Geom. **24** (1986), no. 1, 97–132. MR857378 Cited in Problem 4.130.

[Pow80] Jerome Powell. Homeomorphisms of S^3 leaving a Heegaard surface invariant. *Trans. Amer. Math. Soc.*, 257(1):193–216, 1980. `doi:10.2307/1998131`. Cited in Problem 2.15.

[PP00] Carlo Petronio and Joan Porti, *Negatively oriented ideal triangulations and a proof of Thurston's hyperbolic Dehn filling theorem*, Expo. Math. **18** (2000), no. 1, 1–35. MR1751141 Cited in Problem 3.3.

[PR09] Gopal Prasad and Andrei S. Rapinchuk, *Weakly commensurable arithmetic groups and isospectral locally symmetric spaces*, Publ. Math. Inst. Hautes Études Sci.

109 (2009), 113–184, DOI 10.1007/s10240-009-0019-6. MR2511587 Cited in Problem 3.17.

[PR15] Gopal Prasad and Andrei S. Rapinchuk, *Weakly commensurable groups, with applications to differential geometry*, Handbook of group actions. Vol. I, Adv. Lect. Math. (ALM), vol. 31, Int. Press, Somerville, MA, 2015, pp. 495–524. MR3380340 Cited in Problem 3.17.

[PR21] Mark Powell and Arunima Ray, *The development of topological 4-manifold theory*, The disc embedding theorem, Oxford Univ. Press, Oxford, 2021, pp. 295–330. MR4519519 Cited in Problem 1.62.

[Pra73] Gopal Prasad, *Strong rigidity of* **Q**-*rank* 1 *lattices*, Invent. Math. **21** (1973), 255–286, DOI 10.1007/BF01418789. MR385005 Cited in Problem 3.0.

[Pri77] T. M. Price. Homeomorphisms of quaternion space and projective planes in four space. *J. Austral. Math. Soc. Ser. A*, 23(1):112–128, 1977. doi:10.1017/s1446788700017407. Cited in Problem 4.37.

[Prz91] Józef H. Przytycki, *Skein modules of* 3-*manifolds*, Bull. Polish Acad. Sci. Math. **39** (1991), no. 1-2, 91–100. MR1194712 Cited in Problem 1.32.

[Prz00] Józef H. Przytycki. Kauffman bracket skein module of a connected sum of 3-manifolds. *Manuscripta Math.*, 101(2):199–207, 2000. doi:10.1007/s002290050014. Cited in Problem 3.65.

[Prz15] Piotr Przytycki, *Arcs intersecting at most once*, Geom. Funct. Anal. **25** (2015), no. 2, 658–670, DOI 10.1007/s00039-015-0320-0. MR3334237 Cited in Problem 2.12.

[PS00a] B. Doug Park and Zoltán Szabó, *The geography problem for irreducible spin four-manifolds*, Trans. Amer. Math. Soc. **352** (2000), no. 8, 3639–3650, DOI 10.1090/S0002-9947-00-02467-3. MR1653371 Cited in Problem 4.3.

[PS00b] Józef H. Przytycki and Adam S. Sikora, *On skein algebras and* $\mathrm{Sl}_2(\mathbf{C})$-*character varieties*, Topology **39** (2000), no. 1, 115–148, DOI 10.1016/S0040-9383(98)00062-7. MR1710996 Cited in Problems 1.36 and 3.65.

[PS01] Józef H. Przytycki and Maxim V. Sokolov, *Surgeries on periodic links and homology of periodic 3-manifolds*, Math. Proc. Cambridge Philos. Soc. **131** (2001), no. 2, 295–307, DOI 10.1017/S0305004101005308. MR1857121 Cited in Problem 1.15.

[PS14] Józef H. Przytycki and Radmila Sazdanović, *Torsion in Khovanov homology of semi-adequate links*, Fund. Math. **225** (2014), no. 1, 277–304, DOI 10.4064/fm225-1-13. MR3205574 Cited in Problems 1.30 and 1.31.

[PT76] J. F. Plante and W. P. Thurston, *Polynomial growth in holonomy groups of foliations*, Comment. Math. Helv. **51** (1976), no. 4, 567–584, DOI 10.1007/BF02568174. MR436167 Cited in Problem 2.34.

[PT25] Priyam Patel and Samuel J. Taylor, *Constructing endperiodic loxodromics of infinite-type arc graphs*, Math. Z. **310** (2025), no. 4, Paper No. 82, 12, DOI 10.1007/s00209-025-03784-w. MR4917177 Cited in Problem 2.37.

[Pup07] Volker Puppe, *Do manifolds have little symmetry?*, J. Fixed Point Theory Appl. **2** (2007), no. 1, 85–96, DOI 10.1007/s11784-007-0021-x. MR2336501 Cited in Problem 4.84.

[Pur08] Jessica S. Purcell, *Slope lengths and generalized augmented links*, Comm. Anal. Geom. **16** (2008), no. 4, 883–905, DOI 10.4310/CAG.2008.v16.n4.a7. MR2471374 Cited in Problem 3.11.

[Put07] Thomas Andrew Putman, *An infinite presentation of the Torelli group*, ProQuest LLC, Ann Arbor, MI, 2007. Thesis (Ph.D.)–The University of Chicago. MR2710199 Cited in Problem 2.9.

[PV96] B. Perron and J. P. Vannier, *Groupe de monodromie géométrique des singularités simples* (French), Math. Ann. **306** (1996), no. 2, 231–245, DOI 10.1007/BF01445249. MR1411346 Cited in Problem 2.29.

[PV18] Priyam Patel and Nicholas G. Vlamis, *Algebraic and topological properties of big mapping class groups*, Algebr. Geom. Topol. **18** (2018), no. 7, 4109–4142, DOI 10.2140/agt.2018.18.4109. MR3892241 Cited in Problem 2.37.

[PW13] Andrew Putman and Ben Wieland, *Abelian quotients of subgroups of the mappings class group and higher Prym representations*, J. Lond. Math. Soc. (2) **88** (2013), no. 1, 79–96, DOI 10.1112/jlms/jdt001. MR3092259 Cited in Problem 2.1.

[PW18] Piotr Przytycki and Daniel T. Wise, *Mixed 3-manifolds are virtually special*, J. Amer. Math. Soc. **31** (2018), no. 2, 319–347, DOI 10.1090/jams/886. MR3758147 Cited in Problem 3.37.

[PW24] Tushar Pandey and Ka Ho Wong. Generalized Bonahon-Wong-Yang volume conjecture of quantum invariants of surface diffeomorphisms i: the figure eight knot complement, 2024. arXiv:2402.04483. Cited in Problem 2.50.

[PY07] Gopal Prasad and Sai-Kee Yeung, *Fake projective planes*, Invent. Math. **168** (2007), no. 2, 321–370, DOI 10.1007/s00222-007-0034-5. MR2289867 Cited in Problem 4.91.

[PY09] Jongil Park and Ki-Heon Yun. Nonisomorphic Lefschetz fibrations on knot surgery 4-manifolds. *Math. Ann.*, 345(3):581–597, 2009. doi:10.1007/s00208-009-0366-0. Cited in Problem 4.98.

[PY10] Gopal Prasad and Sai-Kee Yeung, *Addendum to "Fake projective planes" Invent. Math. 168, 321–370 (2007) [MR2289867]*, Invent. Math. **182** (2010), no. 1, 213–227, DOI 10.1007/s00222-010-0259-6. MR2672284 Cited in Problem 4.91.

[PY15] Jongil Park and Ki-Heon Yun, *Families of nondiffeomorphic 4-manifolds with the same Seiberg-Witten invariants*, J. Symplectic Geom. **13** (2015), no. 2, 279–303, DOI 10.4310/JSG.2015.v13.n2.a2. MR3338893 Cited in Problem 4.8.

[PY17] Jongil Park and Ki-Heon Yun. Lefschetz fibrations on knot surgery 4-manifolds via Stallings twist. *Michigan Math. J.*, 66(3):481–498, 2017. doi:10.1307/mmj/1497513628. Cited in Problem 4.98.

[PZ05] R. Piergallini and D. Zuddas, *A universal ribbon surface in B^4*, Proc. London Math. Soc. (3) **90** (2005), no. 3, 763–782, DOI 10.1112/S0024611504015072. MR2137830 Cited in Problem 4.122.

[PZ16] Carlo Petronio and Adolfo Zanellati, *Algorithmic simplification of knot diagrams: new moves and experiments*, J. Knot Theory Ramifications **25** (2016), no. 10, 1650059, 30, DOI 10.1142/S0218216516500590. MR3548477 Cited in Problems 1.76 and 1.79.

[Qin25] Qianhe Qin, *An RBG construction of integral surgery homeomorphisms*, Algebr. Geom. Topol. **25** (2025), no. 6, 3755–3774, DOI 10.2140/agt.2025.25.3755. MR4972726 Cited in Problem 4.3.

[Qiu24] Haochen Qiu. Surgery formulas for Seiberg-Witten invariants and family Seiberg-Witten invariants, 2024. arXiv:2411.10392. Cited in Problem 4.68.

[QR16] Hoel Queffelec and David E. V. Rose, *The $\mathfrak{sl}_n$ foam 2-category: a combinatorial formulation of Khovanov-Rozansky homology via categorical skew Howe duality*, Adv. Math. **302** (2016), 1251–1339, DOI 10.1016/j.aim.2016.07.027. MR3545951 Cited in Problem 1.37.

[QR18] Hoel Queffelec and David E. V. Rose, *Sutured annular Khovanov-Rozansky homology*, Trans. Amer. Math. Soc. **370** (2018), no. 2, 1285–1319, DOI 10.1090/tran/7117. MR3729501 Cited in Problem 1.36.

[QRS18] Hoel Queffelec, David E. V. Rose, and Antonio Sartori. Annular evaluation and link homology, 2018. arXiv:1802.04131. Cited in Problem 1.34.

[QRSW21] You Qi, Louis-Hadrien Robert, Joshua Sussan, and Emmanuel Wagner. A categorification of the colored Jones polynomial at a root of unity, 2021. arXiv:2111.13195. Cited in Problem 3.67.

[Qui79] Frank Quinn, *Open book decompositions, and the bordism of automorphisms*, Topology **18** (1979), no. 1, 55–73, DOI 10.1016/0040-9383(79)90014-4. MR528236 Cited in Problem 4.121.

[Qui82] Frank Quinn, *Ends of maps. III. Dimensions* 4 *and* 5, J. Differential Geometry **17** (1982), no. 3, 503–521. MR679069 Cited in Problems 4.5 and 4.119.

[Qui86] Frank Quinn, *Isotopy of* 4*-manifolds*, J. Differential Geom. **24** (1986), no. 3, 343–372. MR868975 Cited in Problems 4.48, 4.68, 4.72, 4.76, 4.77, 4.80, 4.81, 4.82, and 4.119.

[QW21] Hoel Queffelec and Paul Wedrich, *Khovanov homology and categorification of skein modules*, Quantum Topol. **12** (2021), no. 1, 129–209, DOI 10.4171/qt/148. MR4233203 Cited in Problem 1.36.

[Ran98] Andrew Ranicki, *High-dimensional knot theory*, Springer Monographs in Mathematics, Springer-Verlag, New York, 1998. Algebraic surgery in codimension 2;

With an appendix by Elmar Winkelnkemper, DOI 10.1007/978-3-662-12011-8. MR1713074 Cited in Problem 4.121.

[Ras03] Jacob Andrew Rasmussen, *Floer homology and knot complements*, ProQuest LLC, Ann Arbor, MI, 2003. Thesis (Ph.D.)–Harvard University. MR2704683 Cited in Problems 1.19, 2.47, and 3.47.

[Ras07] Jacob Rasmussen. Lens space surgeries and L-space homology spheres, 2007. `arXiv:0710.2531`. Cited in Problems 1.9 and 1.20.

[Ras10] Jacob Rasmussen, *Khovanov homology and the slice genus*, Invent. Math. **182** (2010), no. 2, 419–447, DOI 10.1007/s00222-010-0275-6. MR2729272 Cited in Problems 1.28, 1.60, 4.1, and 4.43.

[Ras15] Jacob Rasmussen. Some differentials on Khovanov-Rozansky homology. *Geom. Topol.*, 19(6):3031–3104, 2015. `doi:10.2140/gt.2015.19.3031`. Cited in Problems 1.31 and 1.34.

[Ras17] Sarah Dean Rasmussen, *L-space intervals for graph manifolds and cables*, Compos. Math. **153** (2017), no. 5, 1008–1049, DOI 10.1112/S0010437X16008319. MR3705248 Cited in Problem 3.48.

[Ras21] Jacob Rasmussen, *Knots, polynomials, and categorification*, Quantum field theory and manifold invariants, IAS/Park City Math. Ser., vol. 28, Amer. Math. Soc., Providence, RI, [2021] ©2021, pp. 77–170. MR4380678 Cited in Problem 1.28.

[Rat81] John G. Ratcliffe, *On the ends of higher-dimensional knot groups*, J. Pure Appl. Algebra **20** (1981), no. 3, 317–324, DOI 10.1016/0022-4049(81)90066-9. MR604323 Cited in Problem 4.36.

[Rav22] Huygens C. Ravelomanana, *Knot complement problem for L-space* $\mathbb{Z}HS^3$, J. Knot Theory Ramifications **31** (2022), no. 4, Paper No. 2250028, 7, DOI 10.1142/S0218216522500286. MR4445220 Cited in Problem 1.12.

[Rei91] Alan W. Reid, *Arithmeticity of knot complements*, J. London Math. Soc. (2) **43** (1991), no. 1, 171–184, DOI 10.1112/jlms/s2-43.1.171. MR1099096 Cited in Problems 1.88, 2.19, and 3.10.

[Rei92] Alan W. Reid, *Isospectrality and commensurability of arithmetic hyperbolic 2- and 3-manifolds*, Duke Math. J. **65** (1992), no. 2, 215–228, DOI 10.1215/S0012-7094-92-06508-2. MR1150584 Cited in Problem 3.17.

[Rei06] Alan W. Reid, *Surface subgroups of mapping class groups*, Problems on mapping class groups and related topics, Proc. Sympos. Pure Math., vol. 74, Amer. Math. Soc., Providence, RI, 2006, pp. 257–268, DOI 10.1090/pspum/074/2264545. MR2264545 Cited in Problems 2.23 and 4.95.

[Rei07] Alan Reid. The geometry and topology of arithmetic hyperbolic 3-manifolds. In *Proc. Symposium Topology, Complex Analysis and Arithmetic of Hyperbolic Spaces, Kyoto 2006*, volume 1571 of *RIMS Kokyuroku Series*, pages 31–58. Research Institute for Mathematical Sciences, Kyoto University, 2007. Cited in Problem 3.10.

[Rei14] Alan W. Reid, *Traces, lengths, axes and commensurability* (English, with English and French summaries), Ann. Fac. Sci. Toulouse Math. (6) **23** (2014), no. 5, 1103–1118, DOI 10.5802/afst.1438. MR3294604 Cited in Problem 3.17.

[Rei15] Alan W. Reid, *Profinite properties of discrete groups*, Groups St Andrews 2013, London Math. Soc. Lecture Note Ser., vol. 422, Cambridge Univ. Press, Cambridge, 2015, pp. 73–104. MR3445488 Cited in Problem 3.8.

[Reu23] David Reutter, *Semisimple four-dimensional topological field theories cannot detect exotic smooth structure*, J. Topol. **16** (2023), no. 2, 542–566, DOI 10.1112/topo.12288. MR4577071 Cited in Problem 3.78.

[Rob01a] Rachel Roberts, *Taut foliations in punctured surface bundles. I*, Proc. London Math. Soc. (3) **82** (2001), no. 3, 747–768, DOI 10.1112/plms/82.3.747. MR1816696 Cited in Problem 3.41.

[Rob01b] Rachel Roberts, *Taut foliations in punctured surface bundles. II*, Proc. London Math. Soc. (3) **83** (2001), no. 2, 443–471, DOI 10.1112/plms/83.2.443. MR1839461 Cited in Problem 3.41.

[Roh51] V. A. Rohlin, *A three-dimensional manifold is the boundary of a four-dimensional one* (Russian), Doklady Akad. Nauk SSSR (N.S.) **81** (1951), 355–357. MR48808 Cited in Problem 3.68.

[Rol90] Dale Rolfsen, *Knots and links*, Mathematics Lecture Series, vol. 7, Publish or Perish, Inc., Houston, TX, 1990. Corrected reprint of the 1976 original. MR1277811 Cited in Problems 1.0 and 1.8.

[Rom12] Vitaliĭ Roman'kov, *Equations over groups*, Groups Complex. Cryptol. **4** (2012), no. 2, 191–239, DOI 10.1515/gcc-2012-0015. MR3043434 Cited in Problem 2.36.

[Ros07a] Stephan Rosebrock, *The Whitehead conjecture—an overview*, Sib. Èlektron. Mat. Izv. **4** (2007), 440–449. MR2465444 Cited in Problem 4.38.

[Ros07b] Jonathan Rosenberg, *Manifolds of positive scalar curvature: a progress report*, Surveys in differential geometry. Vol. XI, Surv. Differ. Geom., vol. 11, Int. Press, Somerville, MA, 2007, pp. 259–294, DOI 10.4310/SDG.2006.v11.n1.a9. MR2408269 Cited in Problem 4.128.

[Ros13] Christian Rosendal, *Global and local boundedness of Polish groups*, Indiana Univ. Math. J. **62** (2013), no. 5, 1621–1678, DOI 10.1512/iumj.2013.62.5133. MR3188557 Cited in Problems 2.33 and 2.46.

[Ros22] Christian Rosendal, *Coarse geometry of topological groups*, Cambridge Tracts in Mathematics, vol. 223, Cambridge University Press, Cambridge, 2022. MR4327092 Cited in Problem 2.46.

[Roz10] Lev Rozansky. A categorification of the stable SU(2) Witten-Reshetikhin-Turaev invariant of links in $S^2 \times S^1$, 2010. `arXiv:1011.1958`. Cited in Problem 1.32.

[Roz14] Lev Rozansky, *An infinite torus braid yields a categorified Jones-Wenzl projector*, Fund. Math. **225** (2014), no. 1, 305–326, DOI 10.4064/fm225-1-14. MR3205575 Cited in Problem 1.31.

[RR11] E. J. Janse van Rensburg and A. Rechnitzer, *Generalized atmospheric sampling of knotted polygons*, J. Knot Theory Ramifications **20** (2011), no. 8, 1145–1171, DOI 10.1142/S0218216511009170. MR2834121 Cited in Problem 1.73.

[RR24] Julie Rana and Sönke Rollenske, *Standard stable Horikawa surfaces*, Algebr. Geom. **11** (2024), no. 4, 569–592. MR4773392 Cited in Problem 4.94.

[RS72] C. P. Rourke and B. J. Sanderson, *Introduction to piecewise-linear topology*, Ergebnisse der Mathematik und ihrer Grenzgebiete [Results in Mathematics and Related Areas], Band 69, Springer-Verlag, New York-Heidelberg, 1972. MR350744 Cited in Problems 4.46 and 4.56.

[RS77] Frank Raymond and Leonard L. Scott, *Failure of Nielsen's theorem in higher dimensions*, Arch. Math. (Basel) **29** (1977), no. 6, 643–654, DOI 10.1007/BF01220468. MR467773 Cited in Problem 4.84.

[RS96] Hyam Rubinstein and Martin Scharlemann, *Comparing Heegaard splittings of non-Haken 3-manifolds*, Topology **35** (1996), no. 4, 1005–1026, DOI 10.1016/0040-9383(95)00055-0. MR1404921 Cited in Problem 3.34.

[RS97] Daniel Ruberman and Ronald J. Stern, *A fake smooth* $\mathbf{C}\mathrm{P}^2\#\mathbf{R}\mathrm{P}^4$, Math. Res. Lett. **4** (1997), no. 2-3, 375–378, DOI 10.4310/MRL.1997.v4.n3.a6. MR1453067 Cited in Problem 4.20.

[RS00] Daniel Ruberman and Sašo Strle, *Mod 2 Seiberg-Witten invariants of homology tori*, Math. Res. Lett. **7** (2000), no. 5-6, 789–799, DOI 10.4310/MRL.2000.v7.n6.a11. MR1809302 Cited in Problem 4.92.

[RS01] Jonathan Rosenberg and Stephan Stolz, *Metrics of positive scalar curvature and connections with surgery*, Surveys on surgery theory, Vol. 2, Ann. of Math. Stud., vol. 149, Princeton Univ. Press, Princeton, NJ, 2001, pp. 353–386. MR1818778 Cited in Problem 4.128.

[RS05] Daniel Ruberman and Nikolai Saveliev, *Casson-type invariants in dimension four*, Geometry and topology of manifolds, Fields Inst. Commun., vol. 47, Amer. Math. Soc., Providence, RI, 2005, pp. 281–306, DOI 10.1090/fic/047/18. MR2189939 Cited in Problem 4.22.

[RS07] Daniel Ruberman and Nikolai Saveliev, *Dirac operators on manifolds with periodic ends*, J. Gökova Geom. Topol. GGT **1** (2007), 33–50. MR2386534 Cited in Problem 4.128.

[RS09] Yann Rollin and Michael Singer. Constant scalar curvature Kähler surfaces and parabolic polystability. *J. Geom. Anal.*, 19(1):107–136, 2009. `doi:10.1007/s12220-008-9053-8`. Cited in Problem 4.130.

[RS11] Kasra Rafi and Saul Schleimer, *Curve complexes are rigid*, Duke Math. J. **158** (2011), no. 2, 225–246, DOI 10.1215/00127094-1334004. MR2805069 Cited in Problem 2.41.

[RS19] Daniel Ruberman and Laura Starkston, *Topological realizations of line arrangements*, Int. Math. Res. Not. IMRN **8** (2019), 2295–2331, DOI 10.1093/imrn/rnx190. MR3942162 Cited in Problem 4.102.

[RT91] N. Reshetikhin and V. G. Turaev, *Invariants of 3-manifolds via link polynomials and quantum groups*, Invent. Math. **103** (1991), no. 3, 547–597, DOI 10.1007/BF01239527. MR1091619 Cited in Problems 2.49, 3.4, and 3.67.

[RT00] John G. Ratcliffe and Steven T. Tschantz, *The volume spectrum of hyperbolic 4-manifolds*, Experiment. Math. **9** (2000), no. 1, 101–125. MR1758804 Cited in Problems 4.19 and 4.124.

[RT05] John G. Ratcliffe and Steven T. Tschantz, *Some examples of aspherical 4-manifolds that are homology 4-spheres*, Topology **44** (2005), no. 2, 341–350, DOI 10.1016/j.top.2004.10.006. MR2114711 Cited in Problem 4.124.

[Rub82] Daniel Ruberman, *Imbedding four-manifolds and slicing links*, Math. Proc. Cambridge Philos. Soc. **91** (1982), no. 1, 107–110, DOI 10.1017/S0305004100059168. MR633261 Cited in Problem 4.60.

[Rub84] Daniel Ruberman, *Invariant knots of free involutions of* S^4, Topology Appl. **18** (1984), no. 2-3, 217–224, DOI 10.1016/0166-8641(84)90011-7. MR769292 Cited in Problems 4.20 and 4.87.

[Rub87] Daniel Ruberman, *Mutation and volumes of knots in* S^3, Invent. Math. **90** (1987), no. 1, 189–215, DOI 10.1007/BF01389038. MR906585 Cited in Problem 3.61.

[Rub88] Daniel Ruberman, *Rational homology cobordisms of rational space forms*, Topology **27** (1988), no. 4, 401–414, DOI 10.1016/0040-9383(88)90020-1. MR976583 Cited in Problem 4.59.

[Rub95] Joachim H. Rubinstein, *An algorithm to recognize the* 3*-sphere*, Proceedings of the International Congress of Mathematicians, Vol. 1, 2 (Zürich, 1994), Birkhäuser, Basel, 1995, pp. 601–611. MR1403961 Cited in Problems 3.27 and 4.53.

[Rub96] Daniel Ruberman, *The minimal genus of an embedded surface of non-negative square in a rational surface*, Turkish J. Math. **20** (1996), no. 1, 129–133. MR1392668 Cited in Problem 4.44.

[Rub97] J. H. Rubinstein, *Polyhedral minimal surfaces, Heegaard splittings and decision problems for* 3*-dimensional manifolds*, Geometric topology (Athens, GA, 1993), AMS/IP Stud. Adv. Math., vol. 2, Amer. Math. Soc., Providence, RI, 1997, pp. 1–20, DOI 10.1090/amsip/002.1/01. MR1470718 Cited in Problem 3.27.

[Rub98] Daniel Ruberman, *An obstruction to smooth isotopy in dimension* 4, Math. Res. Lett. **5** (1998), no. 6, 743–758, DOI 10.4310/MRL.1998.v5.n6.a5. MR1671187 Cited in Problems 4.61, 4.68, 4.70, and 4.80.

[Rub99a] Daniel Ruberman. Mutation and gauge theory. I. Yang-Mills invariants. *Comment. Math. Helv.*, 74(4):615–641, 1999. doi:10.1007/s000140050108. Cited in Problem 3.61.

[Rub99b] Daniel Ruberman, *A polynomial invariant of diffeomorphisms of 4-manifolds*, Proceedings of the Kirbyfest (Berkeley, CA, 1998), Geom. Topol. Monogr., vol. 2, Geom. Topol. Publ., Coventry, 1999, pp. 473–488, DOI 10.2140/gtm.1999.2.473. MR1734421 Cited in Problems 4.75 and 4.81.

[Rub01] Daniel Ruberman, *Positive scalar curvature, diffeomorphisms and the Seiberg-Witten invariants*, Geom. Topol. **5** (2001), 895–924, DOI 10.2140/gt.2001.5.895. MR1874146 Cited in Problem 4.61.

[Rud76] Lee Rudolph. How independent are the knot-cobordism classes of links of plane curve singularities? *Notices Amer. Math. Soc*, 23:410, 1976. Cited in Problem 1.39.

[Rud83] Lee Rudolph, *Algebraic functions and closed braids*, Topology **22** (1983), no. 2, 191–202, DOI 10.1016/0040-9383(83)90031-9. MR683760 Cited in Problems 1.64, 1.68, 1.69, 1.70, and 4.104.

[Rud84] Lee Rudolph, *Some topologically locally-flat surfaces in the complex projective plane*, Comment. Math. Helv. **59** (1984), no. 4, 592–599, DOI 10.1007/BF02566368. MR780078 Cited in Problems 1.46 and 4.32.

[Rud87] Lee Rudolph. Isolated critical points of mappings from $\mathbf{R}^4$ to $\mathbf{R}^2$ and a natural splitting of the Milnor number of a classical fibered link. I. Basic theory; examples. *Comment. Math. Helv.*, 62(4):630–645, 1987. doi:10.1007/BF02564467. Cited in Problem 1.87.

[Rud92] Lee Rudolph, *Constructions of quasipositive knots and links. III. A characterization of quasipositive Seifert surfaces*, Topology **31** (1992), no. 2, 231–237, DOI 10.1016/0040-9383(92)90017-C. MR1167166 Cited in Problem 1.69.

[Rud93] Lee Rudolph, *Quasipositivity as an obstruction to sliceness*, Bull. Amer. Math. Soc. (N.S.) **29** (1993), no. 1, 51–59, DOI 10.1090/S0273-0979-1993-00397-5. MR1193540 Cited in Problem 1.5.

[Rud95] Lee Rudolph, *An obstruction to sliceness via contact geometry and "classical" gauge theory*, Invent. Math. **119** (1995), no. 1, 155–163, DOI 10.1007/BF01245177. MR1309974 Cited in Problem 1.62.

[Rud99] Lee Rudolph, *Positive links are strongly quasipositive*, Proceedings of the Kirbyfest (Berkeley, CA, 1998), Geom. Topol. Monogr., vol. 2, Geom. Topol. Publ., Coventry, 1999, pp. 555–562, DOI 10.2140/gtm.1999.2.555. MR1734423 Cited in Problems 1.5, 1.69, and 1.82.

[RW] Oscar Randal-Williams. Topological embeddings of S^2 in S^4. https://www.dpmms.cam.ac.uk/~or257/notes/Embeddings.pdf. Cited in Problem 4.72.

[RW92] Yong Wu Rong and Shi Cheng Wang. The preimages of submanifolds. *Math. Proc. Cambridge Philos. Soc.*, 112(2):271–279, 1992. doi:10.1017/S030500410007095X. Cited in Problem 3.33.

[RW98] J. Hyam Rubinstein and Shicheng Wang, *π_1-injective surfaces in graph manifolds*, Comment. Math. Helv. **73** (1998), no. 4, 499–515, DOI 10.1007/s000140050066. MR1639876 Cited in Problem 3.36.

[RW20] Louis-Hadrien Robert and Emmanuel Wagner, *Symmetric Khovanov-Rozansky link homologies* (English, with English and French summaries), J. Éc. polytech. Math. **7** (2020), 573–651, DOI 10.5802/jep.124. MR4086581 Cited in Problem 1.34.

[RW23] Jonathan Rosenberg and Shmuel Weinberger, *Positive scalar curvature on manifolds with boundary and their doubles*, Pure Appl. Math. Q. **19** (2023), no. 6, 2919–2950, DOI 10.4310/pamq.2023.v19.n6.a12. MR4703039 Cited in Problem 4.128.

[RW24] Qiuyu Ren and Michael Willis. Khovanov homology and exotic 4-manifolds, 2024. arXiv:2402.10452. Cited in Problem 4.66.

[Ryf23] Levi Ryffel, *Curves intersecting in a circuit pattern*, Topology Appl. **332** (2023), Paper No. 108522, 15, DOI 10.1016/j.topol.2023.108522. MR4575135 Cited in Problem 2.29.

[Sab23] Joshua M. Sabloff, *On a refinement of the non-orientable 4-genus of torus knots*, Proc. Amer. Math. Soc. Ser. B **10** (2023), 242–251, DOI 10.1090/bproc/166. MR4604871 Cited in Problem 1.47.

[Sad04] Rustam Sadykov, *Elimination of singularities of smooth mappings of 4-manifolds into 3-manifolds*, Topology Appl. **144** (2004), no. 1-3, 173–199, DOI 10.1016/j.topol.2004.04.006. MR2097135 Cited in Problem 5.9.

[Sae93a] Osamu Saeki, *Topology of special generic maps into $\mathbf{R}^3$*, Mat. Contemp. **5** (1993), 161–186. Workshop on Real and Complex Singularities (São Carlos, 1992). MR1305180 Cited in Problem 5.9.

[Sae93b] Osamu Saeki, *Topology of special generic maps of manifolds into Euclidean spaces*, Topology Appl. **49** (1993), no. 3, 265–293, DOI 10.1016/0166-8641(93)90116-U. MR1208678 Cited in Problem 5.9.

[Sae03] Osamu Saeki. Fold maps on 4-manifolds. *Comment. Math. Helv.*, 78(3):627–647, 2003. doi:10.1007/s00014-003-0758-9. Cited in Problem 5.9.

[Sae04] Osamu Saeki. *Topology of singular fibers of differentiable maps*, volume 1854 of *Lecture Notes in Mathematics*. Springer-Verlag, Berlin, 2004. doi:10.1007/b100393. Cited in Problem 5.9.

[Sae06] Osamu Saeki, *Stable mapping class groups of 4-manifolds with boundary*, Trans. Amer. Math. Soc. **358** (2006), no. 5, 2091–2104, DOI 10.1090/S0002-9947-05-03748-7. MR2197449 Cited in Problems 4.75 and 4.83.

[Sae10] Osamu Saeki, *Special generic maps on open 4-manifolds*, J. Singul. **1** (2010), 1–12, DOI 10.5427/jsing.2010.1a. MR2671763 Cited in Problem 5.9.

[Sak86] Makoto Sakuma, *On strongly invertible knots*, Algebraic and topological theories (Kinosaki, 1984), Kinokuniya, Tokyo, 1986, pp. 176–196. MR1102258 Cited in Problem 1.49.

[Sak01] Makoto Sakuma, *Surgery description of orientation-preserving periodic maps on compact orientable 3-manfolds*, Rend. Istit. Mat. Univ. Trieste **32** (2001), no. suppl. 1, 375–396 (2002). Dedicated to the memory of Marco Reni. MR1893406 Cited in Problem 1.15.

[Sal15] Nick Salter, *Surface bundles over surfaces with arbitrarily many fiberings*, Geom. Topol. **19** (2015), no. 5, 2901–2923, DOI 10.2140/gt.2015.19.2901. MR3416116 Cited in Problem 2.24.

[Sal19] Nick Salter. Monodromy and vanishing cycles in toric surfaces. *Invent. Math.*, 216(1):153–213, 2019. doi:10.1007/s00222-018-0845-6. Cited in Problem 2.6.

[San12] Ramanujan Santharoubane, *Limits of the quantum SO(3) representations for the one-holed torus*, J. Knot Theory Ramifications **21** (2012), no. 11, 1250109, 13, DOI 10.1142/S021821651250109X. MR2969639 Cited in Problem 2.49.

[San20] Abdoul Karim Sane. Unicellular maps and filtrations of the mapping class group, 2020. arXiv:2006.15880. Cited in Problem 2.13.

[San21] Abdoul Karim Sane, *Curves on surfaces and surgeries*, European J. Combin. **93** (2021), Paper No. 103281, 20, DOI 10.1016/j.ejc.2020.103281. MR4186621 Cited in Problem 2.13.

[Sat24a] Kouki Sato. Counterexamples to Allen's conjectures, 2024. arXiv:2407.12049. Cited in Problem 1.47.

[Sat24b] Kouki Sato. An unoriented analogue of slice-torus invariant, 2024. arXiv:2404.04056. Cited in Problem 1.47.

[Şav20] Oğuz Şavk, *More Brieskorn spheres bounding rational balls*, Topology Appl. **286** (2020), 107400, 10, DOI 10.1016/j.topol.2020.107400. MR4179129 Cited in Problems 3.72 and 4.58.

[Şav24] Oğuz Şavk, *A survey of the homology cobordism group*, Bull. Amer. Math. Soc. (N.S.) **61** (2024), no. 1, 119–157, DOI 10.1090/bull/1806. MR4678574 Cited in Problems 3.68 and 5.7.

[Say09] Nabil Sayari, *Reducible Dehn surgery and the bridge number of a knot*, J. Knot Theory Ramifications **18** (2009), no. 4, 493–504, DOI 10.1142/S0218216509007038. MR2514544 Cited in Problem 1.11.

[SC24] Anschel Schaffer-Cohen, *Graphs of curves and arcs quasi-isometric to big mapping class groups*, Groups Geom. Dyn. **18** (2024), no. 2, 705–735, DOI 10.4171/ggd/751. MR4729823 Cited in Problem 2.46.

[Sca15] Christopher W. Scaduto, *Instantons and odd Khovanov homology*, J. Topol. **8** (2015), no. 3, 744–810, DOI 10.1112/jtopol/jtv012. MR3394316 Cited in Problem 1.37.

[Sch76] Martin Scharlemann, *Constructing strange manifolds with the dodecahedral space*, Duke Math. J. **43** (1976), no. 1, 33–40. MR402760 Cited in Problem 4.22.

[Sch84] Martin Scharlemann, *The four-dimensional Schoenflies conjecture is true for genus two imbeddings*, Topology **23** (1984), no. 2, 211–217, DOI 10.1016/0040-9383(84)90040-5. MR744851 Cited in Problem 4.23.

[Sch85a] Martin Scharlemann, *Smooth spheres in* $\mathbf{R}^4$ *with four critical points are standard*, Invent. Math. **79** (1985), no. 1, 125–141, DOI 10.1007/BF01388659. MR774532 Cited in Problems 1.59 and 4.29.

[Sch85b] Martin G. Scharlemann, *Unknotting number one knots are prime*, Invent. Math. **82** (1985), no. 1, 37–55, DOI 10.1007/BF01394778. MR808108 Cited in Problems 1.3 and 1.80.

[Sch88] Uwe Schöning, *Graph isomorphism is in the low hierarchy*, J. Comput. System Sci. **37** (1988), no. 3, 312–323, DOI 10.1016/0022-0000(88)90010-4. MR975447 Cited in Problem 3.27.

[Sch90] Martin Scharlemann, *Producing reducible* 3*-manifolds by surgery on a knot*, Topology **29** (1990), no. 4, 481–500, DOI 10.1016/0040-9383(90)90017-E. MR1071370 Cited in Problem 1.11.

[Sch96] Paul Schmutz. Compact Riemann surfaces with many systoles. *Duke Math. J.*, 84(1):191–198, 1996. doi:10.1215/S0012-7094-96-08406-9. Cited in Problem 2.12.

[Sch04] Martin Scharlemann, *Automorphisms of the 3-sphere that preserve a genus two Heegaard splitting*, Bol. Soc. Mat. Mexicana (3) **10** (2004), no. Special Issue, 503–514. MR2199366 Cited in Problem 2.15.

[Sch08] Martin Scharlemann, *Generalized property R and the Schoenflies conjecture*, Comment. Math. Helv. **83** (2008), no. 2, 421–449, DOI 10.4171/CMH/131. MR2390052 Cited in Problem 4.23.

[Sch11] Saul Schleimer, *Sphere recognition lies in NP*, Low-dimensional and symplectic topology, Proc. Sympos. Pure Math., vol. 82, Amer. Math. Soc., Providence, RI, 2011, pp. 183–213, DOI 10.1090/pspum/082/2768660. MR2768660 Cited in Problem 3.27.

[Sch20] Martin Scharlemann, *One Powell generator is redundant*, Proc. Amer. Math. Soc. Ser. B **7** (2020), 138–141, DOI 10.1090/bproc/58. MR4165895 Cited in Problem 2.15.

[Sch25] Martin Scharlemann, *Powell's conjecture on the Goeritz group of S^3 is stably true*, Algebr. Geom. Topol. **25** (2025), no. 6, 3775–3787, DOI 10.2140/agt.2025.25.3775. MR4972727 Cited in Problem 2.15.

[Sch24] Martin Scharlemann, *Generating the Goeritz group of S^3*, J. Assoc. Math. Res. **2** (2024), no. 2, 209–335. MR4863334 Cited in Problem 2.15.

[Sco73] G. P. Scott, *Compact submanifolds of 3-manifolds*, J. London Math. Soc. (2) **7** (1973), 246–250, DOI 10.1112/jlms/s2-7.2.246. MR326737 Cited in Problem 3.37.

[Sco78] Peter Scott, *Subgroups of surface groups are almost geometric*, J. London Math. Soc. (2) **17** (1978), no. 3, 555–565, DOI 10.1112/jlms/s2-17.3.555. MR494062 Cited in Problem 2.11.

[Sco03] Alexandru Scorpan, *Existence of foliations on 4-manifolds*, Algebr. Geom. Topol. **3** (2003), 1225–1256, DOI 10.2140/agt.2003.3.1225. MR2026332 Cited in Problem 4.120.

[Scu21] Joe Scull. The homeomorphism problem for hyperbolic manifolds I, 2021. `arXiv:2108.00779`. Cited in Problems 3.27 and 3.30.

[SDS89] Herman Servatius, Carl Droms, and Brigitte Servatius, *Surface subgroups of graph groups*, Proc. Amer. Math. Soc. **106** (1989), no. 3, 573–578, DOI 10.2307/2047406. MR952322 Cited in Problem 2.30.

[Sei33] H. Seifert. Topologie Dreidimensionaler Gefaserter Räume. *Acta Math.*, 60(1):147–238, 1933. `doi:10.1007/BF02398271`. Cited in Problem 5.7.

[Sei35] H. Seifert. Über das Geschlecht von Knoten. *Math. Ann.*, 110(1):571–592, 1935. `doi:10.1007/BF01448044`. Cited in Problem 1.8.

[Sei97] Paul Seidel. *Floer homology and the symplectic isotopy problem.* PhD thesis, University of Oxford, 1997. Cited in Problem 2.48.

[Sel02] Z. Sela, *Diophantine geometry over groups and the elementary theory of free and hyperbolic groups*, Proceedings of the International Congress of Mathematicians, Vol. II (Beijing, 2002), Higher Ed. Press, Beijing, 2002, pp. 87–92. MR1957023 Cited in Problems 2.35 and 2.36.

[Ş11] Mehmet Haluk Şengün, *On the integral cohomology of Bianchi groups*, Exp. Math. **20** (2011), no. 4, 487–505, DOI 10.1080/10586458.2011.594671. MR2859903 Cited in Problem 3.6.

[Ş12] Mehmet Haluk Şengün, *On the torsion homology of non-arithmetic hyperbolic tetrahedral groups*, Int. J. Number Theory **8** (2012), no. 2, 311–320, DOI 10.1142/S1793042112500182. MR2890481 Cited in Problem 3.6.

[Ser06] Caroline Series, *Thurston's bending measure conjecture for once punctured torus groups*, Spaces of Kleinian groups, London Math. Soc. Lecture Note Ser., vol. 329, Cambridge Univ. Press, Cambridge, 2006, pp. 75–89. MR2258745 Cited in Problem 3.25.

[Sey80] P. D. Seymour, *Disjoint paths in graphs*, Discrete Math. **29** (1980), no. 3, 293–309, DOI 10.1016/0012-365X(80)90158-2. MR560773 Cited in Problem 1.86.

[Sha11] Peter B. Shalen, *A generic Margulis number for hyperbolic 3-manifolds*, Topology and geometry in dimension three, Contemp. Math., vol. 560, Amer. Math. Soc., Providence, RI, 2011, pp. 103–109, DOI 10.1090/conm/560/11094. MR2866926 Cited in Problem 3.23.

[Sha12] Kenneth J. Shackleton, *Tightness and computing distances in the curve complex*, Geom. Dedicata **160** (2012), 243–259, DOI 10.1007/s10711-011-9680-2. MR2970053 Cited in Problem 2.10.

[Sha13] Peter B. Shalen, *Small optimal Margulis numbers force upper volume bounds*, Trans. Amer. Math. Soc. **365** (2013), no. 2, 973–999, DOI 10.1090/S0002-9947-2012-05657-1. MR2995380 Cited in Problem 3.23.

[She00] Vsevolod Shevchishin. Pseudoholomorphic curves and the symplectic isotopy problem, 2000. arXiv:math/0010262. Cited in Problem 4.101.

[She04] Vsevolod V. Shevchishin, *On the local Severi problem*, Int. Math. Res. Not. **5** (2004), 211–237, DOI 10.1155/S1073792804211163. MR2038713 Cited in Problem 4.102.

[Shi73] Arnold C. Shilepsky, *Homogeneity by isotopy for simple closed curves*, Duke Math. J. **40** (1973), 463–472. MR319179 Cited in Problem 1.81.

[Shi97] Hiroshige Shiga, *On monodromies of holomorphic families of Riemann surfaces and modular transformations*, Math. Proc. Cambridge Philos. Soc. **122** (1997), no. 3, 541–549, DOI 10.1017/S0305004197001825. MR1466656 Cited in Problem 2.20.

[Sht05] M. A. Shtan'ko. On Markov's theorem on the algorithmic nonrecognizability of manifolds. *Fundam. Prikl. Mat.*, 11:257–259, 2005. doi:10.1007/s10958-007-0375-z. Cited in Problem 4.53.

[Shu07] Alexander N. Shumakovitch, *Rasmussen invariant, slice-Bennequin inequality, and sliceness of knots*, J. Knot Theory Ramifications **16** (2007), no. 10, 1403–1412, DOI 10.1142/S0218216507005889. MR2384833 Cited in Problems 1.69 and 1.70.

[Shu14] Alexander N. Shumakovitch, *Torsion of Khovanov homology*, Fund. Math. **225** (2014), no. 1, 343–364, DOI 10.4064/fm225-1-16. MR3205577 Cited in Problem 1.30.

[SIA$^+$09] R. Scharein, K. Ishihara, J. Arsuaga, Y. Diao, K. Shimokawa, and M. Vazquez, *Bounds for the minimum step number of knots in the simple cubic lattice*, J. Phys. A **42** (2009), no. 47, 475006, 24, DOI 10.1088/1751-8113/42/47/475006. MR2765342 Cited in Problem 1.73.

[Sie65] Laurence Carl Siebenmann, *THE OBSTRUCTION TO FINDING A BOUNDARY FOR AN OPEN MANIFOLD OF DIMENSION GREATER THAN FIVE*, ProQuest LLC, Ann Arbor, MI, 1965. Thesis (Ph.D.)–Princeton University. MR2615648 Cited in Problem 4.49.

[Sie70] L. C. Siebenmann, *Are nontriangulable manifolds triangulable?*, Topology of Manifolds (Proc. Inst., Univ. of Georgia, Athens, Ga., 1969), Markham Publishing Co., Chicago, IL, 1970, pp. 77–84. MR271956 Cited in Problem 5.8.

[Sik03] Jean-Claude Sikorav, *The gluing construction for normally generic J-holomorphic curves*, Symplectic and contact topology: interactions and perspectives (Toronto, ON/Montreal, QC, 2001), Fields Inst. Commun., vol. 35, Amer. Math. Soc., Providence, RI, 2003, pp. 175–199, DOI 10.1090/fic/035/12. MR1969276 Cited in Problem 4.101.

[Sik08] Adam S. Sikora. Quantizations of character varieties and quantum knot invariants, 2008. arXiv:0807.0943. Cited in Problem 1.25.

[Sil92] D. S. Silver, *On knot-like groups and ribbon concordance*, J. Pure Appl. Algebra **82** (1992), no. 1, 99–105, DOI 10.1016/0022-4049(92)90013-6. MR1181096 Cited in Problem 1.53.

[Sim02] Jonathan Simon, *Physical knots*, Physical knots: knotting, linking, and folding geometric objects in $\mathbb{R}^3$ (Las Vegas, NV, 2001), Contemp. Math., vol. 304, Amer. Math. Soc., Providence, RI, 2002, pp. 1–30, DOI 10.1090/conm/304/05181. MR1953008 Cited in Problem 1.71.

[Sim21] Jonathan Simone, *Using rational homology circles to construct rational homology balls*, Topology Appl. **291** (2021), Paper No. 107626, 16, DOI 10.1016/j.topol.2021.107626. MR4215138 Cited in Problem 3.72.

[Sin25] Oliver Singh, *Pseudo-isotopies and diffeomorphisms of 4-manifolds*, J. Topol. **18** (2025), no. 4, Paper No. e70043, DOI 10.1112/topo.70043. MR5006904 Cited in Problem 4.74.

[SJ23] Stefan Schnabel and Wolfhard Janke. Monte Carlo simulation of long hard-sphere polymer chains in two to five dimensions. *Macromolecular Theory and Simulations*, 32:2200080, 2023. doi:10.1002/mats.202200080. Cited in Problem 1.73.

[Sma61] Stephen Smale. Generalized Poincaré's conjecture in dimensions greater than four. *Ann. of Math. (2)*, 74:391–406, 1961. doi:10.2307/1970239. Cited in Problem 5.9.

[Sma62a] S. Smale, *On the structure of manifolds*, Amer. J. Math. **84** (1962), 387–399, DOI 10.2307/2372978. MR153022 Cited in Problems 4.1, 4.46, 4.56, and 5.9.

[Sma62b] Stephen Smale, *On the structure of* 5*-manifolds*, Ann. of Math. (2) **75** (1962), 38–46, DOI 10.2307/1970417. MR141133 Cited in Problem 4.118.

[Smi41] P. A. Smith, *Periodic and nearly periodic transformations*, Lectures in Topology, Univ. Michigan Press, Ann Arbor, MI, 1941, pp. 159–190. MR5302 Cited in Problem 5.3.

[Smi01a] Ivan Smith, *Geometric monodromy and the hyperbolic disc*, Q. J. Math. **52** (2001), no. 2, 217–228, DOI 10.1093/qjmath/52.2.217. MR1838364 Cited in Problem 2.22.

[Smi01b] Ivan Smith, *Torus fibrations on symplectic four-manifolds*, Turkish J. Math. **25** (2001), no. 1, 69–95. MR1829080 Cited in Problem 2.21.

[Smy66] N Smythe. Boundary links. In *Topology Seminar, Wisconsin, 1965*, volume 60 of *Annals of Mathematics Studies*, pages 69–72. Princeton University Press, Princeton, N.J., 1966. Edited by R. H. Bing and R. J. Bean. Cited in Problem 1.62.

[SS94] Masa-Hiko Saitō and Ken-Ichi Sakakibara. On Mordell-Weil lattices of higher genus fibrations on rational surfaces. *J. Math. Kyoto Univ.*, 34(4):859–871, 1994. doi:10.1215/kjm/1250518890. Cited in Problem 2.21.

[SS97] Paul Schmutz Schaller, *Extremal Riemann surfaces with a large number of systoles*, Extremal Riemann surfaces (San Francisco, CA, 1995), Contemp. Math., vol. 201, Amer. Math. Soc., Providence, RI, 1997, pp. 9–19, DOI 10.1090/conm/201/02617. MR1429190 Cited in Problem 2.12.

[SS99] Osamu Saeki and Kazuhiro Sakuma. Special generic maps of 4-manifolds and compact complex analytic surfaces. *Math. Ann.*, 313(4):617–633, 1999. doi:10.1007/s002080050275. Cited in Problem 5.9.

[SS06] Paul Seidel and Ivan Smith, *A link invariant from the symplectic geometry of nilpotent slices*, Duke Math. J. **134** (2006), no. 3, 453–514, DOI 10.1215/S0012-7094-06-13432-4. MR2254624 Cited in Problem 1.35.

[SS14] Peter Scott and Hamish Short, *The homeomorphism problem for closed 3-manifolds*, Algebr. Geom. Topol. **14** (2014), no. 4, 2431–2444, DOI 10.2140/agt.2014.14.2431. MR3331689 Cited in Problem 3.27.

[SS25] Hirofumi Sasahira and Matthew Stoffregen, *Seiberg-Witten Floer spectra for* $b_1 > 0$, Memoirs of the European Mathematical Society, vol. 17, EMS Press, Berlin, 2025, DOI 10.4171/mems/17. MR4867561 Cited in Problem 3.47.

[SS23] Nick Salter and Abdoul Karim Sane. Connected components of the topological surgery graph of a unicellular collection, 2023. arXiv:2308.09165. Cited in Problems 2.6 and 2.13.

[SS24a] Saul Schleimer and Henry Segerman, *From loom spaces to veering triangulations*, Groups Geom. Dyn. **18** (2024), no. 2, 419–462, DOI 10.4171/ggd/742. MR4729814 Cited in Problem 3.21.

[SS24b] András I. Stipsicz and Zoltán Szabó, *Definite four-manifolds with exotic smooth structures*, J. Reine Angew. Math. **817** (2024), 267–290, DOI 10.1515/crelle-2024-0072. MR4834887 Cited in Problem 4.3.

[SS24c] András I. Stipsicz and Zoltán Szabó, *On the minimal genus problem in four-manifolds*, Frontiers in geometry and topology, Proc. Sympos. Pure Math., vol. 109, Amer. Math. Soc., Providence, RI, [2024] ©2024, pp. 215–232. MR4772954 Cited in Problem 4.12.

[SSvdM13] Paweł Strzelecki, Marta Szumańska, and Heiko von der Mosel, *On some knot energies involving Menger curvature*, Topology Appl. **160** (2013), no. 13, 1507–1529, DOI 10.1016/j.topol.2013.05.022. MR3091327 Cited in Problem 1.75.

[ST88] Martin Scharlemann and Abigail Thompson, *Unknotting number, genus, and companion tori*, Math. Ann. **280** (1988), no. 2, 191–205, DOI 10.1007/BF01456051. MR929535 Cited in Problem 1.4.

[ST89] Martin Scharlemann and Abigail Thompson, *Link genus and the Conway moves*, Comment. Math. Helv. **64** (1989), no. 4, 527–535, DOI 10.1007/BF02564693. MR1022995 Cited in Problem 1.80.

[ST05] Bernd Siebert and Gang Tian, *On the holomorphicity of genus two Lefschetz fibrations*, Ann. of Math. (2) **161** (2005), no. 2, 959–1020, DOI 10.4007/annals.2005.161.959. MR2153404 Cited in Problem 4.101.

[ST18] Jonathan Spreer and Stephan Tillmann, *The trisection genus of standard simply connected PL 4-manifolds*, 34th International Symposium on Computational Geometry, LIPIcs. Leibniz Int. Proc. Inform., vol. 99, Schloss Dagstuhl. Leibniz-Zent. Inform., Wadern, 2018, pp. Art. No. 71, 13. MR3824315 Cited in Problem 4.113.

[ST19] Alessandro Sisto and Samuel J. Taylor, *Largest projections for random walks and shortest curves in random mapping tori*, Math. Res. Lett. **26** (2019), no. 1, 293–321, DOI 10.4310/MRL.2019.v26.n1.a14. MR3963985 Cited in Problem 2.11.

[ST20] Nick Salter and Bena Tshishiku, *Arithmeticity of the monodromy of some Kodaira fibrations*, Compos. Math. **156** (2020), no. 1, 114–157, DOI 10.1112/s0010437x19007668. MR4036450 Cited in Problem 2.24.

[ST22] Jonathan Spreer and Stephan Tillmann, *Determining the trisection genus of orientable and non-orientable PL 4-manifolds through triangulations*, Exp. Math. **31** (2022), no. 3, 897–907, DOI 10.1080/10586458.2020.1723744. MR4477412 Cited in Problem 4.113.

[Sta65] John R. Stallings, *On infinite processes leading to differentiability in the complement of a point*, Differential and Combinatorial Topology (A Symposium in Honor of Marston Morse), Princeton Univ. Press, Princeton, NJ, 1965, pp. 245–254. MR180983 Cited in Problem 4.18.

[Sta67] John R. Stallings, *Lectures on polyhedral topology*, Tata Institute of Fundamental Research Lectures on Mathematics, No. 43, Tata Institute of Fundamental Research, Bombay, 1967. Notes by G. Ananda Swarup. MR238329 Cited in Problems 4.46 and 4.56.

[Sta89] Richard P. Stanley, *Log-concave and unimodal sequences in algebra, combinatorics, and geometry*, Graph theory and its applications: East and West (Jinan, 1986), Ann. New York Acad. Sci., vol. 576, New York Acad. Sci., New York, 1989, pp. 500–535, DOI 10.1111/j.1749-6632.1989.tb16434.x. MR1110850 Cited in Problem 1.7.

[Sta20] Laura Starkston, *A new approach to the symplectic isotopy problem*, J. Symplectic Geom. **18** (2020), no. 3, 939–960, DOI 10.4310/JSG.2020.v18.n3.a11. MR4142492 Cited in Problem 4.101.

[Ste06] Ronald J. Stern, *Will we ever classify simply-connected smooth 4-manifolds?*, Floer homology, gauge theory, and low-dimensional topology, Clay Math. Proc., vol. 5, Amer. Math. Soc., Providence, RI, 2006, pp. 225–239. MR2249255 Cited in Problem 4.57.

[Sti95] Robert Paul Stingley, *Singularities of maps between 4-manifolds*, ProQuest LLC, Ann Arbor, MI, 1995. Thesis (Ph.D.)–State University of New York at Stony Brook. MR2692922 Cited in Problem 5.9.

[Sti99] András I. Stipsicz, *On the number of vanishing cycles in Lefschetz fibrations*, Math. Res. Lett. **6** (1999), no. 3-4, 449–456, DOI 10.4310/MRL.1999.v6.n4.a7. MR1713143 Cited in Problem 2.21.

[Sti01] András I. Stipsicz, *Spin structures on Lefschetz fibrations*, Bull. London Math. Soc. **33** (2001), no. 4, 466–472, DOI 10.1017/S0024609301008232. MR1832559 Cited in Problem 2.6.

[Sti02] András I. Stipsicz, *Singular fibers in Lefschetz fibrations on manifolds with* $b_2^+ = 1$, Topology Appl. **117** (2002), no. 1, 9–21, DOI 10.1016/S0166-8641(00)00105-X. MR1874001 Cited in Problem 2.21.

[Sti10] András I. Stipsicz, *Ozsváth-Szabó invariants and 3-dimensional contact topology*, Proceedings of the International Congress of Mathematicians. Volume II, Hindustan Book Agency, New Delhi, 2010, pp. 1159–1178. MR2827836 Cited in Problem 3.41.

[Sti15] András I. Stipsicz, *Symplectic 4-manifolds, Stein domains, Seiberg-Witten theory and mapping class groups*, Interactions between low-dimensional topology and mapping class groups, Geom. Topol. Monogr., vol. 19, Geom. Topol. Publ., Coventry, 2015, pp. 173–200, DOI 10.2140/gtm.2015.19.173. MR3609908 Cited in Problem 2.22.

[Sto78] Neal W. Stoltzfus, *Algebraic computations of the integral concordance and double null concordance group of knots*, Knot theory (Proc. Sem., Plans-sur-Bex, 1977),

Lecture Notes in Math., vol. 685, Springer, Berlin-New York, 1978, pp. 274–290. MR521738 Cited in Problem 1.48.

[Sto93] Richard Stong, *Uniqueness of π_1-negligible embeddings in 4-manifolds: a correction to Theorem 10.5 of Topology of 4-manifolds [Princeton Univ. Press, Princeton, NJ, 1990; MR1201584 (94b:57021)] by M. H. Freedman and F. Quinn*, Topology **32** (1993), no. 4, 677–699, DOI 10.1016/0040-9383(93)90046-X. MR1241868 Cited in Problem 4.41.

[Sto05] Alexander Stoimenow, *Newton-like polynomials of links*, Enseign. Math. (2) **51** (2005), no. 3-4, 211–230. MR2214886 Cited in Problem 1.7.

[Sto09] Marko Stošić, *Khovanov homology of torus links*, Topology Appl. **156** (2009), no. 3, 533–541, DOI 10.1016/j.topol.2008.08.004. MR2492301 Cited in Problem 1.31.

[Sto13] Matthew Stover, *On the number of ends of rank one locally symmetric spaces*, Geom. Topol. **17** (2013), no. 2, 905–924, DOI 10.2140/gt.2013.17.905. MR3070517 Cited in Problem 5.14.

[Sto15] A. Stoimenow, *Application of braiding sequences III: Concordance of positive knots*, Internat. J. Math. **26** (2015), no. 7, 1550050, 36, DOI 10.1142/S0129167X15500500. MR3357039 Cited in Problem 1.53.

[Sto16] Alexander Stoimenow, *Diagram genus, generators, and applications*, Monographs and Research Notes in Mathematics, CRC Press, Boca Raton, FL, 2016. MR3467592 Cited in Problem 1.7.

[Sto21] A. Stoimenow, *Independence polynomials and Alexander-Conway polynomials of plumbing links*, J. Combin. Theory Ser. A **183** (2021), Paper No. 105487, 18, DOI 10.1016/j.jcta.2021.105487. MR4268723 Cited in Problem 1.7.

[Sto25] Matthew Stover. A hyperbolic 4-orbifold with underlying space $\mathbb{P}^2$, 2025. `arXiv:2506.11667`. Cited in Problem 4.95.

[Str00] Ileana Streinu, *A combinatorial approach to planar non-colliding robot arm motion planning*, 41st Annual Symposium on Foundations of Computer Science (Redondo Beach, CA, 2000), IEEE Comput. Soc. Press, Los Alamitos, CA, 2000, pp. 443–453, DOI 10.1109/SFCS.2000.892132. MR1931841 Cited in Problem 1.72.

[STWZ19] Vivek Shende, David Treumann, Harold Williams, and Eric Zaslow, *Cluster varieties from Legendrian knots*, Duke Math. J. **168** (2019), no. 15, 2801–2871, DOI 10.1215/00127094-2019-0027. MR4017516 Cited in Problem 1.67.

[SU82] J. Sacks and K. Uhlenbeck, *Minimal immersions of closed Riemann surfaces*, Trans. Amer. Math. Soc. **271** (1982), no. 2, 639–652, DOI 10.2307/1998902. MR654854 Cited in Problem 3.12.

[Sul76] Dennis Sullivan, *Cycles for the dynamical study of foliated manifolds and complex manifolds*, Invent. Math. **36** (1976), 225–255, DOI 10.1007/BF01390011. MR433464 Cited in Problem 4.107.

[Sul79a] Dennis Sullivan, *A homological characterization of foliations consisting of minimal surfaces*, Comment. Math. Helv. **54** (1979), no. 2, 218–223, DOI 10.1007/BF02566269. MR535056 Cited in Problem 3.18.

[Sul79b] Dennis Sullivan, *Hyperbolic geometry and homeomorphisms*, Geometric topology (Proc. Georgia Topology Conf., Athens, Ga., 1977), Academic Press, New York-London, 1979, pp. 543–555. MR537749 Cited in Problems 4.125 and 4.131.

[Sul85] Dennis Sullivan, *Quasiconformal homeomorphisms and dynamics. II. Structural stability implies hyperbolicity for Kleinian groups*, Acta Math. **155** (1985), no. 3-4, 243–260, DOI 10.1007/BF02392543. MR806415 Cited in Problem 3.25.

[Sul87] Dennis Sullivan, *Quasiconformal homeomorphisms in dynamics, topology, and geometry*, Proceedings of the International Congress of Mathematicians, Vol. 1, 2 (Berkeley, Calif., 1986), Amer. Math. Soc., Providence, RI, 1987, pp. 1216–1228. MR934326 Cited in Problem 4.61.

[Sul95] Dennis Sullivan, *Exterior d, the local degree, and smoothability*, Prospects in topology (Princeton, NJ, 1994), Ann. of Math. Stud., vol. 138, Princeton Univ. Press, Princeton, NJ, 1995, pp. 328–338. MR1368667 Cited in Problem 4.131.

[Sul96] D. P. Sullivan, *Triangulating and smoothing homotopy equivalences and homeomorphisms. Geometric Topology Seminar Notes*, The Hauptvermutung book, *K*-Monogr. Math., vol. 1, Kluwer Acad. Publ., Dordrecht, 1996, pp. 69–103, DOI 10.1007/978-94-017-3343-4_3. MR1434103 Cited in Problem 4.46.

[Sul99] Dennis Sullivan, *On the foundation of geometry, analysis, and the differentiable structure for manifolds*, Topics in low-dimensional topology (University Park, PA, 1996), World Sci. Publ., River Edge, NJ, 1999, pp. 89–92. MR1883619 Cited in Problems 4.61 and 4.131.

[Sum75] D. W. Sumners, *Smooth Z_p-actions on spheres which leave knots pointwise fixed*, Trans. Amer. Math. Soc. **205** (1975), 193–203, DOI 10.2307/1997199. MR372893 Cited in Problems 1.49 and 4.87.

[Sun85] Toshikazu Sunada, *Riemannian coverings and isospectral manifolds*, Ann. of Math. (2) **121** (1985), no. 1, 169–186, DOI 10.2307/1971195. MR782558 Cited in Problems 2.16 and 3.17.

[Sun15] Nathan S. Sunukjian, *Surfaces in 4-manifolds: concordance, isotopy, and surgery*, Int. Math. Res. Not. IMRN **17** (2015), 7950–7978, DOI 10.1093/imrn/rnu187. MR3404006 Cited in Problem 4.45.

[Suz69] Shin'ichi Suzuki, *Local knots of 2-spheres in 4-manifolds*, Proc. Japan Acad. **45** (1969), 34–38. MR248851 Cited in Problem 4.42.

[Suz76] Shin'ichi Suzuki, *Knotting problems of 2-spheres in 4-sphere*, Math. Sem. Notes Kobe Univ. **4** (1976), no. 3, 241–371. MR445508 Cited in Problem 4.29.

[SW02] Daniel S. Silver and Susan G. Williams, *Mahler measure, links and homology growth*, Topology **41** (2002), no. 5, 979–991, DOI 10.1016/S0040-9383(01)00014-3. MR1923995 Cited in Problem 2.47.

[SW07] Jennifer Schultens and Richard Weidman, *On the geometric and the algebraic rank of graph manifolds*, Pacific J. Math. **231** (2007), no. 2, 481–510, DOI 10.2140/pjm.2007.231.481. MR2346507 Cited in Problem 3.35.

[SW10] Sucharit Sarkar and Jiajun Wang, *An algorithm for computing some Heegaard Floer homologies*, Ann. of Math. (2) **171** (2010), no. 2, 1213–1236, DOI 10.4007/annals.2010.171.1213. MR2630063 Cited in Problem 3.60.

[SY82] Richard Schoen and Shing Tung Yau, *Complete three-dimensional manifolds with positive Ricci curvature and scalar curvature*, Seminar on Differential Geometry, Ann. of Math. Stud., No. 102, Princeton Univ. Press, Princeton, NJ, 1982, pp. 209–228. MR645740 Cited in Problem 3.12.

[SY06] Osamu Saeki and Takahiro Yamamoto, *Singular fibers of stable maps and signatures of 4-manifolds*, Geom. Topol. **10** (2006), 359–399, DOI 10.2140/gt.2006.10.359. MR2224462 Cited in Problem 5.9.

[SY17] András I. Stipsicz and Ki-Heon Yun, *On the minimal number of singular fibers in Lefschetz fibrations over the torus*, Proc. Amer. Math. Soc. **145** (2017), no. 8, 3607–3616, DOI 10.1090/proc/13480. MR3652812 Cited in Problem 2.21.

[SY23] Steven Sivek and Mehdi Yazdi, *Thurston norm and Euler classes of tight contact structures*, Bull. Lond. Math. Soc. **55** (2023), no. 6, 2976–2990, DOI 10.1112/blms.12905. MR4698323 Cited in Problem 3.15.

[SZ22] Steven Sivek and Raphael Zentner, *A menagerie of $SU(2)$-cyclic 3-manifolds*, Int. Math. Res. Not. IMRN **11** (2022), 8038–8085, DOI 10.1093/imrn/rnaa330. MR4425831 Cited in Problem 3.52.

[Sze73] G. Szekeres, *Polyhedral decompositions of cubic graphs*, Bull. Austral. Math. Soc. **8** (1973), 367–387, DOI 10.1017/S0004972700042660. MR325438 Cited in Problem 1.86.

[Sze10] Błażej Szepietowski, *Embedding the braid group in mapping class groups*, Publ. Mat. **54** (2010), no. 2, 359–368, DOI 10.5565/PUBLMAT_54210_04. MR2675928 Cited in Problem 2.29.

[Tag19] Keiji Tagami, *On the Lagrangian fillability of almost positive links*, J. Korean Math. Soc. **56** (2019), no. 3, 789–804, DOI 10.4134/JKMS.j180399. MR3949585 Cited in Problem 1.68.

[Tan12] Shunsuke Tanaka, *On sections of hyperelliptic Lefschetz fibrations*, Algebr. Geom. Topol. **12** (2012), no. 4, 2259–2286, DOI 10.2140/agt.2012.12.2259. MR3020206 Cited in Problems 2.21 and 2.22.

[Tan23] Martin Tancer. Simpler algorithmically unrecognizable 4-manifolds, 2023. `arXiv:2310.07421`. Cited in Problem 4.53.

[Tap23] Chaitanya Tappu. A moduli space of marked hyperbolic structures for big surfaces, 2023. `arXiv:2311.01551`. Cited in Problem 2.40.

[Tau87] Clifford Henry Taubes, *Gauge theory on asymptotically periodic 4-manifolds*, J. Differential Geom. **25** (1987), no. 3, 363–430. MR882829 Cited in Problems 4.5, 4.58, and 4.61.

[Tau92] Clifford Henry Taubes, *The existence of anti-self-dual conformal structures*, J. Differential Geom. **36** (1992), no. 1, 163–253. MR1168984 Cited in Problem 4.130.

[Tau94] Clifford Henry Taubes, *The Seiberg-Witten invariants and symplectic forms*, Math. Res. Lett. **1** (1994), no. 6, 809–822, DOI 10.4310/MRL.1994.v1.n6.a15. MR1306023 Cited in Problems 4.90, 4.95, and 4.99.

[Tau95] Clifford Henry Taubes, *More constraints on symplectic forms from Seiberg-Witten invariants*, Math. Res. Lett. **2** (1995), no. 1, 9–13, DOI 10.4310/MRL.1995.v2.n1.a2. MR1312973 Cited in Problems 4.90 and 4.99.

[Tau96] Clifford H. Taubes, SW $\Rightarrow$ Gr*: from the Seiberg-Witten equations to pseudo-holomorphic curves*, J. Amer. Math. Soc. **9** (1996), no. 3, 845–918, DOI 10.1090/S0894-0347-96-00211-1. MR1362874 Cited in Problems 4.62, 4.90, 4.96, and 4.100.

[Tau00] Clifford Henry Taubes, *Seiberg Witten and Gromov invariants for symplectic 4-manifolds*, First International Press Lecture Series, vol. 2, International Press, Somerville, MA, 2000. Edited by Richard Wentworth. MR1798809 Cited in Problem 3.47.

[Tau10a] Clifford Henry Taubes, *Embedded contact homology and Seiberg-Witten Floer cohomology I*, Geom. Topol. **14** (2010), no. 5, 2497–2581, DOI 10.2140/gt.2010.14.2497. MR2746723 Cited in Problem 3.47.

[Tau10b] Clifford Henry Taubes, *Embedded contact homology and Seiberg-Witten Floer cohomology II*, Geom. Topol. **14** (2010), no. 5, 2583–2720, DOI 10.2140/gt.2010.14.2583. MR2746724 Cited in Problem 3.47.

[Tau10c] Clifford Henry Taubes, *Embedded contact homology and Seiberg-Witten Floer cohomology III*, Geom. Topol. **14** (2010), no. 5, 2721–2817, DOI 10.2140/gt.2010.14.2721. MR2746725 Cited in Problem 3.47.

[Tau10d] Clifford Henry Taubes, *Embedded contact homology and Seiberg-Witten Floer cohomology IV*, Geom. Topol. **14** (2010), no. 5, 2819–2960, DOI 10.2140/gt.2010.14.2819. MR2746726 Cited in Problem 3.47.

[Tau10e] Clifford Henry Taubes, *Embedded contact homology and Seiberg-Witten Floer cohomology V*, Geom. Topol. **14** (2010), no. 5, 2961–3000, DOI 10.2140/gt.2010.14.2961. MR2746727 Cited in Problem 3.47.

[Tau13] Clifford Henry Taubes. Compactness theorems for SL(2;C) generalizations of the 4-dimensional anti-self dual equations, 2013. `arXiv:1307.6447`. Cited in Problem 1.32.

[Tau16] Clifford Henry Taubes. Some 4-manifold geometry from hyperbolic knots in S^3, 2016. `arXiv:1602.01687`. Cited in Problem 4.130.

[Tau18] Clifford Henry Taubes. Sequences of Nahm pole solutions to the SU(2) Kapustin-Witten equations, 2018. `arXiv:1805.02773`. Cited in Problem 1.32.

[Tau19] Clifford Henry Taubes. The $\mathbb{R}$ invariant solutions to the Kapustin Witten equations on $(0,\infty) \times \mathbb{R}^2 \times \mathbb{R}$ with generalized Nahm pole asymptotics, 2019. `arXiv:1903.03539`. Cited in Problem 1.32.

[Tay79] Laurence R. Taylor, *On the genera of knots*, Topology of low-dimensional manifolds (Proc. Second Sussex Conf., Chelwood Gate, 1977), Lecture Notes in Math., vol. 722, Springer, Berlin, 1979, pp. 144–154. MR547461 Cited in Problem 1.46.

[Tei97] Peter Teichner, *On the star-construction for topological 4-manifolds*, Geometric topology (Athens, GA, 1993), AMS/IP Stud. Adv. Math., vol. 2, Amer. Math. Soc., Providence, RI, 1997, pp. 300–312, DOI 10.1090/amsip/002.1/17. MR1470734 Cited in Problem 4.52.

[Thi87] Morwen B. Thistlethwaite, *A spanning tree expansion of the Jones polynomial*, Topology **26** (1987), no. 3, 297–309, DOI 10.1016/0040-9383(87)90003-6. MR899051 Cited in Problems 1.1 and 1.23.

[Thi88] Morwen B. Thistlethwaite, *Kauffman's polynomial and alternating links*, Topology **27** (1988), no. 3, 311–318, DOI 10.1016/0040-9383(88)90012-2. MR963633 Cited in Problem 1.23.

[Thi01] Morwen Thistlethwaite, *Links with trivial Jones polynomial*, J. Knot Theory Ramifications **10** (2001), no. 4, 641–643, DOI 10.1142/S0218216501001050. MR1831681 Cited in Problem 1.24.

[Tho94] Abigail Thompson, *Thin position and the recognition problem for S^3*, Math. Res. Lett. **1** (1994), no. 5, 613–630, DOI 10.4310/MRL.1994.v1.n5.a9. MR1295555 Cited in Problems 3.27 and 4.53.

[Tho22] Abigail Thompson. The freeness index of a graph, 2022. `arXiv:2206.12939`. Cited in Problem 1.86.

[Tho25] Bryce Thomson. The Legendrian Hopf Link has exactly two Lagrangian fillings, 2025. `arXiv:2506.15111`. Cited in Problem 1.67.

[Thu74a] William Thurston, *Foliations and groups of diffeomorphisms*, Bull. Amer. Math. Soc. **80** (1974), 304–307, DOI 10.1090/S0002-9904-1974-13475-0. MR339267 Cited in Problem 4.85.

[Thu74b] William P. Thurston, *A generalization of the Reeb stability theorem*, Topology **13** (1974), 347–352, DOI 10.1016/0040-9383(74)90025-1. MR356087 Cited in Problems 2.34 and 3.22.

[Thu82] William P. Thurston, *Three-dimensional manifolds, Kleinian groups and hyperbolic geometry*, Bull. Amer. Math. Soc. (N.S.) **6** (1982), no. 3, 357–381, DOI 10.1090/S0273-0979-1982-15003-0. MR648524 Cited in Problems 1.88, 3.0, 3.2, and 4.127.

[Thu86] William P. Thurston, *A norm for the homology of* 3*-manifolds*, Mem. Amer. Math. Soc. **59** (1986), no. 339, i–vi and 99–130. MR823443 Cited in Problems 2.14, 2.24, 3.15, and 4.123.

[Thu97] William P. Thurston, *Three-dimensional geometry and topology. Vol. 1*, Princeton Mathematical Series, vol. 35, Princeton University Press, Princeton, NJ, 1997. Edited by Silvio Levy. MR1435975 Cited in Problem 3.23.

[Thu22] William P. Thurston, *On the geometry and dynamics of diffeomorphisms of surfaces*, Collected works of William P. Thurston with commentary. Vol. I. Foliations, surfaces and differential geometry, Amer. Math. Soc., Providence, RI, [2022] ©2022, pp. 495–509. Reprint of [0956596]. MR4554452 Cited in Problems 3.7 and 4.127.

[Toe11] Otto Toeplitz. Ueber einige Aufgaben der Analysis situs. *Verhandlungen der Schweizerischen Naturforschenden Gesellschaft*, 4:197, 1911. Cited in Problem 2.14.

[Tor53] Guillermo Torres, *On the Alexander polynomial*, Ann. of Math. (2) **57** (1953), 57–89, DOI 10.2307/1969726. MR52104 Cited in Problem 1.8.

[Tor17] Rafael Torres, *Smooth structures on nonorientable four-manifolds and free involutions*, J. Knot Theory Ramifications **26** (2017), no. 13, 1750085, 20, DOI 10.1142/S0218216517500857. MR3730553 Cited in Problem 4.11.

[Tor25] Rafael Torres, *Smoothly knotted and topologically unknotted nullhomologous surfaces in 4-manifolds* (English, with English and French summaries), Ann. Inst. Fourier (Grenoble) **75** (2025), no. 6, 2501–2527, DOI 10.5802/aif.3685. MR4940732 Cited in Problem 4.30.

[Tou01] Godfried Toussaint, *A new class of stuck unknots in* Pol_6, Beiträge Algebra Geom. **42** (2001), no. 2, 301–306. MR1865519 Cited in Problem 1.72.

[Tro73] H. F. Trotter. On S-equivalence of Seifert matrices. *Invent. Math.*, 20:173–207, 1973. `doi:10.1007/BF01394094`. Cited in Problem 1.6.

[TS21] Robert E. Tuzun and Adam S. Sikora, *Verification of the Jones unknot conjecture up to 24 crossings*, J. Knot Theory Ramifications **30** (2021), no. 3, Paper No. 2150020, 6, DOI 10.1142/S0218216521500206. MR4259507 Cited in Problem 1.24.

[Tsa23] Chi Cheuk Tsang, *Veering Triangulations and Pseudo-Anosov Flows*, ProQuest LLC, Ann Arbor, MI, 2023. Thesis (Ph.D.)–University of California, Berkeley. MR4675421 Cited in Problem 3.21.

[Tsu08] Takashi Tsuboi, *On the uniform perfectness of diffeomorphism groups*, Groups of diffeomorphisms, Adv. Stud. Pure Math., vol. 52, Math. Soc. Japan, Tokyo, 2008, pp. 505–524, DOI 10.2969/aspm/05210505. MR2509724 Cited in Problem 4.85.

[Tsu12] Takashi Tsuboi, *On the uniform perfectness of the groups of diffeomorphisms of even-dimensional manifolds*, Comment. Math. Helv. **87** (2012), no. 1, 141–185, DOI 10.4171/CMH/251. MR2874899 Cited in Problem 4.85.

[Tuc74] Thomas W. Tucker, *Non-compact 3-manifolds and the missing-boundary problem*, Topology **13** (1974), 267–273, DOI 10.1016/0040-9383(74)90019-6. MR353317 Cited in Problem 4.49.

[Tuk88] Pekka Tukia, *Homeomorphic conjugates of Fuchsian groups*, J. Reine Angew. Math. **391** (1988), 1–54, DOI 10.1515/crll.1988.391.1. MR961162 Cited in Problem 3.24.

[Tur88] V. G. Turaev. The Conway and Kauffman modules of a solid torus. *Zap. Nauchn. Sem. Leningrad. Otdel. Mat. Inst. Steklov. (LOMI)*, 167(Issled. Topol. 6):79–89, 190, 1988. `doi:10.1007/BF01099241`. Cited in Problem 1.32.

[Tur91] Vladimir G. Turaev, *Skein quantization of Poisson algebras of loops on surfaces*, Ann. Sci. École Norm. Sup. (4) **24** (1991), no. 6, 635–704. MR1142906 Cited in Problem 1.36.

[Tur04] Vladimir Turaev, *Virtual strings* (English, with English and French summaries), Ann. Inst. Fourier (Grenoble) **54** (2004), no. 7, 2455–2525 (2005), DOI 10.5802/aif.2086. MR2139700 Cited in Problem 1.83.

[Tur24] Joshua P. Turner, *Affine Springer fibers and generalized Haiman ideals (with an appendix by Eugene Gorsky and Joshua P. Turner)*, Int. Math. Res. Not. IMRN **16** (2024), 11878–11909, DOI 10.1093/imrn/rnae146. MR4789101 Cited in Problem 1.33.

[TV92] V. G. Turaev and O. Ya. Viro, *State sum invariants of* 3*-manifolds and quantum* $6j$*-symbols*, Topology **31** (1992), no. 4, 865–902, DOI 10.1016/0040-9383(92)90015-A. MR1191386 Cited in Problem 3.4.

[TWY08] Valentino Tosatti, Ben Weinkove, and Shing-Tung Yau, *Taming symplectic forms and the Calabi-Yau equation*, Proc. Lond. Math. Soc. (3) **97** (2008), no. 2, 401–424, DOI 10.1112/plms/pdn008. MR2439667 Cited in Problem 4.93.

[Uhl83] Karen K. Uhlenbeck, *Closed minimal surfaces in hyperbolic* 3*-manifolds*, Seminar on minimal submanifolds, Ann. of Math. Stud., vol. 103, Princeton Univ. Press, Princeton, NJ, 1983, pp. 147–168. MR0795233 Cited in Problem 3.18.

[Usu12] Takuya Usui. Heegaard Floer homology, L-spaces and smoothing order on links II, 2012. `arXiv:1202.3333`. Cited in Problem 3.53.

[vdV08] Roland van der Veen, *Proof of the volume conjecture for Whitehead chains*, Acta Math. Vietnam. **33** (2008), no. 3, 421–431. MR2501851 Cited in Problem 1.27.

[Vig80] Marie-France Vignéras, *Variétés riemanniennes isospectrales et non isométriques* (French), Ann. of Math. (2) **112** (1980), no. 1, 21–32, DOI 10.2307/1971319. MR584073 Cited in Problems 2.16 and 3.17.

[Vir75] O. Ja. Viro, *Positioning in codimension* 2*, and the boundary* (Russian), Uspehi Mat. Nauk **30** (1975), no. 1(181), 231–232. MR420641 Cited in Problem 1.47.

[Vir76] O. Ja. Viro, *Nonprojecting isotopies and knots with homeomorphic coverings* (Russian, with English summary), Zap. Naučn. Sem. Leningrad. Otdel. Mat. Inst. Steklov. (LOMI) **66** (1976), 133–147, 207–208. Studies in topology, II. MR451252 Cited in Problem 1.17.

[Vir84] O. Ya. Viro, *The signature of a branched covering* (Russian), Mat. Zametki **36** (1984), no. 4, 549–557. MR771233 Cited in Problems 4.122 and 5.6.

[VKF74] I. A. Volodin, V. E. Kuznecov, and A. T. Fomenko, *The problem of the algorithmic discrimination of the standard three-dimensional sphere* (Russian), Uspehi Mat. Nauk **29** (1974), no. 5(179), 71–168. Appendix by S. P. Novikov. MR405426 Cited in Problem 4.53.

[Vla24a] Nicholas G. Vlamis. Homeomorphism groups of telescoping 2–manifolds are strongly distorted, 2024. `arXiv:2403.03887`. Cited in Problem 2.38.

[Vla24b] Nicholas G. Vlamis, *Homeomorphism groups of self-similar 2-manifolds*, In the tradition of Thurston III. Geometry and dynamics, Springer, Cham, [2024] ©2024, pp. 105–167, DOI 10.1007/978-3-031-43502-7_5. MR4759043 Cited in Problem 2.38.

[Vog82] Pierre Vogel, *On the obstruction group in homology surgery*, Inst. Hautes Études Sci. Publ. Math. **55** (1982), 165–206. MR672183 Cited in Problem 4.59.

[Vor97] Ya. B. Vorobets, *Ergodicity of billiards in polygons* (Russian, with Russian summary), Mat. Sb. **188** (1997), no. 3, 65–112, DOI 10.1070/SM1997v188n03ABEH000211; English transl., Sb. Math. **188** (1997), no. 3, 389–434. MR1462024 Cited in Problem 2.42.

[VS99] Luis Gerardo Valdez Sánchez, *Dehn fillings of 3-manifolds and non-persistent tori*, Topology Appl. **98** (1999), no. 1-3, 355–370, DOI 10.1016/S0166-8641(99)00038-3. II Iberoamerican Conference on Topology and its Applications (Morelia, 1997). MR1720012 Cited in Problem 1.11.

[Vv11] Siniša T. Vrećica and Rade T. Živaljević. Fulton-MacPherson compactification, cyclohedra, and the polygonal pegs problem. *Israel J. Math.*, 184:221–249, 2011. doi:10.1007/s11856-011-0066-9. Cited in Problem 2.14.

[Waj99] Bronislaw Wajnryb, *Artin groups and geometric monodromy*, Invent. Math. **138** (1999), no. 3, 563–571, DOI 10.1007/s002220050353. MR1719815 Cited in Problems 2.26 and 2.29.

[Wal64a] C. T. C. Wall, *Diffeomorphisms of 4-manifolds*, J. London Math. Soc. **39** (1964), 131–140, DOI 10.1112/jlms/s1-39.1.131. MR163323 Cited in Problems 4.30, 4.70, 4.75, and 4.76.

[Wal64b] C. T. C. Wall, *On simply-connected 4-manifolds*, J. London Math. Soc. **39** (1964), 141–149, DOI 10.1112/jlms/s1-39.1.141. MR163324 Cited in Problems 4.7 and 4.14.

[Wal65] C. T. C. Wall, *Finiteness conditions for CW-complexes*, Ann. of Math. (2) **81** (1965), 56–69, DOI 10.2307/1970382. MR171284 Cited in Problem 4.49.

[Wal68a] Friedhelm Waldhausen, *Heegaard-Zerlegungen der 3-Sphäre* (German), Topology **7** (1968), 195–203, DOI 10.1016/0040-9383(68)90027-X. MR227992 Cited in Problem 4.112.

[Wal68b] Friedhelm Waldhausen, *On irreducible 3-manifolds which are sufficiently large*, Ann. of Math. (2) **87** (1968), 56–88, DOI 10.2307/1970594. MR224099 Cited in Problem 3.32.

[Wal70] Friedhelm Waldhausen, *On mappings of handlebodies and of Heegaard splittings*, Topology of Manifolds (Proc. Inst., Univ. of Georgia, Athens, Ga., 1969), Markham Publishing Co., Chicago, IL, 1970, pp. 205–211. MR271941 Cited in Problem 3.33.

[Wal78] Friedhelm Waldhausen, *Some problems on 3-manifolds*, Algebraic and geometric topology (Proc. Sympos. Pure Math., Stanford Univ., Stanford, Calif., 1976), Proc. Sympos. Pure Math., XXXII, Amer. Math. Soc., Providence, RI, 1978, pp. 313–322. MR520549 Cited in Problem 4.117.

[Wal79] C. T. C. Wall (ed.), *Homological group theory*, London Mathematical Society Lecture Note Series, vol. 36, Cambridge University Press, Cambridge-New York, 1979. MR564417 Cited in Problem 3.38.

[Wal99] C. T. C. Wall, *Surgery on compact manifolds*, 2nd ed., Mathematical Surveys and Monographs, vol. 69, American Mathematical Society, Providence, RI, 1999. Edited and with a foreword by A. A. Ranicki, DOI 10.1090/surv/069. MR1687388 Cited in Problems 4.46, 4.87, and 5.5.

[Wan02] Shicheng Wang, *Non-zero degree maps between 3-manifolds*, Proceedings of the International Congress of Mathematicians, Vol. II (Beijing, 2002), Higher Ed. Press, Beijing, 2002, pp. 457–468. MR1957056 Cited in Problem 3.62.

[Wan12] Andy Wand, *Mapping class group relations, Stein fillings, and planar open book decompositions*, J. Topol. **5** (2012), no. 1, 1–14, DOI 10.1112/jtopol/jtr025. MR2897047 Cited in Problem 4.105.

[Wan15] Andy Wand, *Tightness is preserved by Legendrian surgery*, Ann. of Math. (2) **182** (2015), no. 2, 723–738, DOI 10.4007/annals.2015.182.2.8. MR3418529 Cited in Problems 3.43 and 3.44.

[Wan22] Joshua Wang, *The cosmetic crossing conjecture for split links*, Geom. Topol. **26** (2022), no. 7, 2941–3053, DOI 10.2140/gt.2022.26.2941. MR4540900 Cited in Problems 1.21 and 1.56.

[Wan25] Zhihao Wang, *Kauffman bracket intertwiners and the volume conjecture*, Algebr. Geom. Topol. **25** (2025), no. 4, 2143–2177, DOI 10.2140/agt.2025.25.2143. MR4950901 Cited in Problem 2.50.

[Wat10] Liam Watson, *A remark on Khovanov homology and two-fold branched covers*, Pacific J. Math. **245** (2010), no. 2, 373–380, DOI 10.2140/pjm.2010.245.373. MR2608442 Cited in Problem 1.16.

[Wat16] Yohsuke Watanabe, *Intersection numbers in the curve graph with a uniform constant*, Topology Appl. **204** (2016), 157–167, DOI 10.1016/j.topol.2016.03.009. MR3482712 Cited in Problem 2.10.

[Wat19] Tadayuki Watanabe. Some exotic nontrivial elements of the rational homotopy groups of Diff(S^4), 2019. arXiv:1812.02448. Cited in Problems 4.67, 4.70, and 4.73.

[Wat23] Tadayuki Watanabe. Theta-graph and diffeomorphisms of some 4-manifolds, 2023. arXiv:2005.09545. Cited in Problem 4.75.

[Web07] Ben Webster, *Khovanov-Rozansky homology via a canopolis formalism*, Algebr. Geom. Topol. **7** (2007), 673–699, DOI 10.2140/agt.2007.7.673. MR2308960 Cited in Problem 1.34.

[Web08] Claude Weber, *On the topology of singularities*, Singularities II, Contemp. Math., vol. 475, Amer. Math. Soc., Providence, RI, 2008, pp. 217–251, DOI 10.1090/conm/475/09285. MR2454369 Cited in Problem 1.87.

[Web15] Richard C. H. Webb, *Combinatorics of tight geodesics and stable lengths*, Trans. Amer. Math. Soc. **367** (2015), no. 10, 7323–7342, DOI 10.1090/tran/6301. MR3378831 Cited in Problem 2.10.

[Web17] Ben Webster, *Knot invariants and higher representation theory*, Mem. Amer. Math. Soc. **250** (2017), no. 1191, v+141, DOI 10.1090/memo/1191. MR3709726 Cited in Problem 1.37.

[Wen87] Hans Wenzl, *On sequences of projections*, C. R. Math. Rep. Acad. Sci. Canada **9** (1987), no. 1, 5–9. MR873400 Cited in Problem 1.31.

[Wen10] Chris Wendl, *Strongly fillable contact manifolds and J-holomorphic foliations*, Duke Math. J. **151** (2010), no. 3, 337–384, DOI 10.1215/00127094-2010-001. MR2605865 Cited in Problem 4.105.

[Wes90] James E. West, *Open problems in infinite-dimensional topology*, Open problems in topology, North-Holland, Amsterdam, 1990, pp. 523–597. MR1078666 Cited in Problem 5.4.

[Whi40] J. H. C. Whitehead, *On C^1-complexes*, Ann. of Math. (2) **41** (1940), 809–824, DOI 10.2307/1968861. MR2545 Cited in Problem 4.119.

[Whi41] J. H. C. Whitehead, *On adding relations to homotopy groups*, Ann. of Math. (2) **42** (1941), 409–428, DOI 10.2307/1968907. MR4123 Cited in Problems 4.38 and 5.11.

[Wil00] Burkhard Wilking, *On fundamental groups of manifolds of nonnegative curvature*, Differential Geom. Appl. **13** (2000), no. 2, 129–165, DOI 10.1016/S0926-2245(00)00030-9. MR1783960 Cited in Problem 5.15.

[Wil21] Michael Willis. Khovanov homology for links in $\#^r(S^2 \times S^1)$. *Michigan Math. J.*, 70(4):675–748, 2021. doi:10.1307/mmj/1594281620. Cited in Problem 1.32.

[Win73] H. E. Winkelnkemper, *Manifolds as open books*, Bull. Amer. Math. Soc. **79** (1973), 45–51, DOI 10.1090/S0002-9904-1973-13085-X. MR310912 Cited in Problems 3.39 and 4.121.

[Wis21] Daniel T. Wise, *The structure of groups with a quasiconvex hierarchy*, Annals of Mathematics Studies, vol. 209, Princeton University Press, Princeton, NJ, [2021] ©2021. MR4298722 Cited in Problems 3.7, 3.26, 3.29, 3.37, and 5.13.

[Wit89] Edward Witten. Quantum field theory and the Jones polynomial. *Comm. Math. Phys.*, 121(3):351–399, 1989. http://projecteuclid.org/euclid.cmp/1104178138. Cited in Problems 2.49 and 3.67.

[Wit94] Edward Witten, *Monopoles and four-manifolds*, Math. Res. Lett. **1** (1994), no. 6, 769–796, DOI 10.4310/MRL.1994.v1.n6.a13. MR1306021 Cited in Problems 3.47, 4.61, 4.62, 4.64, and 4.128.

[Wit12] Edward Witten, *Fivebranes and knots*, Quantum Topol. **3** (2012), no. 1, 1–137, DOI 10.4171/QT/26. MR2852941 Cited in Problem 1.32.

[Woo71] John W. Wood, *Bundles with totally disconnected structure group*, Comment. Math. Helv. **46** (1971), 257–273, DOI 10.1007/BF02566843. MR293655 Cited in Problem 3.15.

[Wu14] Hao Wu, *A colored $\mathfrak{sl}(N)$ homology for links in S^3*, Dissertationes Math. **499** (2014), 217, DOI 10.4064/dm499-0-1. MR3234803 Cited in Problem 1.37.

[WW20] Michael Wolf and Yunhui Wu, *Non-existence of geometric minimal foliations in hyperbolic three-manifolds*, Comment. Math. Helv. **95** (2020), no. 1, 167–182, DOI 10.4171/cmh/484. MR4082897 Cited in Problem 3.18.

[WX17] Guozhen Wang and Zhouli Xu, *The triviality of the 61-stem in the stable homotopy groups of spheres*, Ann. of Math. (2) **186** (2017), no. 2, 501–580, DOI 10.4007/annals.2017.186.2.3. MR3702672 Cited in Problem 4.1.

[WY23] Ka Ho Wong and Tian Yang. Relative Reshetikhin-Turaev invariants, hyperbolic cone metrics and discrete Fourier transforms I. *Comm. Math. Phys.*, 400(2):1019–1070, 2023. doi:10.1007/s00220-022-04613-5. Cited in Problem 3.4.

[WYY96] Masaaki Wada, Yasushi Yamashita, and Han Yoshida, *An inequality for polyhedra and ideal triangulations of cusped hyperbolic 3-manifolds*, Proc. Amer. Math. Soc. **124** (1996), no. 12, 3905–3911, DOI 10.1090/S0002-9939-96-03563-0. MR1346992 Cited in Problem 3.3.

[Xia87] Gang Xiao. Fibered algebraic surfaces with low slope. *Math. Ann.*, 276(3):449–466, 1987. doi:10.1007/BF01450841. Cited in Problem 2.20.

[Xie21] Yi Xie, *Instantons and annular Khovanov homology*, Adv. Math. **388** (2021), Paper No. 107864, 51, DOI 10.1016/j.aim.2021.107864. MR4281775 Cited in Problem 1.28.

[XZ25] Yi Xie and Boyu Zhang, *Instanton Floer homology for sutured manifolds with tangles*, J. Differential Geom. **130** (2025), no. 3, 701–769, DOI 10.4310/jdg/1749496718. MR4918219 Cited in Problem 1.28.

[Yas91] Akira Yasuhara, $(2,15)$*-torus knot is not slice in* $\mathbf{CP}^2$, Proc. Japan Acad. Ser. A Math. Sci. **67** (1991), no. 10, 353–355. MR1151354 Cited in Problem 4.42.

[Yas96] Akira Yasuhara, *Connecting lemmas and representing homology classes of simply connected 4-manifolds*, Tokyo J. Math. **19** (1996), no. 1, 245–261, DOI 10.3836/tjm/1270043232. MR1391941 Cited in Problem 1.47.

[Yas08] Kouichi Yasui, *Elliptic surfaces without 1-handles*, J. Topol. **1** (2008), no. 4, 857–878, DOI 10.1112/jtopol/jtn026. MR2461858 Cited in Problem 4.118.

[Yas12] Kouichi Yasui. Nuclei and exotic 4-manifolds, 2012. arXiv:1111.0620. Cited in Problem 4.6.

[Yas17] Kouichi Yasui. Corks, exotic 4-manifolds and knot concordance, 2017. arXiv:1505.02551. Cited in Problem 1.58.

[Yas19] Kouichi Yasui, *Geometrically simply connected 4-manifolds and stable cohomotopy Seiberg-Witten invariants*, Geom. Topol. **23** (2019), no. 5, 2685–2697, DOI 10.2140/gt.2019.23.2685. MR4019901 Cited in Problem 4.118.

[Yau78] Shing Tung Yau. On the Ricci curvature of a compact Kähler manifold and the complex Monge-Ampère equation. I. *Comm. Pure Appl. Math.*, 31(3):339–411, 1978. doi:10.1002/cpa.3160310304. Cited in Problems 4.90 and 4.91.

[Yaz20] Mehdi Yazdi, *On Thurston's Euler class-one conjecture*, Acta Math. **225** (2020), no. 2, 313–368, DOI 10.4310/acta.2020.v225.n2.a3. MR4205409 Cited in Problem 3.15.

[Yon11] Yasuyoshi Yonezawa, *Quantum* $(\mathfrak{sl}_n, \wedge V_n)$ *link invariant and matrix factorizations*, Nagoya Math. J. **204** (2011), 69–123, DOI 10.1215/00277630-1431840. MR2863366 Cited in Problem 1.37.

[Yos85] Tomoyoshi Yoshida. The η-invariant of hyperbolic 3-manifolds. *Invent. Math.*, 81(3):473–514, 1985. doi:10.1007/BF01388583. Cited in Problem 3.64.

[Yos94] Katsuyuki Yoshikawa, *An enumeration of surfaces in four-space*, Osaka J. Math. **31** (1994), no. 3, 497–522. MR1309400 Cited in Problem 4.37.

[Yos96] Han Yoshida, *Ideal tetrahedral decompositions of hyperbolic 3-manifolds*, Osaka J. Math. **33** (1996), no. 1, 37–46. MR1381617 Cited in Problem 3.3.

[Yos98] Katsuyuki Yoshikawa, *The order of a meridian of a knotted Klein bottle*, Proc. Amer. Math. Soc. **126** (1998), no. 12, 3727–3731, DOI 10.1090/S0002-9939-98-04560-2. MR1468209 Cited in Problem 4.37.

[Yos13] Ken'ichi Yoshida, *The minimal volume orientable hyperbolic 3-manifold with 4 cusps*, Pacific J. Math. **266** (2013), no. 2, 457–476, DOI 10.2140/pjm.2013.266.457. MR3130632 Cited in Problem 3.1.

[Zar29] Oscar Zariski. On the Problem of Existence of Algebraic Functions of Two Variables Possessing a Given Branch Curve. *Amer. J. Math.*, 51(2):305–328, 1929. doi:10.2307/2370712. Cited in Problem 4.102.

[Zee64] E. C. Zeeman, *On the dunce hat*, Topology **2** (1964), 341–358, DOI 10.1016/0040-9383(63)90014-4. MR156351 Cited in Problem 5.12.

[Zee65] E. C. Zeeman, *Twisting spun knots*, Trans. Amer. Math. Soc. **115** (1965), 471–495, DOI 10.2307/1994281. MR195085 Cited in Problems 1.59, 4.1, and 4.26.

[Zem16] Drew Zemke, *The simple loop conjecture for 3-manifolds modeled on Sol*, Algebr. Geom. Topol. **16** (2016), no. 5, 3051–3071, DOI 10.2140/agt.2016.16.3051. MR3572359 Cited in Problem 3.36.

[Zem19a] Ian Zemke, *Knot Floer homology obstructs ribbon concordance*, Ann. of Math. (2) **190** (2019), no. 3, 931–947, DOI 10.4007/annals.2019.190.3.5. MR4024565 Cited in Problems 1.53 and 1.56.

[Zem19b] Ian Zemke, *Link cobordisms and functoriality in link Floer homology*, J. Topol. **12** (2019), no. 1, 94–220, DOI 10.1112/topo.12085. MR3905679 Cited in Problems 1.19 and 1.34.

[Zem25] Ian Zemke, *The equivalence of lattice and Heegaard Floer homology*, Duke Math. J. **174** (2025), no. 5, 857–910, DOI 10.1215/00127094-2024-0044. MR4905538 Cited in Problems 3.57 and 5.18.

[Zen11] Raphael Zentner, *Representation spaces of pretzel knots*, Algebr. Geom. Topol. **11** (2011), no. 5, 2941–2970, DOI 10.2140/agt.2011.11.2941. MR2869448 Cited in Problem 3.53.

[Zen17] Raphael Zentner, *A class of knots with simple $SU(2)$-representations*, Selecta Math. (N.S.) **23** (2017), no. 3, 2219–2242, DOI 10.1007/s00029-017-0314-x. MR3663606 Cited in Problem 3.52.

[Zen18] Raphael Zentner, *Integer homology 3-spheres admit irreducible representations in* $\mathrm{SL}(2,\mathbb{C})$, Duke Math. J. **167** (2018), no. 9, 1643–1712, DOI 10.1215/00127094-2018-0004. MR3813594 Cited in Problems 3.27, 3.37, 3.52, and 3.65.

[Zha23] Yue Zhang. Guts and the minimal volume orientable hyperbolic 3-manifold with 3 cusps, 2023. `arXiv:2304.09950`. Cited in Problem 3.1.

[Zhe07] Hao Zheng, *Proof of the volume conjecture for Whitehead doubles of a family of torus knots*, Chinese Ann. Math. Ser. B **28** (2007), no. 4, 375–388, DOI 10.1007/s11401-006-0373-3. MR2348452 Cited in Problem 1.27.

[Zho21] Hugo Zhou, *Homology concordance and an infinite rank free subgroup*, J. Topol. **14** (2021), no. 4, 1369–1395, DOI 10.1112/topo.12211. MR4406694 Cited in Problem 1.61.

[Zib23] Claudius Zibrowius, *On symmetries of peculiar modules, or δ-graded link Floer homology is mutation invariant*, J. Eur. Math. Soc. (JEMS) **25** (2023), no. 8, 2949–3006, DOI 10.4171/jems/1201. MR4612107 Cited in Problem 3.61.

[Zim98] Bruno Zimmermann. On hyperbolic knots with homeomorphic cyclic branched coverings. *Math. Ann.*, 311(4):665–673, 1998. `doi:10.1007/s002080050205`. Cited in Problem 1.16.

[Zun25] Jonathan Zung, *Pseudo-Anosov representatives of stable Hamiltonian structures*, J. Fixed Point Theory Appl. **27** (2025), no. 4, Paper No. 87, 29, DOI 10.1007/s11784-025-01238-8. MR4955684 Cited in Problem 3.21.

[Zup20] Alexander Zupan, *The Powell conjecture and reducing sphere complexes*, J. Lond. Math. Soc. (2) **101** (2020), no. 1, 328–348, DOI 10.1112/jlms.12272. MR4072496 Cited in Problem 2.15.

Problems carried over from K2

As mentioned in the introduction, some problems from **K2** reappear in this volume, about 50 in all. Some are essentially unchanged from the original, but in many cases there has been some progress in the intervening years and we cite only part of the original problem. Here is a list of those problems, with the **K2** problem listed first, followed by the corresponding problem number (in parentheses) in this volume.

Problem 1.11 (1.18)
Problem 1.22 (1.17) and (1.16)
Problem 1.32 (1.38)
Problem 1.36 (1.62)
Problem 1.48 (4.35)
Problem 1.65 (1.1)
Problem 1.69 (1.3)
Problem 1.80 (1.13)
Problem 1.82 (1.10)
Problem 1.103 (4.38)
Problem 3.25 (1.17) and (1.16)
Problem 3.60 (3.1)
Problem 3.63 (3.64) and (3.2)
Problem 3.89 (3.34)
Problem 3.102 (1.15)
Problem 3.105 (3.52)
Problem 4.1 (4.54)
Problem 4.9 (4.56)
Problem 4.18 (4.118)
Problem 4.32 (4.23)
Problem 4.45 (4.1)
Problem 4.89 (4.1)
Problem 4.82 (4.20)
Problem 4.86 (4.2)
Problem 4.97 (4.17)
Problem 4.123 (4.58) and (5.7)
Problem 4.131 (4.62)
Problem 5.2 (5.10) and (5.12)
Problem 5.9 (4.46)

Problem 1.19 (1.58)
Problem 1.31 (1.61)
Problem 1.34 (1.52)
Problem 1.41 (1.57)
Problem 1.53 (3.61)
Problem 1.67 (1.2)
Problem 1.78 (1.20)
Problem 1.81 (1.12)
Problem 1.88 (1.24)
Problem 2.17 (2.25)
Problem 3.58 (3.5)
Problem 3.62 (3.64) and (3.2)
Problem 3.77 (3.38)
Problem 3.96 (3.36)
Problem 3.104 (3.64)
Problem 3.106 (3.50)
Problem 4.6 (4.46)
Problem 4.10 (4.19)
Problem 4.22 (1.50)
Problem 4.34 (4.67)
Problem 4.46 (1.62)
Problem 4.74 (4.21)
Problem 4.84 (4.52)
Problem 4.92 (4.15)
Problem 4.101 (4.94)
Problem 4.126 (4.67)
Problem 4.143 (4.129) and (4.128)
Problem 5.4 (5.11)
Problem 5.27 (5.2)

Index

Selected Published Titles in This Series

295 **R. İnanç Baykur, Robion C. Kirby, and Daniel Ruberman, Editors,** K3: A New Problem List in Low-Dimensional Topology, 2026
294 **Sebastian Schleissinger,** Univalent Functions in Quantum Probability Theory, 2025
293 **Bennett Chow, Brett Kotschwar, and Ovidiu Munteanu,** Ricci Solitons in Dimensions 4 and Higher, 2025
292 **Francisco Urbano,** Morse Index of Minimal Submanifolds, 2025
291 **Tadashi Ochiai,** Iwasawa Theory and Its Perspective, Volume 3, 2025
288 **Anna Geyer and Dmitry E. Pelinovsky,** Stability of Nonlinear Waves in Hamiltonian Dynamical Systems, 2025
287 **Jacques Sauloy,** Basic Modern Theory of Linear Complex Analytic q-Difference Equations, 2024
286 **Zalman Balanov, Wieslaw Krawcewicz, Dmitrii Rachinskii, Hao-Pin Wu, and Jianshe Yu,** Degree Theory and Symmetric Equations Assisted by GAP System, 2025
285 **Niles Johnson and Donald Yau,** Bimonoidal Categories, E_n-Monoidal Categories, and Algebraic K-Theory, 2024
284 **Donald Yau,** Bimonoidal Categories, E_n-Monoidal Categories, and Algebraic K-Theory, 2024
283 **Donald Yau,** Bimonoidal Categories, E_n-Monoidal Categories, and Algebraic K-Theory, 2024
282 **Michael Kapovich and Pranab Sardar,** Trees of Hyperbolic Spaces, 2024
281 **Horst R. Thieme,** Discrete-Time Dynamics of Structured Populations and Homogeneous Order-Preserving Operators, 2024
280 **Tadashi Ochiai,** Iwasawa Theory and Its Perspective, Volume 2, 2024
279 **Ellen Elizabeth Eischen, Wee Teck Gan, Aaron Pollack, and Zhiwei Yun,** Automorphic Forms Beyond GL_2, 2024
278 **Delaram Kahrobaei, Ramón Flores, Marialaura Noce, Maggie E. Habeeb, and Christopher Battarbee,** Applications of Group Theory in Cryptography, 2024
277 **George Metcalfe, Francesco Paoli, and Constantine Tsinakis,** Residuated Structures in Algebra and Logic, 2023
276 **Balázs Bárány, Károly Simon, and Boris Solomyak,** Self-similar and Self-affine Sets and Measures, 2023
275 **Alekos Vidras and Alain Yger,** Multidimensional Residue Theory and Applications, 2023
274 **Willi Freeden and M. Zuhair Nashed,** Recovery Methodologies: Regularization and Sampling, 2023
273 **Gennadiy Feldman,** Characterization of Probability Distributions on Locally Compact Abelian Groups, 2023
272 **Tadashi Ochiai,** Iwasawa Theory and Its Perspective, Volume 1, 2023
271 **Hideto Asashiba,** Categories and Representation Theory, 2022
270 **Leslie Hogben, Jephian C.-H. Lin, and Bryan L. Shader,** Inverse Problems and Zero Forcing for Graphs, 2022
269 **Ralph S. Freese, Ralph N. McKenzie, George F. McNulty, and Walter F. Taylor,** Algebras, Lattices, Varieties, 2022
268 **Ralph S. Freese, Ralph N. McKenzie, George F. McNulty, and Walter F. Taylor,** Algebras, Lattices, Varieties, 2022
267 **Dragana S. Cvetković Ilić,** Completion Problems on Operator Matrices, 2022
266 **Kate Juschenko,** Amenability of Discrete Groups by Examples, 2022

For a complete list of titles in this series, visit the
AMS Bookstore at **www.ams.org/bookstore/survseries/**.

Kolibri 360 GmbH
Ferdinand-Jühlke-Straße 7
99095 Erfurt, Deutschland
produktsicherheit@kolibri360.de